Comparative Mammalian Cytogenetics

Comparative Mammalian Cytogenetics

An International Conference at Dartmouth Medical School Hanover, New Hampshire, July 29—August 2, 1968

edited by Kurt Benirschke
Department of Pathology, Dartmouth Medical School

SPRINGER-VERLAG NEW YORK INC. 1969

ISBN-13:978-3-642-85945-8 **e-ISBN-13:978-3-642-85943-4**
DOI: 10.1007/978-3-642-85943-4

Softcover reprint of the hardcover 1st edition 1969
Library of Congress Catalog Card Number 69-20030

Title Number 1564

ORGANIZERS, SESSION CHAIRMEN AND SPONSORS

Organizers

K. Benirschke and T. C. Jones

Session Chairmen

K. Benirschke, W. Layton, M. Clendenin, D. Hoefnagel, and Doris H. Wurster, Department of Pathology, Dartmouth Medical School, Hanover, New Hampshire

T. C. Jones, New England Regional Primate Research Center, Southborough, Massachusetts

Work Shops

T. C. Hsu, Section of Biology, M. D. Anderson Hospital, Houston, Texas (Bone Marrow)

C. E. Ford, Medical Research Council, Harwell, Didcot, England (Meiosis)

Sponsors of Conference

NATIONAL INSTITUTE OF CHILD HEALTH AND HUMAN DEVELOPMENT (CONTRACT PH 43-68-1006)
BRISTOL-MYERS COMPANY
CHARLES RIVER BREEDING LABORATORIES
LAKEVIEW HAMSTER COLONY
ELI LILLY COMPANY
SANDOZ PHARMACEUTICALS
SCHERING CORPORATION
WALLACE PHARMACEUTICALS

MICHAEL GALTON

IN MEMORIAM

Two days after the closing of this conference my good friend and loyal associate MICHAEL GALTON sustained a tragic road accident to which he succumbed ten days later on August 15, 1968. To him we dedicate this volume.

Born in London, England, he received his professional education at University College in London. Although his training was in obstetrics, he became an experimental pathologist after coming to Boston in 1959. He joined the faculty of Dartmouth Medical School in 1961 and was Associate Professor of Pathology at the time of his death.

He was a productive investigator, having won numerous prizes and awards, of which he coveted the Markle Fellowship the most. His sixty-five professional publications are largely original contributions to an understanding of the pathophysiology of pregnancy. He had recently returned with his family from a one year special leave to Mexico City. There he had studied the coagulation disorders attending late and complicated pregnancy, an interest which was aroused by his active participation in the first of these conferences.

At age 35 he was emerging as an outstanding investigator from whose tenacity and inspiration our field of endeavor would have profited much. His invaluable help in this department, his precise and demanding teaching, and his ever challenging presence will be missed much by this community. His wife and sons, Ian and Kenneth, will always remember him as a loving and devoted father.

Hanover, New Hampshire
August 1968

KURT BENIRSCHKE

Speakers at the Conference on Comparative Mammalian Cytogenetics (Hanover, New Hampshire, July 29–August 2, 1968). *First row:* Benirschke, Sarto, Wurster, Basrur, Klinger, Wahrman, Hsu. *Second row:* Jones, Chang, Hutt, Ohno, Egozcue, Mayr, Taylor, Krooth, Ford. *Third row:* McFeely, Borgaonkar, Carr, Gropp, Nadler, Schwarzacher, Hirschhorn, Hayman.

LIST OF PARTICIPANTS

BASRUR, P. K., University of Guelph, Guelph, Ontario, Canada
BENIRSCHKE, K., Dartmouth Medical School, Hanover, New Hampshire
BORGAONKAR, D. S., Johns Hopkins Medical School, Baltimore, Maryland
BRECKON, G., Radiobiological Research Unit, M.R.C., Harwell, Didcot, Berkshire, England
CARR, D. H., McMaster University, Hamilton, Ontario, Canada
CHANG, M. C., The Worcester Foundation for Experimental Biology, Shrewsbury, Massachusetts
CITOLER, P., Pathologisches Institut, Universität Bonn, Venusberg, Germany
COHEN, M. H., Childrens Hospital, Buffalo, New York
EGOZCUE, J., Oregon Regional Primate Research Center, Beaverton, Oregon
EVANS, E. P., Radiobiological Research Unit, M.R.C., Harwell, Didcot, Berkshire, England
FERM, V. H., Dartmouth Medical School, Hanover, New Hampshire
FORD, C. E., Radiobiological Research Unit, M.R.C., Harwell, Didcot, Berkshire, England
v. GELDER, R. G., American Museum of Natural History, New York, New York
GOITEN, R., Hebrew University of Jerusalem, Jerusalem, Israel
GROPP, A., Pathologisches Institut, Universität Bonn, Venusberg, Germany
HAYMAN, D. L., University of Adelaide, Adelaide, South Australia
HIRSCHHORN, K., Mount Sinai School of Medicine, New York, New York
HSU, T. C., University of Texas, M. D. Anderson Hospital, Houston, Texas
HUNGERFORD, D. A., Institute for Cancer Research, Fox Chase, Philadelphia, Pennsylvania
HUTT, F. B., New York State College of Agriculture, Cornell University, Ithaca, New York
INHORN, S. L., University of Wisconsin Medical Center, Madison, Wisconsin
JONES, T. C., New England Regional Primate Research Center, Southborough, Massachusetts

KLINGER, H. P., Albert Einstein College of Medicine, Yeshiva University, Bronx, New York

KROOTH, R. S., University of Michigan School of Medicine, Ann Arbor, Michigan

LOW, R. J., Dartmouth Medical School, Hanover, New Hampshire

MCFEELY, R. A., University of Pennsylvania, New Bolton Center, Philadelphia, Pennsylvania

MCGAUGHEY, R. W., The Worcester Foundation for Experimental Biology, Shrewsbury, Massachusetts

MARTIN, P. G., University of Adelaide, Adelaide, South Australia

MAYR, E., Museum of Comparative Zoology, Harvard University, Cambridge, Massachusetts

MEAD, R. A., University of Idaho, Moscow, Idaho

NADLER, C. F., Northwestern University Medical School, Chicago, Illinois

NEVO, E., Hebrew University of Jerusalem, Jerusalem, Israel

OHNO, S., City of Hope Medical Center, Duarte, California

OPITZ, J. M., University of Wisconsin Medical Center, Madison, Wisconsin

PERA, F., Anatomisches Institut der Universität, Giessen, Germany

PICKWORTH, S., The Worcester Foundation for Experimental Biology, Shrewsbury, Massachusetts

SARTO, G. E., University of Wisconsin Medical Center, Madison, Wisconsin

SCHWARZACHER, H. G., Anatomisches Institut der Universität, Giessen, Germany

SNYDER, R. L., Penrose Research Laboratories, Philadelphia Zoological Garden, Philadelphia, Pennsylvania

TAYLOR, K. M., San Diego State College, San Diego, California

WAHRMAN, J., Hebrew University of Jerusalem, Jerusalem, Israel

WURSTER, D. H., Dartmouth Medical School, Hanover, New Hampshire

PREFACE

Ten years ago a symposium on Cytotaxonomy was held in London (Proc. Linn. Soc. Lond. *169*:110, 1958) in which a first attempt was made to bring together various disciplines to discuss advances of mammalian cytogenetics and to put them into proper context with the sciences of evolution and taxonomy. The introductory remarks by W. B. Turrill to that symposium, essentially an admonishment to be tolerant of the shortcomings of our respective disciplines, would be a most appropriate beginning to this conference as well. However, the meeting held at Hanover was conceived more along the lines of remarks made by R. B. Seymour Sewell in his presidential address to the same society: "It has been said that scientists in this search for truth are nowadays too much concerned with the accumulation of facts, and make too little use of their imagination in their attempts to explain such facts as they have accumulated." (In "The continental drift theory and the distribution of the Copepoda," *ibid. 166*:149, 1956.)

With this as a background, two years ago we held the first of a series of loosely-structured conferences on reproductive failure in the relaxing atmosphere of this small New England college community. The manuscripts of that meeting have been published (*Comparative Aspects of Reproductive Failure,* Springer-Verlag New York Inc., 1967). The exchange of ideas and information across the boundaries of our relatively rigidly defined disciplines was then a very welcome change from the nature of most of our professional meetings. It encouraged us to plan the meeting which forms the substance of this publication.

In recent years, because of improved methods, wider interest and, particularly, because of the astonishing number of divergent chromosomal changes found in man, there has been much work done in mammalian cytogenetics. Some of it contributes to an understanding of reproductive failure, some to taxonomy. Other work which was sheer fun was undertaken by workers with the ability and the interest to explore areas not trodden before. The result is a mass of rather scattered data, some of it quite raw and seemingly abstract. In looking at all this we concluded that the time had come to bring together some divergent disciplines to sort out findings and put them into their respective contexts, cytologic, taxonomic, evolutionary, or whatever. With this in mind the present contributors willingly agreed to come together, discuss their principal findings (and points of view) with the hope of gradually giving some overall direction to the study of the mammalian genome. As the critical reviewer will spot immediately, this has been only partially successful as there are obvious major gaps. Perhaps some areas of coverage are redundant in this undertaking. This is not easily agreed to, however, by those who sat through this long week of papers and discussions, for all of us felt that, if anything, we needed to hear more rather than less of what was presented. As with the first volume, the reviewer will again deplore the absence of a printed discussion in this book. There was much lively discussion after the papers which, because of its verbal context, was uninhibited and thus, again of greatest value to the participants. Perhaps they profited even more during the long evening sessions which took place in the college dormitories where the visitors lived together. Nevertheless, the printed manuscripts will give much new information of a rapidly developing area of inquiry and may serve as a convenient place for an overview of the general area of mammalian cytotaxonomy and some related subjects.

In addition to these printed pages we heard a stimulating presentation by H. P. Klinger on sex chromatin which, because of his recent paper (Cytogenetics *7*:39, 1968) is not presented here again. Moreover, a majority of registrants were not only entertained by but profited greatly from the two excellent evening workshops on direct bone marrow preparations by T. C. Hsu and on testicular meiosis preparations by C. E. Ford. We had also hoped that we could find an answer to the "Fusion vs. Fission" controversy during this meeting. While "fusion" (Robertsonian translocation of acrocentrics to form metacentrics—J. Morphol. *27*:179, 1916) was generally agreed upon to occur, the converse, "fission" (one metacentric into two acrocentrics), proved a difficult subject even at a late evening round table discussion. The subject is brought up in most of the papers in this volume and clearly is a point of central concern that needs direct cytological and experimental attack.

To have had a professional mammalian taxonomist (Dr. R. G. Van Gelder) as consultant in our midst at all times was most comforting and helpful. To him and all those on our staff we extend our sincere gratitude for their efforts. Dartmouth College made all their facilities available once again in a most generous way and the trusting financial support by the companies and the National Institute of Child Health and Human Development cited at the outset is very warmly acknowledged. Without this help we could not have gathered in such a friendly atmosphere to exchange our materials and views so readily. The week ended in a splendid way. Professor A. Gropp from Germany was able to bring on loan a full-length movie on conservation, "The Last Paradises" by E. Schumacher, which will be remembered by all for its unparalleled photography, beauty and content.

The reader will note the frequent reference to "MCN" (Mammalian Chromosomes Newsletter). This newsletter is an informal publication of great help to workers in this field and is published through the untiring efforts of T. C. Hsu and his staff. It is made available to scientists in this investigative area by him at no cost, but is not usually available in libraries.

Hanover, New Hampshire
August 1968

K. BENIRSCHKE

CONTENTS

Induced Chromosomal Aberrations with Special Reference to Man

K. HIRSCHHORN and M. H. COHEN

Lethal Chromosome Errors

D. H. CARR

Meiosis in Mammals

C. E. FORD

Hybrid Sterility

P. K. BASRUR

Experimental Hybridization

M. C. CHANG, S. PICKWORTH and R. W. McGAUGHEY

Genetic Aspects of Infertility

F. B. HUTT

Gene Action in Human Diploid Cell Strains

R. S. KROOTH

Chromosome Mosaics as Markers in Embryology

H. G. SCHWARZACHER

Multipolar Mitosis and Somatic Segregation in Cell Cultures of Microtus Agrestis

H. G. SCHWARZACHER and F. PERA

Cytogenetics of Marsupials

D. L. HAYMAN and P. G. MARTIN

Insectivore Cytogenetics

D. S. BORGAONKAR

Cytogenetic Studies of Some Armadillos

K. BENIRSCHKE, R. J. LOW and V. H. FERM

Artiodactyl Mammals: Their Chromosome Cytology in Relation to Patterns of Evolution

K. M. TAYLOR, D. A. HUNGERFORD and R. L. SNYDER

Primates

J. EGOZCUE

Considerations of Sex Chromosome Abnormalities in Man

G. E. SARTO, J. M. OPITZ and S. L. INHORN

Anomalies of Sex Chromosomes in Tortoiseshell Male Cats

T. C. JONES

Aneuploidy, Polyploidy and Structural Rearrangement of Chromosomes in Mammals Other Than Man

R. A. McFEELY

Voucher Specimens in Comparative Cytogenetic Studies

R. G. VAN GELDER

Concluding Remarks: Where Do We Go From Here?

S. OHNO

TECHNICAL ADDENDUM

Bone Marrow Preparations for Chromosome Studies

T. C. HSU and J. L. PATTON

Meiotic Preparations from Mammalian Testes

C. E. FORD and E. P. EVANS

A Combined Toluidine Blue Stain and Mounting Medium

G. BRECKON and E. P. EVANS

SPECIES, SPECIATION AND CHROMOSOMES

ERNST MAYR
Harvard University
Cambridge, Massachusetts

From the earliest days of cytogenetics, students of chromosomes have asked what role chromosomal rearrangements play in speciation. Not having a clear idea of the nature of species, some of these authors brought more confusion than light to the subject. It is obvious then that an understanding of the meaning of species is a prerequisite of the understanding of the role of chromosomes in speciation. It is for this reason that Dr. Benirschke has invited me to speak on species and speciation.

Let me begin with a discussion of the species concept. Even though literally hundreds of species definitions have been proposed, they can be reduced to three basic species concepts (Mayr, 1968).

The Essentialist Species Concept This concept, which is also called the typological species concept, goes back to Plato and Aristotle. It recognizes species on the basis of an "essential difference." This is based on the assumption that there are a fixed number of so-called "universals" in the world, and that all observed variability is a manifestation of these universals. Individuals that belong to the same "species" share the same "essence." Each "species" is separated from every other species by a clearcut gap because the underlying universal is likewise separated by a clearcut gap from all other universals. Species of organisms can be defined by an "essential" difference. Acceptance of this species concept induced certain early cytogeneticists to recognize the carrier of each recognizable karyotype as a different species.

Although the essentialist species concept works reasonably well for inanimate objects, for instance "species" of furniture, it corresponds so little to the phenomena of biological variation that it has been universally abandoned (Mayr, 1963).

The Nominalist Species Concept The nominalists, the followers of Occam, went to the extreme of claiming that only individuals exist. According to them, species are arbitrary groupings of individuals brought together by the operations of the human mind. Again, there is some validity in this species concept as far as inanimate objects are concerned. Yet, it is totally invalid when applied to biological species. Every naturalist knows, whether he be a primitive native of the mountains of New Guinea or a trained taxonomist, that the species encountered at a given locality are well defined entities and not an arbitrary figment of human imagination. Why this is so is explained by the third species concept.

The Biological Species Concept The biological species concept was developed to cope with the insufficiency of the two mentioned prebiological concepts. This concept, gradually perfected between 1749 (Buffon) and 1905 (K. Jordan), states, to phrase it in the simplest possible way, that species are "protected gene pools." Species are natural populations which are reproductively isolated by so-called isolating mechanisms from other natural populations. This concept stresses genetic distinctness rather than morphological difference. The isolating mechanisms which protect the gene pool of a species from the admixture of genes from other species are part of the evolved genetic program of the species.

There is no better way to understand the biological significance of species than trying to imagine a universe without species. Let us assume for instance that all living organisms formed a single genetic continuum instead of being organized into species. Since there would be no isolating mechanisms, any individual would mate at random with some other individual and produce with it, through recombination, the zygotes of the next generation. Once in a while one of these zygotes would, by chance, be considerably superior in utilizing the environment than the previously existing zygotes. The chance that such a superior individual would meet a suitable mate during the next breeding season is virtually zero since, as we postulated, the choice of mates is not controlled by isolating mechanisms. The result is that the genotype of the superior zygote would be broken up by recombination and lost. We now see what the biological meaning of species is. Organizing genetic diversity into protected gene pools guarantees the protection of harmonious gene combinations, that is of gene combinations which through a long history of evolutionary selection have become coadapted to each other. It is these protected gene pools which we call species.

There are actually only two mechanisms by which superior genotypes can be protected and both of them are found in the animal kingdom. One is simply the abandonment of sexuality. When this occurs, the superior genotype reproduces asexually (uniparentally). Parthenogenesis, the most frequent form of uniparental reproduction, is widespread in the animal kingdom although not yet substantiated (in the wild) in birds or mammals. Abandonment of sexual reproduction sometimes leads to immense temporary success but it has a fatal Achilles heel, an inability to cope with changes of the environment. A genotype which is now superior almost invariably becomes quite inferior when there is a drastic change of the environment. Unless it returns to sexual reproduction, it is doomed to extinction.

The other mechanism by which superior genotypes can be protected is to organize them into species. There is opportunity for a limited amount of variability within each species but no danger that the essential canalizations and well balanced enzyme systems will be broken up by recombination.

This introduction has given us the necessary background to ask searching questions concerning the biological meaning of chromosomal changes during evolution. Improved cytological techniques have resulted in a wealth of new data. However, as scientists, we are not satisfied with descriptive data gathering, we want to interpret, we want to understand. We are looking for an answer to the question: "What is the selective significance of all forms of chromosomal rearrangements?" And, further, the more specific question: "What role do such chromosomal changes play during speciation?" It is now clear that no single sweeping answer can be given to these questions. The various books on the role of chromosomes in evolution well describe the multiple function of chromosomal rearrangements in helping with the storage of genetic variability, with the maintenance of balanced gene combinations, and with the establishment of isolating mechanisms. I have emphasized the importance of protecting the gene pool against the influx of alien genes. It is now important to emphasize the opposite, the advantage of a certain amount of genetic variability, for it has become evident that too homogeneous a gene pool is likewise a disadvantage. There are a number of different reasons why a certain amount, and as we now suspect a very considerable amount, of genetic variability is favored by selection. One of these is that heterozygosity at many loci seems to improve fitness. A second advantage of genetic variability is that it gives a particular population an increased amplitude in the utilization of the environment; it permits the utilization of more subniches. This was first pointed out by Ludwig and has since been confirmed by Howard Levene, C. C. Li, and many others. The third advantage of genetic variability is that it "preadapts" a population to meet fluctuations in external selection pressures. Every environment is changing all the time, a drought year will be followed by a year with heavy rain, an extremely severe winter will be followed by an unusually mild winter, a previously prevalent disease will be replaced by a new pathogen, etc. Obviously a population which can segregate new gene combinations that can cope with the new selection pressure is selectively superior to a population that lacks such storage mechanisms. Chromosomal mechanisms such as polymorphism for paracentric inversions and any other mechanisms that affect linkage are important. Some of these devices permit the locking up of a whole set of epistatically well adjusted genes into a single "supergene" which, when there is a polymorphism for such supergenes, can combine the advantages of harmonious balance with genetic diversity. It is evident that a great deal of karyotype variation serves to establish a compromise between two very conflicting demands on the genotype: to store as much genetic variability as it can possibly incorporate, and to avoid the production of too many deleterious recombinants. Where the exact balance lies between these two incompatible demands depends on the size of the gene pool as well as on all sorts of aspects of the breeding system (White, 1954; Darlington, 1939; Grant, 1965).

Reproductive Isolation and Chromosomal Mutation

Some early cytogeneticists took it for granted that there was a one to one relation between speciation and chromosomal rearrangement and that neither could occur without the other. Needless to say, this misconception has long since been refuted. In other cases, cytogeneticists found that representatives of two species which co-existed at a locality differed by a given chromosomal rearrangement. They then postulated that one of these species had arisen sympatrically from the other one by a chromosomal mutation. In reality there is a great deal of independence between the two phenomena, the acquisition of reproductive isolation and the occurrence of new chromosome arrangements. This is best illustrated by a diagram (Fig. 1).

		Chromosomal Mutation	
		No	Yes
Reproductive Isolation	No	X	2
	Yes	1	3

Fig. 1. 1 = purely genic isolating mechanisms; 2 = chromosomal polymorphism not affecting isolating mechanisms; 3 = chromosomal isolating mechanisms.

Acquisition of Reproductive Isolation Without Chromosomal Rearrangement Numerous cases have been described in the literature where speciation and the acquisition of isolating mechanisms was apparently purely genic in nature, not involving any visible karyological changes. The most convincing documentation for purely genic speciation was recently given by Carson and co-workers (1967). In a group of 22 Hawaiian species of Drosophila, four groups of species were found within each of which several related species have identical banding patterns in the giant salivary gland chromosomes. In one case no less than three species are identical in their banding patterns. These are unquestionably good species because some co-exist in the same habitat without interbreeding. Clearly, in these cases the isolating mechanisms were established without karyotype mutation.

New Chromosome Arrangements There is now a huge class of phenomena known where chromosomal rearrangements merely lead to chromosomal polymorphism. The paracentric inversions of Drosophila are the best known illustration. It is obvious that this kind of chromosomal mutation has nothing whatsoever to do with speciation.

This category includes all the situations where two closely related species are reproductively isolated and also differ karyologically in such a way that there is a reduction of fitness in the hybrids owing to mechanical difficulties in chromosomal pairing, etc. This category includes various kinds of translocations, as well as any other chromosomal mechanisms that might lead to nondisjunction or difficulties in pairing such as pericentric inversions and Robertsonian changes. It is outside my area of competence to comment on the reduction in fitness of heterozygotes in this class of chromosomal rearrangements. The fact that so often closely related species differ precisely by this kind of mechanisms raises a number of interesting questions which will be discussed below.

We can now summarize that classes 1 and 2 of our diagram prove that many chromosome mutations have nothing to do with speciation and that speciation can occur without chromosome mutations. The early assumption of a one to one relation of the two phenomena is clearly wrong.

This still leaves us with the need for further analysis of class 1.

The Bottleneck of Heterozygosity

The third box in our diagram includes the cases where speciation is associated with chromosomal rearrangement. Astonishingly often we find that a species is distinguished from another species which is apparently its closest relative by a major chromosomal rearrangement. When this would cause mechanical difficulties in a heterozygote, one must ask oneself how such a new gene arrangement could have ever passed through the stage of heterozygosity in the face of counter selection. I have discussed elsewhere (Mayr, 1963, p. 540) the difficulties caused by such incompatibilities. Homozygotes for new arrangements would occur far too rarely in large populations to have any chance to become established and to acquire the other two prerequisites of successful speciation, perfected isolating mechanisms and ecological exclusion. The evolutionist, therefore, believes that such chromosomal shifts can occur only in very small peripherally isolated populations where they are exposed to selection pressures that differ from those in the main portion of the range. Under these conditions of severe inbreeding, homozygotes are frequent and the opportunity is provided for the possibility that the selective superiority of the new homozygous karyotype is so great that it overrides the disadvantage of the heterozygotes. Any chromosomal mechanism will be favored by selection, as shown by Wallace (1959), which will facilitate reproductive isolation between the incipient new species and the parent population.

The periphery of the species range, where species populations are often broken up into small, widely separated, and heavily inbreeding populations is the ideal location for this type of speciation. Simultaneously, in such populations there is often also a veritable "genetic revolution" (Mayr, 1954) of the genes not involved in the chromosomal rearrangement. All the conditions are met for an extraordinarily rapid speciation, complete with the perfection of isolating mechanisms and a shift in niche occupation.

This type of rapid speciation seems to occur most frequently in highly sedentary animals that tend to occur in small colonies. It has been described for newts (Spurway and Callan, 1960), for grasshoppers (White and Chinnick, 1957; White, 1968), beetles (Smith, 1960), Drosophila (Carson, 1965), and plants (Lewis, 1966). I suspect that we will learn of other instances during this conference, for instance in Professor Wahrman's presentation.

I would like to make one final point. Most widespread and successful species are rich in genic and chromosomal polymorphism. Curiously, it has been found that those populations which have speciated with the help of chromosomal mechanisms are generally quite monomorphic with respect to those other kinds of chromosomal variations (like paracentric inversions) which cause no mechanical difficulties. This supports the idea that extreme homozygosity is a temporary condition during the shift from one karyotype to another one. Indeed, it may be a prerequisite for such a shift.

The lesson we have learned from the analysis of chromosomal speciation is that it cannot be understood by merely looking at microscopic slides. All sorts of ecological, behavioral, geographical, and historical factors must be carefully considered. The size of the population in which the chromosomal rearrangement takes place is of the utmost importance; so is the amount of gene flow between this population and others of the species; so is the nature of the environment of the incipient species as compared to that of the remainder of the species. All this must be kept in mind when we attempt to reconstruct the path of speciation of species now living geographically adjacent to each other and reproductively isolated as well as differing in a largely incompatible chromosomal rearrangement.

Chromosomal evolution cannot be fully understood unless analyzed within the context of the population in which it occurs.

References

Carson, H. L.: Chromosomal morphism in geographically widespread species of Drosophila, p. 503–531, in H. G. Baker and G. L. Stebbins (eds.), The Genetics of Colonizing Species, Academic Press, New York, 1965.

———, F. E. Clayton and H. D. Stalker: Karyotypic stability and speciation in Hawaiian Drosophila. Proc. Natl. Acad. Sci. *57*:1280, 1967.

Lewis, H.: Speciation in flowering plants. Science *152*:167, 1966.

Mayr, E.: Animal Species and Evolution. Cambridge: Belknap Press, 1963.

Smith, S. G.: Cytogenetics of insects. Ann. Rev. Entomol. *5*:69, 1960.

———: Cytogenetic pathways in beetle speciation. Canad. Entomol. *94*:941, 1962.

Spurway, H. and H. G. Callan: The vigour and male sterility of hybrids between the species *Triturus vulgaris* and *T. helveticus*. Jour. Genetics *57*:84, 1960.

Wallace, B.: The influence of genetic systems on geographical distribution. Cold Spring Harbor Symp. Quant. Biol. *24*:193, 1959.

White, M. J. D.: Speciation in animals. Austr. Jour. Sci. *22*:32, 1959.

———: The role of chromosomal translocations in urodele evolution and speciation. Amer. Nat. *95*:315, 1961.

———: Models of speciation. Science *159*:1065, 1968.

——— and L. J. Chinnick: Cytogenetics of the grasshopper *Moraba scurra*. III. Austr. Jour. Zool. *5*:338, 1957.

MECHANISMS OF CHROMOSOMAL CHANGES IN MAMMALIAN SPECIATION

T. C. HSU AND RODNEY A. MEAD

Section of Cell Biology, Department of Biology, The University of Texas M. D. Anderson Hospital and Tumor Institute at Houston, Houston, Texas, and Department of Biological Science, University of Idaho, Moscow, Idaho

Ultimately, biologists should be able to describe each species (or subdivisions thereof) in terms of base sequences of all its genes and the order in which the genes are aligned. Evolution can then be studied by comparing the base sequences and arrangement of genes in various taxa. This naturally is not possible at the present time, so that only cruder criteria, phenotypes such as morphology, anatomy and behavior, have been used to elucidate species relationships, population structure, and evolutionary trends.

One of the criteria effectively used by geneticists for many years to study evolution is the vehicle for the genetic material, the chromosomes. Their number, morphology, arrangement, etc., can all be employed to gain information concerning phylogenetic relationships. This type of study has been especially fruitful with plants and with dipterous insects. In mammals, such studies have been a recent adventure.

It is not the purpose of this paper to review the subject, "chromosomes and speciation," which has an enormous amount of literature. Interested readers are urged to refer to several classic books (Patterson and Stone, 1952; White, 1954; Stebbins, 1950) for details. In fact, from the scanty data available, chromosomal evolution in mammals does not seem to deviate from the principles well explored and exemplified in other life forms. However, because of the close association of mammals with human welfare, knowledge concerning their interrelationships and their relationships to man would certainly be useful in many branches of biological sciences. A discussion on the types of chromosomal changes that play important roles in Mammalia should be worthwhile.

It should be emphasized at the beginning that chromosomal constitution, especially metaphase chromosome morphology, represents only one set of phenotypes. Placing too much weight on chromosomal features in deciphering phylogenetic relationships is as dangerous as using any single set of morphological characteristics. A difference between two individuals in terms of chromosome number and morphology does not necessarily justify the conclusion that these two should be considered two separate species, or subspecies, etc. Cytogeneticists have found numerous chromosomal aberrations in human populations, yet no one regards the individuals with these aberrations as separate species. However, as long as the investigators realize the limitations, karyological features should be exploited as fully as possible to complement, supplement, and perhaps occasionally correct, the conclusions derived from the classic morphological and anatomical studies.

A good example to illustrate the above-mentioned point is provided by Hsu and Arrighi (1968) in their studies of the genus *Peromyscus*. Taxonomists have subdivided the species *P. truei* into a number of subspecies. Cytological analysis revealed that the specimens from New Mexico and those from California had entirely different karyotypes, one similar to that of *P. californicus* and the other similar to that of *P. gossypinus*. These two karyotypes have many intermediate steps which are represented by different species. It is rather inconceivable that two identical sequences of chromosomal changes could have occurred, one traversing many species, while the other was within a species. Since the chance for identical or nearly identical chromosomal aberrations to occur twice is considered very small, the chance for many identical changes to occur twice should be negligibly small. The authors therefore seriously questioned the validity of classification of *P. truei*. It should be pointed out that Hoffmeister (1951) had a similar feeling based upon morphological and antomical findings. It is therefore satisfying to see that cytological data complement detailed morphological analyses instead of opposing them.

Another example is the classification of the tree shrew, genus *Tupaia*, which may contain at least three closely related species: *T. glis*, *T. chinensis*, and *T. longipes*. The literature regarding the chromosomes of *T. glis* has been confusing. Hsu and Johnson (1963) reported that the diploid number of *T. glis* is 60. Bender and Chu (1963) agreed with this count. Klinger (1963), however, found a diploid number of 62. Egozcue *et al.* (1968) also found a diploid number of 62 in 4 specimens, which were subsequently identified as *Tupaia chinensis* (Borgaonkar, personal communication). Careful analysis of all three subspecies (Arrighi *et al.*, in press) showed that both *glis* and *longipes* have 60 chromosomes but *chinensis* has 62. This example indicates that karyological characteristics, when observed repeatedly, are useful in taxonomic and evolutionary evaluations.

In using karyological characteristics to study evolution, one must also realize that comparison between distantly related taxa may be meaningless. In most cases, comparing karyotypes of different genera within a family or a subfamily, and sometimes even different species within a genus, can be futile because the chromosomes of these taxa bear no resemblance to one another. Therefore, no interpretations can be made regarding their relations. For this reason it is useless to talk about evolution in mammals as a whole by comparing karyotypes. When applied to taxa within a genus or within a species, this group of characteristics can be useful if large numbers of representatives are analyzed.

We must, at this juncture, stress another point. Suppose for one reason or another one decides that karyotype A is the ancestral form of karyotype B. This does not necessarily mean that species M with karyotype A is more primitive than species N with karyotype B. Evolutionary processes may have taken place at various directions and levels other than changes in chromosome morphology. By the same token, several species with the same karyotype are not necessarily equal in rank in the course of evolution. In Chiroptera, for example, species of one genus often exhibit identical karyotypes (Baker and Patton, 1967). Such examples, though disappointing to the workers who hope to compare karyological relationships, provide excellent argument for using chromosome features as a tool in evolutionary studies.

Numerical Changes

Two major types of changes, numerical and structural, can be detected at the metaphase chromosome level in most mammalian material. Polyploid evolution can be disposed of at the onset since its existence has never been found. Polyploid cells are commonly encountered in somatic tissues, but polyploid species rarely exist in vertebrates. Changes in chromosome number in mammals, therefore, occur only at the diploid level. Strikingly, the diploid number of mammals vary from 10 in *Protemnodon* to 84 in *Rhinoceros,* and practically all the numbers in between have been found. However, the tremendous difference in chromosome number does not suggest a tremendous difference in genetic material. When DNA content per cell or total chromosome length is considered, all eutherian mammals appear to possess similar amounts of DNA, regardless of their diploid numbers (Atkin *et al.,* 1965). Thus the difference is merely in packaging, if we regard chromosomes as packages of genetic material.

The Robertsonian Process

The Robertsonian-type of chromosomal evolution is prevalent in mammals, so that it deserves special attention. When two closely related species are compared, one may find that species A has a diploid number of N, and species B has N-2. The karyotype of species A usually has two pairs of extra acrocentrics but lacking one pair of biarmed chromosomes, and species B, the reverse. Either a biarmed chromosome splits into two acrocentrics in the process of evolution, or two acrocentrics join to form a biarmed element. Thus, if one regards each arm of a metacentric to be equivalent to a single acrocentric, the total number of arms (fundamental number of Matthey, 1945) becomes a good guide in evaluating phylogenetic relationships.

The mechanisms by which a metacentric chromosome splits into two acrocentrics and two acrocentrics join to form a metacentric are controversial. It is unfortunate that cytologists use such misleading terms as "fission" and "fusion" respectively to describe these processes, because we really do not have unequivocal evidence that centromeres do fuse or cleave. The prevailing concept of metacentric formation is the "double break and translocation" hypothesis. That is to say, a break occurs close to the centromere on the long arm of one acrocentric and another break at the submicroscopic "short arm" of the other acrocentric. A reciprocal translocation would result in a biarmed element and a "free" centromere, which is subsequently lost.

The "double break and translocation" hypothesis sounds plausible, but there is no direct proof. One difficulty is that the "free" centromere is supposed to be lost. Thus, whenever a metacentric formation is observed, the "free" centromere is not present to tell the story. However, how do we know that the free centromere must be lost? We should realize that a centromere in light microscopy is observed as an achromatic region. An achromatic region is defined only when there are chromatic regions at both sides. When it does not carry visible chromatin, how can we decide that it is not there?

Another possibility is that two centromeres may simply fuse. If a translocation scheme must be entertained, the breaks may occur on the submicroscopic "short arm" of both acrocentrics, and a translocation results in a biarmed chromosome with two centromeres. These two centromeres are so close to each other that they function as if there were only one. In this case the free ends should be lost since they are acentric.

Even though the reciprocal translocation hypothesis still requires direct proof in the study of chromosomal evolution, some evidence has been obtained in induced aberrations to show that such a process did occur. This was usually found in the immediate cell generation after the translocation had taken place. On the other hand, the fission hypothesis appears to have no experimental evidence. In supporting a direct fission hypothesis, one must assume that one functional centromere can be split into two functional centromeres. If this cannot be accomplished, a "fission" of a metacentric chromosome must acquire an extra centromere to yield two good acrocentrics. Sturtevant and Tan (1939) considered the possibility of "borrowing" the centromere from the Y chromosome. In this case, the process again requires two breaks and a translocation. In *Drosophila melanogaster* the Y chromosome is the only acrocentric of the complement. When many acrocentrics are present in a karyotype, as in many mammals, there is no reason why any acrocentric cannot be a donor. However, if direct fusion from two acrocentrics is possible, a fission should be just as easy when a biarmed element contains two centromeres.

At the present stage of development of biological sciences, when integration of information from molecules to populations is in sight, we really should not use empirical concepts which appeared logical 30 or 40 years ago. Whether a centromere can be split into two functional centromeres, whether a misdivision of centromeres is feasible, whether fusion of centromeres can take place, etc., cannot be answered by observations with light microscopy, and some of the antiquated interpretations should be discarded in the light of more recent developments. The work of Brinkley and Stubblefield (1966) on the fine structure of the kinetochore (centromere) is highly significant as the first step toward our understanding of the molecular composition and the architecture of this chromosomal component. However, much more work is required to resolve the problem of centromere structure, so that we should not, at the present time, make unreasonable interpretations on the role of centromeres in evolution.

Circumstantial evidence for both "fusion" and "fission" appears available in animal evolution. Without any concern of the mechanisms, it is quite possible that chromosomes change in both directions. If we consider both processes as translocations whose existence has ample evidence, no one would challenge the theoretical possibility. However, if we do not accept circumstantial evidence, we find little support for the proposals of Staton (1967a, b) and Todd (1967), who consider "fission" as the main force of chromosome evolution in some mammalian taxa. On the other hand, some rather undeniable evidence for metacentric formation can be cited. The best example is found in human populations. No case of fission (even with the translocation scheme) has been found thus far, but fusions (D/D, D/G and G/G translocations) are plentiful in the human cytogenetic literature.

The karyotype of the least cotton rat, *Sigmodon minimus* Mearns, contains 30 chromosomes, most of which are acrocentric. In a small population of southwestern New Mexico, some individuals were found to show 29 chromosomes, missing two acrocentrics but having a new metacentric. Studies on meiosis revealed that the metacentric was a "fusion" product of two homologous chromosomes (Hsu, in press). It is almost impossible to apply the fission hypothesis in this case.

In *Mus minutoides,* the basic karyotype appears to be 36 acrocentrics, including both sex elements. Matthey (1965) described cases of metacentric X and metacentric Y with a corresponding loss of an autosome. In still other instances, both the X and the Y were metacentric. Since the Y chromosome of the majority of mammals thus far analyzed is an acrocentric, the Matthey report again supports the "fusion" process.

If metacentric chromosome formation is indeed a result of reciprocal translocations, it would be unreasonable to assume that a break always occurs at a place so close to the centromere that little or no chromatin goes with the free centromere. In other words, one should be able to find cases where centromeric fragments were not lost. Such an instance was indeed found in the spotted skunks. The taxonomy of spotted skunks, *Spilogale putorius* (family Mustelidae) has been a matter of debate, but at the present, we shall follow the system of Van Gelder (1959). This species has a large number of subspecies, but from our preliminary study, only a few have been karyotyped. Two eastern U.S. subspecies, *S. p. ambarvalis* Bangs, obtained from Florida, and *S. p. interrupta* (Rafinesque), obtained from Texas, showed a diploid number of 64 with 4 pairs of biarmed autosomes (Fig. 1). However, the karyotype of one subspecies from the western U.S., *S. p. latifrons* Merriam, obtained from Oregon, showed a diploid number of 60, including 6 pairs of biarmed autosomes (Fig. 2). The karyological difference can be explained by two sets of single "fusions" or "fissions."

Of interest is the karyotype of another western subspecies, *S. p. phenax* Merriam, which occurs in California and supposedly integrates with *latifrons.* The diploid number is 64, but the karyotype contains 6 pairs of biarmed autosomes instead of 4. In addition, 4 of the acrocentrics are minutes (Fig. 3). If the chromosome breaks took place not directly under, but somewhat away from the centromere, a minute

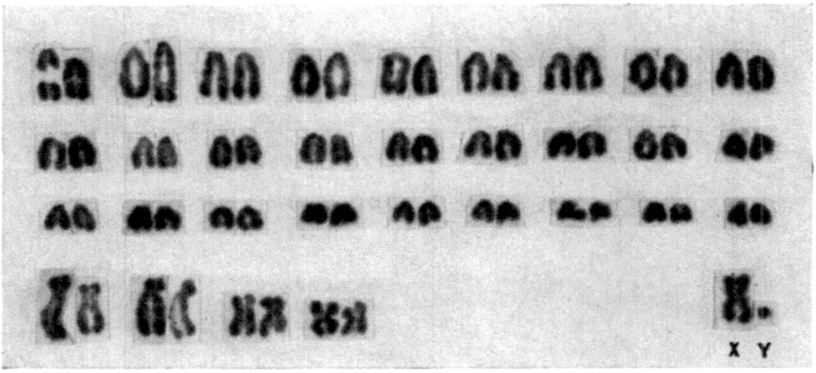

Fig. 1. Karyotype of a male *Spilogale putorius interrupta* (2n = 64.) Note 4 pairs of biarmed autosomes.

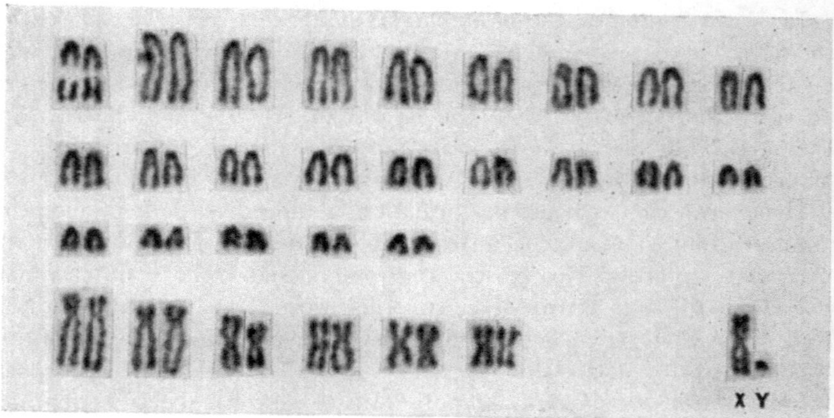

Fig. 2. Karyotype of a male *Spilogale gracilis latifrons* (2n = 60). Note 6 pairs of biarmed autosomes.

chromosome could be produced together with a new metacentric. The karyotype of *phenax* was therefore an intermediate between those of *interrupta* and *latifrons,* and it constitutes one of the few examples to support the translocation hypothesis.

The *Spilogale* story is far from complete, since numerous subspecies have not yet been investigated. The example, however, suggests that minute chromosomes occurring in a number of mammalian species (e.g., in the bats *Myotis, Plecotus,* in the pocket gopher *Thomomys,* etc.) may be explained by the same mechanism, viz., leftover centromeric fragments. If this is true, computing fundamental numbers should be careful not to include the minutes, whether they are biarmed or single-armed.

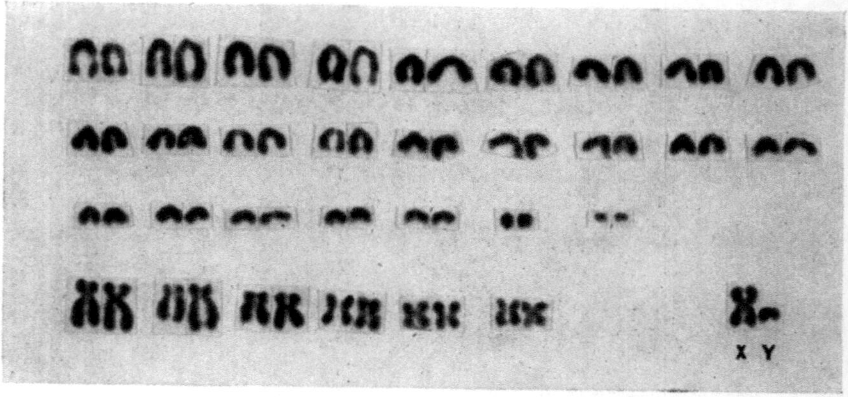

Fig. 3. Karyotype of a male *Spilogale gracilis phenax* (2n = 64). Note 6 pairs of biarmed autosomes and 2 pairs of tiny acrocentrics.

We wish to reiterate that the discussion made thus far does not deny the possibility or even the probability of fission. We only say that the break-and-translocation type of Robertsonian fusion appears to be more frequent than any other type of chromosome changes. In class Mammalia, examples of this type of numerical changes can be cited in practically every order or even every family thus far analyzed. The most amazing case is probably the African pigmy mouse, *Mus minutoides*, of subgenus *Leggada* (cf., Matthey, 1966). The subgenus, unlike the rest of *Mus* ($2n = 40$), has a basic karyotype of 36 acrocentric chromosomes. Various subspecies of *M. minutoides* showed a variety of karyotypes down to a diploid number of 18 with all elements metacentric. The fundamental number, however, remained unaltered in most cases. It is of great interest to note that karyological differences were not only obvious among subspecies, populations, but also within populations. Such polymorphism strongly indicates that new changes occur repeatedly and that the species is at the present time in the crossroads of evolution, experimenting with all possible ways to seek the best combination.

Non-Robertsonian Changes

Changes of chromosome morphology do not have to involve fusion or fission. Pericentric inversion and reciprocal translocation of unequal length may produce drastic changes in karyotype without a reduction or an increase of the diploid number (or the number of centromeres).

While the Robertsonian type of chromosome evolution prevails in *Leggada*, the non-Robertsonian changes are almost exclusive in certain genera of Cricetinae, e.g., *Peromyscus*. From 19 species of *Peromyscus* thus far analyzed (Hsu and Arrighi, 1968), the diploid number is 48. The number of autosomal arms, however, varied from 52 to 92.

When stepwise changes can be found in the karyotypes of a group of related taxa, a phylogenetic tree can be suggested. However, it is a sheer conjecture when one tries to consider which chromosomal phenotype (acrocentric or metacentric) is more ancestral, because a pericentric inversion can turn an acrocentric into a metacentric just as easily as it can turn a metacentric into an acrocentric. Nevertheless, regardless of the direction, it is clear that a karyotype with only one metacentric should be the intermediate form between a karyotype with two acrocentrics and the other with two metacentrics, provided that one of the two metacentrics is homologous to the metacentric of the intermediate type. Homology can only be suggested, not determined, by morphology of the mitotic metaphase chromosomes. In meiosis of the hybrids, the pairing behavior would give more precise information.

Why the number of centromeres must remain constant in some taxa is an open question. It is just as puzzling as the fact that the Robertsonian process operates as a major force in chromosomal evolution in other taxa.

Like *Leggada,* many species of *Peromyscus* have intraspecific polymorphism or even intrapopulational polymorphism. This is especially true in two of the most widely distributed species, *P. leucopus* and *P. maniculatus.* The population structure of these species is extremely complex, so that the problem deserves intensive study by many teams of investigators. Cytological features of *Peromyscus* can be effectively used as a model to elucidate many evolutionary processes when morphological and anatomical characteristics have exhausted their usefulness.

The ultimate aim of taxonomy is not to devise a static classification system, but to interpret evolution, which is a dynamic process of nature. It is understood that morphology and field observational records may be the only information available for museum specimens, but this should not stop naturalists from exploring new ideas and new approaches to the problem, of which cytology is but one. Behavior, physiology, biochemistry or even biophysics should all be included in the repertoire of future taxonomists. Only by effective collaboration between taxonomists and scientists of other fields can the answers to the broad problem of evolution be meaningful.

Acknowledgment

This work was supported by Research Grant GB-1867 from National Science Foundation.

References

Arrighi, F. E., M. W. Sorenson and L. R. Shirley: Chromosomes of the tree shrews (Tupaiidae). Cytogenetics, in press.

Atkin, N. B., G. Mattinson, W. Beçak and S. Ohno: The comparative DNA content of 19 species of placental mammals, reptiles and birds. Chromosoma *17*:1, 1965.

Baker, R. J. and J. L. Patton: Karyotypes and karyotypic variation of North American Vespertilionid bats. J. Mammal. *48*:270, 1967.

Bender, M. A. and E. H. Y. Chu: The chromosomes of primates. In: Evolutionary and Genetic Biology of Primates (J. Buettner-Janusch, ed.) *1*:261–310, New York: Academic Press, 1963.

Brinkley, B. R. and E. Stubblefield: Fine structure of the kinetochore of Chinese hamster cells *in vitro*. Chromosoma *19*:28, 1966.

Egozcue, J., B. Chiarelli, M. Sarti-Chiarelli and F. Hagemenas: Chromosome polymorphism in the tree shrew (*Tupaia glis*). Folia Primat. *8*:150, 1968.

Hoffmeister, D. F.: A taxonomic and evolutionary study of the piñon mouse, *Peromyscus truei*. Ill. Biol. Monogr. *21*:1, 1951.

Hsu, T. C.: Robertsonian fusion between homologous chromosomes in a natural population of the least cotton rat, *Sigmodon minimus* (Rodentia, Cricetidae). Experientia, in press.

——— and F. E. Arrighi: Chromosomes of *Peromyscus* (Rodentia, Cricetidae). Evolutionary trends in 20 species. Cytogenetics *7*:417, 1968.

——— and M. L. Johnson: Karyotypes of two mammals from Malaya. Amer. Nat. *97*:127, 1963.

Klinger, H. P.: The somatic chromosomes of some primates (*Tupaia glis, Nycticebus coucang, Tarsius bancanus, Cercocebus aterrimus, Symphalangus syndactylus*). Cytogenetics *2*:141, 1963.

Matthey, R.: L'evolution de la formule chromosomiale chez les vertébrés. Experientia *1*:50, 78, 1945.

———: Un type nouveau de chromosomes sexuels multiples chez une souris Africaine du groupe *Mus* (*Leggada*) *minutoides* (Mammalia–Rodentia). Chromosoma *16*:351, 1965.

———: Le polymorphisme chromosomique des *Mus* Africains du sous-genre *Leggada*. Révision générale portant sur l'analyse de 213 individus. Rev. Suisse de Zool. *73*:585, 1966.

Patterson, J. T. and W. S. Stone: Evolution in the Genus *Drosophila*. New York: Macmillan, 1952.

Staton, D.: Karyotypes and the phylogeny of New World primates. Mammalian Chromosomes Newsletter *8*:203, 1967.

———: A preliminary analysis of karyotypic evolution in Catarrhines. *ibid.* *8*:220, 1967.

Stebbins, G. L., Jr.: Variation and Evolution in Plants. New York: Columbia University Press, 1950.

Sturtevant, A. H. and C. C. Tan: The comparative genetics of *Drosophila pseudoobscura* and *D. melanogaster*. J. Genet. *34*:417, 1937.

Todd, N. B.: A theory of karyotype fissioning, genetic potentiation and eutherian evolution. Mammalian Chromosomes Newsletter *8*:268, 1967.

Van Gelder, R.: A taxonomic revision of the spotted skunks (genus *Spilogale*). Bull. Amer. Mus. Nat. Hist. *117*:325, 1959.

White, M. J. D.: Animal Cytology and Evolution. London: Cambridge University Press, 1954.

THE MAMMALIAN GENOME IN EVOLUTION AND CONSERVATION OF THE ORIGINAL X-LINKAGE GROUP

SUSUMU OHNO
Department of Biology, City of Hope Medical Center, Duarte, California

I shall take a rather detached standpoint and begin by asking a rather awkward question, "Is it of any importance that mammals display enormous diversity in their karyotypes?"

Each structural gene is a DNA cistron which specifies the amino acid sequence of a polypeptide chain it produces. A mutation is a base substitution which we recognize as an amino acid substitution at one site of a peptide chain. As the complete amino acid sequences of many proteins from diverse species became known, it has become very clear that each structural gene is conservative in that natural selection tolerates only those mutations which do not affect the functionally critical sites of the molecule. For instance, hemoglobin alpha-chain of the horse differs from human alpha-chain at 18 of the 141 sites (Braunitzer and Matsuda, 1963). Yet, as an oxygen carrier, the two are equally effective. Thus, these differences merely reflect tolerable mutations two species have independently accumulated at the alpha-chain locus since they diverged from a common ancestor.

The karyotypic differences can be viewed in the same way. Man has 46 chromosomes, while the horse contains 64 chromosomes in its diploid complement. Yet no one would take the view that $2n = 46$ is a magic formula which made *Homo sapiens* what they are. Man and the horse show markedly different karyotypes, merely because they have followed independent paths of evolution for quite some time.

As the extremely conservative nature of structural genes became known, we began to realize that gene duplication rather than allelic mutations of already existing gene loci has been the major force of evolution.

As long as a single gene locus carries out a vital function, natural selection relentlessly imposes restriction on that gene's freedom to mutate. Only when the genome (haploid set) incorporates two doses of the same gene by duplication is the severe restriction relaxed for a duplicate. An original continues to carry out the function assigned to that locus; thus, a duplicate is now redundant. This redundant locus is permitted to accumulate base substitutions which were formerly forbidden and finally emerges as a new gene locus with a function related to but distinctly different from that assigned to the original gene locus.

Among placental animals, the diploid chromosome number ranges from a high of 84 in the black rhinoceros (Hungerford and Snyder, 1967) to a low of 17 in two species of rodents (Matthey, 1953; Ohno *et al.,* 1963). Yet the genome size remains the same containing about 3.5 mg $\times$ 10^{-9} DNA (Mandel *et al.,* 1950; Atkin *et al.,* 1965). Apparently, much of the gene duplication occurred during earlier parts of the vertebrate evolution (Ohno *et al.,* 1968). As a result, a common ancestor to all the mammals was already endowed with the characteristic degree of genetic redundancy, and subsequent speciation within this class has been carried out by differential use of the fixed amount of redundant genetic materials. In this respect, evolution of mammals is not very interesting.

For instance, mammalian hemoglobin is a tetrameric molecule made of two different classes of subunits; alpha-chains having 141 amino acid residues and beta-like chains having 145 or 146 residues. Duplication of an ancestral gene produced two separate gene loci for the two classes of hemoglobin polypeptides. The very fact that human alpha-chain is more similar to the mouse alpha-chain (17 differences) and the horse alpha-chain (18 differences) than to his own beta-chain (84 differences) reveals that this important duplication occurred before the emergence of the first mammal.

Similarly, mammalian antibody molecules are composed of two different classes of subunits; light-chains and heavy-chains. L-chains are made of about 210 amino acids, while H-chains are twice or three times longer than the L-chain. Man, the mouse and other mammals have two separate gene loci for kappa and lambda light-chains. Here again, the amino acid sequence of human kappa is more similar to that of the mouse kappa than to that of his own lambda. There is little doubt that the primordial mammal, when first emerging on this earth, already had the duplicated gene loci for immunoglobulin light-chains; one for kappa and the other for lambda (Putnam *et al.,* 1967).

In view of the fixed genome size, diverse karyotypes displayed by living mammals appear to mean only one thing; that in this class of vertebrates, there has never been an a priori selection against random fragmentations and regroupings of the original linkage groups during speciation from the common ancestor. Indeed, mammals appear quite capable of coordinating the activities of two unlinked genes so that there has been no need to keep functionally related genes in close linkage. For example, two enzymes, glucose-6-phosphate dehydrogenase (G-6-PD) and 6-phosphogluconate dehydrogenase (6-PGD), catalyze two successive steps of the pentose phosphate shunt, yet the G-6-PD locus is X-linked in man (Childs *et al.,* 1958; Kirkman *et al.,* 1960), in the horse and donkey (Trujillo *et al.,* 1965; Mathai *et al.,* 1966) and in the hare (Ohno *et al.,* 1965), while the 6-PGD locus is autosomally inherited in man and the rat (Parr, 1966), in the deer mouse (Shaw, 1965) and in the cat (Thuline *et al.,* 1967).

In order to produce $\alpha_2\beta_2$ hemoglobin molecules, the equimolecular synthesis of alpha-chains and beta-chains is required of individual erythroblasts, yet the alpha-chain locus is not linked to the beta-chain locus. The equimolecular synthesis of L- and H-chains is also required of individual plasma cells secreting L_2H_2 7S immunoglobulin, yet in man (Kunkel, 1963–64), the rabbit (Oudin, 1966) as well as in the mouse (Herzenberg *et al.,* 1967), the L-chain locus is carried by one autosome and a cluster of H-chain loci by another autosome. Furthermore, it appears that when functionally related genes are clustered together in close linkage, mammals will prefer to use one of them at a time rather than activating all of them concomitantly. A cluster of immunoglobulin H-chain gene loci includes that for gamma-chain of IgG class and that for alpha-chain of IgA class, yet each clone of plasma cells make either IgG or IgA but never both.

In view of the above, if we are uncritical, we may interpret diversity of karyotypes displayed in mammals to mean that they enjoy complete freedom in reorganizing their linkage groups, provided that this reorganization is done within the framework of the constant genome size. However, it will be immediately realized that most reciprocal translocations which unite four broken halves of two non-homologous chromosomes to two new chromosomes are incompatible with successful speciation, for translocation heterozygotes suffer semi-sterility. This semi-sterility has apparently been the means employed by nature to keep the genome size constant. The semi-sterility results from inviability of the gametes carrying duplications and deficiencies. It appears that the only type of translocation which accompanied successful speciation has been the Robertsonian centric fusion which links together two intact chromosomes into one, since heterozygotes for this fusion do not, as a rule, produce unbalanced gametes. The above being the case, it is probable that despite apparent diversity of karyotypes, different mammals of today, by and large, conserved some of the original linkage groups of the primordial mammals. The best example is the conservation of the original X-linkage group. Thus, we shall begin with the

X-linkage group, then we shall proceed to examine the possibility that not only some of the autosomal linkage groups which were not subjected to inversions following centric fusion have been conserved, but also, because of extensive gene duplication in our past, some of the apparent non-homologues in the mammalian genome are in fact homologous.

Conservation of X-linkage

The fact that a majority of fish are already gonochorists reveals that the chromosomal sex-determining mechanism developed at the beginning of vertebrate evolution. However, with a few notable exceptions, the X and the Y (male heterogamety) or the Z and the W (female heterogamety) have remained homomorphic up to the reptiles. The X and the Y or the Z and the W of cold-blooded vertebrates differ from each other by having opposing sex-determining alleles at a few gene loci, otherwise they are homologous. Thus, both sexes are endowed with two doses of every sex-linked gene.

Apparently, miniaturization of the Y chromosome and that of the W chromosome occurred independently to the common ancestor of mammals and to that of avian species. Although miniaturization of the W chromosome forced most of the avian Z-linked genes to exist in the hemizygous state in the heterogametic female sex, the effective dosage compensation mechanism apparently failed to develop in this class of higher vertebrates. Thus, avian species show very clear dosage effect of their Z-linked genes. This point is well illustrated in Figure 1 which shows a male and female pigeon of the same breed carrying a Z-linked plumage color allele *Faded* (B^{of}). A homozygous male ($Z^{B^{of}}Z^{B^{of}}$) carrying two doses of *Faded* is almost pure white, while a hemizygous female ($Z^{B^{of}}W$) is merely grayish resembling the phenotype of a heterozygous male ($Z^{B^{of}}Z^{+}$). If there is a dosage compensation mechanism, homozygous males and hemizygous females should present the same phenotype.

Heterochromatinization of one of the two X-chromosomes in somatic cells of the homogametic female sex endowed mammals with the effective dosage compensation mechanism for their X-linked genes. XX somatic cells of the female and XY somatic cells of the male are equalized in that in either case, X-linked genes remain functionally hemizygous.

The fact that heterochromatinization of one female X occurs not only in all the placental mammals but also in marsupials suggests that this dosage compensation mechanism developed at the dawn of the Cenozoic era more than 70 million years ago in the common ancestor to both infraclasses.

This heterochromatinization of one female X enabled us to realize the trend for conservation of the original X-linkage group by diverse mammalian species. A great majority of placental mammals have the

Fig. 1. A clear cut demonstration of the dosage effect of an avian sex-linked gene. A male (left) and a female (right) pigeon belong to the same breed. A male is homozygous for a sex-linked gene *Faded* (B^{of}), while a female is hemizygous. The double dose of this gene gives almost pure white plumage to the male, while a single dose slightly dilutes the wild-type plumage color of the female. These pigeons belong to Mr. Martin Tidwell.

X chromosome of about the same size comprising of nearly 5% of the genome (Ohno and Beçak, 1964). However, the simple observation that the X's of some species are acrocentrics while those of others are metacentrics reveals that internal rearrangements by inversions have occurred.

The only form of X-autosome translocation compatible with the successful speciation has been the linking of one entire X to one entire autosome. We see this form of X-autosome translocation preserved in the shrew (*Sorex araneus*) (Bovey, 1949; Ford *et al.,* 1957), and the gerbil (*Gerbillus gerbillus*) (Wahrman and Zahavi, 1955) among placentals as well as in a few species of Australian marsupials (Sharman, 1961). They are said to have XY_1Y_2/XX scheme of sex-determining mechanism, whereas, in fact, they have not invented a new sex-determining mechanism. Their X is a combination of one whole X and one whole autosome, and their Y_2 is an intact autosome homologous to the autosomal part of the X. The Y_1 is the original Y. The important fact is that even these species have preserved the original X-linkage in its entirety. Accordingly, only the original X-part in female somatic cells becomes heterochromatinized, and the autosomal part behaves like any other autosome (Hayman and Martin, 1965).

Considerable number of rodent species possess the duplicate-, triplicate- and quadruplicate-type of X chromosomes which comprise of 10%, 15% and 20% of the genome, respectively. For a while it appeared that these unusually large X's occur only among rodents, but recently the duplicate-type X has been found in three species of deer (Wurster *et al.*, 1968; Fraccaro *et al.*, 1968). In each of these unusually large X's, the part in excess of the original X part is rendered permanently heterochromatic (Galton and Holt, 1964; Taylor, 1960; Wolf *et al.*, 1965, 1966). The heterochromatinization of the excessive part appears to suggest that these large X-chromosomes are not the products of X-autosome linkage mentioned above, but rather they arose by literal longitudinal duplication, triplication and quadruplication of the original X.

Conservation of the original X-linkage group in its entirety suggests the homology of X-linked genes among all the placental mammals. Genetic evidence in favor of alleged homology have been slowly accumulating. Although 59 X-linked mutations are known in man (McKusick, 1962) and more than 16 in the mouse, most of these are pathological syndromes in the case of man, and coat color mutations in the case of mice. True homology of the gene loci in different species can be determined only if direct products of the gene loci in question are known. The direct product of a structural gene is invariably a polypeptide chain which is a component of an enzyme or a nonenzymatic protein.

The four well defined structural gene loci known to be X-linked in man are listed below:

(a) The locus for Xg blood group antigen.
(b) The loci for two plasma proteins; antihemophilic factors VIII and IX.
(c) The locus for an enzyme, glucose-6-phosphate dehydrogenase, which catalyzes the first step of the pentose phosphate shunt of the carbohydrate metabolism.
(d) The locus for an enzyme, hypoxanthine-guanine phosphoribosyl transferase, which scavenges guanine (purine) from extracellular fluid and incorporates it to nucleic acids.

The X-linkage of these gene loci has been established in man, because there exists allelic polymorphism at each of these gene loci in human populations. If other species do not exhibit allelic polymorphism at these gene loci, there is no way of testing the possible X-linkage. Fortunately, the X-linkage of three of these gene loci has been established in mammalian species other than man.

(a) The X-linkage of the Xg locus has been determined in man, by the use of anti Xg (a) sera. Erythrocytes of a person carrying an Xg (a+) allele would be agglutinated by this antiserum, while the red cells of a person homozygous or hemizygous for an Xg (a−) allele would not. When chimpanzees, gorillas, orangutans and baboons were examined, they uniformly typed as Xg (a−), suggesting that if there exists allelic polymorphism at this locus in these primate species, their allelic differences are not detectable by available anti human Xg (a) sera. However, gibbons (*Hylobates lar lar*) typed either as Xg (a+) or as Xg (a−) in the same manner as the human and it became clear that the Xg locus is indeed X-linked in gibbons (Gavin *et al.,* 1964). When the X-linked histocompatibility antigen of the mouse (Bailey, 1963) is shown to be detectable on erythrocyte surface, this locus of the mouse can be considered as homologous to the Xg locus of man and gibbons.

(b) In man, a hereditary deficiency of Factor VIII causes hemophilia A and that of Factor IX, hemophilia B. Both are well established X-borne mutations. Both hemophilias have been shown to be inherited as X-borne mutations in the dog (Hutt *et al.,* 1948; Mustard *et al.,* 1960). Thus, two structural gene loci for plasma antihemophilic Factors VIII and IX are X-linked in man and the dog.

(c) In man, several alleles coexist at the gene locus for an enzyme, glucose-6-phosphate dehydrogenase (G-6-PD). Some are amorphic alleles which cause enzyme deficiency, while others code for electrophoretic variants of this enzyme. The X-linkage of this enzyme locus has been well established. By the use of interspecific hybrids, the X-linkage of this locus has been shown on the horse and the donkey (Trujillo *et al.,* 1965); Mathai *et al.,* 1966) as well as on the European hare (Ohno *et al.,* 1965). Recently, Cooper (personal communication) has demonstrated that the G-6-PD locus is autosomally inherited in avian species. Thus, there is no a priori reason for this locus to be on the sex chromosome; the fact that the G-6-PD locus is on the X-chromosome of diverse mammalian species merely reflects conservation of the original X-linkage group of the common ancestor by these mammals of today.

(d) An X-borne neurological mutation known as the Lesch-Nyhan syndrome of man has been shown to be due to an inherited deficiency of an enzyme, hypoxanthine-guanine phosphoribosyl transferase. Mammalian cells are quite capable of *de novo* synthesis of purines and pyrimidines which are building blocks of their nucleic acids, but the possession of this enzyme enables mammalian cells to take in already-made guanine and utilize it. The cells deficient in this enzyme become incapable of incorporating guanine available in extracellular fluid.

When such a deficiency is induced in tissue cultured cells, resistance to 8-azaguanine (guanine analogue) is conferred, for the cells which cannot take in a poison shall not be killed by a poison. To induce an enzyme deficiency on hyperdiploid cells *in vitro* should be a very difficult task, if that enzyme locus is on an autosome. First, a clone of heterozygotes for an enzyme deficiency has to be isolated. Only among heterozygotes can one hope to find a clone of homozygous deficient cells. If the locus is X-linked, on the other hand, a single mutation suffices to produce a clone of hemizygous deficient cells. Due to the dosage compensation mechanism, only one dose of the hypoxanthine-guanine phosphoribosyl tranferase gene is functioning in male as well as female somatic cells of man (Migeon *et al.*, 1968). The rate of mutation which causes 8-azaguanine resistance is already known on the mouse (Littlefield, 1963) as well as the Syrian hamster (Davidson *et al.*, 1966) cell lines. If this mutation occurs at the same rate in human cell lines, it must mean that the gene locus for hypoxanthine-guanine phosphoribosyl transferase is also X-linked in rodents.

Homology Between Non-Homologues

There are only three kinds of ribosomal RNA, 5S, 18S and 28S components, so the genome needs to contain only three different kinds of DNA cistrons for ribosomal RNA. Yet almost half of the total RNA contained in the cell are ribosomal RNA. There exists a need to produce an enormous amount of ribosomal RNA which are few in variety.

In the African water frog (*Xenopus laevis*) and in *Drosophila*, it was shown that it is the nucleolar organizer of the chromosome which contains a hundred duplicates of each of the three cistrons for ribosomal RNA (Wallace and Birnstiel, 1966; Ritossa and Spiegelman, 1965). The mammalian genome which is considerably larger than either the *Xenopus* or the *Drosophila* genome should contain a greater number of duplicates of these three ribosomal cistrons. Indeed, among 46 chromosomes of man, three pairs of D-chromosomes and two pairs of G-chromosomes carry nucleolar organizers. What should be realized here is that the five pairs of acrocentric chromosomes of man are homologous to each other insofar as their nucleolar organizing regions are concerned.

The above example on ribosomal cistrons nicely introduces the subject of homology that exists among apparently non-homologous autosomes of the mammalian genome. Because of extensive gene duplication that has occurred in our prereptilian ancestor, our genome contains a great deal of genetic redundancy. For instance, a part of one autosome which carries the hemoglobin alpha-chain gene locus can be considered as homologous to that part of another autosome which contains the beta-chain locus. There are three gene loci for an enzyme, lactate dehydrogenase (LDH) scattered about in the mammalian genome, and so are two loci for pyruvate kinase, three unlinked gene loci for phosphoglucomutase. These are but a few of the numerous known examples of duplicated genes the mammalian genome contains.

Our study (Ohno *et al.*, 1968) on chromosome complements, genome sizes and degrees of gene duplication of various primitive chordates and fishes indicated that the original diploid complement of vertebrates was probably made of 48 acrocentric chromosomes. During the initial aquatic stage of vertebrate evolution, this complement tended to be conserved despite sometimes great increase in the genome size accomplished by a series of unequal crossings-over and other means. Crossopterygian fish which slithered out of the water to become terrestrial vertebrates at the end of the Devonian period, roughly 280 million years ago, were probably of several different kinds. It is our proposal that one of these belonged to the lineage which eventually gave rise to mammals. This crossopterygian fish attained the genome size characterized by 3.5 mg $\times 10^{-7}$ DNA by becoming a tetraploid. If the above assumption is, by and large, true, all the mammals are remote descendants of a tetraploid species. It might be that the primordial mammal of more than 70 million years ago had the diploid complement which reflected his tetraploid ancestry. Ninety-six acrocentric chromosomes (2×48) should represent the unaltered tetraploid complement. However, by this time, at least two acrocentrics were already linked together into one chromosome which was the original X-linkage group. For this reason, the diploid chromosome number of this primordial mammal could not have been higher than 94. The known highest chromosome number among placentals is 84 which is not too far off this expectation. Successive Robertsonian translocations (centric fusions), interspersed by pericentric inversions, appear to have reduced the diploid chromosome number of a majority of mammals to the range of 44–48, and in many instances still further to lower numbers. Figure 2 schematically illustrates these assumptions. It is then quite possible that two morphologically similar pairs of smaller acrocentrics often included in the diploid complement of mammals are in fact homologous in that they represent four homologues of our remote tetraploid ancestor. Indeed, Penrose (1967) has recently concluded that the Down's syndrome of man is due to the trisomic condition of autosome 21 as well as autosome 22. Similarly, two of the three pairs of D-chromosomes of man may be ancient homologues, so that the so-called D-trisomy syndrome may involve either autosome 14 or 15.

Summary

In recent years, direct products of structural genes came under close scrutiny, and as a result, it was realized that genetic redundancy created by gene duplication rather than allelic mutations at already existing gene loci have been the major force of evolution.

We have proposed that all the mammals descended from a remote tetraploid ancestor (Ohno *et al.*, 1968). Not only the original X-linkage, but some of the autosomal linkage groups of the common ancestor may have been preserved by diverse mammals of today. Thus, diversity of karyotypes displayed by placental mammals can be regarded as more apparent than real.

Because of our tetraploid ancestry, two pairs of smaller acrocentrics in our karyotype may be homologous in the phylogenetic sense in that they represent four original homologues of a tetraploid ancestor.

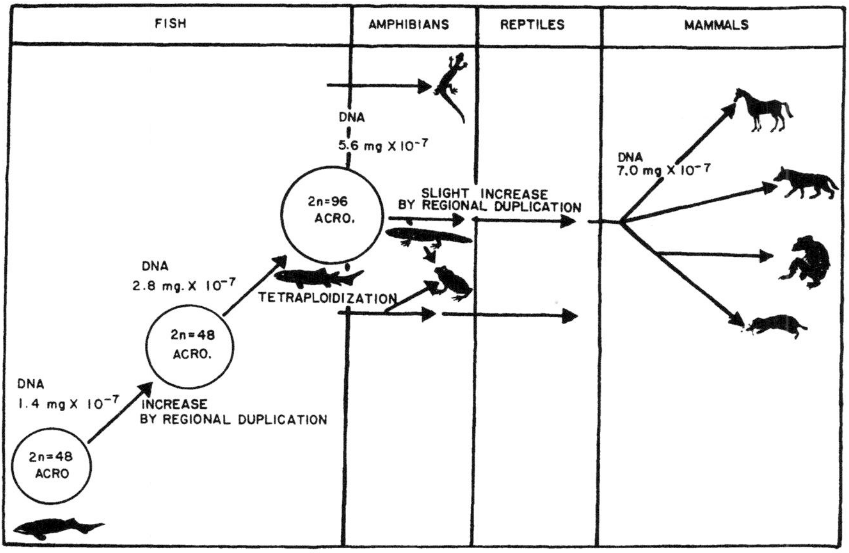

Fig. 2. A schematic illustration of successive changes in the genome size which accompanied vertebrate evolution. An ancient jawless fish (bottom left) probably had 48 acrocentrics in the diploid chromosome complement, and its genome size was only 20% of that of mammals. During the aquatic stage of evolution, extensive experiments with gene duplication took place. Some have increased their genome size exclusively by regional duplication of chromosomal segments. The extreme example of this is seen in lungfish and salamanders. Although some of them attained the genome size 36 times greater than that of mammals, their redundant genetic material have not contributed to the creation of new gene loci. This group apparently hits the evolutional dead end. Those which became direct ancestors of mammals must have gone through tetraploid evolution either shortly before or after they slithered out of water to become amphibians. The basic karyotype of this group can be regarded as having 96 acrocentrics.

Future development of cytogenetics may depend on the use of newer tools such as DNA-DNA hybridization between isolated chromosomes and genetic studies of enzymes and protein, for they would reveal ancient homology between apparently non-homologous chromosomes.

Acknowledgment

This work was supported in part by a grant (CA-05138) from the National Cancer Institute, U.S. Public Health Service, and in part by a research fund established in honor of General James H. Doolittle, Contribution No. 11-68, Department of Biology, City of Hope Medical Center.

References

Bailey, D. W.: Mosaic histocompatibility of skin grafts from female mice. Science *141*:631, 1963.

Bovey, R.: Les chromosomes des Chiroptères et des Insectivores. Rev. suisse Zool. *56*:341, 1949.

Braunitzer, G. and G. Matsuda: Primary structure of the α-chain from horse hemoglobin. J. Biochem. (Japan) *53*:262, 1963.

Childs, B., W. H. Zinkham, E. A. Browne, E. L. Kimbro and J. V. Torbert: A genetic study of a defect in glutathione metabolism of the erythrocyte. Bull. Johns Hopkins Hosp. *102*:21, 1958.

Davidson, R. L., B. Ephrussi and K. Yamamoto: Regulation of pigment synthesis in mammalian cells as studied by somatic hybridization. Proc. Nat. Acad. Sci. *56*:1437, 1966.

Ford, C. E., J. L. Hamerton and G. B. Sharman: Chromosome polymorphism in the common shrew. Nature *180*:392, 1957.

Fraccaro, M., I. Gustavsson, M. Hultén, M. J. Lindsten and L. Tiepolo: Chronology of DNA replication in the sex chromosomes of the reindeer (*Rangifer tarandus L.*). Cytogenetics *7*:196, 1968.

Galton, M. and S. F. Holt: DNA replication patterns of the sex chromosomes in somatic cells of the Syrian hamster. Cytogenetics *3*:97, 1964.

Gavin, J., J. Noades, P. Tippett, R. Sanger and R. R. Race: Blood group antigen Xg (a) in gibbons. Nature *204*:1322, 1964.

Hayman, D. L. and P. G. Martin: An autoradiographic study of DNA synthesis in the sex chromosomes of two marsupials with an XX/XY_1Y_2 sex chromosome mechanism. Cytogenetics *4*:209, 1965.

Herzenberg, L. A., J. D. Minna and L. A. Herzenberg: The chromosome region for immunoglobulin heavy-chains in the mouse. Cold Spring Harbor Symposia for Quant. Biol. *32*:181, 1967.

Hungerford, D. A. and R. L. Snyder: Somatic chromosomes of a black rhinoceros (*Diceros bicornis* Gray 1821). Amer. Naturalist *101*:357, 1967.

Hutt, F. B., C. G. Rickard and R. A. Field: Sex-linked hemophilia in dogs. J. Hered. *39*:2, 1948.

Kirkman, H. N., H. D. Riley, Jr., and B. B. Crowell: Different enzymic expressions of mutants of human glucose-6-phosphate dehydrogenase. Proc. Nat. Acad. Sci. (U.S.) *46*:938, 1960.

Kunkel, H. G.: Myeloma proteins and antibodies. The Harvey Lectures *59*, 1963–1964.

Littlefield, J. W.: The isosinic acid pyrophosphorylase activity of mouse fibroblasts partially resistant to 8-azaguanine. Proc. Nat. Acad. Sci. (U.S.) *50*:568, 1963.

Mathai, C. K., S. Ohno and E. Beutler: Sex-linkage of the glucose-6-phosphate dehydrogenase gene in the family *Equidae*. Nature *210*:115, 1966.

Matthey, R.: La formule chromosomique et le problème de la détermination sexuelle chez *Ellobius lutescens* Thomas. *Rodentia-Muridae-Microtinae*. Arch. Klaus-Stift. Vererb.-Forsch. *28*:65, 1953.

McKusick, V. A.: On the X-chromosome of man. Quart. Rev. Biol. *37*:69, 1962.

Migeon, B. R., V. M. Der Kaloustian, W. L. Nyhan, W. J. Young and B. Childs:

X-linked hypoxanthine-guanine phosphoribosyl transferase deficiency: heterozygote has two clonal populations. Science *160*:425, 1968.

Mustard, J. F., H. C. Roswell, G. A. Robinson, T. D. Hoeksema and H. G. Downie: Canine hemophilia B (Christmas disease). Brit. J. Haemat. *6*:259, 1960.

Ohno, S., J. Jainchill and C. Stenius: The creeping vole (*Microtus oregoni*) as a gonosomic mosaic. I. The OY/XY constitution of the male. Cytogenetics *2*:232, 1963.

———, J. Poole and I. Gustavsson: Sex-linkage of erythrocyte glucose-6-phosphate dehydrogenase in two species of wild hares. Science *150*:1737, 1965.

———, U. Wolf and N. B. Atkin: Evolution from fish to mammals by gene duplication. Hereditas *59*:169, 1968.

Oudin, J.: Genetic regulation of immunoglobulin synthesis. J. Cell Physiol. *67* (Suppl. 1): 77, 1966.

Parr, C. W.: Erythrocyte phosphogluconate dehydrogenase polymorphism. Nature *201*:487, 1966.

Patnum, F. W., K. Titani, M. Wikler and T. Shinoda: Structure and evolution of kappa and lambda light chains. Cold Spring Harbor Symposia Quant. Biol. *32*:9, 1967.

Penrose, L. S.: CIBA Foundation Study Group No. 25: Mongolism. Ed. G. E. Wolstenholme, R. J. Porter and A. Churchill, Ltd., Lond., 1967.

Ritossa, F. M. and S. Spiegelman: Localization of DNA complementary to ribosomal RNA in the nucleolus organizer region of *Drosophila melanogaster*. Proc. Nat. Acad. Sci. (Wash.) *53*:737, 1950–1951.

Sharman, G. B.: The mitotic chromosomes of marsupials and their bearing on taxonomy and phylogeny. Austr. J. Zool. *9*:38, 1961.

Shaw, C. R.: Electrophoretic variation in enzymes. Science *149*:936, 1965.

Taylor, J. H.: Duplication of chromosomes and related events in the cell cycle. In: Cell Physiology of Neoplasia. Austin: University of Texas Press, 1960; pp. 547–575.

Thuline, H. C., A. C. Morrow and A. G. Motulsky: Autosomal phosphogluconic dehydrogenase polymorphism in the cat (*Felis catus* L.) Science *157*:431, 1967.

Trujillo, J. M., B. Walden, P. O'Neil and H. B. Anstall: Sex-linkage of glucose-6-phosphate dehydrogenase in the horse and donkey. Science *148*:1603, 1965.

Wahrman, J. and Z. Zahavi: Cytological contributions to the phylogeny and classification of the rodent genus *Gerbillus*. Nature (Lond.) *175*:600, 1955.

Wallace, H. and M. L. Birnstiel: Ribosomal cistrons and the nucleolar organizer. Biochem. Biophys. Acta *114*:296, 1966.

Wolf, U., G. Flinspach, R. Böhm and S. Ohno: DNS-Reduplikations-muster bei den Riesen-Geschlechtschromosomen von *Microtus agrestis*. Chromosoma *16*:609, 1965.

———, and D. Hepp: DNS-Reduplikationsmuster der somatischen Chromosomen von *Cricetus cricetus* (L.). Chromosoma *18*:438, 1966.

Wurster, D. H., K. Benirschke and H. Noelke: Unusually large sex chromosomes on the sitatunga (*Tragelaphus spekei*) and the black buck (*Antilope cervicapra*). Chromosoma *23*:317, 1968.

GEOGRAPHIC VARIATION OF CHROMOSOME FORMS IN SPALAX, A SUBTERRANEAN MAMMAL OF RESTRICTED MOBILITY

J. Wahrman, Ruth Goitein and E. Nevo
Laboratory of Genetics, The Hebrew University, Jerusalem, Israel

Several different types of rodents are found among the specialized mammals exploiting the subterranean ecological niche. Tunnels dug by them can often be detected through the mounds of earth thrown up during the burrowing. Striking anatomical and behavioral adaptations are associated with their fossorial mode of life (Ellerman, 1956; Pearson, 1959). The mole rats of the genus Spalax are an extreme example of such rodents; they are largely confined to a subterranean existence. Their adaptations include the great reduction in the development of their eyes, which are not visible externally.

Spalax is found in southeastern Europe, the eastern Mediterranean region from Greece to Cyrenaica and in Iraq. A distributional gap in Sinai interrupts the continuity of the range along the Mediterranean coast, between northwestern Egypt and Israel. Some of the populations of Israel are thus marginal, at the southeastern boundary of the range. Spalax has apparently not succeeded in penetrating into true desert. Its limit of distribution in Israel coincides, approximately, with the 100 mm equal rainfall line (Nevo, 1961).

Spalax is the only genus in its family. The general morphology is quite uniform throughout the whole range, with some characters changing in a clinal fashion. Corbet (1966) prefers to look upon the entire assemblage of named taxa as one biological species, while other authorities split it into a number of species (e.g., Ognev, 1947). The standard checklist of Palearctic mammals recognizes two species, while a third has been kept with considerable hesitation (Ellerman and Morrison-Scott, 1951).

About ten years ago it was found that individuals of mole rats from

Israel, indistinguishable morphologically and classified as the south-eastern Mediterranean species *Spalax ehrenbergi* Nehring 1898, differ in their karyotypes. Moreover, animals trapped in one locality, Sa'ar in western Upper Galilee, showed a difference of as many as 6 chromosomes (Wahrman, 1958). These initial observations are now being expanded to a much larger sample. In a systematic survey of chromosome constitutions of mole rats from many localities, bone marrow cells are primarily utilized. The karyotype of each animal is determined with the help of photographs of an average of 10 chromosome plates.

Four diploid chromosome numbers, namely 52, 54, 58 and 60, have been found among the first 100 animals whose chromosomes were analyzed. Figure 1 shows all sampled localities and the chromosome numbers of their populations. Figures 2 to 5 are karyotypes of the four chromosome forms, in which the groups of chromosomes have been labelled. One individual with 59 chromosomes is discussed below. It is not our intention to describe in detail, in this report, the different chromosome sets. However, an attempt is made to compare them (Figure 6 and Table 1). Three groups of chromosomes, designated A,

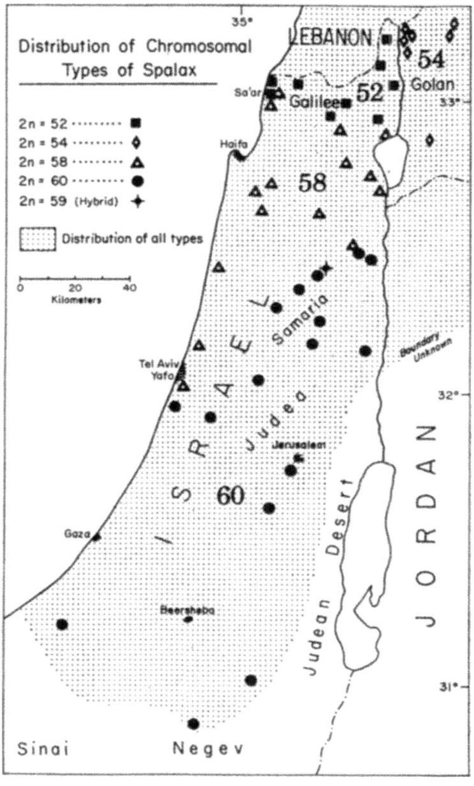

Fig. 1. Distribution map of the four forms of Spalax described in this paper. The boundary of distribution east of the river Jordan is unknown.

Figs. 2–5. Karyotypes of the four forms of Spalax from dividing bone marrow cells, treated hypotonically. The designation of individual metacentric chromosomes can be seen in Fig. 6. Note that in Figs. 2–4 the metacentrics of group B are in a reverse order to that in Fig. 6. × 2250

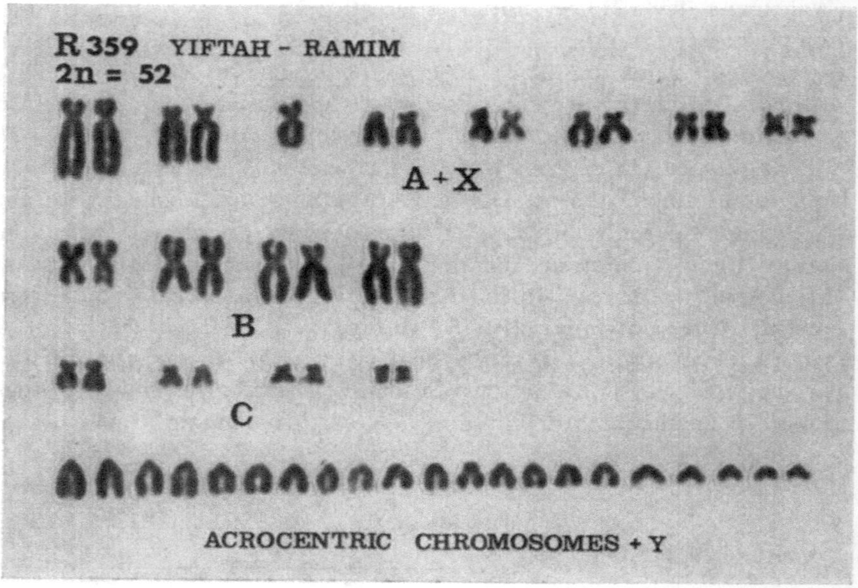

Fig. 2. Karyotype of male (number R359) from Yiftah-Ramim in Upper Galilee. Diploid number of 52.

B and C, are distinguished. Group A is common to all four forms, while groups B and C differ from form to form and are defined on the basis of their postulated rearrangement history. In spite of the obvious limitations of the method of comparative chromosome morphology, the interpretation presented appears to us to be generally acceptable.

Group A comprises the *unchanged chromosomes.* This merely means that all chromosomes which are not affected by cytologically detectable changes, within this particular series of forms, were lumped together into one group, regardless of their possible derivation. Some of these chromosomes have perhaps been produced by Robertsonian conversions, and others might have achieved their present form by pericentric inversions, or by other means. It is not implied that any degree of stability has been reached. On the contrary, a few individuals have been found to possess karyotypes slightly deviating from those in Figures 2 and 3. In all such known cases the modification consists of a decrease in the number of acrocentrics accompanied by an increase in the number of inverted chromosomes. It can be shown that chromosomes of group A are contributing, in this manner, to population polymorphism. Group A includes 7 pairs of metacentric autosomes (A1 to A7) and the metacentric X chromosome, which are individually recognizable. Of the 10 pairs of acrocentric autosomes and the Y chromosome, only a few can be identified and matched consistently in pairs. The metacentrics and acrocentrics are arranged, respectively, in decreasing length. An intensive study of the Spalax complex will, presumably, result in the reduction of group A.

Group B consists of the *Robertsonian chromosomes.* These are the chromosomes which have been involved in whole-arm rearrangements. The metacentrics (B1 to B4) can be characterized individually, while the acrocentrics, which either have been derived from them, or have contributed to their formation were arbitrarily assigned to this group (see Figure 6). Chromosome B4 has an almost symmetrical shape and is readily distinguished in $2n = 52$, 54 and 58. Chromosomes B2 and B3 are present in both $2n = 52$ and $2n = 54$, whereas chromosome B1 is found only in $2n = 52$. All the variability in the number of centromeres can be attributed to changes in which these chromosomes have taken part.

Group C includes the *inversion chromosomes.* These are either chromosomes possessing subterminal centromeres due, most likely, to pericentric inversions, or their equivalent acrocentrics. The metacentric

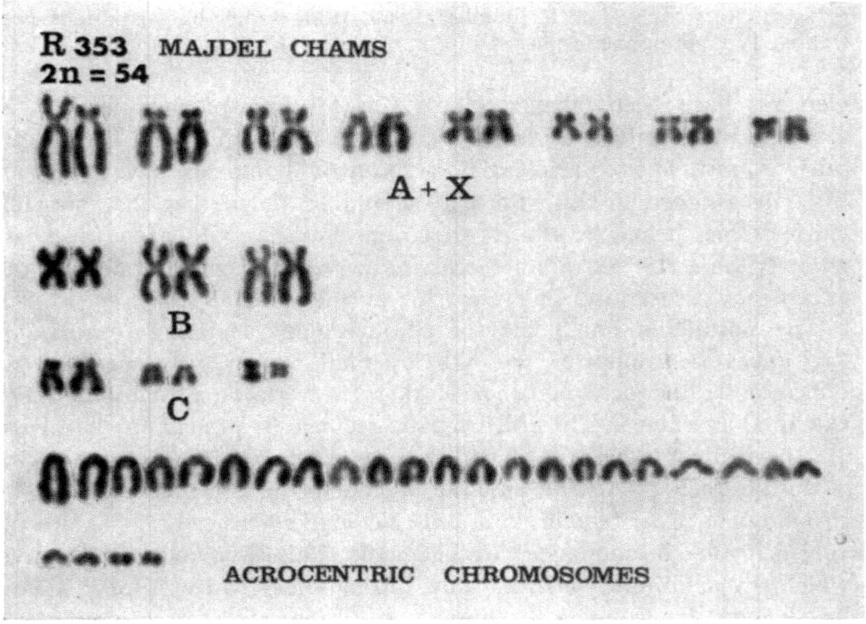

Fig. 3. Karyotype of female (number R353) from Majdel Chams in the Golan. Diploid number of 54.

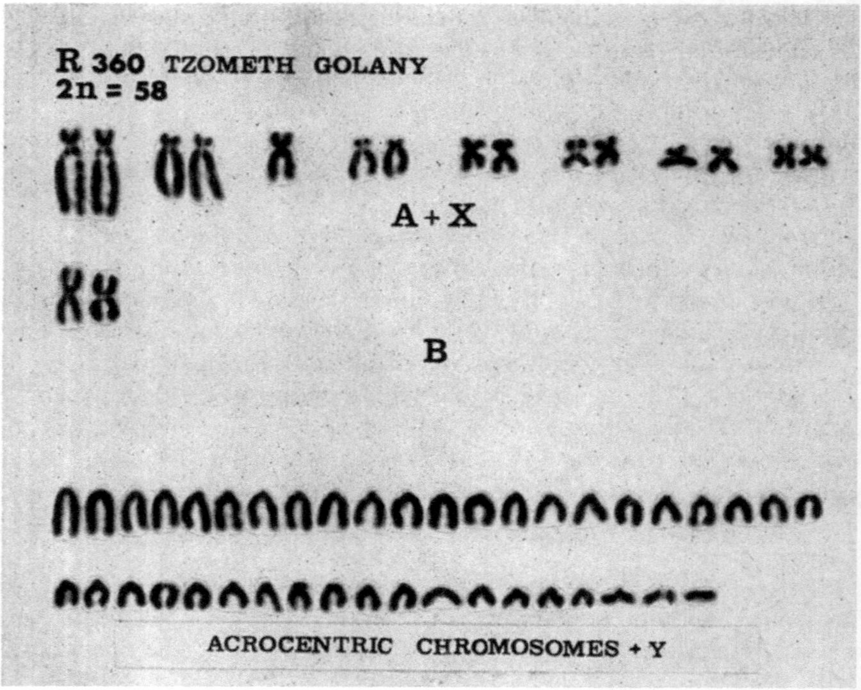

Fig. 4. Karyotype of male (number R360) from Tzometh Golani in Lower Galilee. Diploid number of 58.

elements have been labelled C1 to C4. All four of them appear in $2n = 52$, while only three are present in $2n = 54$, and none in the two other forms. The corresponding acrocentric homologs have been arbitrarily assigned to this group. It should be noted that the inversion chromosomes belong to the shortest quarter of the whole set. In a few animals there are "extra" metacentrics in group C which appear at the expense of acrocentrics of either this or the other groups.

The postulated homologies of chromosomes belonging to different karyotypes are supported not only by their similar gross morphology, but also by the occurrence of markers in certain chromosomes. For example, an extensive heterochromatic segment is present in the proximal part of the long arm of chromosome A1 in all forms.

Within each group the amount of genetic material, as reflected by chromosome mass, remains constant. No marked losses or gains are involved in the chromosome rearrangements. This is obvious for group A which, by definition, includes only chromosomes which display almost no changes. In group B the number of chromosome arms is constant. Though the number of arms is variable in group C, due to shifts in centromere positions, the total chromosome length is similar in all forms. Table 1 summarizes the numerical relationships. The column before the last one indicates the actual number of chromosome arms in each form. This number is the same, namely 76, for $2n = 60$ and $2n = 58$, as these differ only by a Robertsonian change. However, $2n = 52$ and $2n = 54$ have higher numbers of chromosome arms, 82 and 84 respectively. These figures may be corrected since the whole increase is due to pericentric inversions which alter neither the number of chromosomes nor their length. So correcting, one arrives at one and the same number of elements for the entire series—last column in Table 1. (There is no way to pick out chromosomes with a history of centromere shifts, if any, from among the metacentrics of group A. Consequently this last number is not to be considered as the fundamental number of "unmodified arms," and no particular significance should be attached to it.) It may be concluded that all forms retain the same total length of chromosomes, notwithstanding the drastic changes which some of their chromosome sets have undergone.

The four karyotypes, Figures 2 to 5, have been chosen as representative by reason of being the prevalent ones within each diploid number. In fact, very little chromosomal variability was seen in the $2n = 58$ and $2n = 60$ populations, to which most individuals examined belonged

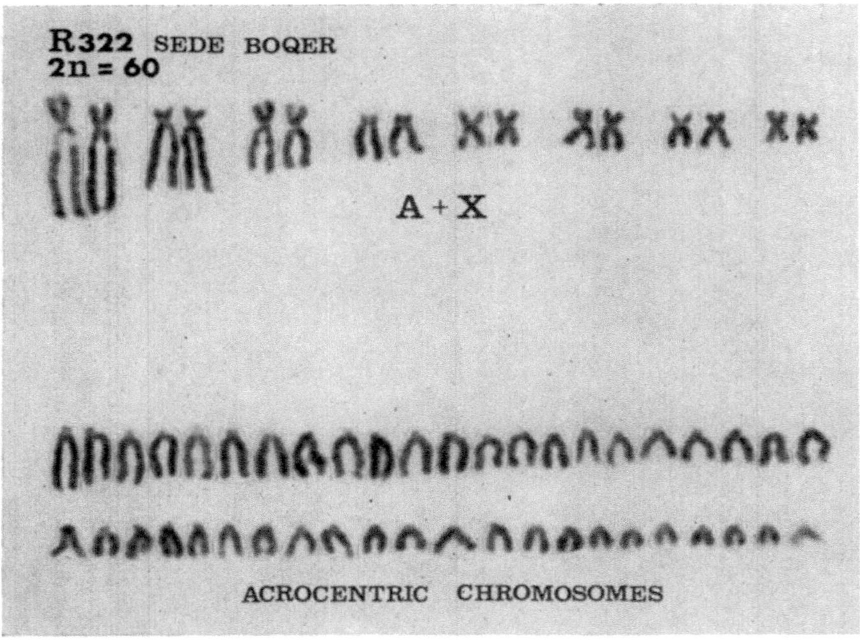

Fig. 5. Karyotype of female (number R322) from Sede Boqer in the Negev. Diploid number of 60.

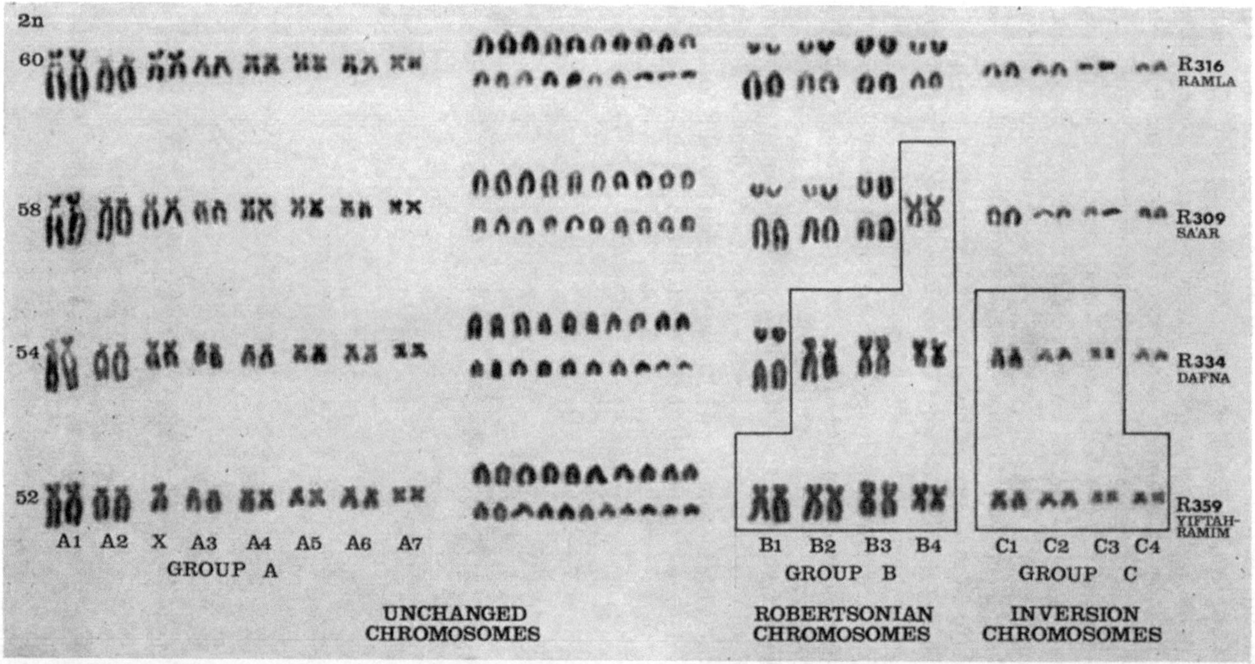

Fig. 6. Comparison between the chromosomes of the four most common karyotypes of Spalax discussed in this paper. Metacentrics of groups B and C in boxes. Acrocentrics were arbitrarily assigned to groups B and C. Further explanations in the text. Localities from where the animals originated, and their serial numbers, at the right hand side.

Table 1: A Breakdown of Chromosome Types of the Different Forms of Spalax Described in the Present Paper

The classification refers to females; the Y chromosome is omitted. $2n$ = diploid chromosome number, M = Metacentric chromosomes, A = Acrocentric chromosomes. The names of the regions given are for general orientation only. Not all of the chromosome form regions accurately overlap the geographic regions named.

Region	$2n$	Group A Unchanged Chromosomes		Group B Robertsonian Chromosomes		Group C Inversion Chromosomes		Total			
		M	A	M	A	M	A	M	A	Arms actual	Arms corrected *
Upper Galilee	52	16	20	8	—	8	—	32	20	84	76
Golan	54	16	20	6	4	6	2	28	26	82	76
Lower Galilee, Central Israel	58	16	20	2	12	—	8	18	40	76	76
Samaria, Judea, Negev	60	16	20	—	16	—	8	16	44	76	76
Samaria	59 †	16	20	1	14	—	8	17	42	76	76

* Metacentric chromosomes of group C, thought to have arisen through pericentric inversions, are counted as if they were one-armed elements

† Hybrid between 58 and 60.

(about 30 in each case). However, among the $2n = 52$ and $2n = 54$ individuals a few animals were found possessing, in each case, an "extra" inverted chromosome replacing a small acrocentric. The number of C group metacentrics is increased, usually by one, in these cases. Such centromere displacements, often in the heterozygous conditions, are interesting in pointing to the dynamic nature of the system which is harboring a certain amount of population polymorphism, especially involving inverted chromosomes in the lower numbered karyotypes.

Occasional chromosomal variations were encountered within the cells of a single individual. These included a centrically fused chromosome in one cell of an animal with 60 chromosomes, and an inverted chromosome in one cell of an animal with 58 chromosomes. Thus, the two main processes of chromosome change may also occur sporadically in the development of the individual.

The following scheme of chromosomal evolution is based on the gradual nature of the observed differences between the karyotypes and on an assumed trend towards increase in chromosome number. The site of origin of ancestral Spalax may have been somewhere in southeastern Europe or Asia Minor. The karyotype of this Proto-Spalax supposedly consisted of a relatively small number of chromosomes, mostly of the larger metacentric type. From this original stock other forms have been derived, spreading out radially. The stepwise repatterning of the chromosome set occurred in association with the extension of the range. Only the southward expansion, eventually reaching the arid parts of Israel, is considered here.

Along the tip of the southward extension which we have studied, there is an increase in chromosome number, culminating, at the margin of the distribution, in the highest number known for Spalax, namely 60. It is suggested that the increase in number is due to the dissociation of metacentrics into acrocentrics. Apart from changes in chromosome number, alterations in chromosome morphology, caused by shifts in centromere position, have also taken place. The generally small size of the inverted chromosomes may point to their origin from Robertsonian acrocentrics. Whether or not these small metacentrics are "secondary," they apparently tend to be preserved in certain populations more than in others.

On general grounds, to be discussed elsewhere, there is every reason to believe that the Robertsonian type of chromosome change may proceed in both directions, that is from acrocentrics to metacentrics, and vice versa. It is also likely that this process is independent of inversions and other rearrangements. Yet there appears to exist in the present series of chromosome forms a predominant preference for a particular kind of chromosome morphology. The types observed fall into a *gradual* series, at the two extremes are the "most metacentric" and the "most acrocentric" karyotypes. Such a "linear" situation might have been achieved, for example, through selective advantages fostering the survival of particular chromosome lengths and centromere positions and the elimination of other chromosome types produced.

The postulated sequence and the minimal number of changes which would account for the observed differences are:

$$52 \xrightarrow[1\text{In}]{1\text{R}} 54 \xrightarrow[3\text{In}]{2\text{R}} 58 \xrightarrow{1\text{R}} 60$$

where the number of Robertsonian steps (R) is indicated above the arrows, and the number of inversions (In) below it. It is possible that the divergence evolved originally by single steps (like the 58—60 change). The multiple karyotypic differences in this series may thus point to extinct or still undetected forms.

We are compelled by the chromosome data to accept the proposed linear relationship between the karyotypes (that is, 52—54—58—60). However, it should be made clear that the suggested *direction* of change is mainly supported by circumstantial evidence and by some highly speculative arguments on the relative selective significance of chromosome size and form (Wahrman, to be presented in another publication). Unfortunately, there is no direct cytological corroboration for either this or the opposite evolutionary sequence. Depending on the direction of change, the Robertsonian steps are either dissociations of metacentrics or centric fusions. Pericentric inversions would convert acrocentrics into metacentrics, or vice versa. It is also plausible that not all evolutionary arrows are pointing in the same direction (consider the possibility, $52 \leftarrow 54 \leftarrow 58 \rightarrow 60$).

The geographic distribution of the various karyotypes in Israel and adjacent territories deserves special attention. Spalax lives in suitable biotopes throughout the area indicated in Figure 1. Each karyotypic form has a distinct range. These ranges are mutually exclusive, but contiguous. There are no significant barriers to the movement of mole rats within the region occupied by Spalax today. Factors which could provide geographic isolation in the classic sense, such as deserts and moving sand, limit their distribution at the periphery.

Definite hybrid zones have not yet been found between any of the forms. Only one possibly hybrid individual was captured. This animal, which had 59 chromosomes, is most likely a natural hybrid between the $2n = 58$ and the $2n = 60$ forms (compare Fig. 7 with Figs. 4 and 5). There can hardly be any doubt that chromosome B4, which is the only Robertsonian metacentric in $2n = 58$, is found heterozygously in the exceptional animal. This animal originated in an area (see Fig. 1) which could well belong to a hybrid zone, though this needs further study. A second animal from the same locality was of the $2n = 60$ form. No further cases of heterozygosity for any of the Robertsonian chromosomes have been found

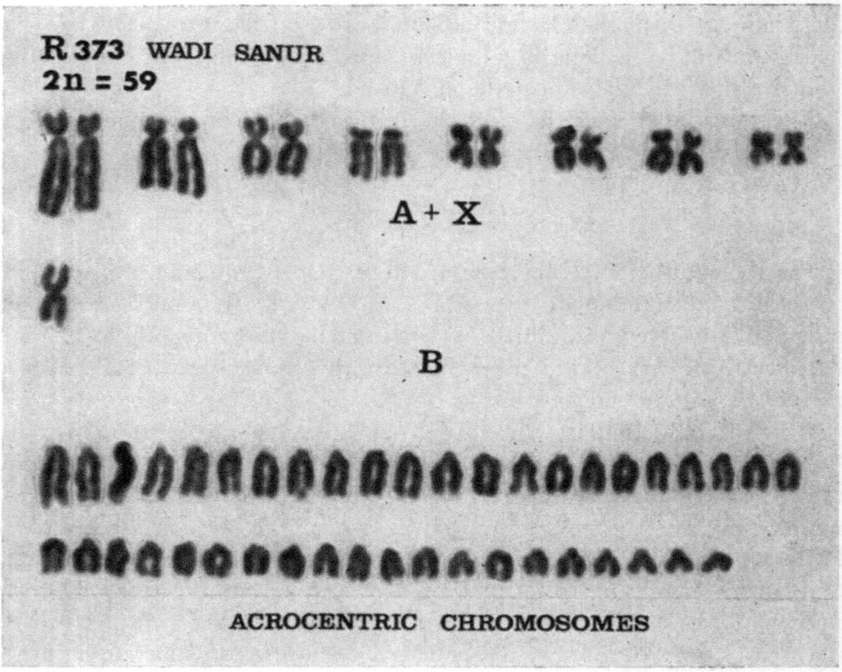

Fig. 7. Karyotype of female (number R373) from Wadi Sanur in Samaria. Chromosome number is 59. This is, presumably, a hybrid between the forms with 58 and 60 chromosomes. Compare with Figs. 4–5. × 2250

so far. Repeated sampling in the vicinity of Sa'ar, a locality situated on the 52–58 border, yielded no hybrids. It should, however, be stressed that hybrid zones, especially if narrow, could easily have been overlooked. A systematic search may reveal such zones in the future, particularly between the more closely related karyotypes.

Unfortunately, no conclusive answer can be given to the question whether any of these forms are reproductively isolated. Eighty crossing experiments were performed (Nevo, 1968). These indicated that the duration of the premating behavior leading to copulation takes about twice the time in pairs with a dissimilar chromosome constitution than it takes in pairs of the same constitution. Furthermore, copulation was found to be achieved more often in chromosomally uniform pairs than in mixed ones (approximately 40 percent, versus 15 percent, of successes). The conclusion that ethological isolating mechanisms have arisen between the chromosome forms will need to be substantiated by a larger number of experiments. A clarification of the existence of developmental compatibility is also necessary as a single pregnant female ($2n = 58$), inseminated by a male of another chromosomal form ($2n = 52$), died at an early stage of embryo development. Further crossings are being planned in the hope of elucidating these crucial issues.

The establishment of homozygosity for most of the chromosomal changes, an apparent rarity of hybrids in nature, as well as the indications of isolating mechanisms at the behavioral level, all seem to indicate phylogenetic divergence that may have reached species rank. The evolution of such a cluster of sibling species can be understood in a number of ways. The following working hypothesis is independent of the direction the chromosome changes may have taken.

The range extension of Spalax was dependent on the building up of adaptations to new environments. It is, for instance, plausible that the occupation by mole rats of the southernmost parts of their present distribution, was linked with progressive adaptation to more and more arid habitats. In this process the preservation of blocks of coadapted genes may be more important than the presence of particular single genes. Therefore, stabilization of favorable gene combinations is useful, while in another set of circumstances a release of crossing over may be advantageous. In considering the possible role of chromosome rearrangement in speciation, consequences of this kind should be appreciated. From among the continually arising chromosome rearrangements, those which had an effect in the "right" direction, under a particular set of circumstances, were selected for. The special conditions of life of subterranean organisms, discussed below, may assist in rapidly achieving a relatively homozygous fixation. (The possible advantage of heterozygotes, though it may play a role in the polymorphism of inverted chromosomes of Spalax, will not be dealt with.) Such a selection may be especially powerful at the margins of distributions, where the establishment of new chromosome combinations will contribute to the colonization of new biotopes. Meiotic drive has also been suggested as acting to ensure the spread of chromosome changes (White, 1968), but there is no proof for this. This part of the hypothesis explains the selection and spread of chromosome changes. However, in a situation of incipient evolutionary divergence reproductive isolation must follow.

How reproductive isolation is built up in a chromosomally conditioned extension of range is unknown. There may be, however, circumstances which will make the achievement of reproductive isolation feasible in the absence of fully fledged physical separation. The most evident of these are restricted mobility and a narrow adaptation to some limiting factor in the environment. Thus, members of true subterranean groups are more or less sedentary and will live only where they can dig. They are, therefore, frequently found in localized, sometimes small, "colonies" or "groups of individuals." Commonly, their small-scale distribution is disjunct as a result of discontinuities in the micro-environment. The probability of the meeting of two gametes including a similar, or even the same, chromosome rearrangement is increased in populations which are relatively isolated and highly inbred. Chromosome configurations which enhanced the adaptation to new environments may thus become fixed. The chromosome differences will function more and more as genetic isolating mechanisms, and sooner or later selection will start to act against the hybrids. Once colonization has occurred, dispersal is diminished in groups of low mobility; only narrow borderline populations remain exposed to the hazards of interbreeding. Ultimately, the gene pools will be practically separated. In fact, the establishment of homozygosity for most of the chromosomal changes in mole rats is most striking, and perhaps indicates the relatively long period of time which has elapsed since the divergence became effective on the specific level. This notion is merely a modification of current models of geographic speciation to fit groups with restricted power of dispersal.

All observers agree in considering mole rats to be strongly territorial and not very mobile. They are solitary, one individual occupies a territory, the average area of which has been estimated, in Israel, as 670 square meters (Nevo, 1961). No measurements of actual distances travelled by Spalax are available for Israel. However, these distances presumably will not exceed those of pocket gophers of the genus Thomomys in North America. Careful experiments with these rodents have shown that the maximal distance travelled by the form with the greater motility was around 750 meters, while the average for the least motile form was 60 meters (Vaughan, 1963). Shifts of mole rats from cultivated fields to nearby uncultivated areas are on record (cited by Ognev, 1947). Young animals usually settle between territories of adults. These animals are, on the whole, sedentary and very vulnerable when on the surface. This distinguishes them from many other mammals that are active above the ground, possess considerable powers of movement and are much more likely to disperse, or even migrate.

The way in which the distribution of the chromosomal types (Fig. 1) can be seen to match the biogeographical zonation of the area supports the idea that the chromosomal differences are adaptive. The areas held by the different karyotypes are climatically well characterizable (Table 2). This correlation between genetic structure and geographic conditions presumably reflects the adaptive nature of the system. An overall precise correspondence between *clinal* changes in the environment and *gradual* chromosome changes is, however, not to be expected. For example, the environmental gradient from central Israel southwards may correspond to the 58 → 60 transition. Likewise, the West-East 52 → 54 transition may perhaps be correlated with another environmental cline. But, a multiple step change, from 52 to 58, occurs between certain areas in Galilee, where the environmental changes are not very abrupt.

Table 2: Climatic Data for the Geographic Regions Inhabited by the Different Chromosome Forms of Spalax (Atlas of Israel, 1963)

The names of the regions given are for convenience only. Not all of the chromosome form regions accurately overlap the geographic regions named.

Region	2*n*	Mean Annual Values of: Precipitation mm	Number of Rainy Days	Evaporation mm	Temperature °C	Moisture Index (Thornthwaite, 1948)	Climatic Type (Thornthwaite, 1948)
Golan	54	400–1200			10–20		Subhumid to humid
Upper Galilee	52	600–1000	50–65	130–150	15–21	–20 to +60	Dry subhumid to humid
Lower Galilee, Central Israel	58	400–700	45–55	130–170	17–21	–40 to +20	Semiarid to dry subhumid
Samaria, Judea, Negev	60	100–600	15–45	140–200	17–23	–60 to +20	Arid & semiarid, with dry subhumid enclaves

In correlating the environment with the chromosome constitution it may be worth noting that cytologically detectable chromosome polymorphism has been found so far mainly in populations inhabiting more diverse environments. On the other hand, populations which are cytologically uniform live in areas judged to be more uniform. A thorough analysis of factors of special importance for a subterranean organism, including, in this case, edaphic and biotic factors, together with intensive sampling in neighboring countries may, hopefully, lead to a better understanding of the detailed evolutionary relationships within this complex.

Fragmentary information on Spalax from other localities makes it likely that similar situations prevail in other parts of the Spalax distribution. Crude data are provided in Table 3. Some of the published karyotypes are hardly amenable to a detailed comparative analysis. Several figures in Table 3 are based on reinterpretation of the original karyotypes. The diploid numbers have only limited significance, as the same numbers can obviously be due to different chromosome combinations. The differences between the numbers of chromosome arms reflect,

Table 3: Major Distinct Chromosome Constitutions of Spalax Tabulated in Increasing Chromosome Number. Number of Chromosome Arms is Approximate for Most Non-Israeli Populations

$2n$ = diploid chromosome number, M = number of metacentric chromosomes including 2 X chromosomes, A = number of acrocentric chromosomes. The Y chromosome is not included.

Origin	$2n$	M	A	Arms	Source
?CAUCASUS, ?UKRAINE	48			*	Matthey (1959)
YUGOSLAVIA: Banat	48	36	12	84	Soldatović *et al.* (1966a, 1967)
ISRAEL: Upper Galilee	52	32	20	84+ †	Present authors
YUGOSLAVIA: Hercegovina	54	34	20	88 ‡	Soldatović *et al.* (1967)
BULGARIA	54	32	22	86 §	Walknowska (1963)
ISRAEL: Golan	54	28	26	82+ †	Present authors
YUGOSLAVIA:					
Beograd, Zlatibor	56	38	18	94	Soldatović *et al.* (1967)
YUGOSLAVIA: Vlasina	56	24	32	80	Soldatović *et al.* (1966b)
YUGOSLAVIA: Čigota	56	26	30	82	" " "
ISRAEL:					
Lower Galilee, Central	58	18	40	76	Present authors
ISRAEL: Samaria	59‖	17	42	76	" "
ISRAEL:					" "
Samaria, Judea, Negev	60	16	44	76	" "

* Centromere position cannot be determined for many chromosomes. Number of arms probably around 80.

† The + sign denotes the existence of individuals in the respective population with a higher number of chromosome arms.

‡ Reinterpretation of original Fig. 3.

§ Very poor karyotypes. Numbers given are based on original Fig. 6.

‖ Hybrid between the forms with 58 and 60 chromosomes.

however, the pericentric inversions distinguishing certain karyotypes. The karyotype of the population at Banat, in Yugoslavia, has been most fully described (Soldatović *et al.,* 1966a). It has the lowest chromosome number known, namely 48. This and the other karyotypes from Yugoslavia have been compared, as far as possible, with the karyotypes from Israel. It can be tentatively concluded that all chromosome sets of Spalax display a basic similarity. The differences have been arrived at by changes of the kinds described, molding the same stock of chromosome material.

The non-Israeli material included in Table 3 appears in the original publications under the name *Spalax leucodon* Nordmann 1840. Little importance should, in our opinion, be attached to this designation in the light of the present findings. It should, however, be pointed out that both the X and Y chromosomes are claimed to be large metacentrics in all animals from Yugoslavia (Soldatović *et al.,* 1966a, 1967). In our series of forms, and apparently also in the specimen studied by Matthey (1959), the Y chromosome is a minute element, either acrocentric or nearly so (possibly having a very short second arm). As the sex chromosomes tend to preserve their size and form in some series of closely related mammals, it would be of interest to study this difference in more detail.

Among the subterranean rodents a pertinent situation has recently been described in pocket gophers of the genus Thomomys, to which two specific names have been applied (Patton and Dingman, 1968). Despite exhibiting two constant chromosome numbers, extensive chromosome variability exists. The variation must be chiefly due to pericentric inversions. Hybrids are produced, but though these are fertile, gene exchange is regarded to be low. In the burrowing species of Ctenomys, in South America, a wide karyotypic variability between closely related taxa is known (Kiblisky and Reig, 1966; Reig and Kiblisky, 1968). A detailed population analysis would seem to be worthwhile within this group.

Six chromosome forms have been discovered in the pocket mouse, *Perognathus goldmani* (Patton, 1967; 1968). These forms appear to be separated by geographical barriers and no hybrids have been found. The possibility of their being distinct biological species has apparently not yet been tested experimentally. Pocket mice, though burrowing rodents, are not subterranean in the strict sense. Nevertheless, their mobility has been judged to be weak (Patton, 1968).

Faced with a comparable situation in the *viatica* group of morabine grasshoppers of Australia, White has recently invoked a stasipatric model of speciation (White *et al.,* 1967; White, 1968). In his discussion, White has entertained some of the possibilities also suggested and preferred here (see, especially, 1968, Fig. 2). The common denominator which emerges from a comparison of those cases, including these grasshoppers and the mammalian examples, showing a cytogenetic mechanism leading to the creation of sibling species, is their relatively restricted mobility and narrow adaptive specialization

Separate designations should, in principle, be assigned to reproductively isolated gene pools. This policy was recently applied to sibling species in the insectivore genus Sorex (Ott, 1968). After a search had failed to reveal any hybrids at the contact zones between two chromosomal forms, the two forms were formally separated. This was done even though the slight morphological differences detected between the siblings would not alone suffice for their differentiation.

It may be said, in conclusion, that whatever further experiments and observations will bring to light, and no matter what other interpretations will be adopted, one thing is abundantly clear even now; mole rats embody a remarkable amount of chromosome variability. If this cytogenetic variability leads to any changes at the anatomical or physiological levels, these are masked by the extraordinary convergent adaptations associated with their special ecological requirements and unique modes of life.

Summary

Four main chromosome forms have been found among rodents conforming to the classical species *Spalax ehrenbergi* from Israel and adjoining areas. Though their diploid numbers are 52, 54, 58 and 60, the chromosome mass is about the same throughout the series. The karyotypic differences are due to two kinds of chromosome changes: whole-arm rearrangements (also known as Robertsonian changes) and pericentric inversions. An attempt was made to break up the chromosome sets into two groups of chromosomes of like history, corresponding to these two processes. A third group includes the chromosomes common to all four karyotypes (Fig. 6). It is tentatively suggested that the ancestral karyotype of Spalax was mostly metacentric and that progressive dissociation of metacentrics has taken place.

One 59 chromosome hybrid was found in an area which might belong into a $2n = 58 \times 2n = 60$ hybrid zone. Chromosome polymorphism, involving only inverted chromosomes, is present primarily in populations of 52 and 54 chromosomes. Robertsonian changes have reached homozygous fixation in the populations examined, which may indicate that they diverged a relatively long time ago.

The four chromosome forms are distributed in a definite geographic pattern, corresponding, in general, to biogeographical zones. They are not separated by any considerable physical barriers. It is considered possible that these forms, as well as those recorded from different parts of Yugoslavia, are indeed sibling species displaying few, if any, morphological differences. The existence of reproductive isolation between some of them is suspected on the basis of preliminary crossing experiments, observations on premating behavior, the homozygosity for most chromosome rearrangements, and the rarity of hybrids in population samples.

Penetration of Spalax into new environments may have profited from the establishment of novel chromosome configurations, conferring adaptive advantages in these regions. The frequency of chromosome and gene mutations may increase because of the isolated and inbred nature of the "colonies" of these almost totally subterranean rodents. Following the expansion into new areas, partial or complete reproductive isolation develops, based on chromosomal and perhaps ethological mechanisms. The overall correlation between karyotype and geographic region may reflect the adaptive nature of the situation.

In general, speciation processes in groups characterized by limited mobility and narrow environmental specialization may proceed in the absence of distinct geographical barriers required by most classical models of allopatric speciation.*

Acknowledgment

We thank Mrs. M. Bronstein for help in the cytological work.

References

Corbet, G. B.: The terrestrial mammals of western Europe. London: Foulis, 1966.

Ellerman, J. R.: The subterranean mammals of the world. Trans. Roy. Soc. S. Afr. *35*:11–20, 1959.

——— and T. C. S. Morrison-Scott: Checklist of Palaearctic and Indian Mammals 1758–1946. London: British Museum, 1951.

Kiblisky, P. and O. A. Reig: Variation in chromosome number within the genus Ctenomys and description of the male karyotype of *Ctenomys talarum talarum* Thomas. Nature (Lond.) *212*:436–438, 1966.

Matthey, R.: Formules chromosomiques de Muridae et de Spalacidae. La question du polymorphisme chromosomique chez les mammifères. Rev. Suisse Zool. *66*:175–209, 1959.

Nevo, E.: Observations on Israeli populations of the mole rat *Spalax e. ehrenbergi* Nehring 1898. Mammalia *25*:127–144, 1961.

———: Mating behavior of the mole rat, *Spalax ehrenbergi*. Manuscript, 1968.

Ognev, S. I.: Mammals of the U.S.S.R. and adjacent countries. Volume 5: Rodents. Moskva-Leningrad, 1947. English translation, Jerusalem, 1963.

Ott, J.: Nachweis natürlicher reproduktiver Isolation zwischen *Sorex gemellus*

* Since this paper was presented, 5 further hybrid individuals have been found. Some information on the chromosomes of Spalax from Romania is now available (Raicu, P., S. Bratosin and M. Hamar: Study on the karyotype of *Spalax leucodon* Nordm. and *S. microphthalmus* Guld. Caryologia *21*:127–135, 1968).

sp. n. und *Sorex araneus* Linnaeus 1758 in der Schweiz (Mammalia, Insectivora). Rev. Suisse Zool. *75*:53–75, 1968.

Patton, J. L.: Chromosome evolution in the pocket mouse, *Perognathus goldmani* Osgood (Rodentia: Heteromyidae). 43rd annual meeting of the Southwestern and Rocky Mountain Division of the AAAS (Abstract), 1967.

———: Personal communication, 1968.

——— and R. E. Dingman: Chromosome studies of pocket gophers, genus Thomomys. I. The specific status of *Thomomys umbrinus* (Richardson) in Arizona. J. Mammal. *49*:1–13, 1968.

Pearson, O. P.: Biology of the subterranean rodents, Ctenomys, in Peru. Memor. Mus. Hist. Nat. Javier Prado, Lima *9*:1–56, 1959.

Reig, O. A. and P. Kiblisky: Chromosomes in four species of rodents of the genus Ctenomys (Rodentia, Octodontidae) from Argentina. Experientia *24*:274–276, 1968.

Soldatović, B., B. Garzičić and S. Živković: Chromosome analysis of *Spalax leucodon*. Caryologia *19*:261–265, 1966a.

———, S. Živković, I. Savić and M. Milošević: Comparative karyotype analysis of two populations of *Spalax leucodon* Nordmann, 1840. Arhiv Biolokih Nauka *18*:127–128, 1966b.

———, S. Živković, I. Savić and M. Milošević: Vergleichende Analyse der Morphologie und der Anzahl der Chromosomen zwischen verschiedenen Populationen von *Spalax leucodon* Nordmann, 1840. Z. f. Säugetierkunde *32*:238–245, 1967.

State of Israel, Surveys Department: Atlas of Israel. 1956.

Thornthwaite, C. W.: An approach toward a rational classification of climate. Geogr. Rev. *38*:55–94, 1948.

Vaughan, T. A.: Movements made by two species of pocket gophers. Amer. Midl. Nat. *69*:367–372, 1963.

Wahrman, J.: Unpublished work on chromosome forms and meiosis in Spalax. 1958.

Walknowska, J.: Les chromosomes chez *Spalax leucodon* Nordm. Folia Biologica *11*:293–307, 1963.

White, M. J. D., R. E. Blackith, R. M. Blackith and J. Cheney: Cytogenetics of the *viatica* group of morabine grasshoppers. I. The "coastal" species. Aust. J. Zool. *15*:263–302, 1967.

———: Models of speciation. Science *159*:1065–1070, 1968.

INDUCED CHROMOSOMAL ABERRATIONS WITH SPECIAL REFERENCE TO MAN

KURT HIRSCHHORN AND MAIMON H. COHEN

Division of Medical Genetics, Department of Pediatrics, Mt. Sinai School of Medicine, New York, New York, and Division of Human Genetics, Department of Pediatrics, State University of New York at Buffalo Medical School and Buffalo Children's Hospital, Buffalo, New York

Induced mutagenesis has long been an active and fruitful area of genetic research. The recent technical advances in tissue culture techniques, as well as in chromosome preparations, have greatly facilitated the expansion of this field to the chromosomal level. Four general classes of exogenous agents have been described which are capable of causing chromosome aberrations and, therefore, possibly gene mutations as well: irradiation, viruses, physical stimuli, and chemicals. This last group can be further subdivided into those compounds affecting biosynthesis of macromolecules (DNA, RNA, and protein): antitumor agents, antibiotics, alkylating agents, nitroso-compounds, and a heterogeneous miscellaneous group (Lea, 1955; Revell, 1953; Kihlman, 1961, 1966; Ostertag, 1966; Cohen and Shaw, 1965).

More recently, attention has been focused on several relatively common environmental agents which has led to the rather intensive cytogenetic investigation of hallucinogens (Cohen *et al.*, 1967a, b; Irwin and Egozcue, 1967; Egozcue *et al.*, 1968; Jarvik and Kato, 1968; Lisco and Lisco, 1968), tranquilizers (Cohen *et al.*, 1968b; Stenchever *et al.*, 1968), oral contraceptives (Carr, 1967), herbicides (Wuu and Grant, 1966a), pesticides (Wuu and Grant, 1966b) and air pollutants (Fetner, 1962; Tough and Court Brown, 1965). Although much controversy surrounds the actual significance of chromosomal damage induced in the circulating leukocytes of man, or in plant and animal cells grown *in vitro*, several analogies have been drawn which suggest the possible consequences of such lesions. These include the well documented association between neoplasia and chromosomal imbalance, the possible teratogenicity to fetuses exposed *in utero,* and a true genetic effect, should the chromosomes of functional gametes be similarly affected. Although such outcomes are no more than suggested by the observations at present, they should serve as sufficient warning that these agents must be treated with caution until more definitive data are available.

It should be emphasized that it is not necessarily the breaks themselves but the mechanisms by which they arise which are of greatest interest and which will possibly yield information concerning chromosome structure and function. Although morphologically chromosome breaks might appear identical and their nonrandom distribution among and within chromosomes similar (Cohen and Shaw, 1965), it is quite possible, and even probable, that the various agents act by different mechanisms. Although most mechanisms of chromosome breakage are unknown, several suggestions have been made based on the known biochemical action of some of these compounds. One hypothesis concerns the synthesis and metabolism of nucleic acids. A basic consideration of the latter involves the inhibition of incorporation of DNA and RNA precursors (ribonucleotides) by such agents as adenine and azaserine. Since these compounds act at the level of purine synthesis, both DNA and RNA are affected (Kihlman, 1966). A second class of compounds may act after the separation of the common pathways of DNA and RNA synthesis, e.g., between purine or pyrimidine biosynthesis and the formation of deoxyriboside triphosphates. Such appears to be the case for FUDR and AdR which have an effect on DNA synthesis but not RNA synthesis. Yet another mechanism is suggested by those agents which interact with primer DNA. Since primer DNA will serve in the biosynthesis of both DNA and RNA through polymerase reactions, such agents will inhibit the synthesis of both nucleic acids (Elliot, 1963; Reich, 1964; Warig, 1964).

A second group of mechanisms concerns the degradation and denaturation of DNA after its synthesis. This concept is supported by the action of a group of agents which cause chromosome damage subsequent to the S period, that is, in early prophase. It is obvious that in such cases the cytological damage cannot be an effect on DNA synthesis per se, which is completed before prophase begins. Therefore, such breaks necessitate the interaction of the agent and the DNA-protein complex. The chemicals streptonigrin, caffeine and Actinomycin D have yielded evidence for such interaction. Streptonigrin initiates a rapid breakdown of E. coli DNA (Radding, 1963). Caffeine combines with DNA to alter its physical properties by denaturation (T'so *et al.,* 1962; T'so and Lu, 1964) and Actinomycin D inhibits strand separation caused by the impairment of the DNA template (Reich, 1964).

The action of most chromosome breaking agents, however, is thought to be the alkylation of bases or the incorporation of base analogues leading to the production of abnormal DNA. It is generally considered that

DNA, rather than its precursors, is affected by alkylation of the heterocyclic nitrogen in the bases (Stacey *et al.,* 1958). Cross-linking of DNA moieties by agents such as mitomycin C (Schwartz *et al.,* 1963; Iyer and Szybalski, 1963) is also considered to be important in this process.

Additionally, several mechanisms, not integrally related to DNA or precursor synthesis, have been proposed. It has long been contended that the cytological effects of X-irradiation are due to the formation of free radicals (Lea, 1955). Faulty ATP and protein synthesis requisite for the repair of chromosomal breaks have also been implicated (Wolff, 1960; Wolff and Luippold, 1955). In this context it may be possible that the mutagenicity of substances such as caffeine is due not to actual chromosome breakage but to the prevention of successful enzymatic repair of breaks. Therefore, there would be an actual potentiation of the chromosome breaking potential of many agents when used in combination with caffeine. A similar synergistic relationship exists for 5-BUDR and X-ray sensitivity (Somers and Humphrey, 1963; Humphrey and Dewey, 1965).

Another hypothesis based on enzyme digestion has been suggested by various investigators (Kersten, 1962; Kersten *et al.,* 1964). These investigators have demonstrated that the degraded DNA resulting from mitomycin C treatment is produced by DNAase released from the microsomes. Fragmentation by DNAase of isolated dipteran lampbrush chromosomes grown *in vitro* (MacGregor and Callan, 1962) has been described. It is reasonable, therefore, to suggest the release of large amounts of DNAase during a period of the cell cycle when the chromosomes are quite vulnerable, leading to the production of chromosomal abnormalities.

A curious paradox exists in the observation of chromosome breakage in relation to the cell cycle. Many agents utilized during the DNA synthetic period (S) yield aberrations which appear to be induced during G_2, e.g., chromatid breaks. Although some agents are active in almost all phases of the cell cycle, G_1, S, and G_2, others, although present during G_1 and S, seem to cause most of the chromosome damage during G_2. Several explanations can be suggested to account for this phenomenon, but one of the more engaging hypotheses involves a "delayed reaction" related to the DNAase release described above. It may be that the primary action of the drug or agent in question is not on the chromosome per se but on some other cellular organelle or component. This initial action may induce a metabolic imbalance involving pathways which will not be expressed until a later stage in the cell cycle. Therefore, the quantitative "end point" of chromosome breakage may indeed be only a secondary response to a metabolic insult occurring prior to mitosis but capable of expression only during G_2. Supportive evidence for this hypothesis has been offered by recent studies on lysosomes. Changes in various lysosomal enzymes have been implicated in viral pathogenicity (Chessin, 1963; Allison and Mallucci, 1965; Mallucci and Allison, 1965).

It has also been specifically suggested that the release of lysosomal DNAase caused by the destabilization of lysosomal membranes in human diploid cells produces chromosomal aberrations (Allison and Paton, 1965). Additional experiments also indicate that chromosome breakage may at times be a secondary response, as mediated by changes in the environmental milieu. For example, cells grown in plasma derived from previously irradiated individuals or in medium containing irradiated sucrose have exhibited such damage (Hollowell and Littlefield, 1967; Shaw and Hayes, 1966).

Recently, we have had the opportunity to investigate the effects of the hallucinogen LSD on human chromosomes both *in vivo* and *in vitro* (Cohen *et al.*, 1967a, b). One of the first effects of LSD noted *in vitro* was a marked suppression of mitosis. The rate of mitosis in control cultures was approximately 6% and a significant mitotic inhibition was apparent only upon the addition of the drug at any concentration ranging from 10–0.001 μg/ml. However, the longer the exposure time to the drug, the greater was the suppression of mitosis. Induction of chromosomal damage in the *in vitro* experiments was also clearly demonstrated. These experiments consisted of cultures derived from six normal individuals (three males and three females). Among the controls, 3.9% of the cells exhibited chromosomal abnormalities, while among the treated cultures, the lowest frequency of abnormal cells was almost twice the control value (7.7%); and it ranged to over four times control values (17.5%). The vast majority of these abnormalities were simple chromatid and isochromatid breaks, although a few chromatid exchanges and dicentric chromosomes were observed. It should be mentioned that all of the morphological structural rearrangements were seen in the treated cultures and none in the control. An *in vivo* study of 18 LSD users corroborated the *in vitro* findings. These individuals had a history of LSD ingestion ranging from two to over 300 exposures. The treated group consisted of 18 patients who were matched for age and sex with a group of 12 controls. The control subjects showed a mean breakage rate of 3.8% of the cells with chromosomal abnormalities (range 2–5.5%). Among the LSD users there was a wide range of chromosomal abnormalities ranging from control levels of 5% to in excess of 25%. As in the *in vitro* studies, the majority of the abnormalities were chromatid and isochromatid breaks; however, 12 dicentric chromosomes and five exchange figures were observed in the LSD group whereas no structural rearrangements were noted in the control group. There was no apparent correlation between the number of doses, the amount per dose, nor the interval between the last dose and the time of bleeding for leukocyte cultures and the frequency of chromosome breaks. The vast majority of LSD users are inveterate experimenters with other drugs as well. However, again there was apparently no relationship in the chromosomal break frequency between the type and number of other drugs used. Elevated frequencies of chromosomal damage were seen regardless of which drugs were ingested in combination with LSD.

We also performed cytogenetic investigations on nine children exposed to LSD *in utero,* four children of other LSD users not exposed *in utero,* the mothers of these children, and six additional adults taking the drug (Cohen *et al.,* 1968b). Significantly elevated frequencies of chromosomal damage were observed in all individuals exposed when compared to matched controls. However, the frequency of aberrations was much lower in those children not exposed *in utero.* Eight structural rearrangements were observed, six dicentric chromosomes and two quadriradial formations, all among those individuals exposed to the drug and none in the controls. In spite of the obvious chromosomal aberrations, all of the individuals in this study are apparently healthy and show no obvious birth defects. Recently, Skakkebaek *et al.* (1968), as well as Cohen and Mukherjee (1968), have demonstrated that LSD injected into male mice can cause meiotic chromosomal abnormalities. These effects were observed both in cells in active meiosis as well as cells of somatic origin (e.g., bone marrow and spermatogonial metaphases). It is interesting to note that structural rearrangements such as biarmed chromosomes, evidence for translocations in mice, were obtained in spermatogonial metaphases.

Observations of chromosomal abnormalities induced by LSD both *in vitro* and *in vivo* have now been corroborated by a number of other laboratories throughout the United States. Additionally, there have been some negative reports of chromosomal abnormalities among LSD users (Loughman *et al.,* 1967; Bender and Sankar, 1968; Sparkes *et al.,* 1968). At this time cells from approximately 220 LSD "users" have been examined for chromosomal abnormalities and an increase in the frequency of chromosomal breakage has been observed in approximately 70–80%. There is obviously a wide range of reaction among the individuals who do take LSD. This is by no means an all or nothing phenomenon as was evident even in our own sample of 18 LSD users (Cohen *et al.,* 1967b). Some of the exposed individuals exhibited frequencies well within the normal range while the majority showed elevated breakage frequencies.

The possible significance and dangers of chromosomal breakage must be viewed from several aspects. First, there are the possible consequences to the individual exposed to the agent. From this point of view we have no definitive evidence derived from man, but one may, however, discuss the problem from our knowledge of an analogous situation which could give rise to similar types of chromosome breakage and rearrangement. Chromosomal aberrations similar to those induced by LSD have been noted to occur spontaneously in three diseases known to be inherited in an autosomal recessive pattern: Bloom's syndrome, Fanconi's anemia and ataxia telangiectasia. Patients with these syndromes show a high propensity to leukemia and other neoplasms. In this context, cells of neoplastic origin frequently illustrate a variety of chromosomal aberrations, many of which are not unlike those seen after LSD exposure. Moreover, many of the agents known to produce similar aberrations (such as radiation and benzene in man and other species, as well as viruses in laboratory animals) are known carcinogens. Whether or not chromosomal damage and the formation of neoplasia are causally related is not known; however, the association of chromosome anomalies and cancer is well documented. One of the obvious potential dangers, therefore, of exposure to agents such as LSD is the possible future increase in the incidence of leukemia and other neoplasms in the persons exposed. A recent case of subacute leukemia in an LSD user demonstrates a Ph_1-like chromosome (Grossbard *et al.*, 1968).

In addition, other less evident dangers exist. In view of the demonstration of chromosome damage in children exposed *in utero*, the production of congenital defects by this and possibly other drugs is a distinct possibility. Although again evidence in man is not yet available, the teratogenic properties of LSD, through the production of congenital abnormalities and increased abortions, have been observed in animal studies. When injected into rats, hamsters, or mice early in pregnancy, a wide range of birth defects was noted from overall stunting of development in some animals to severe malformations of the central nervous system as well as other morphologic anomalies (Alexander *et al.*, 1967; Geber, 1967; Auerbach and Rugowski, 1967). These effects were most dramatic if the drug was administered early in pregnancy but they were observed to a limited extent if given later. Streptonigrin and mitomycin C, two additional drugs capable of somatic chromosome damage (Cohen *et al.*, 1963; Cohen, 1963; Cohen and Shaw, 1964; Shaw and Cohen, 1965), and meiotic chromosome damage (Jagiello, 1967) also possess teratogenic potential (Warkany and Takacs, 1965; Kury and Craig, 1967) and can induce sex linked lethal mutations in Drosophila (Ostertag and Haake, 1966).

Perhaps the greatest danger of such agents lies in the potential chromosome damage to the gametes. Both streptonigrin (Jagiello, 1967) and LSD (Skakkebaek *et al.*, 1968; Cohen and Mukherjee, 1968) can apparently cause chromosomal damage in rodent meiotic cells. Therefore, it is important to regard these drugs with extreme caution. If chromosomal damage similar to that observed in the leukocytes and occurring in spermatogonial cells of the mouse give rise to viable gametes, then the effect of LSD may indeed be a genetic one. Ample evidence already exists that segregation of translocations in man results in chromosomal imbalance, leading to fetal wastage and a high percentage of offspring with congenital abnormalities and mental retardation. Since the carriers of such balanced translocations are usually clinically normal, as are many of their offspring, the consequences of this type of chromosomal abnormality may not be apparent for several generations.

The possibility that LSD might similarly produce translocations in human germ cells has become a probability with the recent investigation of an abnormal child, born to a couple, both of whom had ingested the drug prior to, but not during pregnancy (Hirschhorn *et al.*, 1968a). The infant, who died shortly after birth, manifested many stigmata of trisomy 13. Cytogenetic analysis revealed a modal number of 46 chromosomes including five members of group D and an extra chromosome resembling a #3 (Fig. 1). This marker element was interpreted as a translocation ("centric fusion") between two D group chromosomes with subsequent adjacent segregation during meiosis leading to the trisomic condition. Both parents were karyotypically normal in their somatic cells, and therefore, the child most likely illustrates a *de novo* translocation produced by chromosomal breakage and recombination in a parental germ cell.

The literature of human cytogenetics contains numerous examples of such aberrations which have been termed "centric fusions" between two acrocentric chromosomes. This phenomenon is similar to that of Robertsonian fusion, a process of great evolutionary importance in many species.

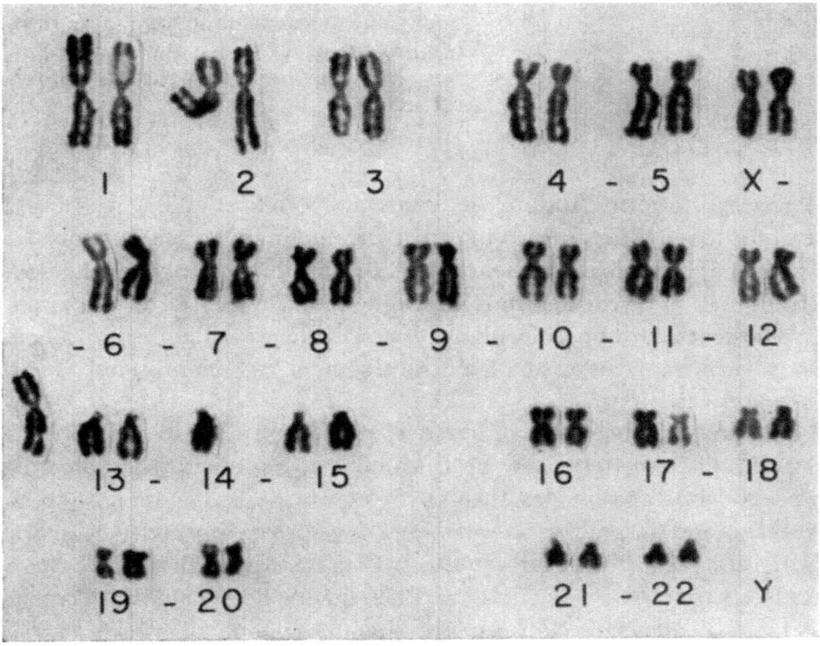

Fig. 1. Karyotype of trisomy 13 with D/D translocation.

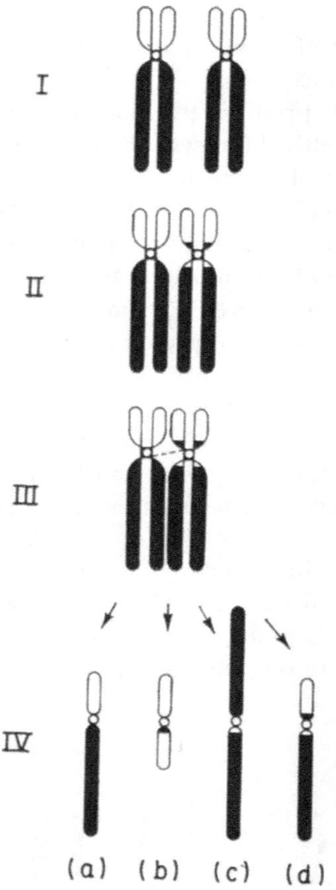

Fig. 2. Isochromosome formation resulting from a pericentric inversion (II—chromosome on the right) followed by crossing over (III—dotted line indicates region of crossing over). Four gametes may be produced: (a) normal, (b) isochromosome of short arm, (c) isochromosome of long arm, (d) pericentric inversion carrier.

However, the term "fusion" as discussed by White (1961) may actually be a misnomer when describing such chromosomal recombinations. Like all other morphological alterations, "fusion" most likely results from a recombination of chromatin following a break in each of two acrocentric chromosomes, on opposite sides of the centromeres. The reciprocal product of such an event, a small centric fragment, is generally lost at subsequent cell divisions.

In many instances, such altered elements are almost perfectly metacentric and have frequently been assumed to represent isochromosomes. However, isochromosomes themselves may actually be the products of breakage and recombination, or even crossover events rather than the tacitly accepted, but never proven, hypothesis of misdivision of the centromere proposed by Darlington (1939, 1940). For example, a crossover near the centromere between two homologs, one of which carries a small pericentric inversion, can result in an isochromosome of either arm.

The inversion itself, of necessity, is the result of a previous two-break event (Fig. 2).

There is, therefore, the possibility of continuous Robertsonian chromosomal evolution in man. Such phenomena require the participation of two pairs of chromosomes for stability and could only result from one event with four breaks, or a two-break event in a carrier of a "centric fusion" type of translocation. Heterozygous translocation carriers occur frequently in the human population, with an estimated incidence of perhaps 1/500 to 1/1000. Even a four-break event, as described above, is not impossible. A recent case describes an infant with clinical features of Trisomy 13 but with only 45 chromosomes, including three normal D group elements and two extra chromosomes resembling #3 (Fig. 3). Autoradiographic analysis revealed that the patient was a double translocation carrier representing one 13/14 and one 13/15 translocation product (Cohen *et al.*, 1968a).

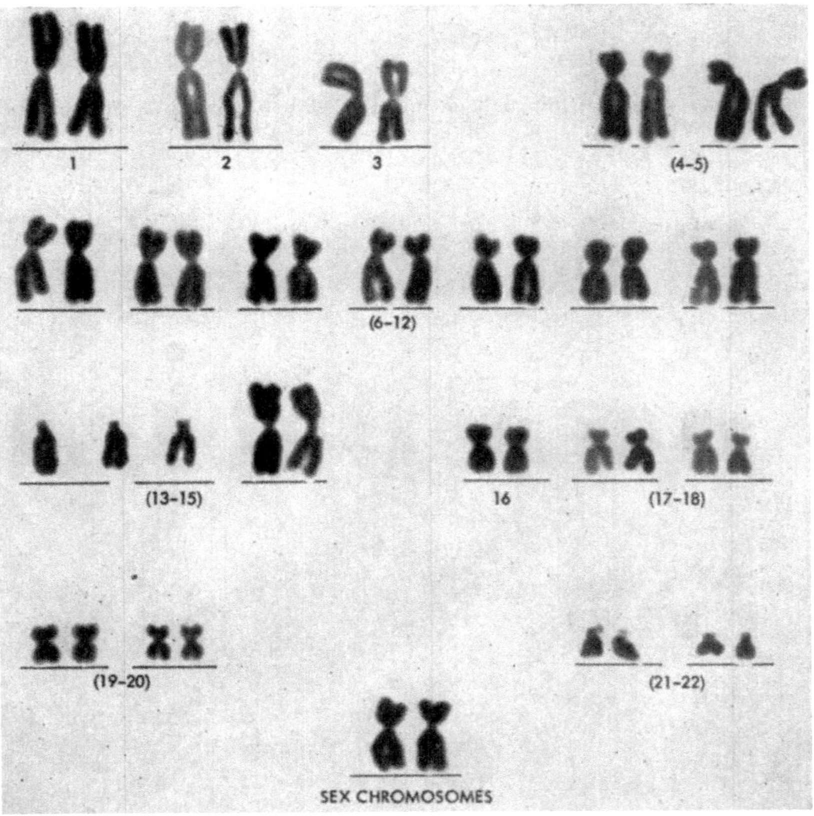

Fig. 3. Karyotype of trisomy 13 with two D/D translocations.

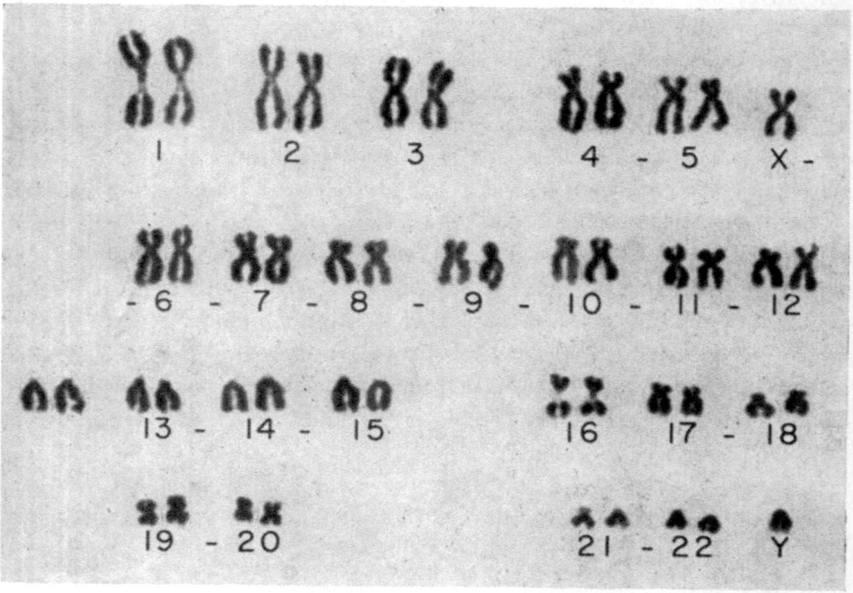

Fig. 4. Karyotype of human leukemia line with two extra D chromosomes.

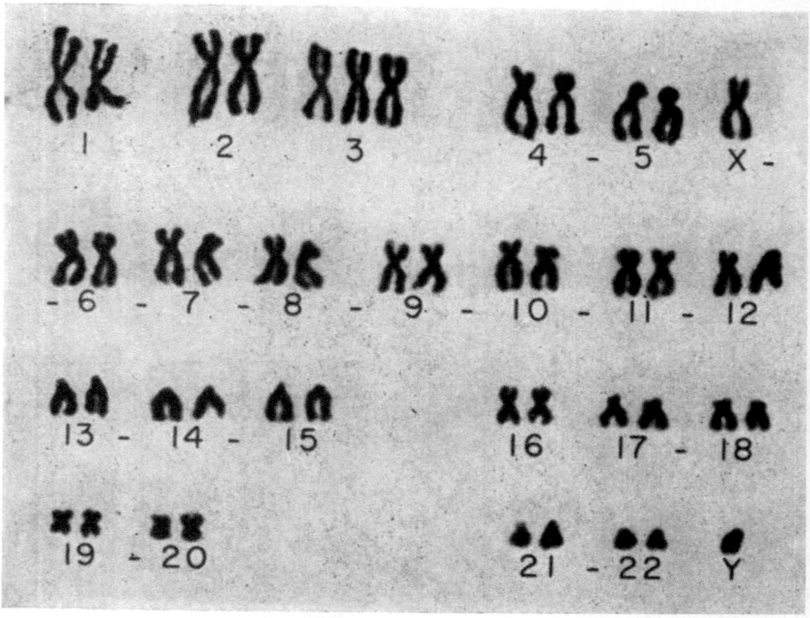

Fig. 5. Karyotype of same leukemia line shown in Fig. 4 but with "fusion" of two D chromosomes forming a chromosome resembling No. 3.

A similar type of translocation was recently observed *in vitro* (Hirschhorn *et al.*, 1968b) in a human cell line derived from a leukemic bone marrow. Two karyotypically distinguishable sublines evolved at different times, one of which exhibited a modal number of 48 chromosomes with eight D group elements (Fig. 4), while the other had 47 chromosomes with six normal D's and one extra chromosome resembling a #3 (Fig. 5). No other chromosomal abnormalities were observed. Such tissue culture models may prove extremely useful in the stuly of the cytogenetic effects of various agents as well as of the basic mechanisms of chromosomal breakage.

Many other types of chromosomal translocations, both balanced and unbalanced, have been described in man. Homozygosity for such balanced translocations has also been found to be an important mechanism

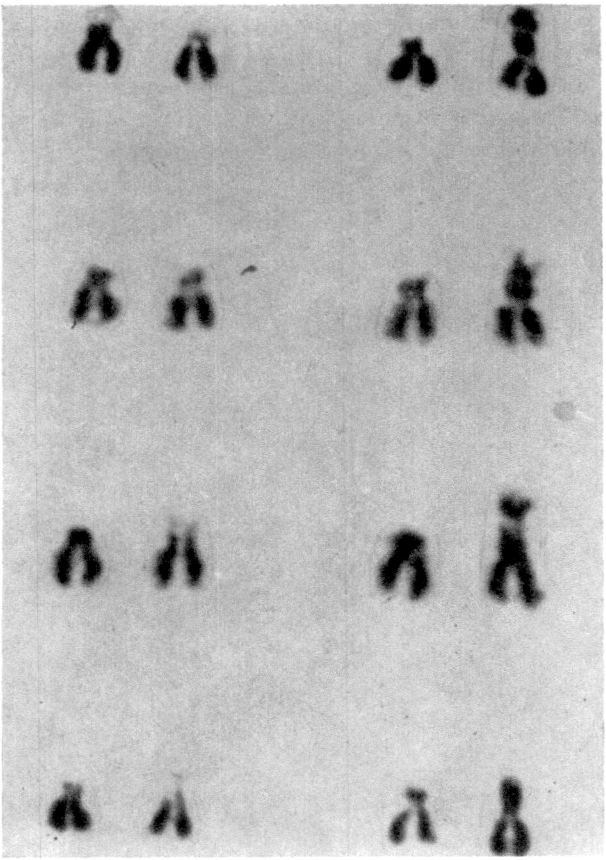

Fig. 6. Four sets of G chromosomes from an individual showing an extended short arm with secondary constriction.

in mammalian evolution (Borgaonkar and Gould, 1968). Some of these morphological changes have been termed "normal variants" and most frequently describe alterations in the short arms of acrocentric chromosomes (Hirschhorn, 1965; Nusbacher and Hirschhorn, 1968). Such changes are generally stable and may occur in multiple generations of a family. Elongation or enlargement of the short arm of an acrocentric, at times with marked secondary constrictions in addition to the usual satellite stalk, are quite common. In a patient with Turner's syndrome and $46,XX_i/45,X$ mosaicism, enlargement of the short arm of a G group chromosome occurred which occasionally simulated a second centromere (Fig. 6). That this element was not truly a dicentric chromosome was evidenced by the fact that it occurred in every cell without variation.

The karyotype of a young boy with congenital heart disease exhibited the simultaneous similar enlargement of the short arms of both a G and a D group chromosome (Fig. 7). A family study in this case revealed that these two abnormal chromosomes were present singly in each of the parents of the propositus as well as in his sibs (Fig. 8). It is quite likely that these "variant" chromosomes represent duplications of short arm material, resulting from translocations due to breaks near the centromere of one and in the satellite stalk of the other acrocentric chromosome involved. Such exchanges may be analogous to "unequal homolo-

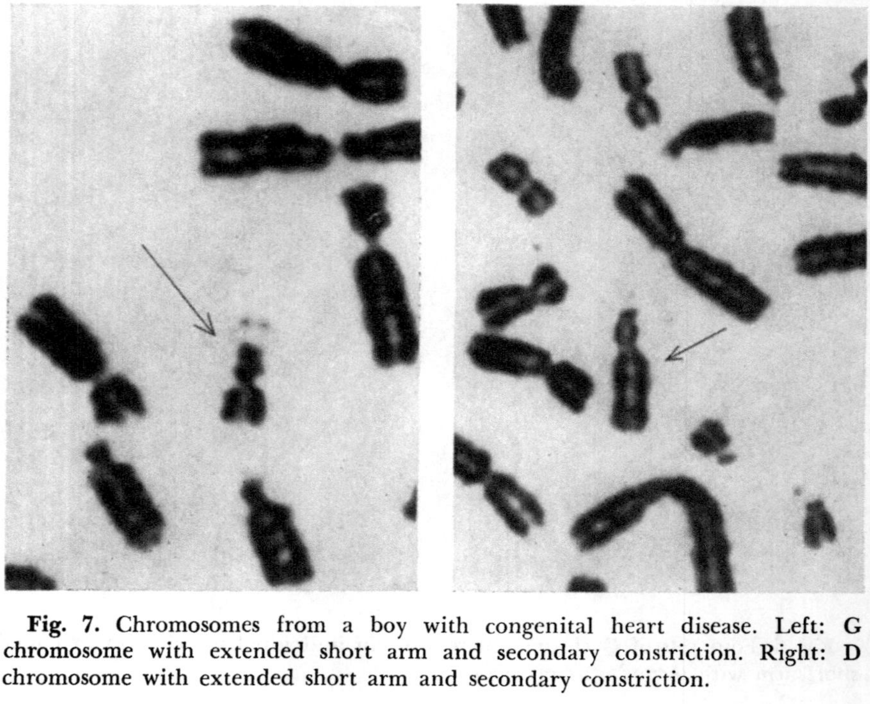

Fig. 7. Chromosomes from a boy with congenital heart disease. Left: G chromosome with extended short arm and secondary constriction. Right: D chromosome with extended short arm and secondary constriction.

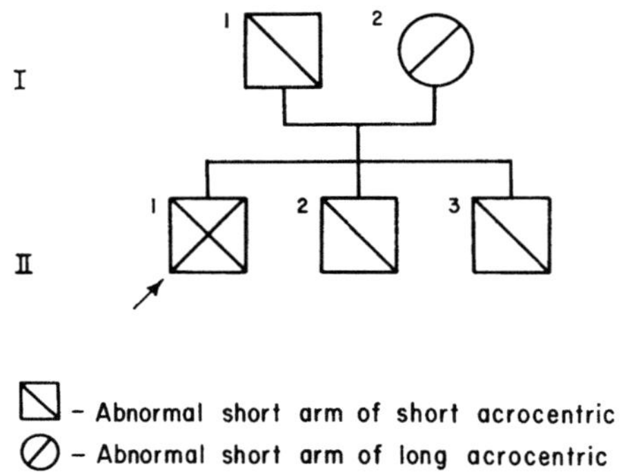

Fig. 8. Pedigree of family of boy whose chromosomes are shown in Fig. 7.

gous crossing over" which in itself is likely due to breakage and recombination.

Another family, ascertained through one of two sisters with primary amenorrhea, possessed a very long short arm of a D chromosome without the additional secondary constriction but with a normal satellite region (Fig. 9). In this case, the breaks probably occurred near the centromere of one chromosome and towards the distal end of the euchromatic part of the short arm of another acrocentric chromosome. These two sisters were otherwise chromosomally normal, and had a brother carrying this abnormal chromosome. The mother of the family, herself the product of a consanguineous marriage, was homozygous for this chromosome (Fig. 10). Although size measurements speak against it, an alternative interpretation is that of a pericentric inversion. In effect, this homozygote represents a new cytogenetic "subspecies." Her brother, a hetero-

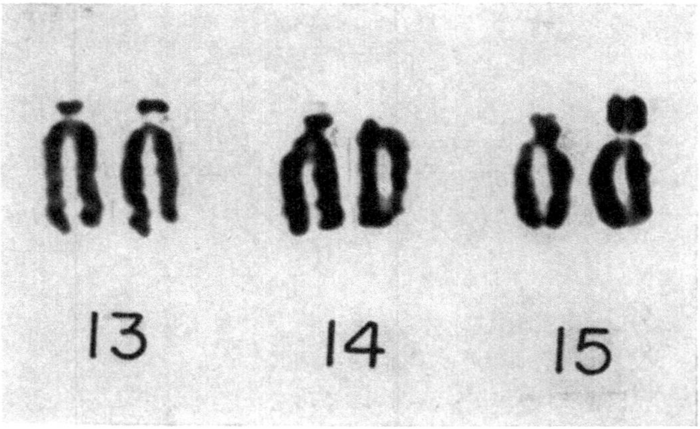

Fig. 9. D chromosomes of girl with primary amenorrhea with extended short arms of one D chromosome.

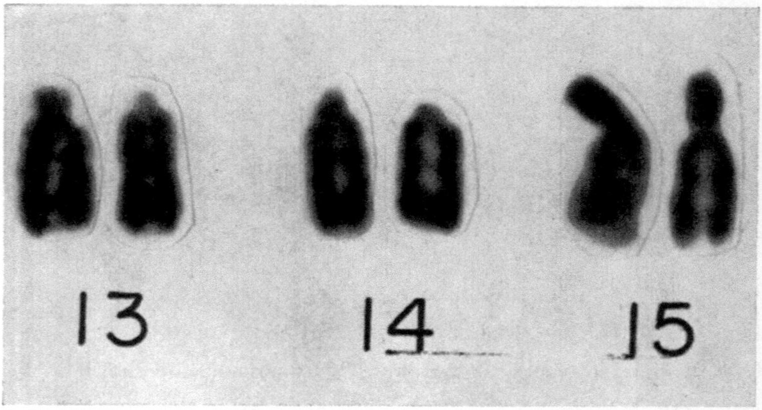

Fig. 10. D chromosomes of mother of patient whose chromosomes are shown in Fig. 9. Note two similar abnormal D chromosomes.

zygous carrier, had two daughters, the younger of which began menstruating at age 12 (Fig. 11). In this family the chromosomal duplication is therefore compatible with both normality and fertility in both sexes.

A recent investigation reveals that the mechanism of translocation may indeed be responsible for at least some (if not all) of these so-called "variants" (Lieber *et al.,* 1968). The propositus was a small child with failure to thrive and a large cartilaginous mass on his head. This child died at one month of age. Chromosome investigation revealed a #13 chromosome with an apparently missing short arm (Fig. 12). Familial segregation of this marker chromosome over four generations was dem-

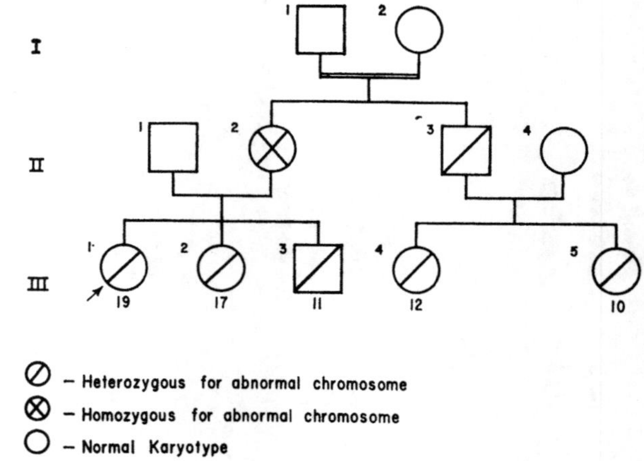

Fig. 11. Pedigree of a family with hereditary transmission of an abnormally long short arm on a chromosome in 13–15 group. Note consanguinity in generation I. Proposita indicated by arrow.

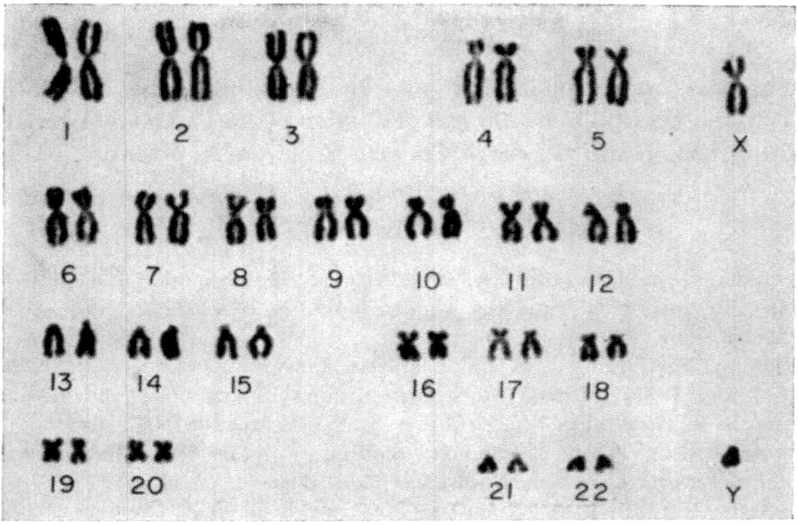

Fig. 12. Karyotype of boy with apparent deletion of short arms of chromosome 13.

onstrated. Many similar "telocentric" chromosomes described in the literature are assumed to represent deletions. However, meiotic studies in one of the normal males in this family who possessed the "variant chromosome" revealed 20 autosomal bivalents, an XY bivalent and one quadrivalent, probably involving another D chromosome pair. There was an apparent association of the long arm of a number 14 with the satellites of the other acrocentric chromosomes, indicating that this family may exhibit a t(13p;14q) translocation (The Chicago Conference, 1966). This observation clearly demonstrates that reciprocal translocation, resulting from breakage and recombination may be quite common in the human population.

It would seem, therefore, that the chemical agents described at the beginning of this paper, as well as an unknown number of other environmental influences, are capable of producing chromosomal alterations in man. The possible mechanisms by which such aberrations may be produced are multiple, but current experimental evidence is insufficient to assign relative importance to any postulated mechanism. However, an undetermined proportion of such chromosomal change is detrimental and assumes the form of fetal loss, congenital malformations, mental retardation and possibly neoplasia (Allison, 1968). It remains to be seen, however, to what extent such environmental factors contribute to the potential acceleration of a subtle form of human evolution as produced by stable alterations of the human chromosome complement.

Acknowledgment

This work was supported in part by grants from the USPHS(HD-02552), the US Children's Bureau (417), the American Heart Association and the NYC Health Research Council (K.H., Career Scientist, I-513).

References

Alexander, G. J., B. E. Miles, G. M. Gold and R. B. Alexander: LSD: injection early in pregnancy produces abnormalities in offspring of rats. Science *157*:459, 1967.

Allison, A. C: Chromosomes and neoplasia. European J. Cancer *3*:481, 1968.

——— and L. Mallucci: Histochemical studies of lysosomes and lysosomal enzymes in virus injected cell cultures. J. Exper. Med. *121*:463, 1965.

——— and G. R. Paton: Chromosome damage in diploid cells following activation of lysosomal enzymes. Nature *207*:1170, 1965.

Auerbach, R. and J. A. Rugowski: Lysergic acid diethylamide: effect on embryos. Science *155*:1325, 1967.

Bender, L. and D. V. S. Sankar: Chromosome damage not found in leukocytes of children treated with LSD-25. Science *159*:749, 1968.

Borgaonkar, D. S. and E. Gould: Homozygous reciprocal translocation as a mode of speciation in *Microgale* Thomas 1883 (Tenrecidae-Insectivora). Experientia *24*:506, 1968.

Carr, D. H.: Chromosomes after oral contraceptives. Lancet *ii*:830, 1967.

Chessin, L. N.: Inherited variation of human somatic cells to poliovirus infection. M. D. Honors Thesis, New York University, 1963.

Cohen, M. M.: The specific effects of streptonigrin activity on human chromosomes in culture. Cytogenetics *2*:271, 1963.

——— and M. W. Shaw: Effects of mitomycin C on human chromosomes. J. Cell Biol. *23*:386, 1964.

——— and M. W. Shaw: In vitro, Vol. *1*:50. Baltimore, Md.: Waverly Press, 1965.

———, M. W. Shaw and A. P. Craig: The effects of streptonigrin on cultured human leukocytes. Proc. Nat. Acad. Sci. U.S.A. *50*:16, 1963.

———, M. J. Marinello and N. Back: Chromosomal damage in human leukocytes induced by lysergic acid diethylamide. Science *155*:1417, 1967a.

———, K. Hirschhorn and W. A. Frosch: *In vivo* and *in vitro* chromosomal damage induced by LSD-25. New Eng. J. Med. *277*:1043, 1967b.

———, N. Takagi and E. K. Harrod: Trisomy D_1 with two D/D translocation chromosomes. Am. J. Dis. Child. *115*:185, 1968a.

———, K. Hirschhorn, S. Verbo, W. A. Frosch and M. M. Groeschel: The effect of LSD-25 on the chromosomes of children exposed *in utero*. Ped. Res. *2*:486, 1968b.

———, K. Hirschhorn and W. A. Frosch: Cytogenetic effects of tranquilizing drugs *in vivo* and *in vitro*. J.A.M.A. *207*:2425, 1969.

——— and A. B. Mukherjee: Meiotic chromosome damage induced by LSD-25. Nature *219*:1072, 1968.

Darlington, C. D.: Misdivision and the genetics of the centromere. J. Genet. *37*:341, 1939.
———: The origin of isochromosomes. J. Genet. *39*:351, 1940.
Egozcue, J., S. Irwin and C. A. Maruffo: Chromosomal damage in LSD users. JAMA *204*:214, 1968.
Elliot, W. H.: The effects of antimicrobial agents on deoxyribonucleic acid polymerase. Biochem. J. *86*:562, 1963.
Fetner, R. H.: Ozone-induced chromosome breakage in human cell cultures. Nature *194*:793, 1962.
Geber, W. F.: Congenital malformations induced by mescaline, lysergic acid diethylamide and bromolysergic acid diethylamide in the hamster. Science *158*:265, 1967.
Grossbard, L., D. Rosen, E. McGilvray, A. DeCapoa, O. J. Miller and A. Bank: Acute leukemia with Ph_1-like chromosomes in an LSD user. JAMA *205*:791, 1968.
Hirschhorn, K.: Abnormalities of the acrocentric chromosomes (abs). In: Genetics Today, Vol. 1. Proc. XI Int. Cong. Genetics. S. J. Geerts, ed., pp. 304–305, New York: Pergamon Press, 1965.
———, L. Hsu and L. Strauss: Translocation-trisomy 13 in a child conceived after ingestion of LSD. (In preparation.) 1968a.
———, N. Gersht and C. Friend: Chromosomal evolution in a human leukemic cell line by translocation. (In preparation.) 1968b.
Hollowell, J. G. and L. G. Littlefield: Chromosome aberrations induced by plasma from irradiated patients. J. S. Carolina Med. Assoc. *63*:437, 1967.
Humphrey, R. M. and W. C. Dewey: Radiosensitivity of normal and 5-bromodeoxyuridine treated mammalian cells during different phases of the cell cycle. Exp. Cell Res. *39*:483, 1965.
Irwin, S. and J. Egozcue: Chromosomal abnormalities in leukocytes from LSD-25 users. Science *157*:313, 1967.
Iyer, V. N. and W. Szybalski: A molecular mechanism of mitomycin action: linking of complementary DNA strands. Proc. Nat. Acad. Sci. (Wash.) *50*:355, 1963.
Jagiello, G.: Streptonigrin: effect of first meiotic metaphase of the mouse egg. Science *157*:453, 1967.
Jarvik, L. G. and T. Kato: Is lysergide a teratogen? Lancet *i*:250, 1968.
Kersten, H.: Action of mitomycin C in nucleic acid metabolism in tumor and bacterial cells. Biochim. Biophys. Acta *55*:558, 1962.
———, W. Kersten, G. Leopold and B. Schnieders: Effect of mitomycin C on DNAse and RNA in *Escherichia coli*. Biochim. Biophys. Acta *80*:521, 1964.
Kihlman, B. A.: Biochemical aspects of chromosome breakage. Advances Genet. *10*:1, 51, 1961.
———: Actions of Chemicals on Dividing Cells. Englewood Cliffs, N.J.: Prentice-Hall, Inc., 1966.
Kury, G. and J. M. Craig: The effect of mitomycin C on developing chicken embryos. J. Embryol. Exp. Morph. *17*:229, 1967.
Lea, E. C.: Actions of Radiations on Living Cells. 2nd ed. Cambridge: University Press, 1955.
Lieber, E., W. Seegers, K. Hirschhorn, P. Allderdice, A. DeCapoa and O. J.

Miller: Apparent familial deletion of the short arm of chromosome 13 due to translocation. (In preparation.) 1968.

Lisco, H. and E. Lisco: (Personal communication), 1968.

Loughman, W. D., T. W. Sargent and D. M. Israelstam: Leukocytes of humans exposed to lysergic acid diethylamide: lack of chromosomal damage. Science *158:*508, 1967.

Mallucci, L. and A. C. Allison: Lysosomal enzymes in cells infected with cytopathic and non-cytopathic viruses. J. Exper. Med. *121:*477, 1965.

MacGregor, H. C. and H. G. Callan: The actions of enzymes on lampbrush chromosomes. Quart. J. Microscop. Sci. *103:*173, 1962.

Nusbacher, J. and K. Hirschhorn: Autosomal abnormalities. Adv. in Teratology (in press), 1968.

Ostertag, W.: Chemische Mutagenese an menschlichen Zellen in Kultur. Abh. Math. Naturw. Kl. *1:*5, 1966.

——— and J. Haake: The mutagenicity in *Drosophila melanogaster* of caffeine and of other compounds which produce chromosome breakage in human cells in culture. Z. Vererbungsl. *98:*299, 1966.

Radding, C. M.: Incorporation of H^3-thymidine by K12 (λ) induced by streptonigrin. In: Genetics Today. S. J. Geerts, ed. Oxford: Pergamon Press, p. 22, 1963.

Reich, E.: Actinomycin: correlation of structure and function of its complexes with purines and DNA. Science *143:*684, 1964.

Revell, S. H.: Chromosome breakage by X-rays and radiomimetic substances in *Vicia.* In: Symposium on breakage. Heredity (Suppl.) *6:*107, 1953.

Schwartz, H. S., J. E. Sodergren and E. S. Philips: Mitomycin C: Chemical and biological studies on alkylation. Science *142:*1181, 1963.

Shaw, M. W. and M. M. Cohen: Chromosome exchanges in human leukocytes induced by mitomycin C. Genetics *51:*181, 1965.

——— and E. Hayes: Effects of irradiated sucrose on the chromosomes of human lymphocytes *in vitro.* Nature *211:*1254, 1966.

Skakkebaek, N. E., J. Phillip and O. J. Rafaelson: LSD in mice: abnormalities in meiotic chromosomes. Science *160:*1246, 1968.

Somers, C. E. and R. M. Humphrey: A chromosome study of radiation sensitization by 5-bromodeoxyuridine. Exp. Cell Res. *30:*208, 1963.

Sparkes, R. S., J. Melnyk and L. P. Bozzetti: Chromosomal effect *in vivo* of exposure to lysergic acid diethylamide. Science *160:*1343, 1968.

Stacey, K. A., M. Cobb, S. F. Cousens and P. Alexander: The reactions of the "radiomimetic" alkylating agents with macromolecules *in vitro.* Ann. N.Y. Acad. Sci. *68:*682, 1958.

Stenchever, M. A., R. Frankel, J. A. Jarvis and K. Veress: Some effects of Diazepam (Valium) on human cells *in vitro.* Amer. J. Obs. & Gyn. (in press), 1968.

The Chicago Conference: Standardization in Human Cytogenetics. In: Birth Defects—Original Article Series Vol. II. Bergsma, Hamerton and Klinger, eds. Published by the National Foundation March of Dimes, 1966.

Tough, I. M. and W. M. Court Brown: Chromosome aberrations and exposure to ambient benzene. Lancet *i:*684, 1965.

T'so, P. O. P., G. K. Helkamp and C. Sander: Interaction of nucleosides and related compounds with nucleic acids as indicated by the change of helix-coli transition temperature. Proc. Nat. Acad. Sci. (Wash.) *48*:686, 1962.

——— and P. Lu: Interaction of nucleic acids. I. Physical binding of thymidine, adenine, steroids, and aromatic hydrocarbons to nucleic acids. Proc. Nat. Acad. Sci. (Wash.) *51*:17, 1964.

Warig, M. J.: Complex formation with DNA and inhibition of *Escherichia* coli RNA polymerase by ethidium bromide. Biochim. Biophys. Acta *87*:358, 1964.

Warkany, J. and E. Takacs: Congenital malformations in rats from streptonigrin. Arch. Path. *79*:65, 1965.

White, M. J. D.: The Chromosomes. Methuen's Monographs on Biclogical Subjects. 5th ed. New York: John Wiley and Sons, 1961.

Wolff, S.: Radiation studies on the nature of chromosome breakage. Amer. Nat. *94*:85, 1960.

——— and H. E. Luippold: Metabolism and chromosome break rejoining. Science *122*:231, 1955.

Wuu, K. D. and W. F. Grant: Induced abnormal behavior in a barley plant (*Hordeum vulgare, L.*) with the herbicide Lorox. Phyton *23*:63, 1966a.

——— and W. F. Grant: Morphological and somatic chromosomal aberrations induced by pesticides in barley (*Hordeum volgare*). Can. J. of Gen. and Cyt. *8*:418, 1966b.

LETHAL CHROMOSOME ERRORS

D. H. CARR
Department of Anatomy, McMaster University, Hamilton, Ontario, Canada

The word "lethal" is used by geneticists in a relative sense. They regard a genetic defect as lethal if it prevents the possessor from reproducing itself through premature death. If the defective state is sufficiently severe, it leads to prenatal death. The severity of lethality due to a chromosomal defect may be expressed in degrees. Chromosome anomalies may conveniently be divided into three groups according to life expectancy of the zygote as listed in Table 1. There is some overlapping and, as usual with biological phenomena, the divisions are not absolute. The time of death varies considerably among individuals and chromosome anomalies which are generally associated with early intrauterine death may occasionally be found in the living.

There are many examples of individuals with a mixture of chromosomally normal and abnormal cell lines. The effect of this mosaicism on the developing organism is very variable as one would expect. Not only the proportion of normal cells but their distribution in the body must surely affect the phenotype. This should vary between the extremes of

Table 1: Classification of Lethal Chromosome Anomalies

Early abortion or resorption.	Most autosomal monosomies. YO anomaly. Pentaploids and other high polyploids. ?Some trisomies.
Recognized abortions in man—usually incompatible with live birth.	Most trisomies. Triploids. Tetraploids.
Compatible with live birth—semi- or sub-lethal.	Trisomy 18. Trisomy 13. 45,X anomaly. Many structural anomalies.

the individual's constituent cell make-up. One mosaic condition for which considerable evidence is available is the normal/21-trisomy defect. The first report of this genotype concerned a girl with an I.Q. of 100 and minimal signs of Down's syndrome (Clarke, Edwards and Smallpiece, 1961). A review of this subject has shown that the phenotype of this mosaicism ranges from normal to individuals with typical Down's syndrome (Finley *et al.,* 1966). Although the effect of a normal cell line is variable, there is a tendency for a moderating influence on the phenotype. Thus, a chromosomal abnormality which usually produces a conceptus in category 2 may be modified in such a way that it falls in category 3 if mosaicism occurs.

Information on chromosome anomalies before birth comes principally from the 17 published series of cytogenetic studies of spontaneous abortions. The results of these investigations are summarized in Table 2. The reports are divided into two categories; those in which there was no selection in the collection of aborted material and those in which the

Table 2: Chromosome Studies in Spontaneous Abortions

Unselected Series	No. of Specimens	No. Chromosomal Abnormalities	% Chromosomal Abnormalities
1. Boué, Boué and Lazar, 1967	132	50	37
2. Bowen and Lee, 1966	22	5	23
3. Carr, 1967; Singh and Carr, 1966 & 67	227	50	22
4. Clendenin and Benirschke, 1963	10	3	30
5. Gropp, 1967	19	3	16
6. Hall and Kallen, 1964	8	2	25
7. Inhorn, Therman and Patau, 1964	88	8	9
8. Kadotani and Pergament, 1966	77	16	21
9. Kerr and Rashad, 1966	35	6	17
10. Makino *et al.,* 1967	37	4	11
11. Schlegel *et al.,* 1966	30	3	10
12. Stenchever, Hempel and Macintyre, 1967	101	8	8
13. Waxman, Arakaki and Smith, 1967	25	10	40
TOTAL	811	168	21
Selected Series			
14. Jacobson and Barter, 1967	35	8	23
15. Szulman, 1965	25	16	64
16. Thiede and Metcalfe, 1966	54	24	44
17. Wingate, 1966	50	1	2

investigator collected specific material based on gestational age or type of specimen. Although there is wide variation in the incidence of chromosome anomalies among spontaneous abortions, the mean for 13 unselected series is 21 percent. The incidence of anomalies is highest for first trimester abortions and declines strikingly in the second trimester.

Conceptuses Absorbed or Aborted Early

The commonest type of chromosome disorder in man is associated with the presence of one extra chromosome. Usually this leads to trisomy but in the male, an extra sex chromosome produces XXY or XYY aneuploids. If the presence of an extra element is due to non-disjunction, the reciprocal condition, monosomy, should be equally common. Except for X-monosomy, this disorder has rarely been described even in abortuses (Inhorn, Therman and Patau, 1964; Jacobson and Barter, 1967; Waxman, Arakaki and Smith, 1967). The six autosomal monosomies which these authors describe represent only 4 percent of the chromosome anomalies in abortions or about $\frac{1}{15}$ the incidence of trisomies. Notwithstanding the great rarity of monosomies in abortions, there has been one report of an infant with monosomy-G in whom no evidence of mosaicism could be found in cells from leucocyte bone marrow or skin cultures (Al-Aish *et al.*, 1967). What is one to make of this finding? It appears to make nonsense of the classification suggested in this paper. The explanation of "hidden mosaicism" or "hidden translocation" is frequently advanced to explain any finding which deviates from one's preconceived notions. Surely the explanation is within the bounds of all biological experience. A certain genetic or chromosomal disorder usually has a certain phenotypic equivalent. It would be surprising if there were no exceptions. If this is the case, we can say that autosomal monosomy is usually incompatible with development of the zygote but may occasionally be associated with the birth of an infant at term.

The effect of polyploidy on the developing animal will be discussed in the next section. Suffice it to say here that multiples of the haploid number above 4 are exceedingly rare in mammals at any stage of development. Pentaploidy has been described in a six-day rabbit blastocyst and this is probably the only known example of this defect in a mammal (Shaver and Carr, 1967). The blastocyst was not only normal in appearance but larger than diploid conceptuses of the same stage.

Hexaploidy in pure form probably produces a zygote in category 1. However, two conceptuses have been found to be hexaploid/diploid mosaics. The karyotype from the hexaploid cell is shown in Figure 1. There were exactly 138 elements in a well circumscribed area and karyotyping demonstrated the presence of an XXXXYY sex chromosome complex. Such a cell could not have arisen in culture unless a triploid cell

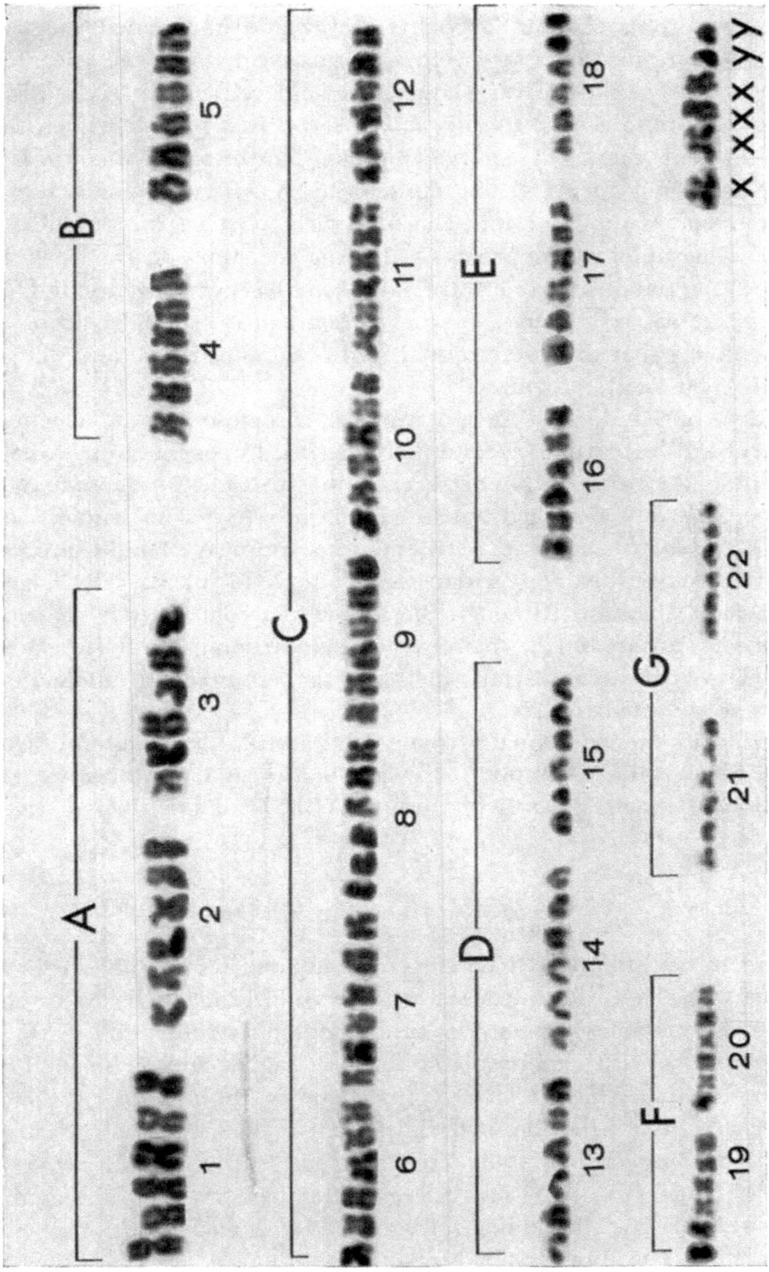

Fig. 1. Hexaploid karyotype with an XXXXYY sex chromosome complex.

line was present and the sex chromosome complement excluded the accidental aggregation of three 46,XY cells. If triploid cells were present in the conceptus, it is unlikely that they would have failed to grow as their proliferative potential *in vitro* appears normal. On the other hand, it seems likely that the very large hexaploid cell might have difficulty undergoing mitosis both *in vivo* and *in vitro*. It is not surprising, therefore, that only one of 16 cells which grew from this conceptus was found to be hexaploid, the rest being diploid (46,XY). It is concluded that the chromosome findings in this abortus reflect its true cytogenetic state, that is a hexaploid/diploid mosaic. During the study of over 6,000 cells from 227 spontaneous abortions, such a mosaic was never found. The only hexaploid cells noted were in predominantly triploid cultures and it is unlikely that they were missed in diploid cultures because they are striking and easily recognized.

Only 4 or 5 individual members of the chromosome complement can be identified by morphology and therefore it is impossible to say whether each individual trisomy has yet been found in man. Chromosome No. 1 can be positively identified but has not been reported in trisomic form, even in an abortus. In general, trisomies for groups A, B and C are much rarer than those involving groups D, E and G (Table 3). The F group is the most aberrant. Although the two chromosome pairs which make up this group are small, their existence in trisomic form is relatively rare. Trisomy-F accounts for only about 2 percent of all anomalies in abortuses reported to date.

If all trisomies are equally common at fertilization (and this is not necessarily a valid assumption) it suggests a greater resorption or early abortion of trisomies involving groups A, B, C and F than of the remainder.

Recognized Abortions in Man

Most of the autosomal trisomies, triploidy and tetraploidy are incompatible with live birth unless a mosaicism with normal cells exists. However, the commonest trisomy in man is not only compatible with live birth but also with fertility of the female (Tagher and Reisman, 1966).

Many trisomies have been recognized among the products of conception, especially during the first trimester of pregnancy. The presence of partial or complete embryonic remnants in relation to karyotype are summarized in Table 3. (The reference numbers are the same as those used in Table 2.) The number of abortuses with each trisomy are listed by authors. The numbers in brackets show how many had embryonic material and a question mark is used if this information is lacking. Frequently the embryo recorded has been severely disorganized or mac-

Table 3: Presence of Embryos in Spontaneous Abortions Related to Karyotype

Ref	Morphology of Trisomy in Groups A	B	C	D	E	F	G	Double Trisomy	45,X	Triploidy	Tetra-ploidy	Other
1	0	1 (0)	5 (3)	7 (3)	5 (2)	1 (0)	8 (6)	2 (2)	8 (6)	10 (7)	1 (0)	2 (2)
2	1 (0)				1 (0)				3 (1)	1 (0)		
3	1 (0)	1 (1)	2 (0)	6 (3)	9 (0)	1 (0)	6 (1)	1 (0)	12 (9)	9 (5)	2 (1)	
4					1 (1)				1 (1)			1 (1)
5		1 (1)			1 (0)						1 (0)	
6	2 (0)											
7						1 (0)			1 (?)	2 (0)	2 (0)	2 (?)
8						no details						
9	2 (0)				1 (0)			1 (0)				2 (1)
10				1 (1)						1 (?)		2 (?)
11										3 (2)		
12				1 (1)					2 (1)	3 (2)		2 (1)
13					3 (?)	1 (?)			1 (?)	1 (?)	1 (?)	3 (?)
14			1 (1)		1 (1)		1 (1)		2 (2)			3 (0)
15			1 (1)		2 (0)		3 (0)		5 (4)	5 (1)		
16	2 (?)	0	0	2 (?)	7 (?)	0	4 (?1)	0	5 (?)	1 (?)	3 (?)	
17							1 (1)					
TOTALS	8 (0)	3 (2)	9 (5)	17 (8)	31 (4)	4 (0)	23 (10)	4 (2)	40 (24)	36 (17)	10 (1)	17 (5)

erated but it has been included in the figure given if identified by the investigator.

Although trisomies for chromosomes 2 and 3 are known, none has been associated with recognized embryonic material. There have been 4 reports of B-trisomy, 3 listed in Table 3 and another recorded by Sato (1965). Two of these 4 abortuses had an embryo, one being apparently normal and measuring 11 mm while the other was severely disorganized.

Trisomy C presents two particularly important aspects as far as development is concerned. In the first place, the X chromosome falls within this group; and secondly, extra C-group chromosomes other than the X, cannot be positively identified. Let us first consider X-polysomy. Although this element is one of the largest members of group C it has a special status. Trisomy-X apparently has little effect on the phenotype, though the incidence of mental retardation may be increased (Carr, 1968). This relatively mild effect of X-trisomy is believed to be due to the inactivation, partial or complete, of X chromosomes in excess of one. The presence of an extra X chromosome does not even produce infertility, as it does in the male, but the extra element is rarely passed on to the offspring.

Trisomy for an autosome in group C has been found in 9 abortuses and 5 of these had some indication of embryonic development. The largest embryo was only 15 mm in length, the expected development at 35–40 days. As the autosomes in group C cannot be separated on morphological grounds, it is not possible to distinguish the various trisomies from one another. This applies equally to the autosomal C group trisomies which have been found in mosaic form in several subjects with varying phenotypes (Jacobs *et al.,* 1961; Pfeiffer, Schellong and Kosenow, 1962; El-Alfi, Powell and Bieselle, 1963; Inhorn *et al.,* 1963; Stalder, Buhler and Weber, 1963; Stolte, Evers and Blankenborg, 1964; Wolf and Reinwein, 1965; Jalbert *et al.,* 1966; Bargman *et al.,* 1967; Sergovich, 1967; Monnet *et al.,* 1967). The presence of mosaicism with a normal cell line is no guarantee of a modified phenotype. Kerr and Rashad (1966) found C-trisomy/normal mosaicism in an acardiac monster, while the other twin had normal chromosomes.

Even gross morphology as seen in published karyotypes of these individuals with C-trisomy suggests that different elements are involved from case to case. It is, therefore, not surprising that no consistent phenotype has emerged. It does in fact range between the extremes of a normal woman and a grossly disorganized fetus. This illustrates well the fact mentioned in the introduction, that a variable degree of expression is to be expected in association with chromosomal mosaicism.

Trisomy-13 is compatible with live birth but D-trisomy is frequently found in cells grown from abortuses. It is not known whether chromosome 13, 14 or 15 is involved in these specimens. The phenotype has been variable including gross disorganization and apparently normal embryos. The single major defect in other embryos with D-trisomy were: gross retardation of lower limb development, facial cleft and anencephaly, and cleft lip. Apart from the last named, none had features suggestive of trisomy-13. They may in fact represent trisomies for one of the other two pairs of chromosomes in group D. However, there was no major defect common to the embryos with D-trisomy.

One of the commonest trisomies among aborted embryos involves chromosome 16. This element can usually be recognized on morphological grounds. The phenotypic effect is variable but many conceptuses consisted of intact empty sacs or a disorganized nodular embryo. In spite of this severe effect on development, trisomy-16 is apparently compatible with grossly normal embryogenesis, at least in the early stages (Clendenin and Benirschke, 1963).

In spite of the common occurrence of trisomy-16 in abortuses, it appears to be rare in the liveborn, even in mosaic form. Lewis and coworkers (1963) described a mentally retarded female, with an extra chromosome which was indistinguishable from No. 16. The same trisomy, in mosaic form, was demonstrated on repeated blood cultures from a boy with metaphyseal dysostosis (Schmidt et al., 1963). Trisomy 17–18, which is compatible with live birth may also lead to abortion (Carr, 1963; Szulman, 1965; Jacobson and Barter, 1967). Only one of these three abortuses contained an embryo. This measured 13 mm and did not show obvious anomalies. It was not established whether chromosome 17 or 18 was involved in these cases. Trisomy-18 has been defined in a 570 gm stillborn female fetus with hydrocephalus, absence of the left ear, flexion of both wrists and a short right arm (El-Alfi, Biesele and Smith, 1964).

Several E-trisomic abortions have not been further specified regarding the chromosome involved. This is not surprising. Though we usually state that chromosome 16 is readily recognizable on morphological grounds, this is not entirely true. When there is an extra element in Group E, chromosomes 16 and 17 can be confused except in ideal cells.

Although the chromosomes in group F are among the smallest in the complement, trisomy involving the members of this group, is rare. This presumably means that both chromosome pairs carry genes which are essential to early embryonic development or which are readily disturbed by a triple dose. None of the F-trisomy abortuses are known to have been associated with embryonic development.

F-trisomy has never been described in pure form in the living but there have been several patients with F-trisomy/normal mosaicism, including one with Cockayne's syndrome (Civantos, 1961). There was no consistent phenotype but it is noteworthy that in three instances, F-trisomy mosaicism was familial (Böök, Santesson and Zetterquist, 1961;

Gagnon *et al.,* 1963; Mosonyi *et al.,* 1966). It is possible, of course, that the extra element was not a chromosome in group F but a product of a translocation as suggested by Böök, Santesson and Zetterquist (1961).

The presence of one extra chromosome in group G is not only the commonest cytogenetic disorder in the liveborn but also one of the commonest lethal trisomies. Only one of the 6 abortuses with G-trisomy studied by Singh and Carr (1967) had an embryo. This showed extreme microcephaly and probable retardation of upper limb development. More recently, G-trisomy mosaicism was found in an embryo with exencephaly. This conceptus was the twin of a normal appearing embryo of an earlier developmental age. Boué and colleagues (1967) found only retarded development rather than gross anomalies in several embryos with G-trisomy which they examined.

The two pairs constituting group G cannot be consistently distinguished from one another. Even autoradiography does not give a repeatable pattern of replication (Back *et al.,* 1967; Fraccaro, Lindsten and Tiepolo, 1967; Hecht and Weleber, 1967). Without a method of distinguishing between chromosomes 21 and 22 it is quite pointless to discuss which member, when present in trisomic form, is responsible for abortion with G-trisomy. The diagnosis of trisomy-21 is clinical and only a fetus with features of Down's syndrome can identify the extra element (Wingate, 1963).

It is indeed disappointing to reflect that no trisomy found among abortuses presents even a trend in its effect on the phenotype. This is no doubt due, in part, to the fact that individual members of the chromosome complement can rarely be identified. There is, therefore, no way of knowing whether we are dealing with a consistent genotype. Further, trisomies are apparently so devastating in their effect on the developing conceptus that the embryo is frequently resorbed or very stunted at the time of abortion. Microscopic study of these very stunted embryos is frequently hopeless and, at best, disappointing. If any tissues are recognizable they show only a vague resemblance to normally developed systems and are grossly disorganized.

Trisomy for mammals other than man has been demonstrated in mouse, rat and rabbit. The 8 trisomies reported in the mouse were all in males (Griffen and Bunker, 1967). The phenotype was indistinguishable from normal except that the animals were sterile or semisterile. The fact that the extra chromosome was an autosome rather than a sex chromosome was based on meiotic studies. The extra element failed to associate with the XY bivalent in the male. The lack of structural defects in the trisomic mouse is in strong contrast to the situation existing in man and this difference is difficult to understand.

Butcher and Fugo (1967) demonstrated presumed autosomal trisomy in embryos of rats examined at 11 days gestation. They were found only in the experimental animals whose ova had been aged. No pure trisomies were discovered among control embryos. On the other hand, Shaver and Carr (1967) found only one trisomic blastocyst among 135 studied in rabbits. This was not in the "delayed mating" group but in a control animal mated immediately after injection of gonadotrophin. Neither of these experiments provides information on the phenotypic effect of trisomy at term in rat and rabbit. It would be interesting to know whether these animals resemble the mouse or man as far as trisomy is concerned.

The effect of triploidy in animals is of considerable interest. The presence of three instead of two chromosomal sets is found, not only as spontaneous aberrations in amphibia but also as a normal state in certain salamanders (Uzzell, 1963). Regular triploidy is rare in sexually reproducing animals. However, it has recently been described in a parthenogenetic species of lizard (Pennock, 1965). Triploidy has been described in an intersex adult chicken (Ohno *et al.*, 1963) and diploid/triploid mosaicism in an underdeveloped chicken embryo (Bloom and Buss, 1966).

Thus, triploidy is tolerated in a variety of animal classes. What is the situation in mammals? Since the improvement in cytogenetic techniques, the precise demonstration of triploid cells has become possible. However, earlier studies on mammalian blastocysts and early embryos suggested that triploid conceptuses did not survive beyond midterm (Beatty, 1957).

By far the majority of triploid conceptuses in man are aborted, usually before the end of the third month of gestation. A large number of such abortuses are now known and the variable embryonic phenotype is striking. Information on the available material is summarized in Table 4. Among 59 triploid conceptuses, 31 had embryos which were classified into one of three groups. There were 13 embryos from 5 to 19 mm in length which were normal as far as the author could judge. There were 7 with focal abnormalities, 5 having meningomyeloceles, one an encephalocele and one with multiple anomalies including cleft lip and palate and a large umbilical hernia. Eleven of the embryos were either so severely macerated that the author could not estimate whether they were normal or not; or, the embryo was grossly disorganized ("nodular, cylindrical or stunted," according to one classification). In no less than 27 specimens (45 percent) the author noted hydatidiform degeneration of the chorionic villi (Fig. 2) or fully developed hydatidiform mole. This was the most characteristic phenotypic feature of the triploid conceptuses and was first noted by Makino, Sasaki and Fukuschima (1964).

There are 3 possible sex chromosome complexes in triploidy XXY, XXX and XYY. The XXY sex chromosome complex was the commonest, being found in almost half the cases (29). The XXX complex was almost

Table 4: Phenotypic and Cytogenetic Findings in Triploid Conceptuses

Gonosomes	Incomplete Specimen	Intact Empty Sac	Normal Embryo	Embryo With Focal Abnormality	Severely Macerated or Disorganized Embryo	* Mole or Hydatidiform Degeneration	Insufficient Details Available	Totals
XXY	10	3	5	2	7	12	2	29
XXX	3	1	5	5	4	10	2	20
XYY	1	0	2	0	0	2	1	4
XXY/XY	1	0	0	0	0	1	0	1
Unknown	2	0	1	0	0	2	2	5
Totals	17	4	13	7	11	27	7	59

* All 27 specimens are included under the other headings.
Additional references not quoted in Table 2: Penrose and Delhanty, 1961; Delhanty, Ellis and Rowley, 1961; Rashad and Kerr, 1966; Makino, Sasaki and Fukuschima, 1964; Atkin and Klinger, 1962; Beischer, Fortune and Fitzgerald, 1967; Edwards *et al.*, 1967; Singh and Carr, 1967; Carr, unpublished data.

Fig. 2. Hydropic or hydatidiform degeneration in a conceptus which showed triploid cells in tissue culture (mm scale).

as common being found in cells grown from 20 specimens. However, XYY gonosomes were found in only 4 of the conceptuses.

The mechanisms underlying triploidy have been discussed by Austin (1960). Basically, triploidy is due to the male or female contributing two haploid sets instead of each contributing single sets of chromosomes. Boué and co-workers (1967) have tried to assess the importance of the different mechanisms from the sex chromosome complexes of known triploids. However, this assumes proportional recovery of the three different sex chromosome types from aborted material. There may be an excess of XYY abortuses in category 1—those conceptuses which are resorbed or aborted early. From the occurrence of XYY abortuses, it is certain that the male may contribute two chromosome sets but the importance of this relative to digyny and ovarian aneugamy is unknown.

As noted elsewhere, there appears to be an increased risk of polyploidy in women who become pregnant soon after discontinuing oral contraceptives (Carr, 1967a). In a study of 18 abortuses from women who became pregnant within six months of discontinuing oral contraceptives, 13 had chromosomal abnormalities. The results are summarized in Table 5. The 13 abnormalities included two trisomies, three 45,X aneuploids and 8 polyploids of different types. Among the polyploids were 5 with triploidy, 1 with tetraploidy and 2 with hexaploid/diploid mosaicism. The incidence of polyploidy under these circumstances represents a 9-fold increase over that found in unselected series. This is all the more striking as the occurrence of trisomy is unchanged from the expected level. One possible mechanism for the increased risk of polyploidy is a temporary rise in L.H. production by the pituitary after stopping synthetic progesterone ingestion. There are two reports of changes in the zona pellucida in primates following injection of luteiniz-

Table 5: Chromosome Anomalies in Abortuses Conceived Soon After Ceasing Oral Contraceptives*

Patient No.	Age	Type of Medication	On Medication (mos.)	No. of Periods Before Pregnancy	Chromosomes	Abortus
1	37	C	18	2	45,X	Normal 15 mm. embryo
2	26	C	20	5	Triploid	Normal 18 mm. embryo
3	25	C, S	30	2	46,XY/138, XXXXYY mosaic	Ruptured sac
4	27	S	3	3	Triploid 69,XXX	Normal 5 mm. embryo hydropic villi
5	24	?	12	2	46,XX	Intact empty sac
6	26	C	11	4	45,X	Normal 15 mm. embryo
7	20	S	2	2	46,XX	Normal female fetus
8	39	C, S	18	5	Triploid 69,XXY	Hydropic villi
9	26	C	20	1	45,X	Normal 17 mm. embryo
10	27	?	22	1	46,XY	Anencephalic embryo
11	29	C	30	3	Triploid 69,XXY	Hydropic villi
12	21	?	12	4	46,XY	Normal embryo
13	24	C	20	3	46,XX	Hydropic villi
14	32	C	15	5	Triploid 69,XXY	Embryo with meningomyelocele
15	20	C	2	1	Tetraploid 92,XXXX	Hydropic villi
16	19	C	13	1	Trisomy G	Ruptured sac
17	30	?	24	3	Diploid/hexaploid mosaic	Normal embryo
18	21	C	4 (15)	6	Trisomy F	Hydropic villi

* C = combined, S = sequential, ? = unknown medication.

ing hormone (Katzberg and Hendrickx, 1966; Maruffo, 1967). It is possible that abnormalities of fertilization and early cell division of the zygote account for the increased risk of polyploidy due to overproduction of L.H.

The association between triploidy and hydatidiform degeneration of the chorionic villi has already been mentioned. The relationship of these two events may be indirect, with an underlying common etiology (Carr, 1968a).

Although pure triploidy is apparently always lethal before full development, there have been two accounts of infants surviving for a few hours after birth in whom only triploid cells were found (Edwards *et al.*, 1967; Bernard *et al.*, 1967). Both infants were underdeveloped, presented systemic defects including abnormal external genitalia and had a 69,XXY chromosome complement. Unfortunately, only one tissue was studied in each case. The finding of triploidy in lymphocyte culture of one infant has established the possibility of this chromosome complement persisting in rapidly dividing cells.

There have now been four reports of humans who have survived to childhood in whom definite triploid/diploid mosaicism was demonstrated. They all presented mental retardation and certain other anomalies including porencephaly (Böök and Santesson, 1960), congenital asymmetry (Ellis *et al.*, 1963; Ferrier *et al.*, 1964) and multiple skeletal disorders (Sacrez *et al.*, 1967). In addition, Lejeune *et al.* (1967) studied a female infant with a 69,XXY cell line. The authors concluded that the infant was a chimera who had received the triploid cells from an incompletely resorbed co-twin. An equally unusual situation was the report of a child who was not only a triploid mosaic but had aneuploidy as well (Schmid and Vischer, 1967). The child was in fact a 48,XXYY/71,XXXYY mosaic and presented multiple anomalies including colobomata of the eyes and syndactyly.

In addition to these examples of 3n/2n mosaicism there have been two examples of near-triploid mosaicism involving localized tissues in the body. Gropp, Jussen and Odunjo (1964) found hypertriploid cells in tissue adjacent to a cleft palate. However, similar studies carried out in a series of such cases was negative (Soukup and Warkany, 1966). Johnston and Penrose (1966) studied tissue from patients with congenital asymmetry and one of the nine had a few hypertriploid cells in tissue grown from the hypertrophied arm. Whether or not these cases did involve local abnormal cell clones is uncertain but their reporting is important as it directs our attention to this possibility.

The early work on triploidy in mammals has already been mentioned. With the introduction of modern techniques for chromosome analysis, additional information has become available. Butcher and Fugo (1967) found no difference in the incidence of triploidy in 11-day rat embryos between those in which ovulation was delayed and normal controls. However, Austin (1967) and Shaver and Carr (1967) both demonstrated an increase in the incidence of triploidy among embryos derived from aged ova in rabbits. If ova were aged by 8 or 9 hours, it was found that 12 percent of blastocysts, recovered on day 6, were triploid. There is probably a high loss of abnormal zygotes either before or early in the implantation stage in mammals. Among 11-day rat embryos, Butcher and Fugo (1967) found only two triploids in 410 examined. On the other hand, nearly 6 percent of 10-day blastocysts in pigs were triploid (McFeeley, 1967). This may represent a species difference but it is more likely due to the difference in the stage of embryonic development.

Other mammals probably resemble man in their tolerance to triploid cells as diploid/triploid chimerism has been described in cat and mink (Chu, Thuline and Norby, 1964; Nes, 1966). In both animals the chromosome complement was $2n$,XX/$3n$,XXY.

The behavior of X-chromosomes in the triploid cell is apparently irregular. This is based on the varied sex chromatin findings in pure triploid cell cultures (Carr, 1965; Boué, Boué and Lazar, 1967). Melander (1963) suggested that "mammalian pure triploids would have a minimum chance of survival because of a necessarily abnormal chromosome balance in the early embryonic cells of males and in the soma of both sexes."

On the other hand, Melander (1963) felt that gonosome : autosome imbalance should not be an obstacle to embryonic development in the tetraploid. Whatever the reason, tetraploidy seems to be highly lethal in man and in mammals generally. There have been many examples of tetraploidy in human aborted conceptuses, with both XXXX and XXYY sex chromosome complexes (Table 3). In spite of this, only one highly disorganized embryo has been found. Thiede and Salm (1964) did find 16 percent tetraploid cells in a 75 mm apparently normal male fetus. Of even greater interest is the finding of about 70 percent tetraploid cells in leucocyte cultures from a grossly underdeveloped term infant (Kohn *et al.,* 1967). This male child had several anomalies and died suddenly at 9 months of age.

Tetraploidy is well known in plants and has been reported sporadically in amphibians. The condition has recently been discovered in an apparently bisexual amphibian (Beçak, Beçak and Rabello, 1966). However, tetraploidy in mammals including man seems to be lethal. The reason for this is not clear.

Chromosome Anomalies Compatible with Live Birth (Semi- or Sub-lethals)

In 1960, Edwards and co-workers and Patau and his colleagues reported simultaneously two different syndromes which they showed were associated with autosomal trisomy. Edward's syndrome, or trisomy-18, is associated with anomalies involving every system in the body. The most constant features: developmental retardation, failure to thrive, low set malformed ears, micrognathia, flexion deformity of the fingers, arches on 3 or more fingers and congenital heart disease, are found in 90 percent of affected infants (Taylor, 1967). Although born at term, the weight tends to be low with a mean of 2,200 gm. The mean age at death is about 10 weeks and there is a marked excess of affected females in a ratio over 3:1 (Weber, 1967). Weber has shown that there is a greatly increased mortality in the first 3 months among males with trisomy-18 syndrome. This may be the continuation of a situation existing before birth.

The other lethal trisomy syndrome (Patau's), which involves chromosome 13, also results in low birth weight, term infants with multiple anomalies. The mean weight at birth is 2,550 gm and the mean survival time is about 14 weeks. Although infants with this disorder live, on the average, longer than those with 18-trisomy, the spread is greater, 25 percent being dead at the end of one week (Magenis and Hecht, 1966). In addition to the developmental retardation already mentioned, twitching and apnea, deafness, malformed ears, microphthalmus or coloboma of the eye, single palmar crease, distal axis triradius and undescended testes are found in 90 percent of affected infants.

Single umbilical artery is an important finding in both these lethal conditions being found in the 18 and 13 trisomy syndromes with an incidence of 80 and 50 percent respectively.

Sex chromosome disorders in man are relatively innocuous and rarely lethal. One sex chromosome anomaly is, however, an exception being highly lethal. This is the absence of one sex chromosome leading to a 45,X constitution. This disorder is found in 1 in 2,500 female infants (Maclean *et al.,* 1964) and is frequently associated with skeletal, urinary and cardio-vascular anomalies. Although perinatal mortality in this condition is difficult to assess, it is probably in the region of 20 percent. Even more striking is the high prenatal mortality. The incidence of the 45,X chromosome disorder in spontaneous abortions is about 5 percent. If we assume that 15 percent of pregnancies end in abortion, this means that no more than 2 of every 100 zygotes which lack one sex chromosome, reach term. This high mortality is puzzling. Various anomalies are found among XO embryos and fetuses. These include cystic hygroma, horseshoe kidney and unicornuate uterus (Singh and Carr, 1966). However, these disorders are all compatible with live birth. Further, many of the 45,X embryos and fetuses which are aborted appear normal in every way. The 22 mm embryo shown in Fig. 3 was well preserved and was completely normal on dissection.

The only mammal in which the XO anomaly has been commonly found is the mouse (Russell, 1962), Its estimated mean incidence in this animal is about the same as that in man at fertilization. However, there

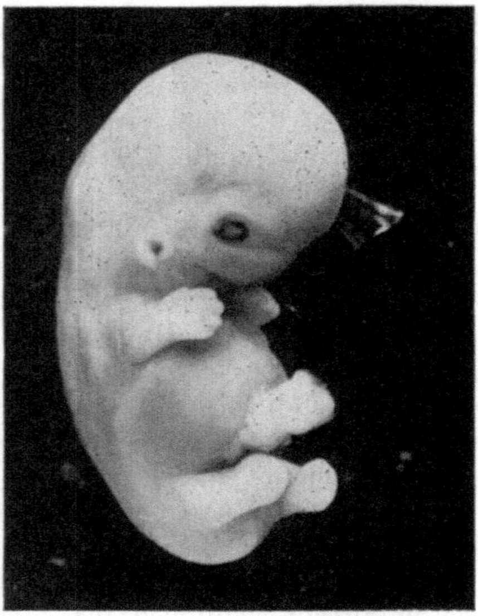

Fig. 3. Apparently normal 22 mm embryo with a 45,X chromosome complement.

is no apparent increased lethality in the mouse. This may perhaps be due to loss of lethal genes on the X chromosome associated with inbreeding of laboratory stocks. In man, on the other hand, lethal genes may be commonly carried on the X chromosome. These would be suppressed in the normal female and male, by the second, relatively inactive X, or by the Y chromosome. It is known that in Drosophila the Y is able to suppress lethal genes carried on the X chromosome (Lindsley, Edington and Von Halle, 1960).

One of the pointers which led to chromosome analysis of abortions was the presumed occurrence of unbalanced products from carriers of reciprocal translocations (Macintyre *et al.*, 1962). So far, in studies of series of spontaneous abortions, these chromosomally unbalanced products of conception from translocation carriers have been notable by their absence. As far as can be judged, the presence of an extra chromosome is usually a sporadic trisomy and not the product of an interchange (Singh, Rubinoff and Carr, 1966). It is also noteworthy that though large numbers of infants with abnormalities have been found to have chromosome deletions or rings, none have yet been described in abortuses.

Conclusions

Lethal chromosome errors have been tentatively divided into three categories. These are not meant to be exclusive. There is no doubt overlapping and what we say about chromosome errors in one category today may not be true tomorrow. We must avoid having too rigid a concept of chromosomal disease. We must avoid making "rules" so that a chromosome error which does not fit with our preconceived ideas has to be explained away by "hidden mosaicism" or some other manoeuvre. We must expect exceptions. For instance, women with only one X chromosome are almost always sterile. An undoubted case of fertility in a woman with 45,X chromosome constitution has been reported, without any evidence of mosaicism (Bahner *et al.,* 1960). We now know that the XO embryo has the germ cells necessary for fertility (Singh and Carr, 1966). Is it surprising, therefore, that these cells which usually die in their abnormal environment occasionally survive?

We have tentatively classified autosomal monosomy in the most severe category of prenatal lethality. But already we have examples of G-monosomy without apparent mosaicism. It may be true that there is some G-chromosome material hidden on another member of the complement. As this is an armchair explanation to cover our preconceived notions let us admit that the living organism is extraordinary. Why should not a chromosome disorder, usually leading to very early abortion, occasionally produce a living child?

It appears that mosaicism, with a normal cell line, modifies the development of the zygote. However, on theoretical grounds we expect, and do indeed find, a wide spectrum of phenotypic effect in mosaic subjects ranging from normal to severely affected. The normal subject will come under the scientist's scrutiny only by accident and a mosaicism with abnormal cell lines is probably commoner than we can ascertain. If an abnormal clone in the early embryo develops in a critical area, a local anomaly may result. The same clone might, on the other hand, remain concealed and the phenotype develop normally.

Many mysteries remain to be explored and, if possible, explained. Why is tetraploidy so devastating in the mammal? The balance of the chromosome complement is grossly unchanged. Perhaps such a large cell simply has mechanical problems in dividing, though this does not appear to be so in culture. Why in man does the absence of an X chromosome cause at least 98 percent of affected zygotes to die? Why is the mouse so different from man in regard to not only the XO anomaly but also autosomal trisomy? And yet we have noted that man and mouse react similarly to polyploidy. These are questions for the future which will continue to interest the scientist whatever his background.

Acknowledgments

The author wishes to thank the nursing and medical staffs of St. Joseph's and Henderson General Hospitals, Hamilton, for their cooperation in collecting spontaneous abortions. The invaluable assistance of my technician, Mrs. G. Feleki, is gratefully acknowledged. This study was made possible by a grant from the Medical Research Council of Canada. Miss P. Godfrey typed this paper with her usual care and expertise.

References

Al-Aish, M. S., F. de la Cruz, L. A. Goldsmith, J. Volpe, G. Mella and J. C. Robinson: Autosomal monosomy in man. Complete monosomy G (21–22) in a four-and-one-half-year-old mentally retarded girl. New Eng. J. Med. *277:*777, 1967.

Atkin, N. B. and H. P. Klinger: The superfemale mole. Lancet *2:*727, 1962.

Austin, C. R.: Anomalies of fertilization leading to triploidy. J. Cell. Comp. Physiol. *56:*Suppl. 1, 1, 1960.

———: Chromosome deterioration in ageing eggs of the rabbit. Nature *213:*1018, 1967.

Back, F., P. Dormer, P. Baumann and E. Albrich: Trisomy 21 or 22 in Down's syndrome. Lancet *1:*1228, 1967.

Bahner, F., G. Schwarz, H. A. Hienz and K. Walter: Turner Syndrom mit voll ausgebildeten sekundären Geschlechtsmerkmalen und Fertilität. Acta Endocr. (Kobenhavn) *35:*397, 1960.

Bargman, G. J., R. L. Neu, T. Kajii, J. C. Leao and L. I. Gardner: Trisomy C mosaicism in a seven month old girl. Humangenetik *4:*13, 1967.

Beatty, R. A.: Parthenogenesis and Polyploidy in Mammalian Development. Cambridge, London, 1957.

Beçak, M. L., W. Beçak and M. N. Rabello: Cytological evidence of constant tetraploidy in the bisexual South American frog *Odontophrynus americanus.* Chromosoma (Berlin) *19:*188, 1966.

Beischer, N. A., D. W. Fortune and M. G. Fitzgerald: Hydatidiform mole and coexistent foetus, both with triploid chromosome constitution. Brit. Med. J. *3:*476, 1967.

Bernard, R., A. Stahl, J. Coignet, F. Giraud, M. Hartung, Y. Brusquet and P. Passeron: Triploidie chromosomique chez un nouveau-né polymalformé. Ann. Genet. *10:*70, 1967.

Bloom, S. E. and G. E. Buss: Triploid-diploid mosaic chicken embryo. Science *153:*759, 1966.

Böök, J. A. and B. Santesson: Malformation syndrome in man associated with triploidy (69 chromosomes). Lancet *1:*858, 1960.

———, B. Santesson and P. Zetterquist: Association between congenital heart malformation and chromosomal variations. Acta Paediat. *50:*217, 1961.

Boué, J. G., A. Boué and P. Lazar: Les aberrations chromosomiques dans les avortements. Ann. Genet. *10:*179, 1967.

Bowen, P. and C. S. N. Lee: Karyotypes of cell strains derived from aborted human fetal tissues. Proc. Int. Conf. Hum. Genet. 3rd Chicago, 1966.

Butcher, R. L. and N. W. Fugo: Overripeness and mammalian ova. II. Delayed ovulation and chromosome anomalies. Fertil. Steril. *18:*297, 1967.

Carr, D. H.: Chromosome studies in abortuses and stillborn infants. Lancet *2:*603, 1963.

———: Chromosome studies in spontaneous abortions Obstet. & Gynec. *26:*308, 1965.

———: Chromosome anomalies as a cause of spontaneous abortion. Amer. J. Obstet. Gynec. *97*:283, 1967.

———: Chromosomes after oral contraceptives. Lancet *2*:830, 1967a.

———: Chromosomal abnormalities in clinical medicine. In press. 1968.

———: Cytogenetic and pathologic studies in hydatidiform degeneration. In press. 1968a.

Chu, E. H. Y., H. C. Thuline and D. E. Norby: Triploid-diploid chimerism in a male tortoiseshell cat. Cytogenetics (Basel) *3*:1, 1964.

Civantos, F.: Human chromosomal abnormalities. Bull. Tulane Univ. Med. Fac. *20*:241, 1961.

Clarke, C. M., J. H. Edwards and V. Smallpiece: 21 trisomy/normal mosaicism in an intelligent child with some mongoloid characters. Lancet *2*:1028, 1961.

Clendenin, T. M. and K. Benirschke: Chromosome studies in spontaneous abortions. Lab. Invest. *12*:1281, 1963.

Delhanty, J. D. A., J. R. Ellis and P. T. Rowley: Triploid cells in a human embryo. Lancet *1*:1286, 1961.

Edwards, J. H., D. G. Harnden, A. H. Cameron, V. M. Cross and O. H. Wolff. A new trisomic syndrome. Lancet *1*:787, 1960.

———, C. Yuncken, D. I. Rushton, S. Richards and U. Mittwoch: Three cases of triploidy in man. Cytogenetics (Basel) *6*:81, 1967.

El-Alfi, O. S., H. C. Powell and J. J. Bieselle: Possible trisomy in chromosome group 6–12 in a mentally retarded patient. Lancet *1*:700, 1963.

———, J. J. Biesele and P. M. Smith: Trisomy 18 in a hydrocephalic fetus. J. Pediat. *65*:67, 1964.

Ellis, J. R., R. Marshall, I. C. S. Normand and L. S. Penrose: A girl with triploid cells. Nature *198*:411, 1963.

Ferrier, P., S. Ferrier, G. Stalder, E. Buhler, F. Bamatter and D. Klein: Congenital asymmetry associated with diploid-triploid mosaicism and large satellites. Lancet *1*:80, 1964.

Finley, W. H., S. C. Finley, C. J. Rosecrans and C. C. Tucker: Normal/21-trisomy mosaicism. Amer. J. Dis. Child. *112*:444, 1966.

Fraccaro, M., J. Lindsten and L. Tiepolo: Trisomy 21 or 22 in Down's syndrome. Lancet *1*:389, 1967.

Gagnon, J., L. Archambault, J. R. Ducharme and N. Katyk-Longtin: Résultats préliminaires sur une étude cytogénétique d' une famille affecté de nephropathie. Rev. Canad. Biol. *22*:133, 1963.

Griffen, A. B. and M. C. Bunker: Four further cases of autosomal primary trisomy in the mouse. Proc. Nat. Acad. Sci. *58*:1446, 1967.

Gropp, A., A. Jussen and F. Odunjo: Near-triploid chromosome constitution in epithelial-cell cultures of palate mucosa from a case of cleft palate. Lancet *1*:1167, 1964.

———: Chromosomenuntersuchungen bei Spontanabortus. Verhandlungen der Deutschen Gesellschaft für Pathologie *51*:278, 1967.

Hall, B. and B. Källén: Chromosome studies in abortuses and stillborn infants. Lancet *1*:110, 1964.

Hecht, F. and R. G. Weleber: Trisomy 21 or 22 in Down's syndrome. Lancet *2*:467, 1967.

Inhorn, S. L., D. W. Smith, K. Patau and E. Therman: Presumed C-group autosomal trisomy/normal mosaicism in a child with multiple anomalies. Second Mammalian Cytogenetics Conference (Abstr.), 1963.

———, E. Therman and K. Patau: Cytogenetic studies in spontaneous human abortions. (Abstr.) Amer. J. Clin. Path. *42*:528, 1964.

Jacobs, P. A., D. G. Harnden, K. E. Buckton, W. M. Court Brown, M. J. King, J. A. McBride, T. N. MacGregor and N. Maclean: Cytogenetic studies in primary amenorrhoea. Lancet *1*:1183, 1961.

Jacobson, C. B. and R. H. Barter: Some cytogenetic aspects of habitual abortion. Amer. J. Obstet. Gynec. *97*:666, 1967.

Jalbert, P., J. Jobert, J. Patet, Cl. Mouriquand and J. Roget: Un nouveau cas de trisomie présumée 6–12. Ann. Genet. *9*:109, 1966.

Johnston, A. W. and L. S. Penrose: Congenital asymmetry. J. Med. Genet. *3*:77, 1966.

Kadotani, T. and E. Pergament: The nature and implications of chromosome aberrations observed in an abortion population. Proc. Int. Conf. Hum. Genet. 3rd Chicago, 1966.

Katzberg, A. A. and A. G. Hendrickx: Gonadotrophin-induced anomalies of the zona pellucida of the baboon ovum. Science *151*:1225, 1966.

Kerr, M. and M. N. Rashad: Chromosome studies on spontaneous abortions. Amer. J. Obstet. Gynec. *94*:322, 1966.

——— and M. N. Rashad: Autosomal trisomy in a discordant monozygotic twin. Nature *212*:726, 1966a.

Kohn, G., B. H. Mayall, M. E. Miller and W. J. Mellman: Tetraploid-diploid mosaicism in a surviving infant. Pediat. Res. *1*:461, 1967.

Lejeune, J., Ch. Salmon, R. Berger, M. O. Réthoré, A. Rossier and J. C. Job: Chimère 46,XX/69,XXY. Ann. Genet. *10*:188, 1967.

Lewis, F. J. W., J. M. Hyman, M. MacTaggart and R. H. Paulding: Trisomy of autosome 16. Nature *199*:404, 1963.

Lindsley, D. L., C. W. Edington and E. S. Von Halle: Sex-linked recessive lethals in Drosophila whose expression is suppressed by the Y chromosome. Genetics *45*:1649, 1960.

Macintyre, M. N., W. I. Staples, A. G. Steinberg and J. M. Hempel: Familial mongolism (trisomy-21 syndrome) resulting from a 15/21 chromosome translocation in more than three generations of a large kindred. Amer. J. Hum. Genet. *14*:335, 1962.

Maclean, N., D. G. Harnden, W. M. Court Brown, J. Bond and D. J. Mantle: Sex chromosome abnormalities in newborn babies. Lancet *1*:286, 1964.

Magenis, R. E. and F. Hecht: D_1 trisomy syndrome: studies on parental age, sex and survival. Proc. Int. Conf. Hum. Genet. 3rd Chicago, 1966.

Makino, S., T. Ikeuchi, M. Sasaki, J. Mumamoto, H. Shimba, S. Fujimoto and S. Matsuda: A preliminary study of the chromosomes in spontaneous abortions. Proc. Japan Acad. *43*:552, 1967.

———, M. S. Sasaki and T. Fukuschima: Triploid chromosome constitution in human chorionic lesions. Lancet *2*:1273, 1964.

Maruffo, C. A.: Zona pellucida of rhesus monkey ovum after gonadotrophin stimulation. Science *157*:1313, 1967.

McFeely, R. A.: Chromosome abnormalities in early embryos of the pig. J. Reprod. Fertil. *13*:579, 1967.

Melander, Y.: A presumed obstacle to mammalian polyploidization. Nature *197*:152, 1963.

Monnet, P., L. Willemin-Clog, J. Gauthier, J. Peytel, C. Laurent, Y. Gay and J. Poncet: La trisomie 6–12. Arch. Franc. Ped. *24*:869, 1967.

Mosonyi, L., D. Schuler, E. Acs and S. Kiss: 19–20 (hereditary?) trisomy in a family with multiple congenital malformations. Acta Med. Acad. Sci. Hung. *23*:41, 1966.

Nes, N.: Diploid-triploid chimerism in a true hermaphrodite mink (*Mustela vison*). Hereditas (Lund) *56*:159, 1966.

Ohno, S., W. A. Kittrell, L. C. Christian, C. Stenius and G. A. Witt: An adult triploid chicken (*Gallus domesticus*) with a left ovotestis. Cytogenetics (Basel) *2*:42, 1963.

Patau, K., D. W. Smith, E. Therman, S. L. Inhorn and H. P. Wagner: Multiple congenital anomaly caused by an extra autosome. Lancet *1*:790, 1960.

Pennock, L. A.: Triploidy in parthenogenetic species of the Teiid lizard genus *Cnemidophorous*. Science *149*:539, 1965.

Penrose, L. S. and J. D. A. Delhanty: Triploid cell cultures from a macerated foetus. Lancet *1*:1261, 1961.

Pfeiffer, R. A., G. Schellong and W. Kosenow: Chromosomenanomalien in den Blutzellen eines Kindes mit multiplen Abartungen. Klin. Wschr. *40*:1058, 1962.

Rashad, M. N. and M. G. Kerr: A human triploid embryo. J. Anat. (Abstr.) *100*:928, 1966.

Russell, L. B.: Chromosome aberrations in experimental mammals. In: Progress in Medical Genetics. A. G. Steinberg, ed. Vol. *2*:230, 1962.

Sacrez, R., J. Clavert, D. Willard, Y. Rumpler, J. M'Bede, R. Meyer and R. Korn: La triploidie chez l'enfant. Pediatrie (Lyon) *22*:267, 1967.

Sato, H.: Chromosome studies in abortuses. Lancet *1*:1280, 1965.

Schlegel, R. J., R. L. Neu, J. C. Leão, E. Farias, M. J. Aspillaga and L. I. Gardner: Observations on the chromosomal, cytological and anatomical characteristics of 75 human conceptuses. Cytogenetics (Basel) *5*:430, 1966.

Schmid, W. and D. Vischer: A malformed boy with double aneuploidy and diploid-triploid mosaicism 48,XXYY/71,XXXYY. Cytogenetics (Basel) *6*:145, 1967.

Schmidt, B. J., W. Beçak, M. L. Beçak, I. Soibelman, A. da S. Queiroz, A. P. Lorga, F. Secaf, C. F. Antonio and A. de A. Carvalho: Metaphyseal dysostosis. J. Pediat. *63*:106, 1963.

Sergovich, F. R.: Cytogenetic practice in a mental retardation clinic. Canad. Psychiat. Ass. J. *12*:35, 1967.

Shaver, E. L. and D. H. Carr: Chromosome abnormalities in rabbit blastocysts following delayed fertilization. J. Reprod. Fert. *14*:415, 1967.

Singh, R. P., A. Rubinoff and D. H. Carr. Chromosomes and abortions. Lancet *2*:445, 1966.

——— and D. H. Carr: Anatomic findings in human abortions of known chromosomal constitution. Obstet. & Gynec. *29*:806, 1967.

Soukup, S. and J. Warkany: Chromosomes in cleft-palate tissues. Lancet *1*:492, 1966.

Stalder, G. R., E. M. Buhler and J. R. Weber: Possible trisomy in chromosome group 6–12. Lancet *1*:1379, 1963.

Stenchever, M. A., J. A. Hempel and M. N. Macintyre: Cytogenetics of spontaneously aborted human fetuses. Obstet. & Gynec. *30:*683, 1967.

Stolte, L., J. Evers and G. Blankenborg: Possible trisomy in chromosome group 6–12 in a normal woman. Lancet *2:*480, 1964.

Szulman, A. E.: Chromosomal aberrations in spontaneous human abortions. New Eng. J. Med. *272:*811, 1965.

Tagher, P. and L. E. Reisman: Reproduction in Down's syndrome (mongolism): Chromosomal study of mother and normal child. Obstet. & Gynec. *27:*182, 1966.

Taylor, A. I.: Patau's, Edward's and cri du chat syndromes: A tabulated summary of current findings. Develop. Med. Child. Neurol. *9:*78, 1967.

Thiede, H. A. and S. B. Salm: Chromosome studies of human spontaneous abortions. Amer. J. Obstet. Gynec. *90:*205, 1964.

——— and S. Metcalfe: Chromosomes and human pregnancy wastage. Amer. J. Obstet. Gynec. *96:*1132, 1966.

Uzzell, T. M.: Natural triploidy in salamanders related to *Ambystoma jeffersonianum*. Science *139:*113, 1963.

Waxman, S. H., D. T. Arakaki and J. B. Smith: Cytogenetics of fetal abortions. Pediatrics *39:*425, 1967.

Weber, W. W.: Survival and the sex ratio in trisomy 17–18. Amer. J. Hum. Genet. *19:*369, 1967.

Wingate, M. B.: A foetus with Down's syndrome. Bristol Med.-chir. J. *78:*121, 1963.

———: The sex ratio of mid-trimester abortions. J. Obstet. Gynaec. Brit. Comm. *73:*296, 1966.

Wolf, U. and H. Reinwein: Chromosomenmosaik C-Trisomie/normal. Humangenetik *1:*686, 1965.

MEIOSIS IN MAMMALS

C. E. Ford
M.R.C. Radiobiological Research Unit, Harwell, Didcot, Berkshire, England

Introduction

The unique features of meiosis are the specific attraction of homologous chromosomes for one another that leads to side-by-side pairing (synapsis) and the subsequent regular occurrence of recombination between them. It has been appreciated for over thirty years that attraction is specific down to small chromosome segments, that it operates over great distances (when considered on the molecular scale) and that it is saturated by a single partner strand (so that, at zygotene in triploids, for example, two of the three strands of a trivalent group are associated at any one point and the third strand is free; Darlington, 1937). Neither attraction nor recombination are understood in molecular terms. Other distinctive but consequential features of meiosis are the exchanges of partner chromatids (chiasmata) seen in stages from diplotene to first metaphase, and the typical occurrence of two successive "divisions" that normally ensure the segregation of the four chromatids of a single bivalent each into a separate telophase nucleus.

The clearest descriptions of the meiotic process have come from flowering plants and certain groups of insects, notably the Acrididae, but sufficient observations have been made on other forms to indicate that the basic principles hold throughout what Whitehouse (1965) has called the "Chromosomal Kingdom" (as opposed to the "Chromonemal Kingdom"). General accounts of normal meiosis and its occasional irregularities have been given by Darlington (1937), John and Lewis (1965), Rhoades (1961), Westergaard (1964) and White (1954).

Studies of meiosis in mammals are as yet few. This is not because of inherent intractability of mammalian material. Preparations can be obtained from some species that are little inferior to those from the most favorable plants and insects, at least as far as the stages of male meiosis from diplotene to diakinesis are concerned. Furthermore, it has been shown recently that preparations of comparable quality may be obtainable from newly-released oocytes of mice and the human female (Henderson and Edwards, 1968; Jagiello *et al.*, 1968). Meiosis in male mammals, so far as my experience goes, conforms well to the classical accounts. To what ends, then, should meiotic studies in mammalian species be directed?

It is unlikely now that repetition and refinement of the old morphological methods, whether on mammalian material or any other, will tell us anything fundamentally new about meiosis. But these methods may still be able to provide valuable information at a more modest level in relation to the accurate identification of structural heterozygosity and its segregational consequences, to the organization of the genome and to the genetic systems of particular species. Examples will be given in the following pages. There remain the issues of hybrid meiosis and of the special behavior of the sex chromosomes. Neither will be discussed here. Too little is yet known of the one, though clear pachytene or later stages from hybrid animals, if obtainable, could potentially reveal something of the extent of structural differentiation between the genomes of the parental species. The other could command a review to itself. Let it only be said that the nature of the association between X and Y chromosomes in species such as the mouse and man, in which only terminal attachments are seen at diplotene, remains unsolved; which means that parasynaptic association of short pairing segments, chiasma formation and early terminalization are not excluded by present evidence.

Studies at Pachytene

Pachytene bivalents in mammalian spermatocytes are shorter and thicker than those in plants such as *Zea mais* (Rhoades, 1950) or *Secale cereale* (Lima-da-Faria, 1952) that have proved favorable for analysis. They are also difficult to spread well and to preserve (fix) with satisfactory detail. The relevant literature is therefore small, notwithstanding the great potential value of observations at this stage.

Schultz and St. Lawrence (1949) identified what they believed to be the nucleolar chromosome in acetic-orcein squashes of human spermatocytes. The nucleolus was located in a sub-median position. This observation was not immediately reconcilable with the recognition, some ten years later, that the acrocentric chromosomes of the D and G groups at mitotic metaphase frequently had satellites appended terminally to their short arms and that the "satellite stalks" were presumptive nucleolar organizers. The apparent discrepancy was resolved by Ferguson-Smith (1964) who showed that the Schultz-St. Lawrence nucleolar chromosome could have been generated by terminal association of a D chromosome and a G chromosome and the attachment of both to the same nucleolus.

A map of the pachytene chromosomes of the mouse was prepared by Slizynski (1949). This map has been criticized on the grounds that detail is shown beyond the theoretical limit of resolution of the microscope. However, I have been informed by Slizynski that he made his observations on stretched regions of the chromosomes and then compiled maps that had been normalized for length. In any case, the maps served to identify the specific chromosomes involved in a series of reciprocal translocation stocks and to locate, at least approximately, the points of interchange as seen in the cross-configurations at pachytene in heterozygotes. Knowledge of the linkage groups involved in several of the translocations then permitted the relation of seven linkage groups to the particular chromosomes that bear them (Slizynski 1957).

A map of the chromosomes of the mouse at pachytene has also been constructed by Griffen (1955). Yerganian (1957) and Eberle (1966) have made contributions towards an eventual map of the human pachytene chromosomes.

Chiasmata

Janssens (1909) was the first to pay careful attention to the intermediate (i.e., non-terminal) points of association between the two members of a bivalent seen in stages from diplotene to diakinesis. He found that where they could be clearly resolved there was a cross-like exchange of partner between the four chromatids, to which he gave the name *chiasma*. Janssens is also renowned as the originator of the hypothesis that chiasmata are the physical basis of genetic recombination, then recently discovered. This, the chiasmatype hypothesis, has had a somewhat controversial history, but doubts and objections have been steadily answered and in view of the great mass of supporting correlative evidence many geneticists today assume its truth without question (e.g., Renwick and Schulze, 1964). An excellent statement of the case for the hypothesis is given by Whitehouse (1968).

Wherever a particular normal bivalent can be recognized unequivocally, the number and position of the *interstitial* chiasmata vary, so that the appearance of the whole structure may differ considerably from cell to cell. Chiasmata are most easily identified and counted at diakinesis. At diplotene, when the bivalents are still long and difficult to fix sharply, some individual chiasmata may be clearly resolvable as unquestioned changes of partner while others are difficult to distinguish from relational coiling. At full metaphase the bivalents are frequently too contracted for the changes of partner to be seen clearly but the numbers can often be counted accurately from an understanding of the three-dimensional form, and this can be built up by studying the earlier stages, when the chromosomes are less contracted. Whatever the stage at which counts are made, it is imperative to use the fine adjustment of the microscope to resolve each structure in depth. Light squashes may be more suitable than air dried preparations, which flatten out the bivalents so that the interpretative advantage of the successive loops that naturally lie in planes at right angles to one-another is lost. This applies particularly to bivalents with three or more chiasmata and, therefore, to species, like man, with a high chiasma frequency. These considerations explain why meiotic stages from diplotene to metaphase are so difficult to illustrate photographically.

The work of Darlington and his co-workers in the 1930's demonstrated that in many species there is a progressive change in the distribution of chiasmata from diplotene onwards to full metaphase, commonly accompanied by a significant reduction of the total count and a significant increase in the number of terminal associations (Darlington, 1937). The major inferences were that there is a movement of interstitial chiasmata away from the centromere, that the terminal associations are produced in consequence of this movement, and that two or more chiasmata in the same arm may reach the end successively, leaving only a single terminal association. Accordingly, the terminal associations are referred to as terminal chiasmata and included in the total count. Different phases from diplotene to metaphase, as well as different species, can be compared in respect of total chiasma frequency and also by the ratio of terminal to total chiasmata, the *terminalization coefficient*.

So far as I am aware there is as yet no comparable body of published data bearing on the questions of the initial numbers of chiasmata and their subsequent movement for any mammalian species other than the mouse. The counts of Crew and Koller (1932) in the male mouse would seem to demonstrate a terminal movement between diplotene and metaphase but it is doubtful how much reliance could now be placed on their data. The more recent counts of Slizynski (1955) give a mean number of chiasmata per autosomal bivalent of 2.02. On the chiasmatype hypothesis this would be equivalent to a total autosomal map length of 19.2 morgans. It is of interest that Carter (1955) obtained an estimate of 16.2 ± 3.5 morgans from linkage test data by calculations based on the assumption that linkage of a new mutant gene to a known marker is a function of total map length.

It is a part of contemporary dogma that natural selection operates on the total effects of gene combinations rather than on the individual effects of particular alleles at specific loci. In any genetic system the rate of production of new gene combinations must be related on the one hand to the haploid number of autosomes and on the other to the total amount of genetic crossing over. If the chiasmatype hypothesis is valid, the mean

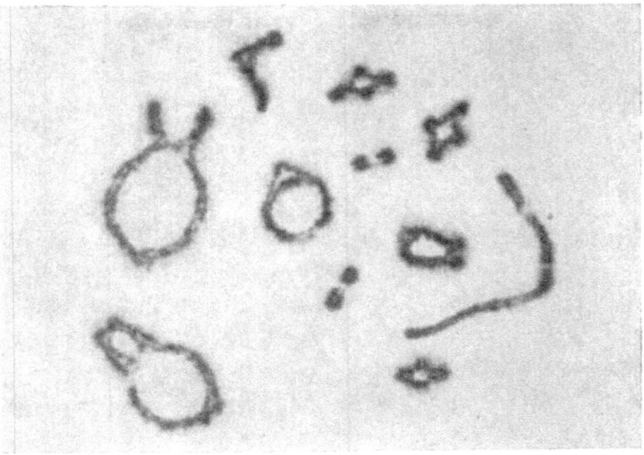

Fig. 1. Chinese hamster (*Cricetulus griseus*). Spermatocyte at early diakinesis. × 1220

chiasma frequency provides a measure of the latter. Darlington (1939, p. 77) has written: "We might (take) the sum of the haploid number of chromosomes and of the average chiasma frequency of all the chromosomes in a meiotic cell as a *recombination index*. Too high an index would be deleterious by breaking up advantageous combinations; too low an index would never achieve the most advantageous combinations."

Typical diakinesis spermatocytes from five mammalian species are shown in Figures 1–5 and the corresponding recombination indexes in Table 1. Since the number of chiasmata does not normally vary greatly from cell to cell within a species, these figures may be taken as reasonably representative of the five species concerned. A sex chromosome was included in the haploid number for the calculation of the recombination

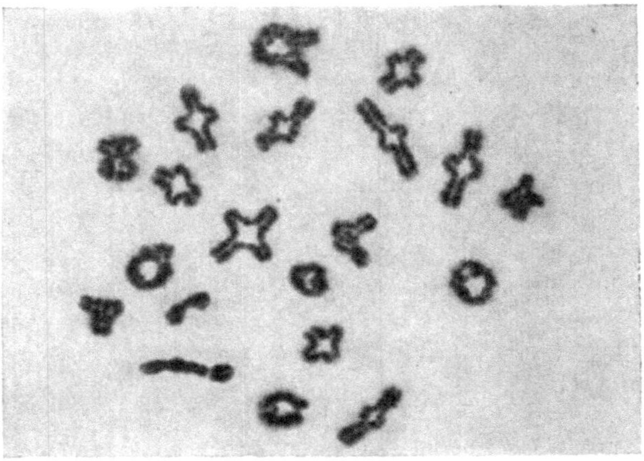

Fig. 2. Mouse (*Mus musculus*). Spermatocyte at diakinesis. × 1400

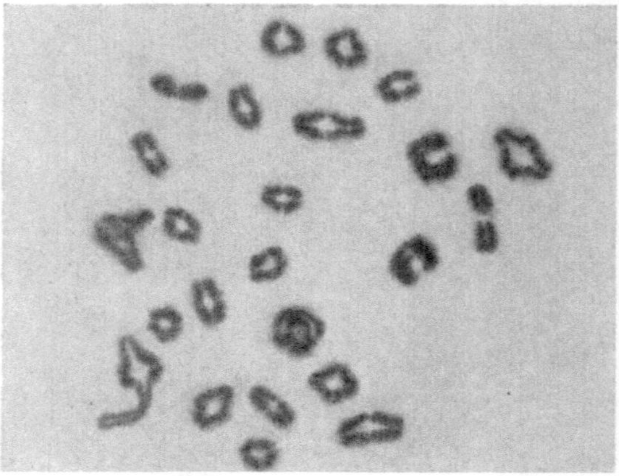

Fig. 3. Wood mouse (*Apodemus sylvaticus*). Spermatocyte at metaphase I. × 1320

indexes given in Table 1, but except in the Chinese hamster (where association between X and Y is frequently by an unquestionable interstitial chiasma) terminal XY associations were not counted as chiasmata. It may, in fact, be better to base comparisons between mammalian species on autosomal number and chiasmata alone. In any case, the figures give an indication of the great range of recombination index within the Class. It remains to be seen whether these differences are of adaptive significance, as Darlington implies, and if so, in what way they are related to the general biology of each species.

Recent studies of meiosis in mammalian oocytes by Henderson and Edwards (1968) have produced two surprising results. The mean chiasma

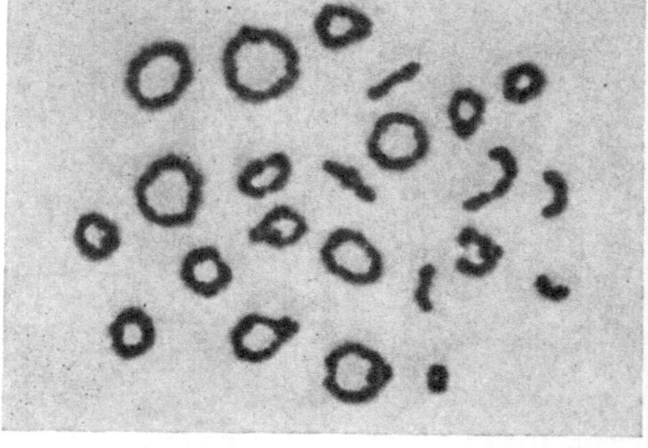

Fig. 4. Marmoset (*Oedipomidas oedipus*). Spermatocyte at diakinesis. × 1240

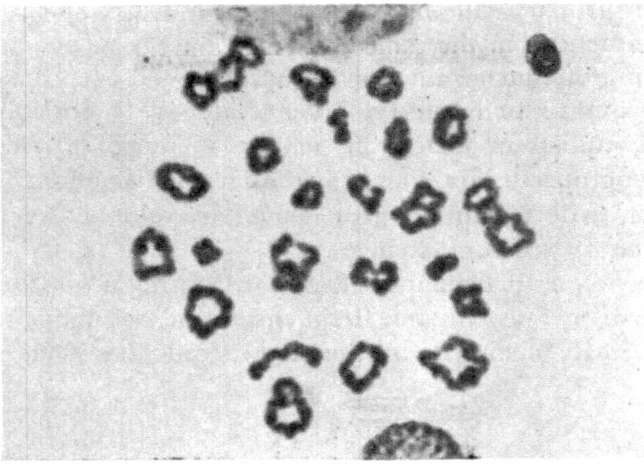

Fig. 5. Man (*Homo sapiens*). Spermatocyte at Diakinesis. × 1050

frequency in oocytes of mice was found to decrease significantly with age, and to be less in human oocytes then spermatocytes, according to published counts. The first is surprising because chiasmata are formed (or at least, first seen) in the diplotene oocytes shortly before birth. These pass into the diffuse dictyate stage and remain in this state until maturation, possibly many months later. The second is surprising because the limited recombinational data available for a comparison between crossing-over in the human male and human female point to an approximately 50% greater figure in the female (Renwick and Schulze, 1965; Renwick, 1968); and though the human data are few, similar and much more numerous comparisons in mice also nearly always reveal greater (usually significantly greater) crossing-over in the female (Dunn and Bennet, 1967). The findings of Henderson and Edwards would appear to be contrary to the chiasmatype hypothesis. Two possible solutions of the paradox were put forward by the authors: (1) Chiasmata are progressively reduced in number during the long dictyate stage by terminalization and later severance of some terminal chiasmata. (2) Diplotene oocytes are produced in sequence; during this

Table 1: Chiasma Counts and Recombination Indexes in Representative Cells of Five Species

Figure	Species	n	Total Chiasmata	Recombination Index
1	Chinese hamster (*Cricetulus griseus*)	11	16	25
2	Mouse (*Mus musculus*)	20	27	46
3	Wood mouse (*Apodemus sylvaticus*)	24	27	51
4	Marmoset (*Oedipomidas oedipus*)	23	40	62
5	Man (*Homo sapiens*)	23	48	71

period the primary chiasma frequency diminishes progressively; the oocytes then mature in the same sequence. Simple (genetic) estimates of recombination frequency in females of different ages should make it possible to confirm or refute the latter hypothesis. Until these questions are settled, studies of meiosis in mature mammalian oocytes will be clearly inappropriate for the calculation of recombination indexes, though they will remain of great potential value in relation to the identification of structural changes.

It is to be hoped that the next decade will bring as much information about normal meiosis, chiasma frequencies and recombination indexes, at least in male mammals, as the last decade has done for somatic karyotypes.

Structural Changes

Although their exact manner of origin is not understood, the great majority of structural changes, or chromosome rearrangements, can be considered to be the product of simple breakage of linear structures followed by rejoining of the broken ends to give new segmental combinations. Structural changes can, therefore, be classified according to the numbers of breaks that must be postulated to account for them. Among the questions that are not satisfactorily answered at the present time are (1) whether stable changes can result from a single break (simple translocations and stable terminal deletions) and (2) to what extent changes that occur exactly at the centromere (or so close to it that the difference cannot be detected by conventional light microscopy) constitute a separate class in respect of their mechanism of origin. Changes of the latter category are inferred from the occurrence of presumptive isochromosomes and the metacentric "fusion" chromosomes of Robertsonian polymorphism.

Heterozygosity for a structural change may be suspected from evidence of hereditary reduced fertility. It may be detected from somatic karyotypic analysis and a formal interpretation given based on the least number of breaks and rejoins necessary to account for the observed differences. But meiotic studies are essential for an assured interpretation. Meiosis in structural heterozygotes was thoroughly investigated in the 1930's and broad, generally applicable rules of chromosome behavior were established.

There are four of these rules. First, zygotene pairing is not a whole-chromosome phenomenon but takes place on a segment-by-segment basis with a high level of specificity. Second, chiasma formation (and genetic crossing-over) involve two of the four chromatids at any one point. Third, chiasma formation may or may not occur in any particular segment that was paired with its partner at pachytene; if a chiasma is formed the segments remain associated through the later stages from diplotene to metaphase; if a chiasma is not formed the segments separate at diplotene and evidence of their previous pairing is unobtainable at later stages. Fourth, chiasmata are arrested in their terminal movement at points where segmental homology changes.

It may be harder today than it used to be to understand and accept these rules because of the widespread practice of thinking of chromosomal phenomena in terms of the DNA double helix, but explanations at the molecular level are far from obvious. In any case, it is very proper for the skeptical observer to doubt them, to examine their validity for himself and to go on doing so. Some limitations of their applicability are indeed already known. For instance, the lower limit of relative length of a displaced segment that can still "find" its partner in zygotene does not appear to have been established. Furthermore, some degree of non-homologous association may occur at zygotene and pachytene that is difficult or impossible to distinguish from specific pairing. At the other extreme, zygotene pairing between some, even all, segments may not take place at all (partial or complete *asynapsis*). If pairing is complete at pachytene it may lapse later without chiasma formation so that the paired chromosomes fall apart as two univalents (*desynapsis*). Finally, although the general principle of arrest of terminalization through change of homology is well-established in cases where large segmental differences are involved, it may require more accurate definition, since arrest at a minor change, like a very small duplication or deficiency, seems unlikely. This test has not yet been made, so far as I am aware.

Verification of inferences made from the application of the four rules may be possible from observations at anaphase I or second metaphase, or from genetic data. All stages of meiosis from pachytene through to second anaphase can therefore be informative. Little, however, has yet appeared in print on meiosis in mammalian structural heterozygotes but what is known is concordant with the four rules of chromosome behavior just discussed. Some features of the principle expected types are considered below.

Inversion

Comparisons of the karyotypes of related species has sometimes identified presumptive homologous chromosomes that differ in the location of their centromeres. Similar differences within the individual have been recognized in clinical cytogenetics. Pericentric inversion is then assumed to be the cause of the difference. It cannot be stressed too strongly that without the evidence of meiotic behavior in the hybrid (or heterozygote), the interpretation is, at best, presumptive. In fact no example of inversion, whether pericentric or paracentric, has yet been identified by meiotic observations in any mammalian species, including man. It is well known that at pachytene a loop is to be expected, but the formation of the loop may be dependent on the length of the inverted segment: if the segment were very short it is possible that it would show non-homologous pairing with its partner. Since crossing-over could not occur in the absence of homologous pairing, the segmental change could then only be recognized genetically. A pronounced reduction of recombination in a particular region would be good evidence of a structural change, though not necessarily of an inversion: the only certain evidence would be from changed order of loci in the linkage map.

It is also well-known that if a chiasma does occur within the inversion loop, a pair of abnormal, unbalanced monocentric chromosomes would be produced if the inversion were pericentric and a dicentric chromosome and an acentric chromosome if the inversion were paracentric. An important diagnostic point is that when a specific paracentric inversion is involved, both the dicentric and acentric fragments are of constant length, and whereas the former may lead to bridges of somewhat variable appearance at anaphase through stretching on the spindle, the latter are invariant.

Duplication and Deficiency

No example of either has been studied by modern methods. At pachytene the appearance expected would be the same for both, namely a kink or "buckle" of varying size depending on the length of the segment concerned. At later stages an unequal bivalent might be detected.

Reciprocal Translocation

The cross-like configurations at pachytene diagnostic of a reciprocal translocation heterozygote have been seen in the mouse by Slizynski and

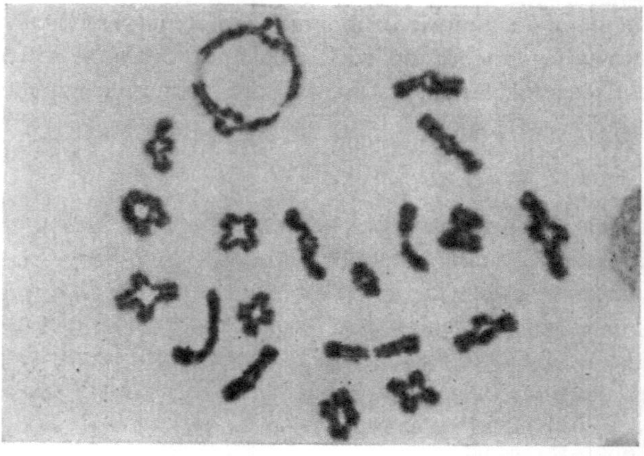

Fig. 6. Mouse (*Mus musculus*). Spermatocyte of an irradiated male at diakinesis showing a ring quadrivalent indicative of a reciprocal translocation. × 1250

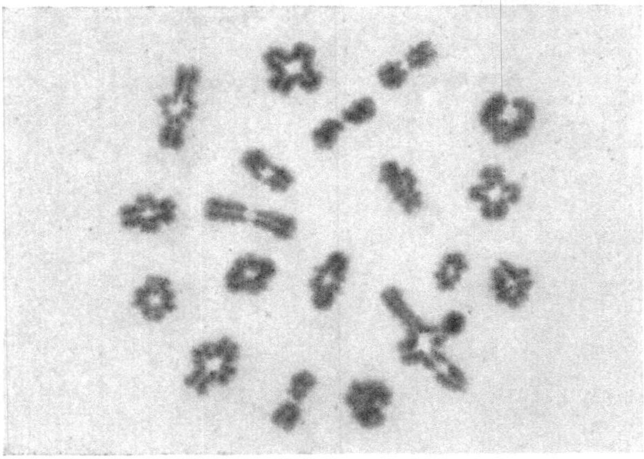

Fig. 7. Mouse (*Mus musculus*). Cattanach's translocation. Spermatocyte of a type I male at diakinesis showing a quadrivalent indicative of an insertion (3-break rearrangement). The chromosomes involved are the X, the Y and an autosome pair. $\times$ 1320

Griffen, as mentioned earlier. At later stages the typical derivative configuration is the ring of four chromosomes (Figs. 6 and 9), but it is possible for failure of chiasma formation to lead to lapse of association in any one, or any combination, of the four arms of the cross to give rise

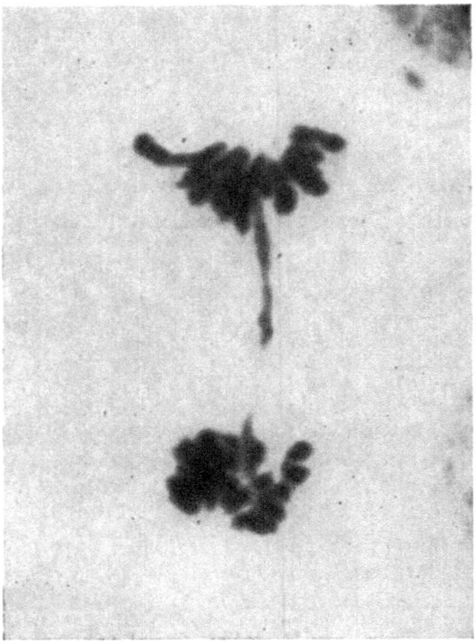

Fig. 8. Mouse (*Mus musculus*). Cattanach's translocation. Spermatocyte at first anaphase showing the dicentric bridge attributable to crossing-over in the inserted segment. $\times$ 2480

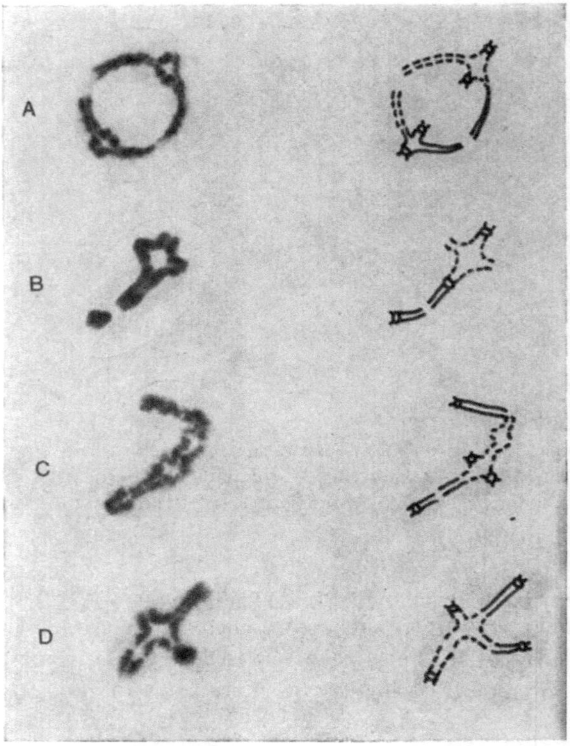

Fig. 9. Mouse (*Mus musculus*). Multivalent configurations. Left: A from Fig. 6; B from Fig. 7; C from a mouse heterozygous for Searle's X-autosomal reciprocal translocation; D from a mouse heterozygous for a "Robertsonian fusion" (T 163). Right: interpretative chromatid diagrams.

to a chain quadrivalent, a trivalent and univalent, two bivalents, a bivalent and two univalents, or, in the extreme, four univalents. Different translocation stocks differ considerably in the frequency of occurrence of these derivative forms of association. Data from three translocation stocks taken from a survey of 12 stocks (Ford, Carter and Hamerton, unpublished) are given in Table 2.

Table 2: Reciprocal Translocations in the Mouse. Primary Spermatocytes in Male Heterozygotes from Three Stocks Classified According to Type of Chromosome Association Observed. Observations Made at Diakinesis and First Metaphase.

Trans-location	18 II Ring IV	18 II Chain IV	18 II III, I	20 II	19 II 2 I	18 II 4 I	Total Cells
T6	12	88	77	4	1	0	182
T264	392	9	0	0	0	0	401
T(5:8)a	190	5	0	957	70	4	1226

The other main form of variation of behavior is in the type of congression on the spindle of the multivalent configurations and its effect on segregation. In the case of the ring quadrivalent, there are three types that lead to two-and-two segregation. These were given the names alternate, adjacent–1 and adjacent–2 by McClintock (1945) and they refer to the manner of disjunction of the four centromeres, the two adjacent types being distinguished by homologous centromeres disjoining to opposite poles (adjacent–1) or going to the same pole (adjacent–2).

These disjunctional arrangements can give rise to 10 possible gametic combinations, 4 of them being contributed exclusively by adjacent–2 disjunction. And when 3:1 and 4:0 disjunctions are also considered, the number is increased to 36 (Table 3). This maximum number, however, requires as a prior condition that chiasmata should be formed in both the *interstitial* segments that lie between centromere and breakpoint. Since the chromosomes of the mouse are all acrocentric the formation of a ring quadrivalent automatically implies chiasma formation in both interstitial segments. This has the interesting (and analytically important) consequence that alternate and adjacent–1 disjunctions are genetically equivalent. This can be verified (and the various gametic combinations enumerated) by the construction of chromatid diagrams and then following through the disjunctional possibilities. The skeptic can check the occurrence of crossing-over in the interstitial segments by examining second metaphase in animals heterozygous for a translocation

Table 3: Tally of the 36 Possible Gametic Combinations that Can Be Formed Following 2:2, 3:1 and 4:0 Anaphase I Disjunctions of the Chromosomes in a Quadrivalent of a Reciprocal Translocation Heterozygote. (Original Chromosomes AB, CD; Rearranged Chromosomes AD, BC.)

Number of Chromosomes Derived from the Quadrivalent that are Included in the Gamete

0	1	2	3	4
	AB	AB AB	AB AB BC	AB AB BC BC
	BC	AB BC	AB AB DC	AB AB BC DC
	DC	AB DC	AB AD BC	AB AB DC DC
	AD	AB AD	AB AD DC	AB AD BC BC
		AD BC	AD AD BC	AB AD BC DC
		AD DC	AD AD DC	AB AD DC DC
		AD AD	BC BC AB	AD AD BC BC
		BC BC	BC BC AD	AD AD BC DC
		BC DC	BC DC AB	AD AD DC DC
		DC DC	BC DC AD	
			DC DC AB	
			DC DC AD	
1	4	10	12	9

in which the segments exchanged are clearly of unequal length: the second metaphase will then be expected to include two chromosomes each with unequal-length chromatids. The extensive information from pedigrees on the behavior of presumptive reciprocal translocations in man has been reviewed recently (Ford and Clegg, 1969).

Robertsonian Translocation

Meiotic studies have been reported in animals heterozygous for a Robertsonian translocation that appeared spontaneously at Harwell (Evans *et al.*, 1967). The expected trivalent configurations were observed in spermatocytes at diakinesis and metaphase I (Fig. 9). It is of interest that in this case, the first of its kind to be investigated, heterozygosity in the male was associated with an approximately 30 percent reduction in fertility. This infertility is at present unexplained: it is not attributable to irregular segregation of the trivalent at first anaphase, since only 1 percent of numerically deviant second metaphase counts were recorded in 600 second spermatocytes from four animals.

Insertion

It is convenient to distinguish two types of 3-break rearrangement, those in which the change involves a single chromosome only and those where two non-homologous chromosomes are involved. White (1954) gives these changes the names *shift* and *insertion* respectively. It is reasonable to suppose that they will be rare in comparison with the two-break rearrangement types. Nevertheless, insertions have been postulated on several occasions to account for unexplained observations in human cytogenetics.

One mammalian example of insertion has been demonstrated unequivocally by observations at meiosis. This is the Cattanach translocation in the mouse. (The generic term translocation is appropriate for all cases where more than one chromosome is involved in a segmental rearrangement). Genetic studies demonstrated that a segment from the chromosome carrying linkage group I had been translocated to the X chromosome. Subsequent observations at meiosis (Ohno and Cattanach, 1962; Ford and Evans, unpublished) then showed that quadrivalents of an unusual kind involving the X and Y chromosomes were present in about five percent of primary spermatocytes. The remaining spermatocytes had 20 bivalents including an XY pair with an X that was appreciably longer than normal, and an unequal autosome bivalent. The form of the quadrivalents could not be reconciled with reciprocal translocation but was compatible with insertion. An example is shown in Figure 9d, and can be compared with the quadrivalent shown in Figure 9c, which is from a mouse heterozygous for Searle's X-autosome reciprocal translocation (Ford and Evans, 1964). Final proof of the insertion interpretation came from the discovery that about five percent of first anaphase figures showed a bridge (Fig. 8) and, usually, an associated fragment. Bridges could only be produced with a frequency equal to the frequency of quadrivalents if (1) the rearrangement was indeed an insertion and (2) the inserted segment had been inverted relative to its position in the donor autosome. All single cross-overs within the inserted segment would, therefore, lead to the formation of unbalanced gametes, so explaining the failure to recover any cross-over progeny in the genetic experiments.

Summary

Relatively few studies of meiosis in mammals have been reported. The potential value of such studies is discussed particularly in relation to chiasma counts and their interpretation and to the identification of different types of structural heterozygosity.

Acknowledgment

I am grateful to Dr. E. P. Evans for permission to use several of his photographs to illustrate this article.

References

Carter, T. C.: The estimation of total genetical map lengths from linkage test data. J. Genet. *53*:21, 1955.

Crew, F. A. E. and P. C. Koller: The sex incidence of chiasma frequency and genetical crossing-over in the mouse. J. Genet. *26*:359, 1932.

Darlington, C. D.: Recent Advances in Cytology. 2nd ed. London: Churchill Ltd., 1937.

———: The Evolution of Genetic Systems. Cambridge: University Press 1939.

Dunn, L. C. and D. Bennet: Sex differences in recombination of linked genes in animals. Genet. Res. *9*:211, 1967.

Eberle, P.: Die Chromosomenstruktur des Menchen in Mitosis und Meiosis. Stuttgart: Fischer, 1966.

Evans, E. P., M. F. Lyon and M. Daglish: A mouse translocation giving a metacentric marker chromosome. Cytogenetics *6*:105, 1967.

Ferguson-Smith, M. A.: The sites of nucleolus formation in human pachytene chromosomes. Cytogenetics *3*:124, 1964.

Ford, C. E. and H. M. Clegg: Reciprocal translocations. Br. med. Bull. *25*:110, 1969.

——— and E. P. Evans: A reciprocal translocation in the mouse between the X chromosome and a short autosome. Cytogenetics *3*:295, 1964.

Griffen, A. B.: A late pachytene chromosome map of the male mouse. J. Morph. *96*:123, 1955.

Henderson, S. A. and R. G. Edwards: Chiasma frequency and maternal age in mammals. Nature *218*:22, 1968.

Jagiello, G., J. Karnicki and R. J. Ryan: Superovulation with pituitary gonadotrophins: method for obtaining meiotic metaphase figures in human ova. Lancet *1*:178, 1968.

Janssens, F. A.: Spermatogenèse dans les batraciens V., La theorie de la chiasmatypie, nouvelle interprétation des cineses de maturation. Cellule *25*:387, 1909.

John, B. and K. R. Lewis: The meiotic system. Protoplasmatologia *VI*:1, 1965.

Lima-da-Faria, A.: Chromosome analysis of the chromosomes of rye. Chromosoma *5*:1, 1952.

McClintock, B.: *Neurospora*. I. Preliminary observations of the chromosomes of *Neurospora crassa*. Am. J. Bot. *32*:671, 1945.

Ohno, S. and B. M. Cattanach: Cytological study of an X-autosome translocation in *Mus musculus*. Cytogenetics *1*:129, 1962.

Renwick, J. H. and J. Schulze: An analysis of some data on the linkage between Xg and color blindness in man. Am. J. hum. Genet. *16*:410, 1964.

——— and ———: Male and female recombination fractions for the nail-patella: ABO linkage in man. Ann. hum. Genet. *28*:379, 1965.

———: Rates of female to male recombination fractions in man. Bull. Eur. Soc. hum. Genet. *2*:7, 1968.

Rhoades, M. M.: Meiosis in Maize. J. Hered. *41*:59, 1950.

———: Meiosis. In: The Cell., 111. J. Brachet and A. E. Mirsky, eds. New York: Academic Press Ltd., *1*, 1961.

Schultz, J. and P. St. Lawrence: A cytological basis for a map of the nucleolar chromosome in man. J. Hered. *40*:31, 1949.

Slizynski, B. M.: A preliminary pachytene chromosome map of the house mouse. J. Genet. *49*:242, 1948–49.

———: Chiasmata in the male mouse. J. Genet. *53*:597, 1955.

———: Cytological analysis of translocations in the mouse. J. Genet. *55*:122, 1957.

Westergaard, M.: Studies on the mechanism of crossing over: 1. Thoretical considerations. C. R. Lab. Carlsberg *34*:359, 1964.

White, M. J. D.: Animal Cytology and Evolution. 2nd ed. Cambridge: University Press, 1954.

Whitehouse, H. L. K.: The Mechanism of Heredity. London: Edward Arnold Ltd., 1965.

Yerganian, G.: Cytologic maps of some isolated human pachytene chromosomes. Am. J. Hum. Genet. *9*:42, 1957.

HYBRID STERILITY

P. K. Basrur

Department of Anatomy, Ontario Veterinary College, University of Guelph, Guelph, Ontario, Canada

Throughout the ages man's attempts to combine the desirable traits of parental species in the progeny through experimental hybridization have been frustrated by several barriers of reproductive isolation. These barriers, which effectively restrict gene exchange between species, are built by a variety of mechanisms, all of which may be interrelated. Depending upon the developmental stage at which a particular isolating mechanism operates, interspecific hybridization may result in hybrid inviability, total or partial sterility of the hybrids, or reduced viability of the back-cross progeny. Examples of the first type of isolating mechanisms are sheep-goat hybrids (Berry, 1938; Hancock, 1964; Warwick *et al.*, 1934) and ferret-mink hybrids (Chang, 1965) all of which perish either in embryonal or fetal stages (Chang and Hancock, 1967). Total sterility of the progeny has been the general trend in most of the equine hybrids (Benirschke *et al.*, 1964; King *et al.*, 1966) despite the exceptions which have been reported from time to time (Benirschke, 1967; Koulischer and Frechkop, 1966). Partial sterility in hybrids is often expressed as sex-restricted malfunctions, which are generally more pronounced in the heterogametic sex (Haldane, 1922). Thus, among the interspecific hybrids, the males generally die *in utero* or grow to be sterile adults in mammals, whereas in birds, with a reversed sex determining mechanism, it is the female hybrids which are either at a reproductive or developmental disadvantage.

The cytogenetic and cytotaxonomic aspects of some of the better known mammalian hybrids have been adequately reviewed by Benirschke (1967), and the experimental data on attempted cross fertilization of several mammalian species have also been presented by Chang and Hancock (1967). This report will be concerned with certain hybrids which have not been previously dealt with in detail. The first part of this paper will review the existing information, including our own, on cattle-bison hybrids and the second part will concern our experiments on chicken-pheasant hybrids which we consider to be an excellent model for studying the problems associated with hybrid sterility.

Cattle-Bison Hybrids (Cattalos)

It has been reported that there were over 300 cattle-bison hybrids in North America at the turn of the century (Boyd, 1914). These hybrids were all females and a majority of them were the progeny of matings between domestic cattle bulls and American bison cows. Mossom Boyd, who was one of the first to report on experimental hybridization of cattle with bison (Boyd, 1908, 1914) defined the word "cattalos" as the progeny of crosses in which both parents were part bison and part cattle. These were distinguished from the F_1 "hybrids" whose parents were pure cattle and pure bison, and from the back-cross progenies which were designated as 25% bison (or 75% cattle) and 12.5% bison (87.5% cattle) depending upon the number of back-crosses to either pure cattle or pure bison. Boyd (1914) noted that domestic cows mated with bison bulls accumulated excessive amounts of fluid in the fetal membranes, with a resultant high mortality of the cows and calves at birth. The reverse cross, which is relatively more successful, results in a disturbed sex ratio with females predominating in the F_1 progeny (Boyd, 1914; Deakin *et al.*, 1935, 1941; Goodnight, 1914; Jones, 1907; Logan and Sylvestre, 1950; Sylvestre *et al.*, 1948).

Although the domestic cattle (*Bos taurus*) and the American bison (*Bison bison*) belong to the same genus, subfamily Bovinae, the two have evolved into ecologically separated species with strikingly different general appearance, natural disposition, nutritional requirements, response to climatic conditions and breeding habits (Baughman, 1951). The cattle-bison hybrids combine some of the characteristics of both parents. For example, the 14 ribs, the tufts of curly hair and the thick hide of the buffaloes are passed on to the hybrids along with their resistance to diseases (Boyd, 1914; Goodnight, 1914). These hybrids, besides being thrifty and winter-hardy like the buffaloes, produce more meat which is low in connective tissue content (Goodnight, 1914). These obviously advantageous traits have prompted hybridization experiments with a view to producing a line of cattalos. This venture, however, has failed on a commercial scale, both in the United States and in Canada, mainly because of the male sterility in F_1 hybrids and the poor viability and low male fertility in the back-cross progenies (Peters, 1964).

Although the F_1 male hybrids and the first back-cross males were consistently sterile, a small percentage of males with varying degrees of fertility was obtained in the subsequent back-cross generations (Deakin *et al.*, 1935, 1941; Logan and Sylvestre, 1950). However, when the proportion of bison parentage was increased to over 14 by mating the cattalos with cows from F_1 or back-cross progenies, the male cattalos were invariably sterile whereas the females from such crosses were generally fertile (Peters *et al.*, 1958).

Male sterility in cattalos has been attributed to a number of different factors, one of which has been the difference in the position and structure of the testes in bison and cattle (Deakin *et al.*, 1935). It was postulated that this was an adaptation against the severe winter conditions for the protection of the bison's germinal epithelium by means of a heavy scrotal wall, which was considered to provide a higher temperature in the intra-scrotal environment. As a consequence, the cattle and the bison genomes were said to be at a disadvantage in the hybrid testes, which would maintain a temperature different from the norm of both species (Deakin *et al.*, 1935). Later it was shown that the intra-testicular temperature of bison, cattle and cattalo was the same (Peters and Newbound, 1957), although at the time of these experiments the average weight of bison testes was less than a third of that of domestic bulls of corresponding age (Peters *et al.*, 1958). It was noted that the rate of post-natal development of the testes in cattle is much greater than that in bison, whereas the cattalos, including the sterile ones, exhibit a growth rate intermediate to those of bison and cattle (Peters, 1964). The data suggest that the developmental stages, when the intra-testicular temperature was taken, were probably not comparable regarding the testicular activities of these animals.

Another cause of male sterility has been postulated to be the differences between bison and cattle chromosomes. Our studies on bison, cattle and cattalos have revealed that the chromosome number in all three groups is 60, and that the only overt difference in the chromosome morphology of cattle and bison is in the Y chromosome which is a metacentric in cattle and an acrocentric, indistinguishable from the autosomes, in bison (Basrur and Moon, 1967).

The degree of sterility in males is reported to vary greatly depending upon the percentage of bison parentage in the male cattalos. Generally, very few males from the second back-cross progeny (12.5% bison) were fertile; however, male fertility steadily increased with repeated back-crossing to domestic cows. If an F_1 hybrid female is back-crossed to a bison male or a domestic bull, the male hybrids carrying 75% bison and 75% cattle respectively, may be fertile (Boyd, 1914; Goodnight, 1914). However, if a fertile bull from this back-cross generation is mated with an F_1 female (50% bison), the males are generally sterile (Boyd, 1914; Goodnight, 1914; Logan and Sylvestre, 1950). These observations led Boyd (1914) to the conclusion that 25% bison is probably the maximum bison parentage in a hybrid female that is compatible with the production of fertile males. In general, cattalos with 14% bison parentage are found to be fertile (Peters, 1964).

Our own studies on the testicular activities of cattalos have been limited to 3 cattalos which were 2.5 years old. The percentage of bison parentage in these cattalos were 31, 22 and 14 (Fig. 1). The testicular activities varied in these cattalos depending upon the percentage of bison parentage. Thus, in the cattalo with 31% bison whose parents were a 14% cattalo bull and an F_1 female (Fig. 1), no indication of spermatogenesis was apparent. The seminiferous tubules, compared with those of a domestic bull of corresponding age, were irregular in outline and were lined by a layer of "indifferent" cells from which a coarse net-like substance projected towards the lumen (Figs. 2, 3 and 4). The most striking features noted in the histological sections, besides the total lack of spermatogenesis, were the excessive thickening of the base-

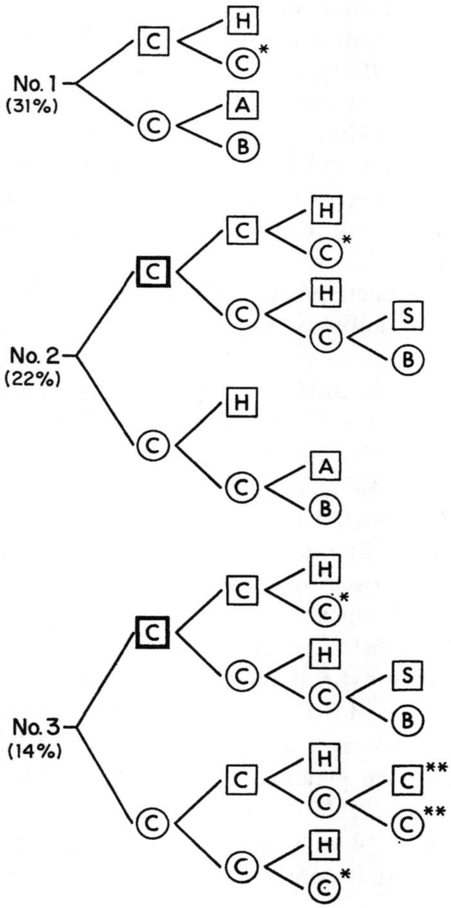

Fig. 1. Pedigrees of the 3 cattalos. A: Angus; B: Bison; C: Cattalo; H: Hereford; S: Shorthorn; □: Sire; ○: Dame; heavier □: Same Sire; *25% Bison; **14% Bison.

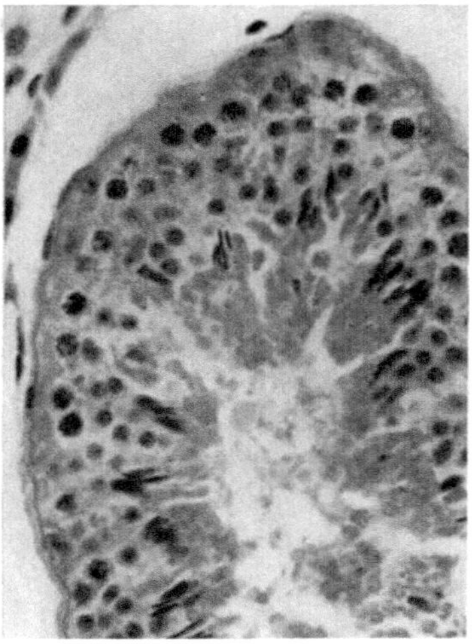

Fig. 2. A section of a seminiferous tubule from a normal Hereford bull (H & E × 280).

ment membrane, the abundance of interstitial cells and the occurrence of large amorphous structures in the lumen (Figs. 3, 4 and 5). The cattalos with 22% and 14% bison were rather interesting from histologic and cytogenetic aspects (Figs. 6–19). Although the number of "active" tubules (Table 1) and the number of epithelial layers lining the tubules were low in cattalo #2, all stages of spermatogenesis including mature spermatozoa were noted in the sections (Figs. 6 and 7). In some tubules the alignment of spermatozoa was such that the head was parallel to the basement membrane. A consistent feature was the accumulation in the lumen of darkly stained bodies surrounded by a thin envelope (Figs. 7 and 8). The dark bodies appeared to be spermatids or sperm-heads which had become altered to highly condensed basophilic bodies. Some of these sperm-like bodies were noted either attached to the Sertoli cells or making their way into the envelope (Fig. 8). The envelope with altered sperms resembled the spermiophages described in the testes and epididymis of sterile men (Phadke and Phadke, 1961) and as a reaction to extravasation of spermatozoa and to granuloma formation (Tjioe

Table 1: Activity of Seminiferous Epithelium as Evidenced by Division Figures

Animal Number	% Bison Parentage	No. of Tubules Showing Division Stages *	Total No. of Tubules Examined
Cattalo #1	31	3	123
Cattalo #2	22	67	104
Cattalo #3	14	81	97

* Mitotic and meiotic stages are included.

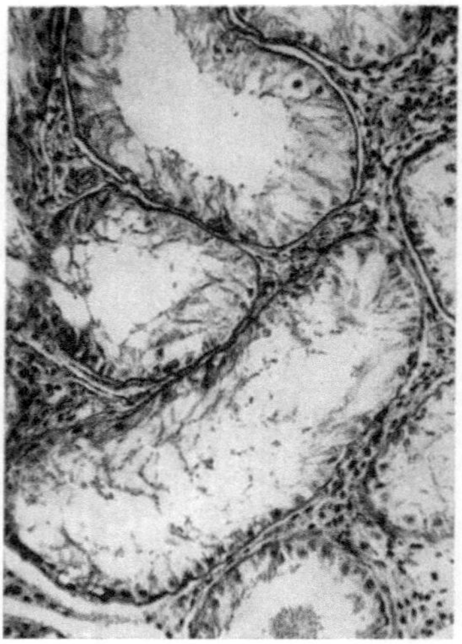

Fig. 3.

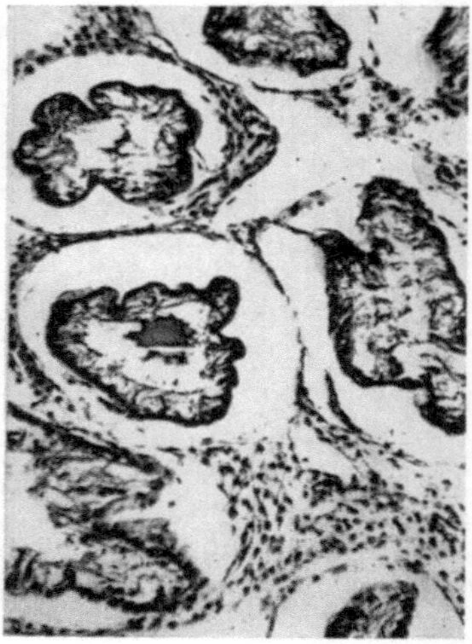

Fig. 4.

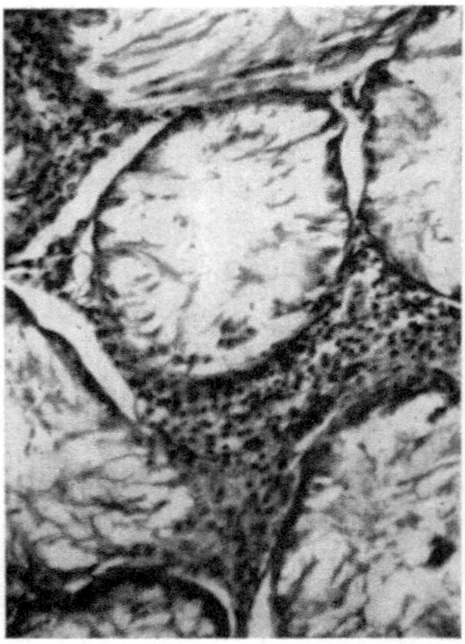

Fig. 5.

Figs. 3–5. Sections of testis from cattalo #1 with 31% bison parentage. Total lack of spermatogenesis is a striking characteristic of the seminiferous tubules from this cattalo. Note the net-like projections in Fig. 3, the shrivelled appearance of the seminiferous tubules and the amorphous deposit in Fig. 4 and the large patches of interstitial cells in Figs. 4 and 5. The connective tissue cell-like appearance of the interstitial cells is more pronounced in Fig. 4 (H & E × 160).

and Steinberger, 1967). In several tubules, nuclear material was also noted in clusters of 8 or 12 without an envelope around it (Fig. 9). These probably were produced by the breaking down of multipolar mitoses (Figs. 10 and 11) which were noted quite frequently. The frequency of meiotic metaphase stages were relatively low in cattalo #2; however, meiotic stages were abundant in cattalo #3 which resembled a normal bull in testicular activity, although the seminiferous tubules were irregular in outline (Fig. 14) and the epididymis was devoid of spermatozoa (Fig. 15). Forty out of 67 metaphase plates analyzed from cattalo

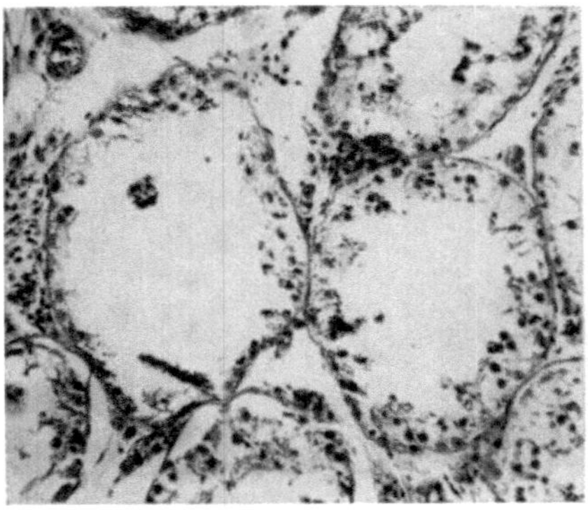

Fig. 6.

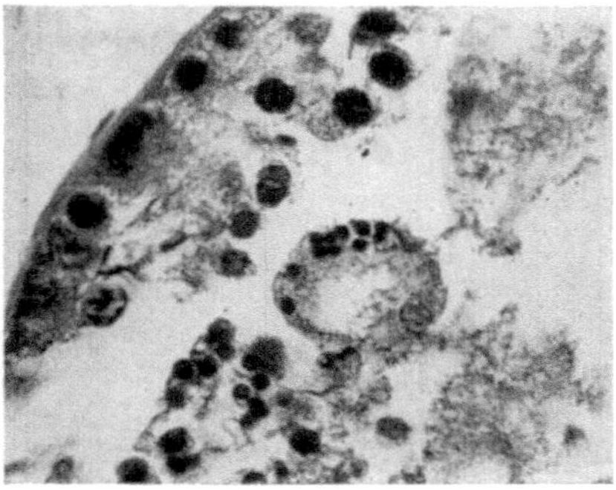

Fig. 7.

Figs. 6–8. Sections of testis from cattalo #2 with 22% bison parentage. The seminiferous tubules are relatively more active than those of cattalo #1 (H & E × 145). Note the large spermiophages with many round sperm heads (Figs. 6 and 7), the spermatozoa oriented parallel to the basement membrane (Fig. 7) and the "altered" sperm heads in groups on the Sertoli cell and on the outer membrane of the spermiophage (Fig. 8).

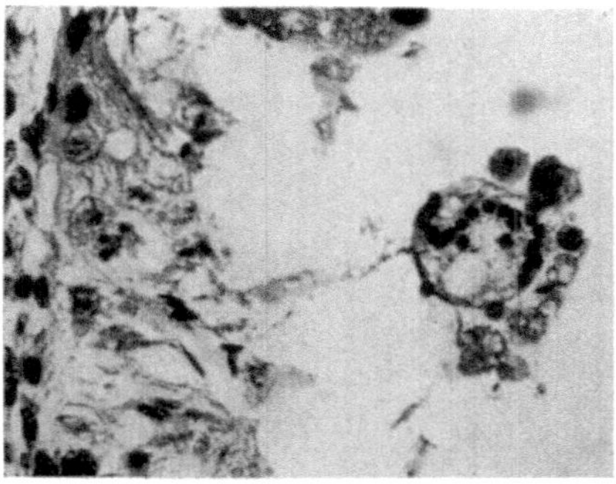

Fig. 8.

#3 (Table 2) exhibited 30 bivalents, whereas 21 plates appeared to have 31 elements (Figs. 12 and 13). In Feulgen squash preparations the pachytene stages revealed no striking difference in synapsis (Figs. 16 and 17) as compared to the normal and included a distinct sex vesicle in a majority of early meiotic stages (Fig. 16). Another striking feature of cattalos #2 and #3 concerns the sperm heads which appeared to be

Table 2: Distribution of Chromosome Elements * in the Meiotic Metaphase Plates of Cattalo #3

	Number of Elements in Each Plate					
	29	30	31	32	33	Total
No. of Cells	2	40	21	1	3	67

* Bivalents and/or univalents.

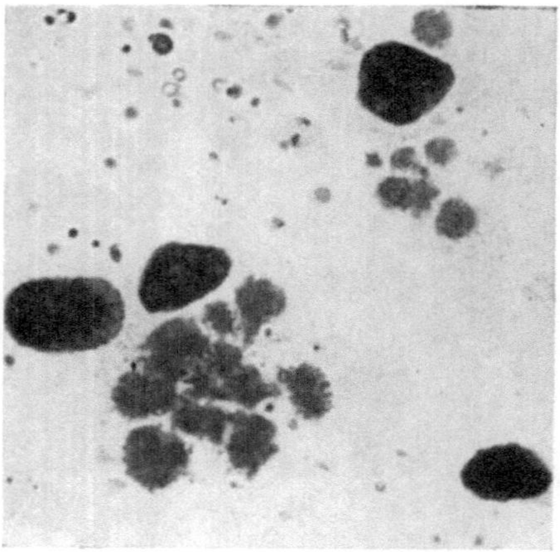

Fig. 9.

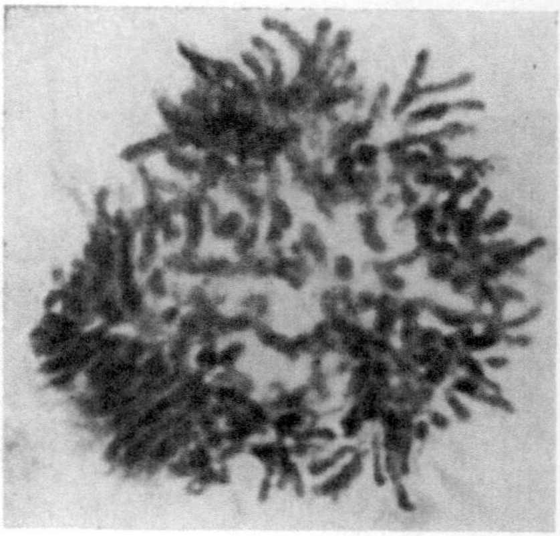

Fig. 10.

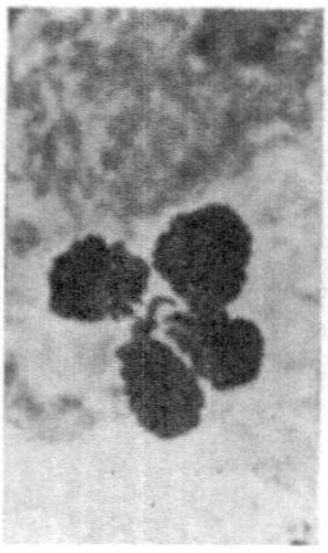

Fig. 11.

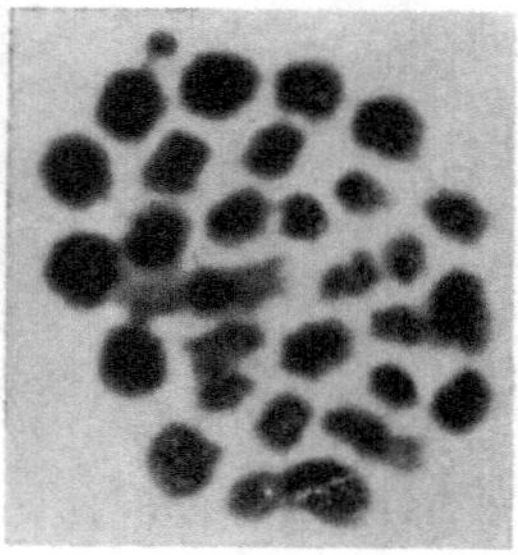

Fig. 12.

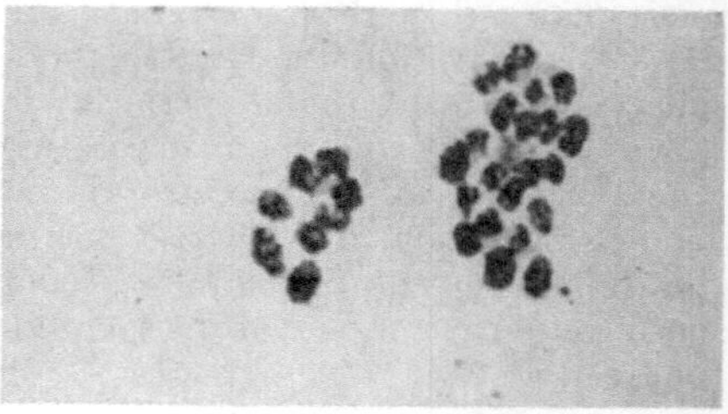

Fig. 13.

Figs. 9–13. Some of the division figures noted in cattalo #2. Clusters of nuclear material probably after the break down of polypolar spindles (Fig. 9), a tetrapolar mitosis still retaining chromosome individuality (Fig. 10), a group of 4 nuclei which appear to be the result of a tetrapolar mitosis (Fig. 11) and meiotic metaphases with 31 bivalents (Fig. 12) and 30 bivalents (Fig. 13) (Feulgen reagent × 1750).

vacuolated, enlarged (Fig. 18) and often bloated to enormous proportions (Fig. 19).

Although all 3 cattalo bulls showed normal libido, the semen samples collected by electro-ejaculation revealed no live, or morphologically normal spermatozoa. The fluid contained multi-nucleated cells and leukocyte-like cells. The levels of testosterone of each of the 3 cattalos were within the range of values for normal domestic bulls (Raeside, personal communication). Cattalo #1, which showed no spermatogenic activity, had more extensive interstitial tissues than normal bulls (Fig. 5).

Our observations seem to indicate that the meiotic activity is relatively uninterrupted in a majority of cells in cattalos #2 and #3,

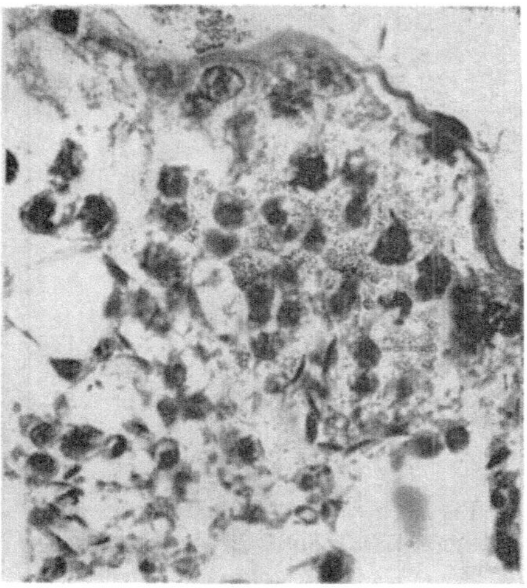

Fig. 14. A section of a seminiferous tubule from cattalo #3 with 14% bison parentage. All stages of meiosis and spermiogenesis were noted in this cattalo (H & E × 415).

although all 3 were sterile. Cattalo #1 resembled the sterile F_1 hybrid in the appearance of seminiferous tubules (Deakin *et al.*, 1935); however, the inability of cattalos #2 and #3 to produce semen, despite the meiotic activity and the presence of mature spermatozoa noted in these animals, is puzzling. At least as far as cattalos #2 and #3 are concerned, the assumption that male sterility in interspecific hybrids is due to pairing abnormalities does not appear to be valid. Meiosis, and even spermiogenesis in the seminiferous tubules, coupled with the total absence of spermatozoa in the epididymis and in the ejaculates, seem to indicate that the majority of the gametes may be destroyed before they get to the epididymis. The presence of spermiophages in the intra-

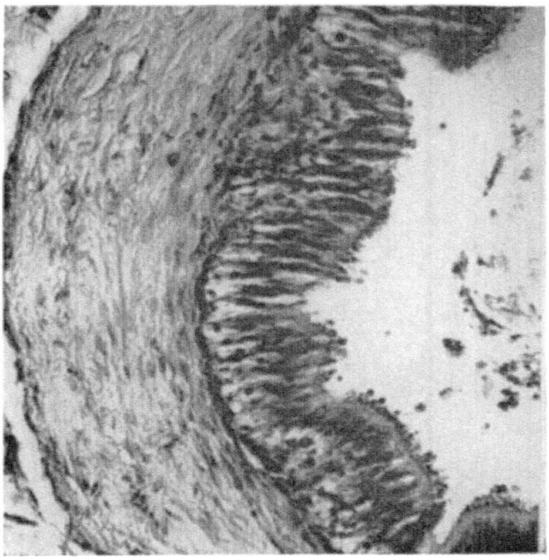

Fig. 15. A section of the epididymis showing absence of spermatozoa. (H & E × 415).

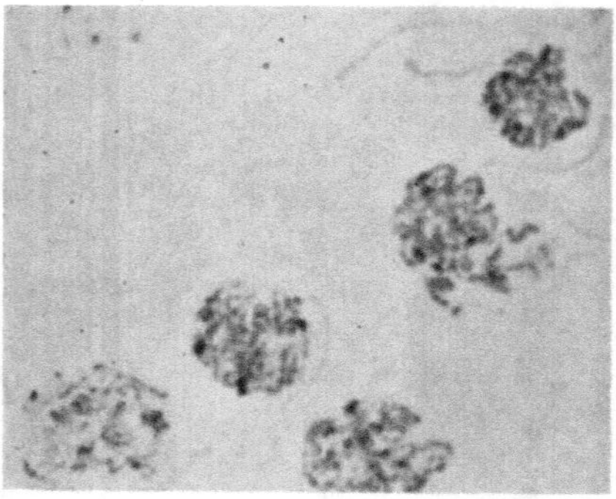

Fig. 16.

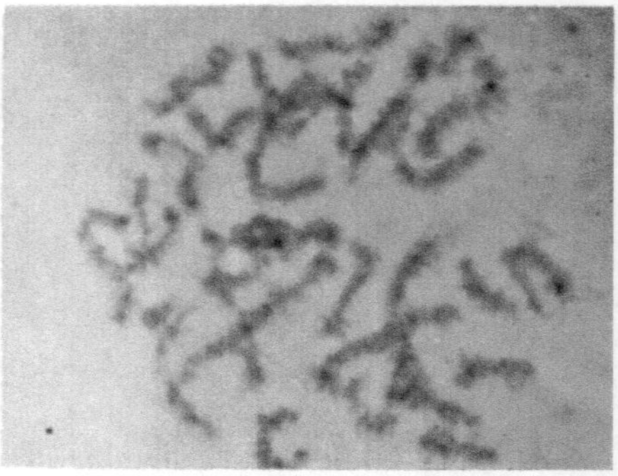

Fig. 17.

Figs. 16 and 17. Pachytene stages showing sex vesicles and apparently complete pairing. Feulgen reagent. × 310 (Fig. 16) and × 1800 (Fig. 17).

tubular space in cattalos #2 and #3 may be related to the fate of these gametes.

Tjioe and Steinberger (1967) have suggested that the spermiophages are lymphocytes of the interstitial area infiltrating the tubular lumen. It has been demonstrated that, under experimental conditions, Sertoli cells may phagocytize injected material (Clegg and MacMillan, 1965) and cell debris (Vilar *et al.*, 1962) and it is conceivable that in the normal seminiferous tubules Sertoli cells generally act as phagocytes. It is conceivable that some of the spermatids, especially those noted in unusual orientation with respect to Sertoli cells, are engulfed by spermiophages. The presence of spermiophages in sterile cattalo testes may thus be a reflection of the impaired activity of Sertoli cells which are also involved in the normal maturation process of spermatids to

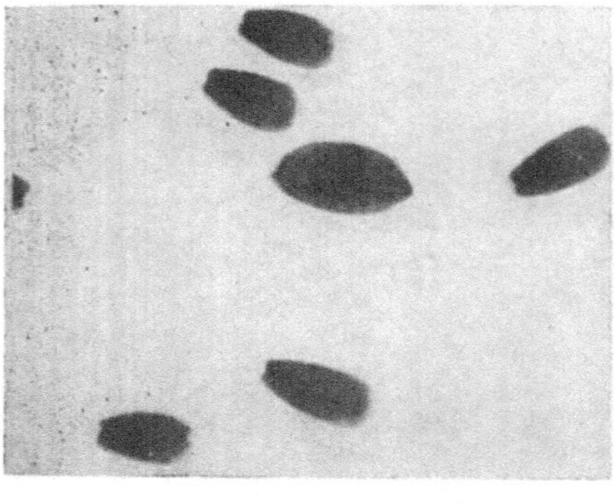

Fig. 18.

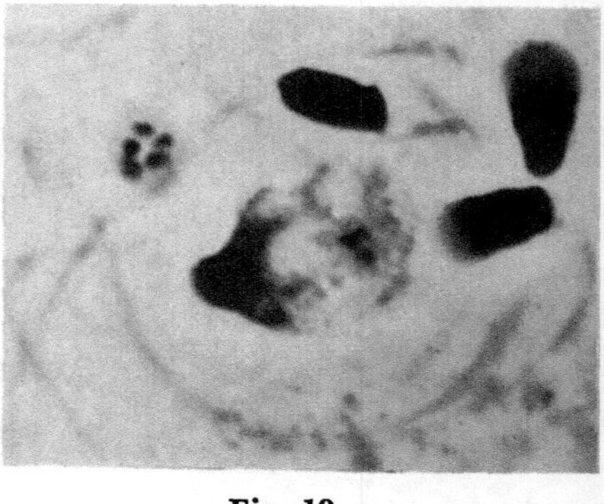

Fig. 19.

Figs. 18 and 19. Spermatozoa showing enlargement of head and vacuolations (Fig. 18) and bloated sperm heads (Rose Bengal × 1800).

spermatozoa (Vilar *et al.,* 1962). Bedford (1965) had demonstrated that only defective spermatozoa, especially those with damaged acrosome and ruptured head membrane, were vulnerable to phagocytosis by leukocytes. Since the frequencies of abnormal sperm heads in cattalos #2 and #3 were relatively high even at the light microscopic level, it would appear that some of the spermatids are defective probably through the segregation of unbalanced chromosome groupings including deficiencies and duplications (Stebbins, 1958). This unbalanced chromosome make-up in the developing gametes may be attributable to undetected structural differences in the chromosome complements of bison and cattle which, however, appeared to differ only in their Y chromosomes. The genetic imbalance caused by deficiencies and duplications may not affect the earlier stages of spermatogenesis whereas it may be deleterious to spermatids and spermatozoa which are haploid.

The European domestic cattle, *Bos taurus,* has been known to hybridize with several species of the family Bovidae (Gray, 1954; Haldane, 1922; Melander, 1959). In all well authenticated cases a fertile progeny has been obtained only when the parental species appeared karyotypically identical. The most successful of the crosses so far has been that with the Afro-Asian zebu, *Bos indicus,* which has been reported to be indistinguishable from European cattle in chromosome morphology (Makino, 1944; Sasaki and Makino, 1962). The complete interfertility of these two species has prompted some investigators to suggest that their classification into separate species may be unjustifiable (Kieffer and Cartwright, 1968). Recently Kieffer and Cartwright (1968) have demonstrated that the Y chromosome of *Bos indicus* is an acrocentric. These investigators postulated that the Y chromosome of *Bos indicus* is similar to that of *Bos taurus* because of the lack of reproductive barriers between these species. A morphologic difference of this magnitude could easily be derived through a pericentric inversion in the Y chromosome without recourse to pronounced changes in the total genomes.

Interbreeding between European cattle and the European wisent, *Bison bonasus,* has been reported to have occurred repeatedly with good fertility in the F_1 progeny (Heck, 1968; Melander, 1959). On the basis of chromosome studies on female European bison, Melander (1959) had postulated that both male and female European bison may be karyotypically identical to European domestic cattle. This was confirmed by Koulischer *et al.* (1967) who noted that the wisent Y chromosome is a small submetacentric. On the basis of immunological studies, European cattle and European bison have been demonstrated to be identical in several blood group factors (Czaja and Gasparski, 1960; Gasparski, 1962, 1965), whereas specific antigenic differences have been noted between European cattle and American bison (Shaw and Patel,

1963). Crosses between European bison and American bison are known to be successful and the male progenies have been reported to be not only fertile but of superior vigor than the parents (Gray, 1954; Heck, 1968). However, on the basis of karyotypic feature alone the European bison is closer to the European cattle than to the American bison which, in turn, more closely resembles the Afro-Asian zebu. American bison-zebu crosses have been reported to produce fertile females (Bartlett, 1884; Baughman, 1951; Shah, 1935), however, no information is available in the literature about the male hybrids. If the karyotypic similarity is not superficial, American bison-zebu matings should produce progenies which are more fertile than the cattalos. On the other hand, the male sterility in cattalos may have no connection with the differences in morphologic character in the Y chromosomes of American bison and American cattle. It would appear from the striking difference in the reciprocal matings that the incompatibility is probably related to a disharmony between the nuclear complement of one parent with the cytoplasm of the other. Such disharmonies have been attributed to cytoplasmic factors which are capable of exerting the influence of the original mother through several back-cross generations (Stebbins, 1958). The observations that there is a high mortality in the F_1 generations and that at least 2 back-crosses are required to produce fertile males in cattle-bison matings (Peters, 1964) seem to suggest that an incompatibility between the cytoplasm of the domestic cow and the male gametic components of the bison are involved in evoking male sterility. The number of back-crosses required may be indicative of the measure of cytoplasmic factors involved in maintaining the integrity of cattle and bison as distinct species.

Chicken-Pheasant Hybrids

Many of the problems associated with mammalian hybrid sterility have received only superficial considerations and have remained poorly understood because of the time and expense involved in conducting such studies on the species concerned. Since the mechanisms responsible for hybrid sterility are basically the same in all higher vertebrates, we have directed our attention to a system that is relatively simple and much less expensive to maintain for hybridization experiments than cattle and bison.

Interspecific and intergeneric hybridizations have been reported to occur quite regularly in class Aves (Gray, 1958). The two avian species we have selected, the domestic chicken (*Gallus domesticus*) and the ring-necked pheasant (*Phasianus colchicus*), belonging to the family *Phasianidae,* have been reported to interbreed occasionally with a pronounced inviability of females (Asmundson and Lorenz, 1967; Cutler, 1918; Shaklee and Knox, 1953). Two-way crosses between Columbian Rock chickens and Ring-Necked pheasants by means of artificial insemination have resulted in a fair percentage of fertility although approximately half of these eggs died within the shell (Bhatnagar, 1968). A majority of the eggs hatched, including 4 females, are living at present (Table 3). Asmundson and Lorenz (1967) had observed that chicken males mated to pheasant females gave a higher fertility than the reciprocal cross. In our experiments the fertility was essentially the same in both crosses; however, hatchability was lower in the chicken male-pheasant female cross. The chicken-pheasant hybrid males exhibit the traits of both parents such as the general plumage of chicken and the feathering of pheasant on the head and neck.

The karyotypes of chicken-pheasant hybrids are presented in Figure 20. Our observations on chicken and pheasant chromosomes are in agreement with those of the previous authors (Krishan and Shoffner, 1966; Stenius *et al.*, 1963) in that the pheasant complements include only one pair of biarmed autosomes compared to the 3 large biarmed autosome pairs in chicken. The Z chromosome of both species are identical, whereas the W chromosomes of the female birds are different (Krishan and Shoffner, 1966). In the hybrids, as in the case of the chicken-quail hybrids (Ohno, 1967), one set of macrochromosomes of each parent can be distinguished in all somatic cells (Fig. 20).

Gonadal differentiation in both male and female hybrids is markedly low as compared with normal birds of the parental species. The hybrid female gonad is totally inactive, while the male gonad exhibits a variety of abnormalities. The size of the hybrid testes and the number of layers of germinal epithelium are much reduced compared to those of chicken. The tubular lumen in the hybrid testes often contained laminated structures derived from the clusters of cells which are similar in morphology to the spermiophages noted in the testes of sterile cattalos. However, no spermatids or spermatozoa were detected in the chicken-pheasant male hybrids even at one year of age.

Electron microscopic studies on the hybrid testes revealed several notable characteristics. A consistent feature was a distortion in the

Table 3: Fertility and Hatchability of the Chicken-Pheasant Crosses and the Gonadal Sex of the F_1 Hybrids Sacrificed

Year	No. of Eggs	Fertile Eggs	Hybrids Hatched ♂	Hybrids Hatched ♀	Incubation Period (in days)	Hybrids Sacrificed	Gonadal Sex
1967	120	5	3	0	22.5	3	3 ♂
1968	335	100	48	4	22.5	2	1 ♂; 1 ♀

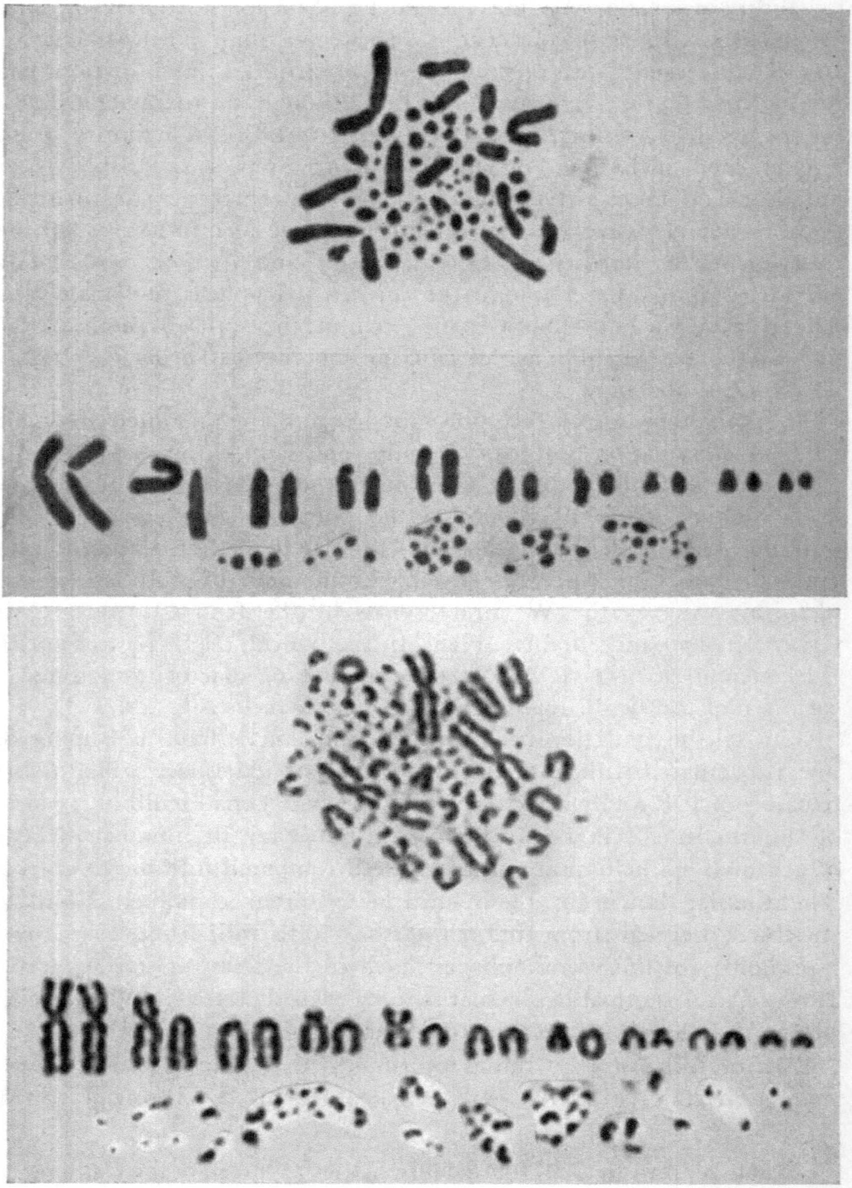

Fig. 20. Metaphase plates and arrays of macrochromosomes of chicken-pheasant hybrids. Note one element each of chicken and pheasant complements. The Z chromosomes of chicken and pheasant are indistinguishable. The W chromosome of the heterogametic female hybrid is derived from the female pheasant parent.

general appearance and in the internal structure of mitochondria (Fig. 21) which appeared to be most pronounced in the Golgi zone. Mitochondria exhibiting various degrees of damage were noted in the Sertoli cells and in those cells which resembled early primary spermatocytes.

Another consistent abnormality was the excessive thickening of the basement membrane and the accumulation of the basement membrane-like material through the germinal epithelium close to the tubular periphery. In some tubules the orientation of the basement lamina was not disturbed, even though the peritubular contractile cell appeared to be traversed by basement membrane-like substance. At other areas, the

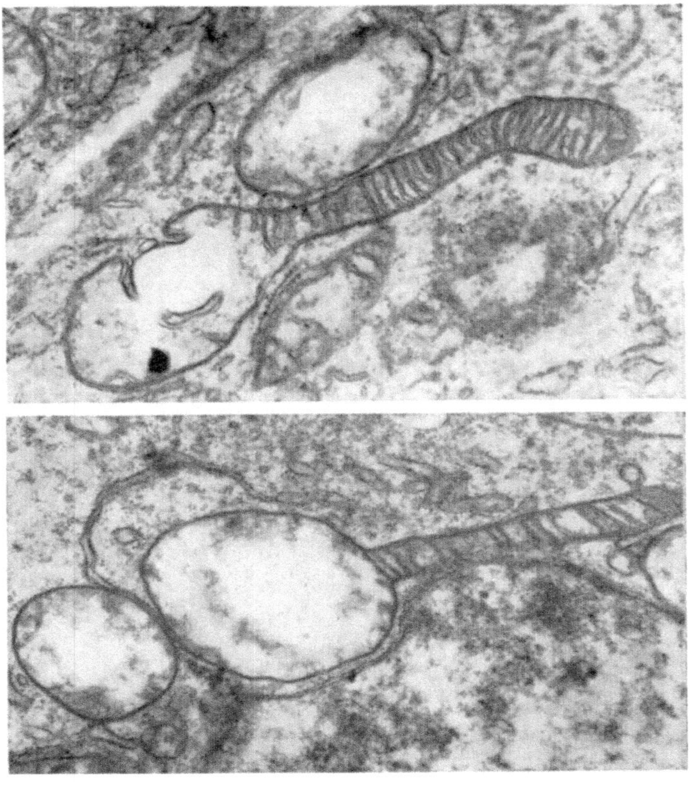

Fig. 21. Mitochondria from chicken-pheasant hybrid testes. Note that both are dilated on one end. The internal cristae are destroyed in most of the mitochondria. A very characteristic feature of these mitochondria is the "tennis racket"-like appearance, with the dilated part showing no internal structure, whereas nondilated area maintains the cristae. The other membrane is distorted in the dilated area in the upper one, whereas it is entire in the lower one. Glutaraldehyde-fixed, osmium post-fixed material stained with uranil acetate × 21,500.

basement lamina was thrown into folds and branches some of which extended amongst the adjacent germ cells (Fig. 22).

With the light microscope, no unequivocal evidence of meiosis was noted in the seminiferous tubules; however, the electron microscopic studies revealed that several layers of primary spermatocytes were present in most tubules (Fig. 23). The ultrastructure of the early meiotic cells was different from that described for normal bivalents (Moses and Coleman, 1964) in that only one of the axial elements of the syn-

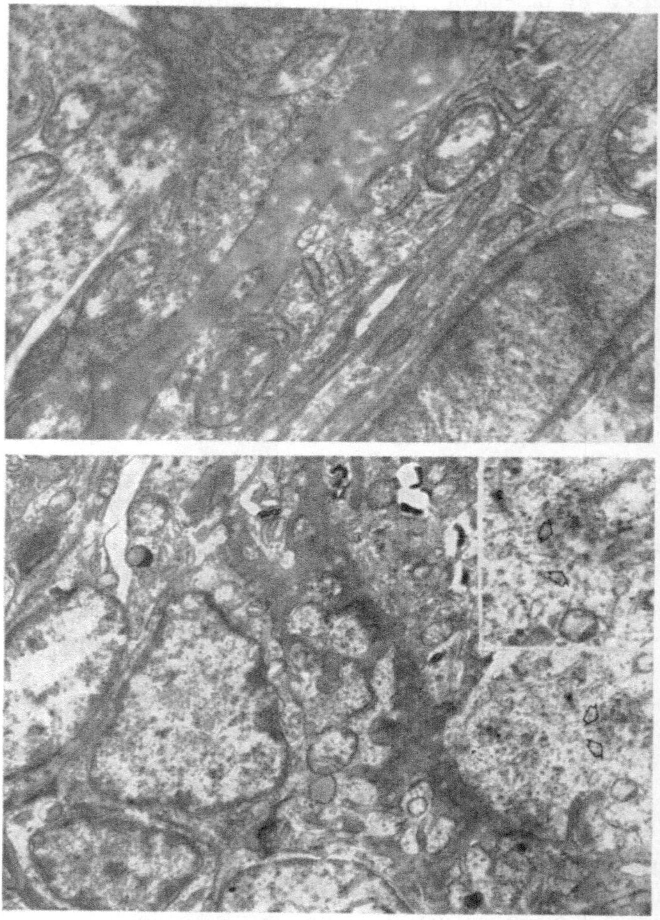

Fig. 22. Basement membrane in chicken-pheasant hybrid testis. Note the extremely thick basement membrane and the parallel streaks of basement membrane-like material towards the peritubular contractile cell in the top figure and the extreme infolding of the basement membrane which extends to the clusters of germ cells in the lower figure. The microtubule-like structures (arrows) in the center right are shown at a higher magnification, above (arrows).

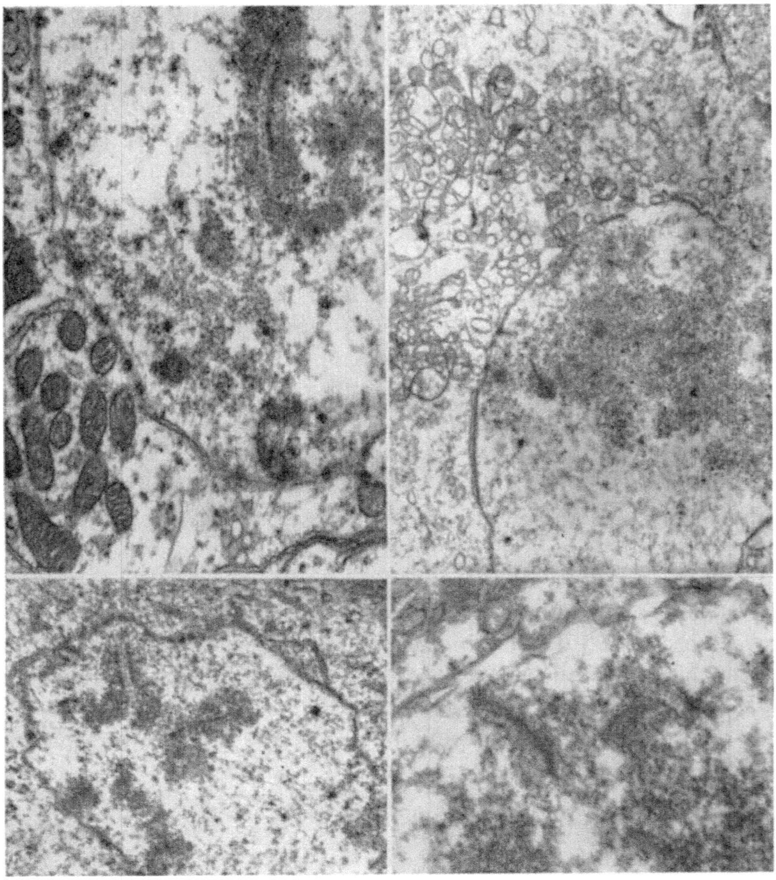

Fig. 23. Ultra structure of the early primary spermatocytes in pheasant (upper left), in chicken (lower left) and in chicken-pheasant hybrids (right). Note the ribbon-like appearance of the synaptinemal complex in pheasant and chicken and the single axial elements in the hybrids at low magnification (upper right) and high magnification (lower right). The synaptinemal complexes have been considered to represent the synapsed homologues during early meiotic prophase. The single axial elements noted in the hybrids may be indicative of lack of homology between chromosomes in the parental species.

aptinemal complex was noted in these cells (Fig. 23). The synaptinemal complexes in chicken and in pheasant were represented by triplets of linear densities similar to those of normal meiosis (Moses and Coleman, 1964).

The degenerative changes in the testes of older birds seemed to be more pronounced which may have been mediated by the thickening of the basement membrane and the resultant interference with the diffusion of nutrients from the intertubular blood vessels. An impairment of diffusion may be causally related to the physical distortions noted in mitochondria, which are known to swell in response to acidification of the medium (Cereijo-Santalo, 1965). Mitochondrial swelling has also been induced with hormones which exhibit estrogenic activity (Dietrich *et al.,* 1961). The type of physical distortions noted in the chicken-pheasant hybrids, whether they were induced by a localized change in intracellular pH or through the accumulation of estrogens produced by the seminiferous tubules (Lacy and Lofts, 1965), will affect the respiratory enzymes. Our studies on two enzymes, glucose-6-phosphate dehydrogenase (G-6-PD) which is common for both glycolysis and respiration and lactate dehydrogenase (LDH) which is unique to glycolysis, showed hybrid patterns in the chicken-pheasant hybrids. It was also revealed that the determinants of G-6-PD may not be located on the Z chromosome in chicken and pheasant since the female progeny in a mating in which chicken was the father failed to exhibit the expected chicken G-6-PD isozyme pattern (Fig. 24). The band pattern for male and female hybrids was identical (Bhatnagar, in press); this would not be the case if this trait was sex linked as suggested by Ohno (1967).

The ultrastructural abnormalities observed in chicken-pheasant hybrid testes may be indicative of a general degenerative change, or they may be associated with the interruption of specific mechanisms which support the normal function of the germinal epithelium. Excessive thickenings, coupled with distortions in the orientation of the basement membranes, have been noted in mammalian testes due to hormonal imbalance (Ross and Grant, 1968) and exposure of testes to irradiation and above-

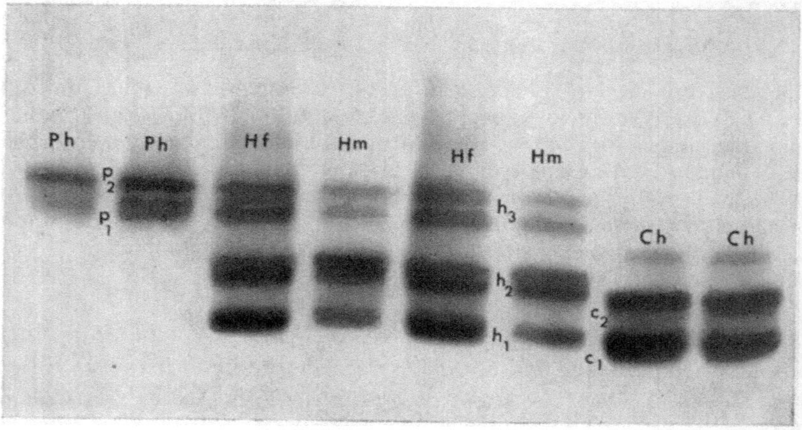

Fig. 24. G6PD zymograms of chicken, pheasant and their hybrids. Ph, Pheasant; p_1 and p_2, Pheasant bands; Ch, Chicken; c_1 and c_2, Chicken bands; Hf, female hybrid; Hm, male hybrid; h_1, h_2 and h_3, hybrid bands. The male and female hybrids show similar patterns indicating that G6PD is not sex-linked in chicken and pheasants.

normal body temperatures (Idänpään-Heikkilä, 1966). In hypophysectomized rats, the basement membrane is thrown into folds apparently in an attempt at reorientation to the reduced surface of the seminiferous epithelium (Ross and Grant, 1968). A collagenous character has been postulated for basement membrane even though histochemically it is composed of mucopolysaccharide and the 640 Å striations characteristic of collagen have not been noted under the electron microscope. Based on the fact that the removal of mucopolysaccharide with pronase will expose the collagen-like fibers in the basement membranes, it has been assumed that the functional integrity of the basement membrane may be dependent upon the ability of collagen molecules to associate with the glycoprotein (Ross and Grant, 1968). The basement membrane of skin has been demonstrated to participate in the guiding of skin regeneration (Ross and Grant, 1968), a phenomenon similar to that which may be operative within the normal seminiferous epithelium. A change in the molecular association of collagen and glycoprotein, probably induced by the altered FSH-LH balance in hypophysectomized rats, may be responsible for the distortion in the peritubular basement membrane. The occurrence of a similar basement membrane distortion in the hybrids could well be indicative of a similar hormonal imbalance.

The change in the structural integrity of the basement membrane may lead to a series of further derangements, especially if this change involves the selective permeability attributed to basement membranes (Ross and Grant, 1968). It has been demonstrated that, in the bovine testes, the normal seminiferous tubule accumulates a large quantity of fluids from the blood vessels of the interstitial tissue (Crabo, 1965). This relatively large volume of fluid consistently passes through the seminiferous tubules along with the spermatozoa into the epididymis, where the water content is resorbed (Crabo and Gustafsson, 1964; Gustafsson, 1964). If structural distortions of the basement membrane affect permeability, these interrelated processes, such as the passage of nutrient fluids from the extra-tubular vascular channel to the seminiferous tubules and the normal "kidney-like" function of the epididymis (Nicander, 1963), may also be adversely affected. At present it is not known whether the basement membrane abnormalities which simulate the conditions noted in hypophysectomized animals (Coombs and Marshall, 1956; Ross and Grant, 1968) are the primary changes that lead to sterility in interspecific hybrids.

A relatively high incidence of teratological defects including an "open skull" was noted in our chicken-pheasant hybrids. The latter abnormality is lethal and is probably associated with the general mechanisms of isolation which are operative in the hybrids at the zygotic, embryonal, prenatal or postnatal stages to prevent introgressive hybridization (Anderson, 1953; Mayr, 1963). A high incidence of chromosome abnormalities has been reported in interspecific hybrids of lower animals (Stebbins, 1958) and their teratological effects have been utilized as indicators of the specific target on which the effect of isolating mechanisms are expressed through the impairment of metabolic processes. The gonads are one of the more consistent target tissue on which these effects are highly pronounced. It is believed that the mechanism which is responsible for reproductive failure is either the same as, or closely related to, that which elicits localized tumors (Gordon, 1937) and teratological abnormalities in the developing embryo (Mayr, 1963; Stebbins, 1958). Our observations on cell cultures of hybrid testis (Basrur and

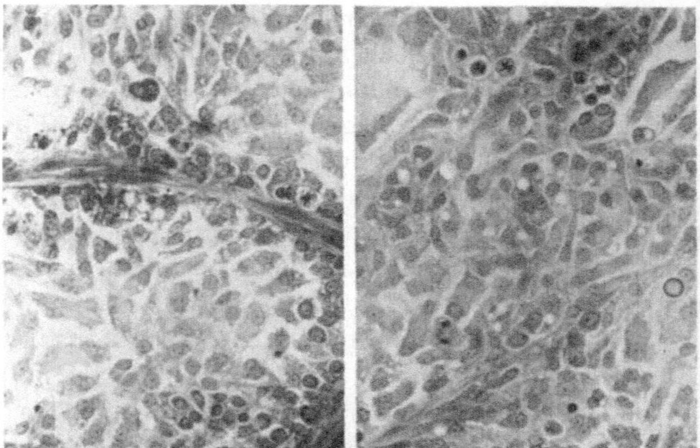

Fig. 25. Hybrid testes cells grown *in vitro* showing attempted reorganization of seminiferous tubules. Note the mitotic abnormalities (upper left) which are shown at higher magnification (lower right). Note the anaphase cell with a "laggard" on the equatorial plate (lower left).

Bhatnagar, 1968) revealed a high percentage of chromosome abnormalities, including spindle malfunction and chromosome breakage (Fig. 25), suggesting that some of the morphologic derangements of the hybrid testis may be related to mitotic abnormalities.

The female hybrids have consistently proved to be infertile and the few females we have observed had follicles devoid of any yolk. The oviduct was consistently filled with a clear liquid simulating a situation commonly noted in sterile mammals (Basrur and Coubrough, 1964).

The thickening of the basement membrane and the spermiophage-like bodies are common to cattalos and chicken-pheasant hybrids. It is possible that the hybrids share a great variety of degenerative features which may not be directly associated with hybrid sterility. Even if such features are not the primary causes of sterility, an understanding of the various changes noted are significant in themselves and should be distinguished as deviations from normal. A thorough understanding of such deviations may bring us close to the problem of infertility. In this respect the avian hybrids would appear to be a very useful experimental model.

Summary

The testicular activities of cattalos with varying percentage of bison parentage were described and the various hypotheses advanced by previous investigators for male sterility in cattle-bison hybrids were discussed.

The histological and ultra structural features of sterile male hybrids resulting from two way crosses between domestic chicken and ring-neck pheasant were described, and the abnormalities noted consistently in the gonads of hybrids were discussed with reference to the mechanisms leading to sterility in interspecific hybrids.

Acknowledgments

This investigation was supported in part by a grant from the National Research Council of Canada.

I am extremely grateful to Dr. H. F. Peters for supplying the cattalos and to Dr. J. P. W. Gilman for his critical evaluation of this manuscript. My thanks are due to Dr. M. K. Bhatnagar for allowing me to use some of his unpublished data and to Dr. H. Kanagawa for his help during this work. My sincere thanks are due to Mr. A. K. Sykes and Mrs. Dorothy Friedman for efficient technical help.

The donation of JEM 6A by the National Cancer Institute of Canada is gratefully acknowledged.

References

Anderson, E.: Introgressive hybridization. Biol. Revs. Cambridge Phil. Soc. *28*:280, 1953.

Asmundson, V. S. and F. W. Lorenz: Hybrids of ring-necked pheasants, turkeys and domestic fowl. Poultry Sci. *36*:1323, 1967.

Bartlett, A. D.: On some hybrid bovine animals bred in the Society's Gardens. Proc. Zool. Soc. Lond. 1884: 399, 1884.

Basrur, P. K. and M. K. Bhatnagar: Testicular cells of chicken-pheasant hybrids. Paper presented at the Canad. Fed. Biol. Soc. Kingston, Ontario, 1968.

——— and R. C. Coubrough: Anatomical and cytological sex of a Saänen goat. Cytogenetics *3*:414, 1964.

——— and J. P. W. Gilman: Testicular activities of cattle-bison hybrids. Am. J. Vet. Res. (in press).

——— and Y. S. Moon: Chromosomes of cattle, bison and their hybrid, the cattalo. Am. J. Vet. Res. *28*:1319, 1967.

Baughman, J. L.: Climate, cattle, and crossbreeding. Beef and milk production in the tropics and subtropics with a bibliography on various phases of the problem. Tex. J. Sci. *3*:253, 1951.

Bedford, J. M.: Effect of environment on phagocytosis of rabbit spermatozoa. J. Reprod. Fertil. *9*:249, 1965.

Benirschke, K.: Sterility and fertility of interspecific mammalian hybrids. In: Comparative Aspects of Reproductive Failure. K. Benirschke, ed. New York: Springer-Verlag, 1967.

———, R. J. Low, L. E. Brownhill, L. B. Caday and J. DeVenecia-Fernandez: Chromosome studies of a donkey-Grevy zebra hybrid. Chromosoma *15*:1, 1964.

———, R. J. Low, M. M. Sullivan and R. M. Carter: Chromosome study of an alleged fertile mare mule. J. Hered. *55*:31, 1964.

Berry, R. O.: Comparative studies on the chromosome numbers in sheep, goat, and sheep-goat hybrids. J. Hered. *29*:343, 1938.

Bhatnagar, M. K.: Glucose-6-phosphate dehydrogenase isozymes of chicken, pheasant and their hybrids. J. Biochem. Genet (in press).

———: Cytogenetic studies on some avian species. Ph.D. Thesis. University of Guelph, 1968.

Boyd, M. M.: A short account on an experiment in crossing the American bison with domestic cows. Rep. Amer. Breed. Ass. *4*:324, 1908.

———: Crossing bison and cattle. J. Hered: *5*:189, 1914.

Cereijo-Santalo, R.: Mitochondrial swelling at acid pH. Canad. J. Biochem. *44*:695, 1966.

Chang, M. C.: Implantation of ferret ova fertilized by mink sperm. J. Exp. Zool. *160*:67, 1965.

——— and J. L. Hancock: Experimental hybridization. In: Comparative Aspects of Reproductive Failure. K. Benirschke, ed. New York: Springer-Verlag, 1967.

Clegg, E. J. and E. W. MacMillan: The uptake of vital dyes and particulate matter by the Sertoli cells in the rat testis. J. Anat. *99*:219, 1965.

Coombs, C. J. F. and A. J. Marshall: The effects of hypophysectomy on the internal testes rhythm in birds and mammals. J. Endocrin. *13*:107, 1956.

Crabo, B.: Studies on the composition of epididymal content in bulls and boars. Acta vet. Scandinavica *6* (Suppl. 5): 1, 1965.
——— and B. Gustafsson: Distribution of sodium and potassium and its relation to sperm concentration in the epididymal plasma of the bull. J. Repr. Fertil. *7*:337, 1964.
Cutler, D. W.: On the sterility of hybrids between the pheasants and the gold Campine fowl. J. Genet. *7*:155, 1918.
Czaja, M. and J. Gasparski: Some red-cell antigens of *Bison bonasus* as compared to red-cell antigens of domestic cattle. Nature *185*:185, 1960.
Deakin, A., G. W. Muir and A. G. Smith: Hybridization of domestic cattle, bison and yak. Report on the Wainwright experiment. Publ. Dep. Agric. Can., No. 479: 30, 1935.
———, G. W. Muir, A. G. Smith and A. S. MacLellan: Hybridization of domestic cattle and buffalo (*Bison americanus*). Progress report of the Wainwright experiment 1935–41. Experimental Farms Service, Department of Agriculture. Dominion of Canada, 10, 1941.
Dietrich, L. S., M. J. Friedland and R. C. Cefaln: Mitochondrial swelling produced by compounds exhibiting estrogenic activity. Proc. Soc. Exp. Biol. Med. *107*:168, 1961.
Gasparski, J.: Studies of the antigenic factors of the blood of Aurochs (*Bison bonasus*). Am. N.Y. Acad. Sci. *97*:285, 1962.
———: Investigations on the blood groups of wisents (*Bison bonasus*) and hybrids in comparison with the blood groups of cattle. Blood Groups of Animals *7*:93, 1964.
Goodnight, C.: My experience with bison hybrids. J. Hered. *5*:194, 1914.
Gordon, M.: The production of spontaneous melanotic neoplasms in fishes by selective matings. Amer. J. Cancer *30*:362, 1937.
Gray, A. P.: Mammalian Hybrids. Techn. Commun. #10 Commonwealth Bureau of Animal Breeding and Genetics. Edinburgh, Bucks, England, Farnham Royal, 1954.
———: Bird Hybrids. Techn. Commun. #13, Commonwealth Bureau of Animal Breeding and Genetics. Edinburgh. Farnham Royal, Bucks, England, 1958.
Gustafsson, B.: The sodium and potassium content in epididymal plasma of bulls with certain reproductive disturbances. Proc. Vth Int. Congr. Anim. Reprod. Trento, *5*:193, 1964.
Haldane, J. B. S.: Sex ratio and unisexual sterility in hybrid animals. J. Genet. *12*:101, 1922.
———: Eland-ox hybrids. Nature (London) *129*:906, 1922.
Hancock, J. L.: Attempted hybridization of sheep and goats. Proc. V. Int. Congr. Anim. Reprod. Trento, *II*:445, 1964.
Heck, H.: Der Bison. A. Ziemsen Verlag. Wittenberg Lutherstadt, 1968.
Idänpään-Heikkilä, P.: Effect of local heat *in vivo* on the fine structure of the basement membrane and the Sertoli cells of the rat testis. Fertil. and Steril. *17*:689, 1966.
Jones, C. J.: Breeding cattalo. Rep. Amer. Breed. Ass. *3*:161, 1907.
Kieffer, N. M. and T. C. Cartwright: Sex chromosome polymorphism in domestic cattle. J. Hered. *59*:35, 1968.
King, J. M., R. V. Short, D. E. Mutton and J. L. Hamerton: The reproductive

physiology of male zebra-horse and zebra-donkey hybrids. In: Comparative Biology of Reproduction in Mammals. I. W. Rowlands, ed. London: Academic Press, 1966.

Koulischer, L. and S. Frechkop: Chromosome complement: A fertile hybrid between *Equus prjewalskii* and *Equus caballus.* Science *151*:93, 1966.

———, J. Tijskens and J. Mortelmans: Mammalian cytogenetics. I. The chromosomes of three species of Bovidae: *Bos taurus, Bison bonasus* and *Cephalopus grimmi.* Acta Zoologica et Pathologica Antverpiensis *43*:135, 1967.

Krishan, A. and R. N. Shoffner: Sex chromosomes in the domestic fowl, *Gallus domesticus,* turkey, *Melagris gallopavo,* and chinese pheasant, *Phasianus colchicus.* Cytogenetics *5*:53, 1966.

Lacy, D. and B. Lofts: Studies on the structure and function of the mammalian testis. I. Cytological and histo-chemical observations after continuous treatment with oestrogenic hormone and the effects of F.S.H. and L.H. Proc. Roy. Soc. B. *162*:188, 1965.

Logan, V. S. and P. E. Sylvestre: Hybridization of domestic beef cattle and buffalo; the 1937–1949 Progress Rep., Anim. Husb. Div., Exper. Farm Service Dept. Agr., Ottawa: 7, 1950.

——— and ———: Hybridization of domestic beef cattle and buffalo. Progress Report, Central Experimental Farm, Canada Dept. Agr., Ottawa, 1950.

Makino, S.: Karyotypes of domestic cattle, zebu and domestic water buffalo. Cytologia *13*:247, 1944.

Mayr, E.: Animal Species and Evolution. Cambridge, Mass.: Harvard University Press, 1963.

Melander, Y.: The mitotic chromosomes of some cavicorn mammals (*Bos taurus* L., *Bison bonasus* L. and *Ovis aries* L.). Hereditas *45*:649, 1959.

Moses, M. J. and J. R. Coleman: Structural pattern and functional organization of chromosomes. In: The Role of Chromosomes in Development. M. Locke, ed. New York: Academic Press Inc., 11–49, 1964.

Nicander, L.: Resorption phenomena in epididymis. Int. J. Fertil. *8*:866, 1963.

Ohno, S.: Sex Chromosomes and Sex-Linked Genes. Berlin: Springer-Verlag, 1967.

———: Cytologic and genetic evidence of somatic segregation in mammals, birds and fishes. In vitro *2*:46–60, 1967.

Peters, H. F.: Experimental hybridization of domestic cattle and American bison. Proc. V. Int. Congr. Anim. Reprod. Trento. VII, 326, 1964.

———, J. A. B. MacArthur, J. E. Lawson, J. A. Vesely and S. Smoliak: Progress report, 1954–1958, Exper. Farm, Manyberries, Alberta, Information Div., Can. Dept. Agr. Ottawa, Ontario, 1958.

——— and K. B. Newbound: Intra-testicular temperature and fertility of bison, cattalo and Herford yearling bulls. Canad. J. Animal Sci. *37*:14, 1957.

Phadke, A. M. and G. M. Phadke: Occurrence of macrophage cells in the semen and in the epididymis in cases of male infertility. J. Reprod. Fertil. *2*:400, 1961.

Ross, M. H. and L. Grant: On the structural integrity of basement membrane. Expt. Cell Res. *50*:277, 1968.

Sasaki, M. S. and S. Makino: A revised study of the chromosomes in somatic cells *in vitro* of *Bos taurus* and *Equus caballus.* J. Hered. *53*:157, 1962.

Shah, S. Z.: A hybrid American bison and domestic cow at the Mysore zoo. J. Bombay Nat. Hist. Soc. *37*:954, 1935.

Shaklee, W. E. and C. W. Knox: Hybridization of pheasant and fowl. Poultry Sci. *32*:923, 1953.

Shaw, D. H. and J. R. Patel: Demonstration of antigenic difference between the American Bison (*Bison bison*) and domestic cattle (*Bos taurus*). Nature *196*:498, 1963.

Stebbins, G. L.: The inviability, weakness and sterility of inter-specific hybrids. Advances in Genetics *9*:147, 1958.

Stenius, C., L. C. Christian and S. Ohno: Comparative cytological study of *Phasianus colchicus, Melagris gallopavo,* and *Gallus domesticus.* Chromosoma (Berlin) *13*:515, 1963.

Sylvestre, P. E., V. S. Logan and G. W. Muir: Hybridization of domestic cattle and the bison. Canada: Department of Agriculture, 7, 1948.

Tjioe, D. Y. and E. Steinberger: Spermiophages in human testes. Fertility and Sterility *18*:807, 1967.

Vilar, O., M. I. Perez del Cerro and R. E. Mancini: The Sertoli cell as a "bridge cell" between the basal membrane and the germinal cells. Exp. Cell Res. *27*:158, 1962.

Warwick, B. L., R. O. Berry and W. R. Horlacher: Cytological and hybridization studies with sheep and goats. Rep. Texas Agric. Exp. Sta. 30, 1934.

EXPERIMENTAL HYBRIDIZATION AND CHROMOSOMES OF HYBRIDS

M. C. Chang, S. Pickworth and R. W. McGaughey
Worcester Foundation for Experimental Biology, Shrewsbury, Massachusetts

Experimental hybridization between the ferret and mink, goat and sheep, and between the domestic rabbit and some other members of the *Leporidae* was discussed at a previous meeting (Chang and Hancock, 1967). In each case successful cross-fertilization was followed by prenatal death of the hybrid. Chromosome studies and sterility of hybrids which survive to adulthood were discussed by Benirschke (1967) and King (1967).

Some results obtained in this field in our laboratory during the past two years are reported in the present paper and progress in experimental mammalian hybridization is reviewed.

Fertilization and Prenatal Development

Peromyscus

The genus *Peromyscus* is divided into many species and subspecies, some of which can interbreed with varying degrees of success and are assigned to the same species group. The chromosome number, 2n = 48, is the same for all *Peromyscus* that have been examined, but karyotypic polymorphism exists both between and within species (Hsu and Arrighi, 1966; Sparkes and Arakaki, 1966; Singh and McMillan, 1966).

Several laboratory studies of hybridization between *Peromyscus maniculatus* and *Peromyscus polionotus* subspecies have been made, using natural mating. Dawson (1965, 1966) recently described an interesting pattern of size inheritance in the hybrid offspring which suggested an interaction between embryonic genotype and a maternal growth factor.

His results also confirmed earlier observations that the cross is more successful when *P. maniculatus* is the female parent (Watson, 1942; Liu, 1953). In the reciprocal cross there appeared to be a high incidence of fetal death in the second half of pregnancy.

Hybridization has recently been attempted in our laboratory between *P. maniculatus bairdii* and *P. leucopus noveboracensis*. The latter belongs to a different species group whose members are inter-fertile in the laboratory (Dice, 1937). Interbreeding between sympatric *P. leucopus* subspecies in the wild occurs only in certain areas (McCarley, 1954). Subspecies of *P. maniculatus* are often sympatric with those of *P. leucopus*, and intermediate morphological types are found, but Dice (1933) obtained no litters from 69 heterospecific pairs in the laboratory. Using induced ovulation and artificial insemination, it has been found that fertilization is successful in both directions of hybridization between these two species, and development of hybrid eggs into blastocysts and, in one animal, to midterm embryos, has been observed in *P. leucopus* females. Some young females have mated with males of the other species under our laboratory conditions, and live embryos were recovered from one *P. maniculatus* female on the eleventh day after mating with a *P. leucopus* male.

Goat–Sheep

Several papers concerning goat × sheep hybridization have appeared in the past two years. Successful development of hybrids in goats has been achieved up to the beginning of the sixth week (Hancock, McGovern and Stamp, 1968) or to approximately two months of pregnancy (Alexander, Williams and Bailey, 1967). Death of the hybrid fetus is not likely to be the direct result of gross chromosomal abnormalities. Alexander and co-workers (1967) found hemolytic antibodies against paternal red blood cells in the serum of female goats after the time of cotyledonary attachment of hybrid fetuses. Whether this occurs during normal pregnancy is not known, but they considered that the conditions of fetal death were consistent with hemolytic disease, developing as a result of an imperfect placental barrier. Both the above group and Hancock *et al.* (1968) observed abnormally large volumes of fetal fluid during hybrid pregnancy. Unusually extensive invasion of trophoblast was found in goats with 5–6-week-old dead fetuses (Hancock *et al.*, 1968).

In the reciprocal cross, where the sheep is the maternal parent, the incidence of fertilization is very low (Bowerman and Hancock, 1963). Development of hybrids, in one case to the 45th day of gestation, has been reported in sheep receiving hybrid eggs by transfer (Hancock *et al.*, 1968).

Ferret–Mink

In recent experiments in our laboratory the use of large numbers of sperm for insemination has greatly increased the proportion of ferret eggs fertilized by mink sperm. A high percentage of fertilized eggs were implanted but few fetuses remained alive after the third week of pregnancy (Chang, 1968). The oldest living hybrid, so far, was recovered 27 days after insemination. Results of chromosome analyses of 23-day-old living hybrids are reported further on in this paper.

In an attempt to increase immunological tolerance toward the development of hybrid embryos, 2 mg cortisone acetate, a suspension of mink spleen cells, or 50r of X-radiation was administered to ferrets (four females per treatment group) 14, 16 and 18 days after insemination with mink sperm. The ferrets were examined 23–27 days after insemination. The proportion of implantation sites containing fetal material, alive or dead, appeared to be higher in the groups treated with cortisone acetate or mink spleen cells as compared with control or irradiated animals (28–29% vs. 7–9%); but the difference was not statistically significant (Table 1).

Table 1: Insemination of Ferrets with Mink Sperm

Treatment of Ferrets	Ferret Females Used	Insem. with Sperm of Mink	Day of Exam.	Total Implantation Sites	Embryos
Injection of 2 mg	F 1	M1	25	8	2 dead
cortisone	F 2	M1	25	7	2 dead ?
acetate on	F11 *	M3	26	9	3 living ?
days 14, 16 & 18	F 6	M2	27	5	1 dead
Spleen cells	F13	M3	23	14	3 living
of mink on	F 3	M1	25	7	4 dead
days 14, 16 & 18	F12	M3	25	0	0
	F 7	M2	27	7	1 dead ?
Irradiation at	F15	M3	24	12	0
50r on	F 4	M1	25	11	2 dead
days 14, 16 & 18	F 9	M2	26	1	0
	F 8	M2	27	2	0
Control	F14	M3	24	8	1 living
	F16	M3	25	9	0
	F 5	M1	26	8	1 living
	F10 *	M2	27	8	1 living ?

* Laparotomy first; then killed 44–45 days after insemination, embryos degenerated.

The reciprocal cross, between the mink female and ferret male, has not been achieved. Female mink were inseminated with ferret sperm via the uterine lumen or the ovarian capsule, but no fertilized eggs were observed (Chang, 1968).

Rabbit—Hare

Three intergeneric crosses have been attempted with the domestic rabbit, *Oryctolagus cuniculus.* Only 30% of rabbit eggs were fertilized by sperm of the cottontail, *Sylvilagus transitionalis* (Chang, 1960), but fertilization was virtually 100% successful after insemination with sperm of the common hare, *Lepus europaeus* (Adams, 1957; Chang and Adams, 1962), and snowshoe hare, *L. americanus* (Chang, Marston and Hunt, 1964). The reciprocal cross was tested in the common hare from which 2-cell hybrid eggs were recovered (Treus and Steklenjov, 1963), and in the snowshoe hare, where only one of ten eggs was fertilized by domestic rabbit sperm (Chang, 1965).

When rabbits were examined 6–7 days after insemination with snowshoe hare sperm, 5 degenerated morulae and 10 early blastocysts, resembling 3–4 day rabbit blastocysts, were found in two animals (Chang *et al.*, 1964). Four rabbits were examined 9–17 days after insemination, but only one small uterine swelling was observed. In order to determine whether implantation of the hybrid does occur and to obtain material for chromosome studies, a total of 23 rabbits were inseminated with snowshoe hare sperm in 1967 and 1968. The results are presented in Table 2. Almost all the rabbit eggs were fertilized by the hare sperm but only 44% of the eggs had developed into early blastocysts in 7 animals examined on days 6 and 7. There were no implantation sites in 5 animals examined on days 11–15.

Formation of the hybrid blastocyst occurred 5–6 days after insemination, 1–2 days later than in the parent species; a comparable delay was observed in rabbits inseminated with sperm of the common hare (Adams, 1957; Chang and Adams, 1962).

Chromosome Studies

Hybrids Reaching Adult Life

Recent karyological studies of the mule (hybrid resulting from the cross, *Equus caballus* ♀ × *E. asinus* ♂) have shown conclusively that its karyotype consists of the donkey haploid (n = 31) plus the horse haploid (n = 32) chromosome sets (Benirschke, Brownhill and Beath, 1962; Trujillo, Stenius, Christian and Ohno, 1962). King, Short, Mutton and Hamerton (1966) discussed the karyology of three zebroids (*Equus cabal-*

Table 2: Insemination of Rabbits with Snowshoe Hare Sperm

Day of Exam.	No. of Rabbits	Total c.l.	Total Eggs Recovered	Total Eggs Fertilized	Stage of Eggs
2	6	73	67 (T.*) (92%)	63 (94%)	15–20 cells
3	3	45	27 (T. U.) (60%)	27 (100%)	12–24 + cells
4	4	37	22 (U.*) (60%)		14 = normal morulae 8 = degenerated morulae
6	3	36	30 (U.) (83%)		8 = early blastocysts 5 = degenerated early blastocysts 17 = degenerated morulae
7	2	26	24 (U.) (92%)		7 = degenerated early blastocysts 17 = degenerated morulae
11–15	5	54	16 (U.) (30%)		No implantations 9 = degenerated early blastocysts 7 = degenerated morulae

* T. = tubal; U. = uterine.

lus ♀ × *E. grevyi* ♂, *E. asinus* ♀ × *E. burchelli böhmi* ♂, and *E. zebra zebra* ♀ × *E. asinus* ♂), the chromosome numbers of which were intermediate between those of the parent species. Another equine hybrid (*Equus caballus* ♀ × *E. prjewalskii* ♂) was recently reported by Koulischer and Frechkop (1966) to possess a karyotype (2n = 65) intermediate between those of its parents. Short (1967) listed the chromosome numbers for several of the Equidae and their hybrids.

Other viable hybrids which recently have been examined cytologically and which have been reported to possess karyotypes intermediate between the parent species include *Thalarctos maritimus* (polar bear) × *Ursus arctos middendorffi* (Alaskan brown bear), *Lama pacos* (Alpaca) × *Lama vicugna* (vicuna) (Benirschke, 1967), and *Mesocricetus newtoni* (Romanian hamster) × *Mesocricetus auratus* (Syrian hamster) (Raicu and Bratosin, 1968).

Hybrids Which Die During Pregnancy

The early report that cells of the goat × sheep hybrid embryo contain a chromosome number intermediate between the number in the parent cells (Berry, 1938) has recently been reexamined. Hancock and Jacobs (1966) found that of 135 cells from 10 hybrid embryos, 125 (93%) contained 57 chromosomes, the intermediate number between the sheep (n = 27) and the goat (n = 30). Similarly, Ilbery, Alexander and Williams (1967) counted 57 chromosomes in each of 21 cells analyzed from four goat × sheep hybrid embryos recovered at 2 months of gestation. These recent investigations led to the conclusion first proposed by Berry (1938) that death of the goat × sheep hybrid embryo, which usually occurs between the first quarter and the midpoint of gestation, cannot be explained in terms of chromosomal abnormalities.

The present studies were undertaken to determine whether or not karyological abnormalities accompanied the death of implanted ferret ♀ × mink ♂ hybrid embryos and/or of domestic rabbit ♀ × snowshoe hare ♂ eggs prior to implantation in the uterus.

Ferret (*Mustela furo*) ♀ × Mink (*Mustela vison*) ♂

Karyotypes of the parent species were verified from squash preparations of adult testis (mink) and embryonic limb-bud (ferret). The mink karyotype (2n = 30) was identical to that reported by Fredga (1961) but differed in the relative size of the Y-chromosome (Fig. 1A) from that reported by Hsu and Benirschke (1968). The ferret karyotype (2n = 40) was similar to that determined by Basrur (1966) and Hsu and Benir-

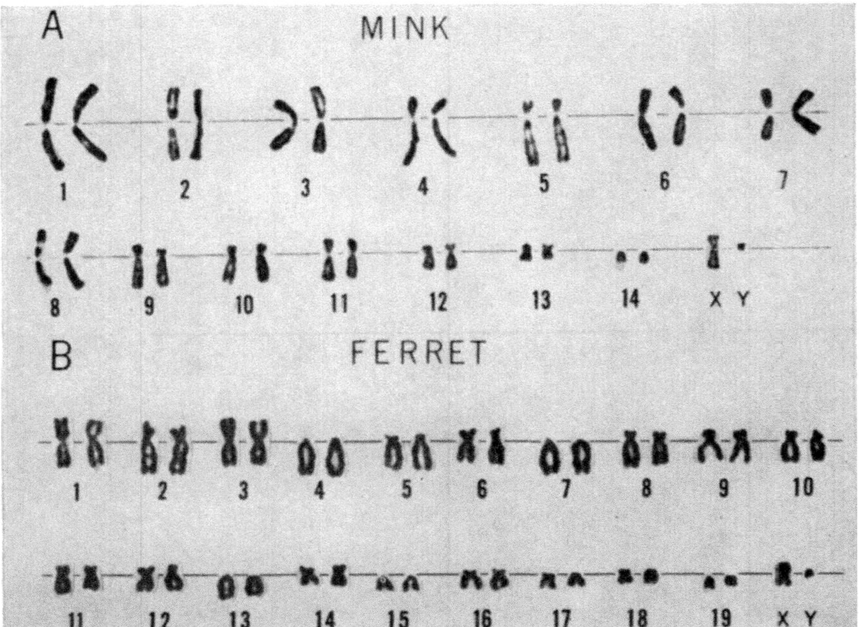

Fig. 1. A. Male mink karyotype (2n = 30). B. Male ferret karyotype (2n = 40). Chromosomes arranged in order of decreasing arm length. (× ca. 1,800) *Note:* The line passes through the centromere of the chromosomes in all figures.

schke (1967). The male displayed a very small metacentric Y-chromosome similar to that observed for the mink in this study (Fig. 1B).

The karyotype of the hybrid was established by means of a squash technique. Two living ferret × mink hybrids, recovered from the uterus of a ferret on day 23 of pregnancy, were cut into small pieces and placed in 0.7% sodium citrate solution for 30 minutes. Following the hypotonic treatment the pieces of embryonic tissue were fixed in 45% acetic acid (aqueous) and squashed under a cover-slip as described by Ohno (1965). The squash preparations were hydrolyzed in 1N HCl and stained in basic fuchsin.

Of 30 well-spread, intact metaphase plates counted from each hybrid embryo, 27 (90%) displayed exactly 35 chromosomes. The chromosomes were found to correspond to mink haploid plus ferret haploid karyotypes (Figs. 2A and B). Both embryonic karyotypes contained a single, very

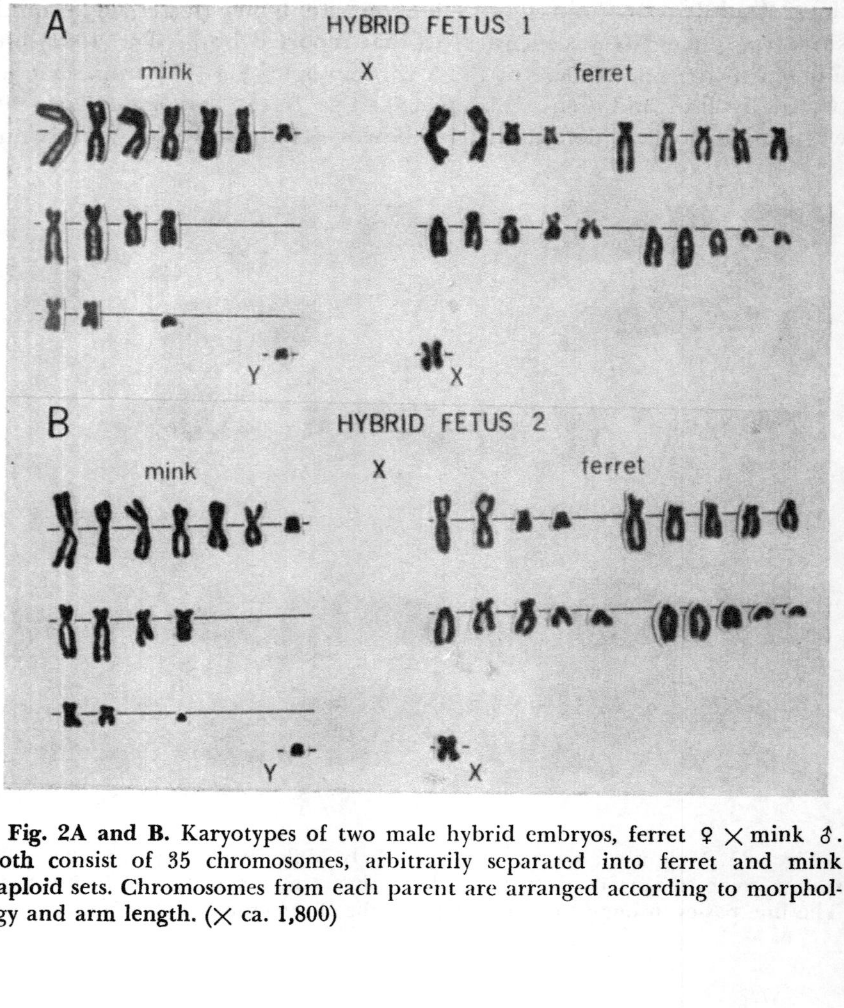

Fig. 2A and B. Karyotypes of two male hybrid embryos, ferret ♀ × mink ♂. Both consist of 35 chromosomes, arbitrarily separated into ferret and mink haploid sets. Chromosomes from each parent are arranged according to morphology and arm length. (× ca. 1,800)

small, metacentric chromosome, corresponding to the mink Y-chromosome, and thus were considered to be males. The relative size of the Y-chromosome in the cells of hybrid embryos corresponded closely with that of the mink karyotype described by Hsu and Benirschke (1968). Since the majority of ferret × mink hybrid embryos are dead by, or shortly after, the midpoint of gestation (normal gestation in the ferret is 42 days), and since the two living embryos examined in this study displayed complete chromosome sets in the majority of cells, it is suggested that in these hybrid embryos death is not necessarily caused by the occurrence of gross chromosomal abnormalities.

Rabbit (*Oryctolagus cuniculus*) ♀ × Snowshoe Hare (*Lepus americanus*) ♂

The hare karyotype (2n = 48) was examined from squashes of adult cornea, and was found to be similar (42 metacentrics, submetacentrics or subtelocentrics, 4 acrocentrics or telocentrics plus the sex chromosome pair) to those described previously for two other members of the *Leporidae* (Hsu and Benirschke, 1967). The rabbit karyotype (2n = 44) was examined from squashes of embryonic lung tissue, and did not differ from that reported by several earlier investigators (Figs. 3A and B).

Due to the early degeneration of rabbit × hare hybrid eggs, chromosomes could be examined only during preimplantation stages of development. An air-dry technique (McGaughey, 1968) was used to examine the chromosomes of hybrid eggs recovered 2 days after insemination. The eggs, flushed from the rabbit oviducts, were treated first with pronase in 2% sodium citrate to remove the mucin coat, then for 5 to 7 minutes in 0.7% sodium citrate, and were transferred to a clean microscope slide. Flattening of the cells was accomplished by dropping 15% acetic acid (in ethanol) over the eggs and allowing it to evaporate at room temperature. The preparations were stained with basic fuchsin. Six-day hybrid blastocysts were recovered from the uterus, treated in hypotonic sodium citrate, squashed and stained with basic fuchsin.

Chromosome counts were made on cells from 12 2- or 6-day hybrid eggs. The majority of cells (5 out of 7) from 2-day eggs contained from 43 to 46 chromosomes, while 17 out of 25 blastocyst cells contained less than 40 chromosomes. Figure 4A shows a karyogram constructed from a cell containing 44 chromosomes. The chromosomes were divided into a complete haploid rabbit karyotype and an incomplete haploid hare karyotype. Figure 4B represents the chromosomes from a cell of a 6-day blastocyst and shows the cell to contain a total of 22 chromosomes from which it was possible to construct a rabbit haploid karyotype. A similar karyotype was prepared from another cell of the same embryo.

Although at the present time only a few of these hybrid eggs have

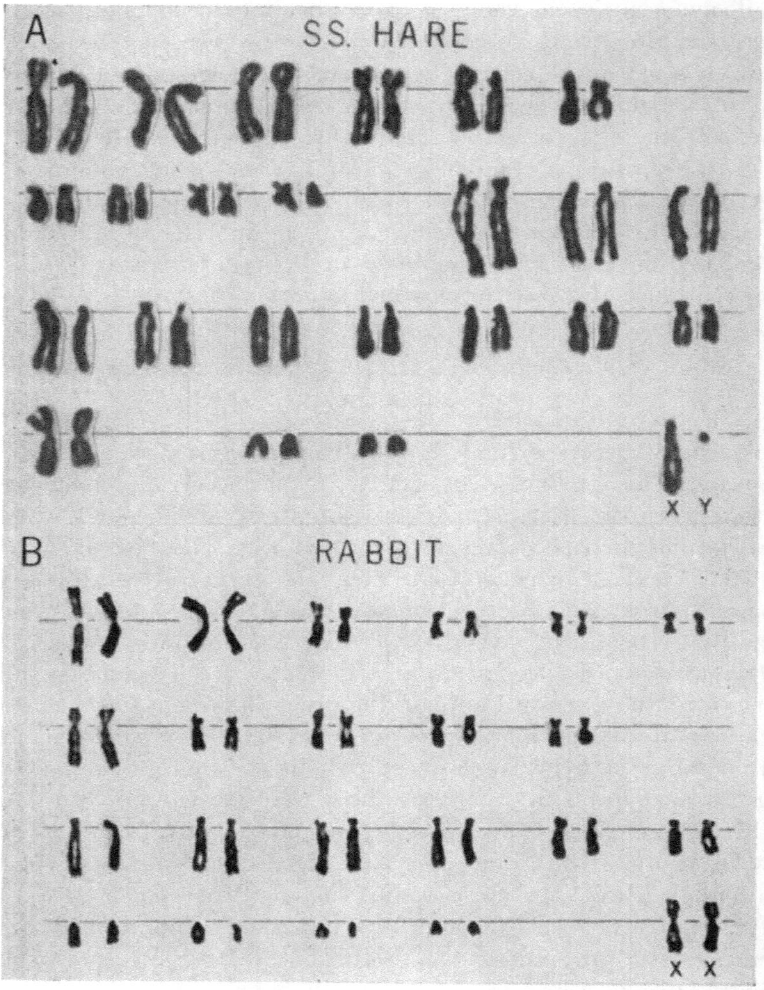

Fig. 3. A. Male Snowshoe hare karyotype (2 = 48). **B.** female rabbit karyotype (2n = 44); chromosomes arranged according to morphology and arm length. (× ca. 1,800)

been analyzed, due to their short developmental life and the small number of cells present, some general implications might be derived from the present results. If the paternal or maternal chromosomes of the zygote were completely lost at the first cleavage division, the expected result would be an egg containing only haploid cells. Two possible haploid 2-day eggs were observed, although only one metaphase was counted from each egg. No 6-day blastocysts were totally haploid, although the majority of cells from these eggs were hypodiploid or haploid for the rabbit (n = 22) or hare (n = 24). In two blastocysts, from which 9 and 11 cells were counted respectively, the largest number of cells (10) were found to contain 22 chromosomes, the rabbit haploid number, although 40–46 chromosomes were present in some cells from each blastocyst (6 cells in all). Thus, in these embryos loss of chromosomes probably did not occur at least until after the first cleavage.

It is suggested that chromosome loss is associated with the early degeneration of these hybrid eggs, although much more information is required to determine whether the observed chromosomal abnormalities are the result of, or the cause of, cellular degeneration at early stages of development.

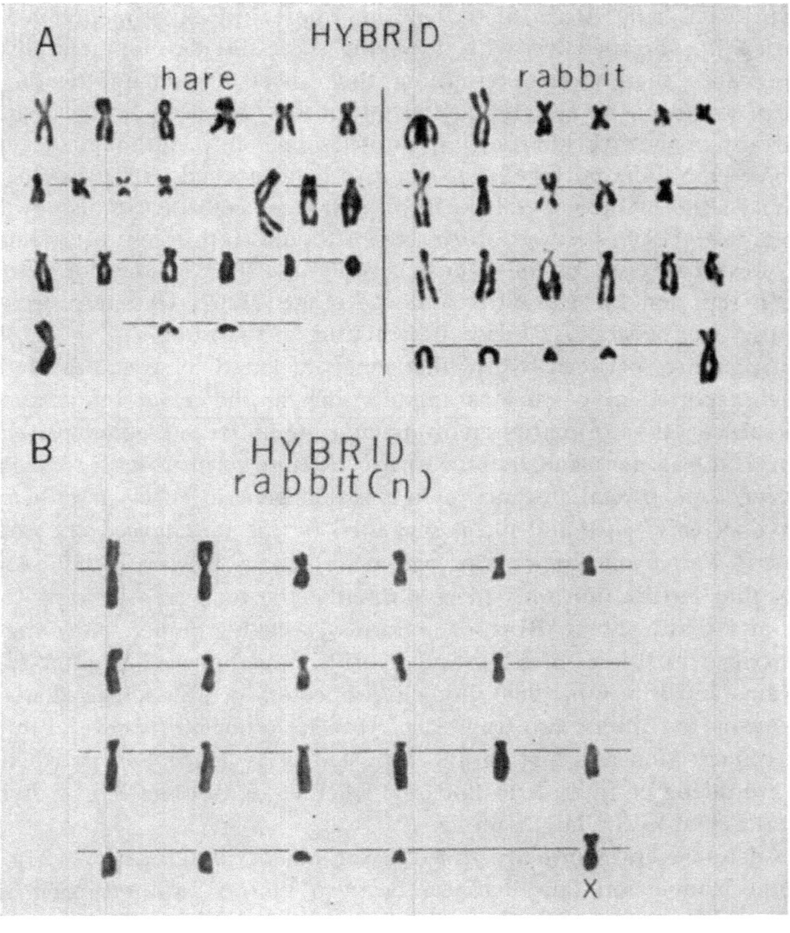

Fig. 4. Karyotypes of cells from domestic rabbit ♀ × Snowshoe hare ♂ hybrid eggs. **A.** A cell containing 44 chromosomes which were arbitrarily divided into a complete rabbit haploid set plus an incomplete hare haploid set (× ca. 1,800). **B.** A cell containing 22 chromosomes which were arbitrarily arranged as a rabbit haploid set. (× ca. 2,100)

Discussion

Cellular material from very different species has been combined to form viable polyploid somatic cell hybrids in tissue culture (Harris *et al.*, 1966). We are concerned, however, with the union of male and female gametes between different species and the development of such diploid zygotes. A variety of isolating mechanisms operates under natural conditions to prevent sexual hybridization between species. Of these, geographical, ethological and mechanical barriers can be overcome by artificial insemination. Many additional factors, however, might be expected to prevent experimental interspecific fertilization and development of hybrid offspring. Much of the current interest in experimental hybridization is concerned less with achieving successful development through pregnancy than with determining the causes of prenatal death. The primary reason for developmental failure has not been firmly identified for any of the hybridizations attempted up to now, although degeneration of rabbit-hare hybrid eggs seems to be associated with gross chromosomal abnormalities. Fetal death of goat-sheep hybrids studied by Hancock *et al.* (1968) occurred during the fifth and sixth weeks of pregnancy. However, survival of the hybrid beyond this stage of development has been reported by Alexander and co-workers (1967) who recovered live fetuses from goats 57–63 days after mating to the ram.

Differences between hybrid littermates in length of prenatal life have been reported in polytocous animals such as the ferret but cannot be identified yet as primarily environmental or individual genotypic effects. Survival of some members of a hybrid litter to term makes investigation easier. One special instance of a possible genetic effect in either prenatal or early postnatal life is suggested by the unbalanced sex ratio of hybrid *Peromyscus* at weaning age, described by Dawson (1965). Assuming that fertilization took place normally, the male hybrid appeared to be at a disadvantage. Birth of abnormally small numbers of young was described in studies of reciprocal hybridization between hamster species (Raicu and Bratosin, 1968), but the proportion of eggs fertilized was not known. Four litters, two from each cross, were born after several months of cohabitation and mating and consisted of 10 males and two females. It would be of interest to find out whether or not the female hybrids are less viable in a large sample.

Adequate and normally timed development and function of a placental connection (and barrier) between mother and offspring is an essential part of mammalian development, and symptoms of placental abnormality have been observed during hybrid pregnancy in ferrets, goats and sheep. Increased antigenic dissimilarity between mother and offspring results in a larger size of the placenta (Billington, 1964; James,

1965) and fetus (James, 1968) in the laboratory mouse. The importance of such factors in other species, however, is not known. Hancock and co-workers observed abnormally extensive trophoblast invasion in goats with dead hybrid fetuses, but the sequence of events has not been determined. The evidence for hemolytic disease in goat-sheep fetuses (Alexander *et al.,* 1967) emphasizes the importance of the placental barrier for successful pregnancy. The hybrid fetus may be doubly vulnerable to immunological attack of this kind, with both an increased chance of placental abnormality and an exaggerated antigenic difference from its mother. Nevertheless, the possibility that the pre-implantation death of rabbit-hare hybrid eggs may be due to chromosome loss during cleavage emphasizes that incompatibility between embryo and maternal environment is not the only factor to account for the death of hybrids.

Summary

Recent experiments have confirmed that rabbit eggs fertilized by snowshoe hare sperm can develop, at a reduced rate, into early blastocysts but do not implant. Severe chromosome loss was observed in the hybrid eggs.

Insemination of ferrets with large numbers of mink sperm resulted in fertilization and implantation of a high proportion of eggs, but few hybrid fetuses lived beyond the third week of pregnancy. The oldest live hybrid was recovered on the 27th day after insemination. The chromosome number of two live 23-day old fetuses was intermediate between those of the parent species.

The longest period of prenatal survival reported from studies of goat-sheep hybridization is approximately two months. Immunological reactions between mother and fetus have been suggested as a possible cause of death. The chromosome number of the hybrids has been verified as intermediate between the goat and sheep.

Recent karyological and morphological studies of hybridization in these and other species are briefly discussed.

Acknowledgment

This work was supported by USPHS grants HD 03472 and HD 03003.

References

Adams, C. E.: An attempt to cross the domestic rabbit (*Oryctolagus cuniculus*) and hare (*Lepus europaeus*). Nature *180:*853, 1957.

Alexander, G., D. Williams and L. Bailey: Natural immunization in pregnant goats against red blood cells of their sheep × goat hybrid foetuses. Aust. J. Biol. Sci. *20:*1217, 1967.

Basrur, P. K.: The somatic chromosomes of the ferret. J. Hered. *57*:110, 1966.

Benirschke, K.: Sterility and fertility of interspecific mammalian hybrids. In: Comparative Aspects of Reproductive Failure. K. Benirschke, ed. New York: Springer-Verlag, p. 218, 1967.

Benirschke, K., L. E. Brownhill and M. M. Beath: Somatic chromosomes of the horse, the donkey and their hybrids the mule and the hinny. J. Reprod. Fert. *4*:319, 1962.

Berry, R. O.: Comparative studies on the chromosome numbers in sheep, goats and sheep/goat hybrids. J. Hered. *29*:343, 1938.

Billington, W. D.: Influence of immunological dissimilarity of mother and foetus on size of placenta in mice. Nature *202*:317, 1964.

Bowerman, H. R. L. and J. L. Hancock: Sheep × goat hybrids. J. Reprod. Fert. *6*:326, 1963. (Abstr.)

Chang, M. C.: Fertilization of domestic rabbit (*Oryctolagus cuniculus*) ova by cottontail rabbit (*Sylvilagus transitionalis*) sperm. J. Exp. Zool. *144*:1, 1960.

———: Artificial insemination of Snowshoe hares (*Lepus americanus*) and the transfer of their fertilized eggs to the rabbit (*Oryctolagus cuniculus*). J. Reprod. Fert. *10*:447, 1965.

———: Reciprocal insemination and egg transfer between ferrets and mink. J. Exp. Zool. *168*:49, 1968.

Chang, M. C. and C. E. Adams: Fate of rabbit ova fertilized by hare spermatozoa. XXV Anno dalla Fondazione dello Instituto Sperimentale Italiano L. Spallanzani. 186, 1962.

Chang, M. C. and J. L. Hancock: Experimental hybridization. In: Comparative Aspects of Reproductive Failure. K. Benirschke, ed. New York: Springer-Verlag, p. 206, 1967.

Chang, M. C., J. H. Marston and D. M. Hunt: Reciprocal fertilization between the domesticated rabbit and the Snowshoe hare with special reference to insemination of rabbits with an equal number of hare and rabbit spermatozoa. J. Exp. Zool. *155*:47, 1964.

Dawson, W. D.: Fertility and size inheritance in a *Peromyscus* species cross. Evolution *19*:44, 1965.

———: Postnatal development in *Peromyscus maniculatus-polionotus* hybrids. I. Developmental landmarks and litter mortality. Ohio J. Sci. *66*:518, 1966.

Dice, L. R.: Fertility relationships between some of the species and subspecies of mice in the genus *Peromyscus*. J. Mammal. *14*:298, 1933.

———: Fertility relations in the *Peromyscus leucopus* group of mice. Contr. Lab. Vert. Biol. U. Mich. *4*:1, 1937.

Fredga, K.: The chromosomes of the mink. J. Hered. *52*:91, 1961.

Hancock, J. L. and P. A. Jacobs: The chromosomes of goat × sheep hybrids. J. Reprod. Fert. *12*:591, 1966.

Hancock, J. L., P. T. McGovern and J. T. Stamp: Failure of gestation of goat × sheep hybrids in goats and sheep. J. Reprod. Fert. Supp. *3*:29, 1968.

Harris, H., J. F. Watkins, C. E. Ford and G. I. Schoefl: Artificial heterokaryons of animal cells from different species. J. Cell Sci. *1*:1, 1966.

Hsu, T. C. and F. E. Arrhighi: Chromosomal evolution in the genus *Peromyscus* (Cricetidae, Rodentia). Cytogenetics *5*:355, 1966.

Hsu, T. C. and K. Benirschke: An Atlas of Mammalian Chromosomes. Vol. 1. New York: Springer-Verlag, 1967.

———: An Atlas of Mammalian Chromosomes. Vol. 2. New York: Springer-Verlag, 1968.

Ilbery, P. L. T., G. Alexander and D. Williams: The chromosomes of sheep × goat hybrids. Aust. J. Biol. Sci. *20:*1245, 1967.

James, D. A.: Effects of antigenic dissimilarity between mother and foetus on placental size in mice. Nature *205:*613, 1965.

———: Some effects of immunological factors on gestation on mice. J. Reprod. Fert. *14:*265, 1967.

King, J. M.: The sterility of two rare equine hybrids. In: Comparative Aspects of Reproductive Failure. K. Benirschke, ed., New York: Springer-Verlag, p. 235, 1967.

King, J. M., R. V. Short, D. E. Mutton and J. L. Hamerton: The reproductive physiology of male zebra-horse and zebra-donkey hybrids. In: Comparative Biology of Reproduction in Mammals. I. W. Rowlands, ed. New York: Academic Press, p. 511, 1966.

Koulischer, L. and S. Frechkop: Chromosome complement: A fertile hybrid between *Equus prjewalskii* and *Equus caballus.* Science *151:*93, 1966.

Liu, T. T.: Prenatal mortality in *Peromyscus* with special reference to its bearing on reduced fertility in some interspecific and intersubspecific crosses. Contr. Lab. Vert. Biol. U. Mich. *60:*1, 1953.

McCarley, W. H.: Natural hybridization in the *Peromyscus leucopus* species group of mice. Evolution *8:*314, 1954.

McGaughey, R. W.: Karyological studies on maturation, fertilization and cleavage in rodent eggs. Ph.D. Dissertation, Boston University, 1968.

Ohno, S.: Direct handling of germ cells. In: Human Chromosome Methodology. J. J. Yunis, ed. New York: Academic Press, 1965.

Raicu, P. and S. Bratosin: Interspecific reciprocal hybrids between *Mesocricetus auratus* and *M. newtoni.* Genet. Res. *11:*113, 1968.

Short, R. V.: Reproduction. Ann. Rev. Physiol. *29:*373, 1967.

Singh, R. P. and D. B. McMillan: Karyotypes of three subspecies of *Peromyscus.* J. Mammal. *47:*261, 1966.

Sparkes, R. S. and D. T. Arakaki: Intrasubspecific and intersubspecific chromosomal polymorphism in *Peromyscus maniculatus* (deer mouse). Cytogenetics *5:*411, 1966.

Treus, V. D. and O. P. Steklenjov: The hybridization of hares and domestic rabbits. Anim. Br. Abstr. *31:*510, 1963.

Trujillo, J. M., C. Stenius, L. C. Christian and S. Ohno: Chromosomes of the horse, the donkey, and the mule. Chromosoma *13:*243, 1962.

Watson, M. L.: Hybridization experiments between *Peromyscus polionotus* and *Peromyscus maniculatus.* J. Mammal. *23:*315, 1942.

GENETIC ASPECTS OF INFERTILITY

F. B. Hutt

Department of Poultry Science, New York State College of Agriculture, Cornell University, Ithaca, New York

As terms like "fecundity," "fertility" and "sterility" have different meanings for different people, it seems desirable to state at the outset what biological territory will be covered in this brief review, and to define clearly its metes and bounds.

For most biologists, the term "fecundity" refers to potential fertility, the ability to produce functional gametes. That capacity is difficult to measure in mammals, but it can be determined with accuracy in avian females. As a result, selection for fecundity has led to the development of highly prolific strains of domestic poultry. Some hens and ducks are known to have produced one gamete daily, without a miss, for a year or more.

Fertility is generally considered as the ability to produce viable offspring, and variations in it are usually rated as high, low, or otherwise, in comparison with the average rate of reproduction for the herd, flock, breed, or species. Normal fertility results when viable gametes are produced in normal numbers, when fusion of such gametes is normal, and when normal viability of zygotes, embryos and fetuses is followed by the birth of viable offspring. This definition is clear so long as we do not have to define also the term "normal."

When this chain of events is interrupted, or when deviations from average performance are great enough to be evident, the result is infertility. Degrees of infertility vary all the way from that which is barely detectable down to complete sterility.

As with fecundity, deviations from normal fertility are more easily studied in birds than in mammals. Infertile birds' eggs can easily be distinguished from fertile ones after a few days of incubation, or even (in opened eggs) at the time of laying. Accordingly, for biologists working with birds the term "infertility" has a special connotation and embryonic mortality is a separate problem. Thus, a hen's fecundity may be

75 percent (of one egg daily for the period measured), infertility of her eggs may range from zero to 100 percent, and hatchability (of her fertile eggs only) may be perfect (100 percent) or poor (70 percent), in relation to a flock average of perhaps 85 percent.

Time permits no complete review here of genetic factors causing infertility in mammals. Instead, I have selected examples to show how, in divers species, mutations have interfered with normal reproduction at different stages all the way from gametogenesis to viability of the newborn. By my terms of reference, I can exclude the infertility of hybrids, but (alas!) must also leave for other contributors such standard examples of infertility as the freemartin and the tortoise-shell male cat.

Anatomical Malformations Causing Infertility

Genetic abnormalities of the reproductive tract are mostly sex-limited. The best known example in domestic animals is probably that commonly referred to as the white-heifer disease of Shorthorn cattle. In that condition, the ovaries and external genitalia are normal, but the uterus and vagina are underdeveloped. Affected females are completely sterile. In one herd of Shorthorns, Rendel (1952) found the frequency of this abnormality during a period of 15 years in red, roan and white heifers to be 1.1, 3.5, and 39.1 percent, respectively.

An extensive study of the same condition was made by Hanset (1965) in the Central and High Belgian breed, in which the gene for white color had been derived from Shorthorns. The proportions of black, blue-roan, and white heifers affected were found to be 0.3, 1 to 5, and 10 to 15 percent, respectively.

Although white-heifer sterility has been known to stockmen for many years, no genetically satisfactory basis for its association with white coat has yet been conclusively established. Remembering that we had to wait over a half a century for the elucidation of two other genetic mysteries—mongolism in man and the tortoise-shell male cat—it is still too early to classify the white-heifer enigma as insoluble.

A similar anatomical abnormality found by Fincher and Williams (1926) in Holstein-Friesian cattle has since been ascribed to the action of a simple recessive gene in the homozygous state. A sire mated to his own daughters had 23 female progeny among which 13 showed on dissection different degrees of arrested development of the Mullerian ducts. These figures fit the assumption that the sire was homozygous for the causative gene (but unaffected by it), that all his daughters were heterozygotes, and, hence, that half his grand-daughters would be defective. Evidently such an abnormality is not necessarily associated with a white coat.

As an example of a genetic anatomical defect causing sterility in males, the impotency of bulls studied by De Groot and Numans (1946) will serve nicely. It is caused by failure of the sigmoid curve of the penis to straighten out, and can be overcome by myectomy of the paired *retractor penis* muscles. After the abnormality was proven to be caused by a recessive gene, the dysgenic nature of that operation was recognized and it was discontinued.

Cryptorchidism is a not uncommon defect in domestic animals. Males with bilateral cryptorchidism are sterile, but those having only one testis retained (often erroneously called "monorchids") are fertile. Although some writers consider cryptorchidism to be a simple recessive defect, the evidence available thus far is hardly conclusive. Like other sex-limited abnormalities that reduce fertility, the condition is somewhat difficult to study genetically. The fact that it is currently a serious problem in Poodles, Boxers, and some other breeds of dogs may be one consequence of another fact—that it can be corrected surgically.

Turning now from sex-limited, anatomical abnormalities, let us consider a few examples of genetic variations that cause infertility in both sexes.

One of these is the dwarfism induced in mice by the autosomal, recessive gene *dw,* which, in homozygotes, prevents normal development of the anterior lobe of the pituitary. Adult dwarfs weigh only 6–8 grams, about a quarter of normal size. In males the testes do not descend, and in females the vagina does not open. Both sexes are sterile. However, when young dwarfs are given daily implantations of rat pituitaries, growth is resumed and the mice attain almost normal size. Males thus treated become fertile, but females do not. Smith and MacDowell (1931) showed that in these dwarfs secretion of gonadotropic hormone from the pituitary is normal, and the action of the causative gene is to prevent or reduce production of the hormone that promotes growth. Heterozygotes of both sexes are fertile and normal in size.

In Swedish Highland cattle, infertility associated with hypoplasia of the gonads was recognized in the 1930's as a major problem in that breed. Lagerlöf found the condition to be equally prevalent in both sexes, with the frequency of cases clearly recognizable reaching 17 percent, and an additional 16 percent that were suspect. The abnormality was usually unilateral, and the proportions affected on the left, right, or both sides were about 82, 3, and 15 percent, respectively.

From matings of hypoplastic animals together, Eriksson (1943) found the proportion of 125 progeny definitely showing the defect to be 49.6 percent, about four times as high as in offspring from normal parents. He considered it to be caused by a recessive gene with average penetrance only about 50 percent. Selection against the defect was begun and its incidence was thus reduced from about 25 percent in 1935 to 8 percent seven years later.

An interesting point is that the hypoplasia of the gonads is associated with white coat color. In an examination of 882 females, Settergren (1954) found that among those 9/10 white, or more, 13.5 percent showed hypoplasia and another 10.3 percent were suspect. Among those less than 9/10 white, there were only two possible cases, both uncertain.

Abnormalities in Gametes

In guinea pigs, a simple recessive gene makes homozygotes of both sexes completely sterile (Van Lone, 1932). In males development of the entire genital system is arrested about the time of birth, and there is no spermatogenesis. In females, the reproductive tract develops normally, and after sexual maturity there is continuous oestrus, but no ovulation. The effect of the gene is thus to prevent gametogenesis in both sexes. Affected guinea pigs attain normal size. Experiments indicated that the causative gene might induce its effects by inhibiting production of gonadotropic hormones. The case thus resembles that of the dwarf mice in that both result from deficiency of a pituitary hormone, but different hormones are deficient in the two species.

Even when gametes are produced in normal numbers, their efficiency can be completely blocked by genetic action. In some Friesian bulls, over 80 percent of the spermatozoa show at high magnification an eccentric thickening of the acrosome, the apical body of the sperm head. Such bulls are completely sterile. This condition, first recognized in Holland by Teunissen, was found by Donald and Hancock (1953) to be caused by an autosomal recessive gene in the homozygous state. Females are apparently unaffected.

The translocations and inversions found by Knudsen (1958) in germ cells of sterile bulls suggest that these and other abnormalities in meiosis may be responsible for infertility in more cases than had been previously recognized. Consideration of such chromosomal aberrations is left for other participants in this program.

Embryonic Mortality and Lethal Genes

Ever since Cuenot showed in 1905 that yellow mice are all heterozygous for a lethal gene, biologists have been discovering other lethal genes—not only in the mouse, but in every species that has been adequately studied, including our own. Some of these cause death of homozygotes soon after implantation or even before, as in the yellow mouse. Others exert their differing deadly effects at birth, or soon thereafter, and sticklers about semantics might even include among the lethal genes those that prevent most of us from becoming nonagenarians. Let us say, simply, that, if fertility refers to the production of viable offspring, then lethal genes contribute greatly to infertility.

Four years ago, a list of known lethal genes in domestic animals included 51 in five species of mammals and 35 in two species of birds (Hutt, 1964). It is also certain that many genes are lethal so early in embryonic development that specific effects of single genes are not recognized among other factors contributing to prenatal mortality. A survey by Hanly (1961) shows estimates of such mortality in the pig to range from 25 to 50 percent. Long ago Corner (1923) found in that species that about 10 percent of the ova reaching the uterus never segment, and another 10 percent die without implantation. Among those implanted, 5 to 10 percent subsequently died.

Much of such mortality is probably due to the action of lethal genes, but it seems equally likely from the recent studies of McFeely (1967) that aneuploidy of various kinds is responsible for some of it.

Maternal Effects

An interesting type of infertility is found in hens homozygous for a recessive gene, *rd,* which prevents them from transferring riboflavin from the feed to the egg, as is normally done by other hens. Birds of that genotype are normal and lay well, and their eggs may all be fertile. However, chick embryos need riboflavin for normal development, hence those in eggs laid by *rd rd* hens never hatch, except when riboflavin is introduced artificially into the incubating eggs. Such hens are thus both fully fertile and completely sterile. Recent studies have shown that this paradoxical situation is caused by deficiency of a protein necessary for binding riboflavin (Winter *et al.,* 1967). The gene *rd* is maintained in the flock by the males and heterozygous females, all of which reproduce normally.

Other maternal effects on fertility and viability are well known in other species. Some are genetic, others not. I know of no case in mammals directly comparable with that of the *rd rd* hens. The incompatibility between Rh+ babies and *rh rh* mothers is not quite the same because the adverse reaction in that case depends on the genotypes of both mother and fetus.

Genetic Indicators of Infertility

It is remarkable that in several cases infertility of one kind or another is associated with white coat color. Apart from the white Shorthorn heifer and the Swedish Highland cattle already considered, other cases include the Heggedal minks and the silver-white guinea pigs.

Malformations of the uterus and vagina varying in degree, and apparently identical to those in white heifers, were found by Nes (1965) in minks heterozygous for the so-called Heggedal (or shadow) gene S^H. That gene is lethal to homozygotes, and in heterozygotes causes extreme dilution of the pigment in the fur. Minks that would be wild type, Platinum, or Sapphire without S^H are converted by that gene to shades known as Heggedal, Blue Shadow, and Sapphire Shadow, respectively. In such S^H females, Nes found the proportions with abnormal reproductive tracts to be 18.5 percent. In older females proven sterile, the proportion was about 42 percent.

In guinea pigs homozygous for *si* and *dm*, two recessive genes each of which causes separately some dilution of coat color, about half are entirely white, and the other half show only small spots of color in the cheeks. These silver-whites, when compared with litter mates not white, are subnormal in viability and fertility (Wright, 1959). Males are sterile, having testes only about 25 percent of normal weight, in which there is no spermatogenesis. About half of the females are sterile, but no consistent abnormalities were found in their reproductive tracts.

Other cases of reduced fertility associated with a gene for dilution of coat color include the Blufrost mink and the platinum fox. All of the latter are heterozygous, and homozygotes are believed to die early in gestation. The Blufrost dilution is lethal to most homozygotes, and the few surviving males are sterile. Dominant white is lethal in canaries and probably also in horses, but, perhaps to keep up from generalizing, it is standard color for the most prolific breed of chickens in the world—the White Leghorns.

It is clear that some kinds of whiteness, and some genes tending to increase whiteness, are incompatible with normal fertility. As the effects of the whitening genes are similar in some cases (the white Shorthorn heifer and the Heggedal mink) but quite different in others, the whites present a nice problem for some biologist to tackle.

Colors other than white sometimes indicate that reduced fertility is to be expected. Litters from yellow mice are reduced by about 20 to 25 percent because of the lethal action of the yellow gene. Black down is lethal during the last two days of incubation to about 3.5 percent of black chicks (Hutt, 1953).

Apart from colors, other genetic variations sometimes indicate reduced fertility. Female goats homozygous for the polled condition are intersexes and sterile. There is also evidence that most homozygous polled bucks are completely infertile (Soller *et al.*, 1963).

Another head appendage sometimes indicating poor fertility is the rose comb of the fowl. It apparently exerts no adverse effects in females, or in heterozygous males, but males of the genotype *RR* do not mate as often as those *Rr* or *rr*, and, as Wehrhahn and Crawford (1965) have shown after a series of studies by Crawford, they are, in effect, sterile. The Wyandotte breed is thus reproduced by heterozygous males and rose-combed females either *RR* or *Rr*.

Other Genetic Variations in Fertility

In addition to the examples given above, there must be many genetic conditions not yet investigated which interfere with normal reproduction. There are indications that in some strains of animals fertility is reduced by an imbalance of endocrine secretions. Inbreeding reduces fertility because it causes homozygosity of lethal genes, but also because it affects behavior. In pigs highly inbred for a few generations, some lines have been lost because the boars refused to mate.

Infertility and sterility of domestic animals has recently been reviewed by Johansson and Rendel (1968). From their discussion, as from the examples cited in this paper, it is clear that, apart even from obvious anatomical abnormalities, genetic influences on fertility can be expected in any species, at any stage from gametogenesis to birth.

Summary

To show that infertility or sterility may be caused by genetic deviations from normal anatomy and by genetically induced abnormalities effective at any stage from gametogenesis to birth, examples of different types of infertility in domestic animals are described.

It is pointed out that sterility and reduced fertility are often associated with white coat color. Examples of that association include the white Shorthorn heifer, the predominantly white Swedish Highland cattle, Heggedal minks, silver-white guinea pigs, and Blufrost dilution in the mink.

References

Corner, G. W.: The problem of embryonic pathology in mammals, with observations upon intra-uterine mortality in the pig. Amer. J. Anat. *31*:523, 1923.

De Groot, T. and S. R. Numans: Over de erfelijkheid der impotentia coeundi bij stieren. Tijdschr. voor Diergeneeskunde. *71*:372, 1946.

Donald, H. P. and J. L. Hancock: Evidence of gene-controlled sterility in bulls. J. Agric. Sci. *43*:178, 1953.

Eriksson, K.: Hereditary Forms of Sterility in Cattle. Biological and Genetical Investigations. I. Lund. Hakan Ohlssons Boktryckeri. 155 pp., 1943.

Fincher, M. G. and W. L. Williams: Arrested development of the Muellerian ducts associated with inbreeding. Cornell Vet. *16*:1, 1926.

Hanly, S.: Prenatal mortality in farm animals. J. Reprod. Fertil. *2*:182, 1961.

Hanset, R.: Recherches sur la White Heifer Disease et son déterminisme génétique. Comptes rend. Inst. Encourage. Recherche Scient. Ind. Agr. Bruxelles. No. 33; 181 pp., 1965.

Hutt, F. B.: Lethal action of the gene for extension of black pigment in the fowl. Genetics *36*:213, 1951.

———: Animal Genetics. New York: The Ronald Press Company. xiv + 546 pp., 1964.

Johansson, I. and J. Rendel: Genetics and Animal Breeding. Edinburgh and London: Oliver and Boyd. xi + 489 pp., Chap. 9, 1968.

Knudsen, O.: Studies on spermiocytogenesis in the bull. Internat. J. Fertil. *3*:389, 1958.

McFeely, R. A.: Chromosome abnormalities in early embryos of the pig. J. Reprod. Fertil. *13*:579, 1967.

Nes, N.: Abnormalities of the female genital organs in mink heterozygous for the Heggedal factor (shadow factor). Acta Vet. Scand. *6*:65, 1965.

Rendel, J. M.: White heifer disease in a herd of dairy Shorthorns. J. Genetics *51*:89, 1952.

Settergren, I.: Cited from Johansson and Rendel, p. 240, 1968.

Soller, M., M. Laor, R. Barnea, Y. Weiss and N. Ayalon: Polledness and infertility in male Saanen goats. J. Hered. *54*:237, 1963.

Smith, P. E. and E. C. MacDowell: The differential effect of hereditary mouse dwarfism on the anterior pituitary hormones. Anat. Rec. *50*:85, 1931.

Van Lone, E. E.: The inheritance, histology, and physiology of a new type of sterility, sexual hypogenesis, in the guinea-pig. J. Exper. Zool. *64*:1, 1932.

Wehrhahn, C. F. and R. D. Crawford: Comb dimorphism in Wyandotte domestic fowl. 2. Population genetics of the rose comb gene. Canadian J. Genetics and Cytology *7*:651, 1965.

Winter, W. P., E. G. Buss, C. O. Clagett and R. V. Boucher. The nature of the biochemical lesion in avian renal riboflavinuria. II. The inherited change of a riboflavin-binding protein from blood and eggs. Comp. Biochem. Physiol. *22*:897, 1967.

Wright, S.: Silvering (si) and diminution (dm) of coat color of the guinea pig, and male sterility of the white or near-white combination of these. Genetics *44*:563, 1959.

GENE ACTION IN HUMAN DIPLOID CELL STRAINS

Robert S. Krooth
*Lawrence D. Buhl Center for Human Genetics,
Department of Human Genetics,
University of Michigan Medical School,
Ann Arbor, Michigan*

Introduction

Human diploid cell strains, biochemically marked by the action of a Mendelian gene, may eventually prove useful in the analysis of the genome, in the study of mutation, and in experiments aimed at producing specific genetic changes. In this essay, however, I shall summarize what we have learned about gene action from the use of such strains. I prefer this topic for two reasons. First, quite a lot has been learned about the subject in the past 10 years, and second, it is the field with which I personally am most familiar.

I shall confine my comments to those cells which are capable of sustained growth outside the body; hence, I shall not be considering cells which can merely be kept alive *in vitro,* or which undergo but two or three generations outside the body. I shall also be concerned mainly with the action of genes in human diploid cells. Finally, I shall limit my discussion chiefly to cases where the Mendelian character of the cellular phenotype has been shown by studies on the families of the donors of the cells. These restrictions distinctly narrow the topic, but they ensure that we need not debate whether the cell phenotype is indeed due to a gene, and not to, say, variation in whole chromosomes, or some epigenetic change. By considering only human cells, we shall cover most (though not all) of the Mendelian markers now known for cultured mammalian cells.

Before examining what we know about specific genes in cultured cells,

I should like to comment on the biology of the cells. To save space, I have included no references to this part of the discussion. The primary sources to support the statements I shall be making can be found in several recent reviews (Harris, 1964; Krooth, 1964b; Eagle, 1965; Krooth, 1965; Gartler and Pious, 1966; Green and Todaro, 1967; Krooth *et al.*, 1968). In general, there are two kinds of mammalian cells now growing *in vitro*. "Heteronuclear" cultures are characterized by the fact that there is marked variation from cell to cell in chromosome complement. If the cells are cloned, each colony will usually regenerate the same spectrum of chromosome complements as was present in the parental mass culture. Hence, the cells of heteronuclear cultures have some kind of hereditary anomaly of mitosis. "Homonuclear" cultures show almost no variation from cell to cell in chromosome complement except when such cultures are developed from chromosomal mosaics. In this latter case the variation is not usually observed between the cells of a clonally derived colony. Except for lines developed from chromosomal mosaics, homonuclear cultures are presumed to contain the initial chromosome complement of the donor—the one present at the time the egg, from which the donor developed, was fertilized.

In some rodents, such as the mouse and Syrian hamster, these two kinds of cultures seem to reflect different stages in the *in vitro* life cycle of the line. Cultures from the normal (i.e., non-neoplastic) tissues of these animals begin as homonuclear cells, stop growing for a time, and then become heteronuclear. In the case of human cells, however, each of the two types of cultures breeds true. Homonuclear human cultures rarely, if ever, give rise to heteronuclear ones. (The converse is also true.) The reason for this remarkable species variation is unknown.

Human heteronuclear lines are only rarely isolated, and it is probably true that they usually arise from tissue involved in a cancerous process. At least some human heteronuclear lines clone with high efficiency and will grow indefinitely in culture. They also have other distinctive features.

A human homonuclear line can easily be isolated from the non-neoplastic tissue of any individual. Such cell lines are difficult to clone, especially after they have grown for some time *in vitro*. Cultures of these cells can be serially propagated *in vitro* for about 1 year with existing methods. At or about 1 year, the generation time of the cells lengthens asymptotically, and growth essentially ceases. In the few mitoses which can be visualized at this time, there is an increased frequency of aneuploidy and tetraploidy. Dicentric and other abnormal chromosomes are sometimes also seen. Even at this stage, however, the *majority* of cells are diploid and have normal chromosome morphology. Although growth is barely detectable, the cultures can be kept alive (in the sense of metabolizing) for at least a year and probably longer.

Gene Action

Cellular Differentiation

Nearly all lines of cultured homonuclear cells, and most heteronuclear lines as well, do not perform most of the tissue-specific functions of intact cells (Levintow and Eagle, 1961, *inter alia*). For example, human diploid cell strains cultured *in vitro* do not seem to synthesize hemoglobin, nor, in all probability, cystathionase (Eagle, 1966)—to give two examples.

The cultured cells do, of course, engage in many general cellular activities, such as the synthesis of protein, RNA, DNA, and a number of small molecules, including eight nucleotides and at least seven amino acids (cf. Eagle, 1959; Krooth, 1965). However, with respect to many loci, the phenotype of the cultured cell is simply unexpressed. Hence, in genetic studies one is presently limited to those loci which affect molecules the cell spontaneously synthesizes. In general, one can probably assume that if a specific protein or other macromolecule is present in all or nearly all the intact tissues of the body, it will also be found in cultured cells (Krooth and Weinberg, 1961). The presence of the molecule would not seem to depend on which way the cell has differentiated. In practice, of course, no enzymes have been looked for in all the tissues of the body. It is probably true that if an enzyme is present in a peripheral blood cell, and a parenchymatous organ other than spleen, it is likely to be ubiquitous and therefore to be present in serially cultured cells (Krooth, 1963, 1964c, 1965). However, the converse of the generalization does not hold: It does not follow that only ubiquitous enzymes will be present in cultured cells (Uhlendorf and Mudd, 1968; Krooth *et al.*, 1968).

The intact cell which gives rise to the serially cultured mammalian cell is not known. In at least some cases, it is logical to suspect that cultured heteronuclear cells are descended from neoplastic cells in the original donor. Human diploid cell lines have a fibroblastic morphology, and there is evidence that such cells synthesize collagen (Green *et al.*, 1966). Hence, it is probable that human diploid lines, and perhaps other cell lines having a fibroblastic morphology, are descended from the fibroblasts of their original donors.

In general, serially cultured diploid cells cannot be shown to resemble the characteristic cell of the organ or tissue from which they are isolated. It is probable, at least in some cases, that the intact cell type which is able to grow under cultural conditions is a ubiquitous cell. It is present in nearly all tissues, but is the characteristic cell in none of them.

Relationship of Method of Obtaining Variant Cells to the Cell Type Employed

There are two methods commonly used to obtain variant cell lines. In one of these, a selective agent is applied to a culture (usually containing cells from a single donor) and the surviving cells are isolated. If the ability to survive in the selective medium is found to persist even after the cells have grown in non-selective media for some time, and if certain other criteria are satisfied, the descendants of the surviving cells are considered hereditary variants. They can then be compared with wild-type cells of the parental line. The second method is to select not cultured cells but rather donors. With this method, one uses individuals known to carry a mutant gene affecting molecules in the cultured cell as a source of variant lines.

In the case of human cells, there exists a relationship between the method of selection employed and the cell type on which the experiments are performed. As a rule, when cells are selected from a population, heteronuclear cultures are employed, whereas when donors are selected, homonuclear cultures are used. Part of the reason for this relationship lies in the rarity with which human heteronuclear lines are isolated. It would require a fortuitous coincidence to develop a heteronuclear line from a donor having a mutant gene affecting cultured cells. Another factor may be the difficulty in cloning homonuclear cells: when homonuclear cells are subjected to a selective agent, the rare surviving cell may not always be able to generate a colony. Finally, the chromosomal homogeneity of homonuclear cultures may tend to restrict the hereditary variation between cells obtained from a single donor.

Genetic Markers in Human Diploid Cell Cultures

Table 1 lists the Mendelian genes which have been shown to confer a distinctive biochemical phenotype on human cell strains.* The Table includes at least twenty-two mutant Mendelian genes, for there are two different electrophoretic variants of glucose-6-phosphate dehydrogenase recognizable in cultured cells, and Hurler's syndrome can be caused by a mutation at either of two genetic loci. If we assume acatalasia I and II are due to different mutations, which is a reasonable possibility, the number becomes twenty-three. It is more difficult to guess the number of

* Table 1 is, of course, a summary of the work of a number of different laboratories. A bibliography for each of the entries in Table 1 (except for the two most recently described cellular phenotypes) is given elsewhere (Krooth *et al.*, 1968). The two new phenotypes are mucoviscidosis (Danes and Bearn, 1968) and homocystinuria (Uhlendorf and Mudd, 1968).

Table 1: Some Genetic Abnormalities that Affect Specific Molecules Present in the Cultured Human Homonuclear Cell

Disease or Variant	Enzyme	Genetic Disease or Variant
Acatalasia I (Japanese variant)	Catalase	Autosomal recessive
Acatalasia II (Swiss variant)	Catalase	Autosomal recessive
Familial non-spherocytic hemolytic anemia	Glucose-6-phosphate dehydrogenase	Sex-linked recessive
Electrophoretic variants of G-6-PD (asymptomatic)	Glucose-6-phosphate dehydrogenase	Sex-linked co-dominant
G-6-PD deficiency (Mediterranean type)	Glucose-6-phosphate dehydrogenase	Sex-linked recessive
G-6-PD deficiency (Negro type)	Glucose-6-phosphate dehydrogenase	Sex-linked recessive
Galactosemia	Uridine diphosphogalactose transferase	Autosomal recessive
Hurler's syndrome	Uncertain	Sex-linked recessive or autosomal recessive
Electrophoretic variants of lactic dehydrogenase	Lactic acid dehydrogenase	Autosomal co-dominant
Branched-chain ketonuria	Uncertain	Autosomal recessive
Niemann-Pick disease	Uncertain	Autosomal recessive
Orotic aciduria	Orotidine-5′-monophosphate pyrophosphorylase and orotidine-5′-monophosphate decarboxylase	Autosomal recessive
Electrophoretic variant of phosphoglucomutase	Phosphoglucomutase	Autosomal co-dominant
Electrophoretic variants of 6-phosphoglucuronic acid dehydrogenase	6-phosphoglucuronic acid dehydrogenase	Autosomal co-dominant
Citrullinemia	Arginosuccinate synthetase	Autosomal recessive
Glycogen storage disease (Cori Type II)	Lysosomal α-1-4-glucosidase	Autosomal recessive
Lesch-Nyhan syndrome	Hypoxanthine-guanine phosphoribosyltransferase	Sex-linked recessive
Chediak-Higashi syndrome	Uncertain	Autosomal recessive
Mucoviscidosis (Pancreatic cystic firbrosis)	Uncertain	Autosomal recessive
Homocystinuria	Cystathionine synthetase	Autosomal recessive

loci, for one must not overlook the possibility that different alleles at the same locus may cause completely different diseases or even enzyme deficiencies, especially if it is argued that a single polypeptide can be incorporated into more than one protein or may have more than one catalytic activity. Similarly, one cannot be sure that all the mutations causing a single condition, or affecting a single enzyme, are at the same locus. In Hurler's syndrome, the two alternative loci where a mutation can cause the disease are even on different chromosomes. It is likely, however, that the genes listed in Table 1 are at about seventeen different loci. The abnormalities include disorders of carbohydrate, amino acid, nucleotide, and lipid metabolism. Seven of the genes in Table 1 are on the X-chromosome; these are the mutations affecting glucose-6-phosphate dehydrogenase (including the two different electrophoretic variants), the mutation causing the Lesch-Nyhan syndrome, and the mutation at one of the loci causing Hurler's syndrome. (The autosomal form of Hurler's syndrome has slightly different clinical manifestations than the sex-linked form. The sex-linked form is sometimes called "Hunter's syndrome.")

Media which select against mutant homozygous cells are effective for three of the loci in Table 1. Under appropriate circumstances, the substitution of galactose for glucose will select against galactosemic cells, but will permit the growth of normal or heterozygous cells (Krooth and Weinberg, 1961; Krooth, 1966). In orotic aciduria (Smith *et al.*, 1966), a double dose of the mutant gene causes about a 100-fold reduction in the final two sequentially acting enzymes of uridine-5′-monophosphate biosynthesis; nonetheless, mutant homozygous cells are able to grow slowly in the absence of exogenous uridine (Krooth, 1964b). If, however, adenosine (or certain other purine nucleosides) are added to the medium, the level of specific activity of at least one of the two enzymes affected by the gene is reduced by another three- to four-fold. Under these circumstances, the mutant homozygous cells develop an absolute nutritional requirement for uridine, whereas normal or heterozygous cells, in the same medium, are able to grow whether uridine is present or not (Krooth, 1964b). In the case of citrullinemia, mutant homozygous cells will not grow, or at least will not grow rapidly, if citrulline is substituted for arginine in the medium. The growth of normal cells is not significantly affected by this substitution (Tedesco and Mellman, 1967).

In the case of five genes listed in Table 1, methods have been developed for determining the phenotype of individual cells. These genes are the ones causing both the sex-linked and autosomal forms of Hurler's syndrome (Danes and Bearn, 1965; Danes and Bearn, 1966a; DeMars and Leroy, 1967), the Chediak-Higashi syndrome (Danes and Bearn, 1967b), the Mediterranean type of glucose-6-phosphate dehydrogenase deficiency (Wajntal and DeMars, 1967) and mucoviscidosis (Danes and Bearn, 1968).

In addition to the genes in Table 1, there are about six other genetic defects, reviewed elsewhere (Krooth *et al.,* 1968), which are probably recognizable in human diploid cell strains. In these conditions, however, the nature of the molecule affected by the gene is less certain.

The enzyme affected in the Lesch-Nyhan syndrome is reported to have reduced activity in some patients with gout (Kelley *et al.,* 1967a). Hence, certain varieties of gout may also be found to be associated with a phenotype recognizable in cultured cells. Since, under appropriate conditions, human diploid cells can use galactose as a substitute for glucose (Krooth and Weinberg, 1961; Krooth, 1966; Baugh *et al.,* 1967), the autosomal recessive gene, described by Gitzelman (1967), for galactokinase deficiency may prove identifiable in diploid cell strains.

The Single Active X-Theory

The Single Active X-Theory (cf. Lyon, 1968), in its simplest form, holds that early in the development of the mammalian female there is inactivation, within each somatic cell, of either the maternal or paternal X-chromosome. The word "inactivation" denotes that the X-chromosome no longer functions. Inactivation is supposed to be random and hereditary over somatic generations. In other words, about half the cells should have the maternal X-chromosome inactive, and about half the paternal one. Moreover, the particular X-chromosome inactivated continues to be the inactive one in all the descendants of the cell.

The theory generates two predictions about diploid cell strains developed from females who are heterozygous for sex-linked genes:

1. Mass cultures should consist of two cellular populations. One population of cells should resemble the wild-type and the other should resemble cells from mutant homozygous females or from mutant hemizygous males.
2. When mass cultures are cloned, two types of colonies should be recovered in about equal proportions. Colonies whose cells have the mutant phenotype and colonies whose cells have the normal phenotype. There should not be any mixed colonies.

The second prediction seems, on the face of it, to be more strict than the first. However, the first prediction contains two assumptions which are not included in the second prediction. The first assumption is that one has a method—say autoradiography or histochemistry—for determining the phenotype of individual cells. The second assumption is that cellular phenotype is autonomous in the sense of not being influenced by the phenotypes of other cells which are adjacent or in the same container. Both predictions assume that neither of the two cell types postulated by the theory has been eliminated by selection *in vivo* or *in vitro.* This latter possibility means that finding only one kind of cell in mass culture or only one kind of colony after cloning does not necessarily invalidate the theory. Finding both phenotypes in a single clone is disastrous for the theory. The two predictions are critical, of course, only if one does *not* find a dimorphism in cell strains developed from hemizygous males, or homozygous females.

Table 2 summarizes the work thus far done to test these predictions. The experiments have used a total of five sex-linked genes, which are probably at three different loci. In every case so far, at least the second prediction has been met. When tested, the first prediction was also satisfied in one case, and an ambiguous result was obtained in another. It has also been shown that the second prediction is *not* met in the case of five autosomal genes tested. In other words, clonal cell populations from females heterozygous for each of these genes displayed the heterozygous phenotype in every case. The mutant genes were the ones causing electrophoretic variants of phosphoglucomutase (Davidson *et al.,* 1965; Sigman and Gartler, 1966), of lactic acid dehydrogenase (Davidson *et al.,* 1966) and of 6-phosphogluconic acid dehydrogenase (Davidson *et al.,* 1966); the other mutant genes were the ones causing branched chain ketonuria (Sigman and Gartler, 1966), and the autosomal recessive form of Hurler's syndrome (Danes and Bearn, 1967a).

The data described above are, in aggregate, the strongest evidence yet collected to show that the Single Active X-Theory is true, or at least that it applies to cultured diploid human cells for the loci studied. Whether the inactivation is absolutely complete and whether it applies to all loci are questions which are presently under study in a number of laboratories.

Some Examples of Gene Action in Human Diploid Cell Strains

The Lesch-Nyhan Syndrome

The Lesch-Nyhan syndrome (Nyhan *et al.,* 1967) is a sex-linked recessive neurological disorder. Affected patients have an abnormally high rate of uric acid synthesis. Diploid cell strains (and erythrocytes) from mutant hemizygous males lack activity for hypoxanthine:guanine:xanthine phosphoribosyltransferase (Seegmiller *et al.,* 1967a). The colons denote that the conversion of each of these three purines to its corresponding ribotide is catalyzed by the same enzyme (Kelley *et al.,* 1967b). The reason the enzyme deficiency leads to over-production of purines is unknown. One possibility is that purine ribotides, which seem to serve as feed-back inhibitors of the first enzyme of the purine synthetic pathway, simply run off as oxypurines in the absence of the enzyme affected by the gene (Seegmiller *et al.,* 1967).* The pathway cannot, as it were, keep track of how much product it has made, because the enzyme which

* It is impossible to read the report of Seegmiller *et al.* (1967) without realizing that they were keenly aware of this possibility. However, I must add, in fairness to them, that they expressed the idea less directly, and much more cautiously, than I.

Table 2: Experimental Tests, Using Heterozygous Human Diploid Cell Strains, of the Single Active X-Theory *

Mutant Phenotypes	Results: Prediction 1 (Cellular Dimorphism)	Results: Prediction 2 (Clonal Dimorphism)	References: Prediction 1	References: Prediction 2
Glucose-6-Phosphate Dehydrogenase Deficiency (Mediterranean Variety)	–†	+	–	Davidson, Nitowsky and Childs (1963)
B-Variant of Glucose-6-Phosphate Dehydrogenase	–	+	–	Davidson, Nitowsky and Childs (1963)
Madison Variant of Glucose-6-Phosphate Dehydrogenase	–	+	–	DeMars and Nance (1964)
Sex-linked Hurler's Syndrome (Tissue culture phenotype: Cellular Metachromasia)	0 ‡	+	Danes and Bearn (1966a)	Danes and Bearn (1967a)
Lesch-Nyhan Syndrome (Tissue culture phenotype: Absence of hypoxanthine: guanine phosphoribosyltransferase *and* failure of individual cells to incorporate ^{3}H-guanine or ^{3}H-hypoxanthine into acid-insoluble material)	+	+	Rosenbloom *et al.* (1967)	Migeon *et al.* (1968)

* The definition of the symbols is as follows: + denotes prediction was found to apply to the specified gene; 0 denotes prediction was found *not* to apply to the specified gene; – denotes prediction has not yet been tested.

† A histochemical method for determining cellular phenotype, with respect to this gene, has just been developed by Wajntal and DeMars (1967). Hence, it should be possible to test Prediction 1 in this case almost at once.

‡ The prediction *does* apply, in one sense, to this locus: about half the cells of female heterozygotes have the metachromatic granules characteristic of the mutant hemizygous genotype. However, cells from one hemizygous male showed some dimorphism in mass culture, which seems to vitiate the theoretical significance of the observation on female cells. Nearly all the cells of two other hemizygous males who were studied contained the granules.

reconverts some of the catabolites of the product back to product is missing. Only the product is inhibitory. When the product can proceed unchecked to its catabolites, not only do the catabolites accumulate, but the synthetic part of the pathway goes too fast.

Whether or not this particular formulation is correct, it is quite likely that the absence of this enzyme has deranged a cellular control mechanism, and thereby causes an abnormally rapid rate of production of purines. The presence of a characteristic biochemical phenotype for this gene in cultured cells should make it far easier to explore the mechanism further. It is of interest that uric acid is not spontaneously synthesized by cultured cells. The enzyme (xanthine oxidase) which catalyzes the conversion of xanthine to uric acid is peculiar to liver. Hence, in this case, it has been possible to determine phenotype in cultured cells even though, because of differentiation, they cannot express the clinically most salient metabolic feature of the disease—overproduction of uric acid.

There are two other features of the mutant hemizygous cells, which also allow their phenotype to be identified. First, because of the enzyme deficiency, they are unable to incorporate tritiated hypoxanthine into RNA (Rosenbloom *et al.*, 1967; Migeon *et al.*, 1968). Hence, one can infer the phenotype of individual cells by autoradiography (see Table 2). Second, the rate of purine synthesis by mutant hemizygous cells is not inhibited by 6-mercaptopurine, for the affected enzyme is apparently the one responsible for converting 6-mercaptopurine to its ribotide (Seegmiller *et al.*, 1967). The 6-mercaptoribotide serves as a feedback inhibitor of purine synthesis in normal cells. It is conceivable that some tumors which become resistant to 6-mercaptopurine owe their resistance to a genetic change at the same sex-linked locus where a mutation can cause the Lesch-Nyhan syndrome.

Hurler's Syndrome

Hurler's syndrome (Dorfman, 1966) is characterized by mental deficiency and a number of morphological abnormalities. As noted above, it can be caused by either an autosomal or a sex-linked locus. In both cases, the mutant gene is recessive. Cells in a number of different tissues of affected individuals contain distinctive metachromatic granules.

The granules cannot be detected in the intact somatic cells of heterozygous persons, but they are readily demonstrable after two or three subcultures when diploid cell strains of such persons are grown *in vitro* (Danes and Bearn, 1966a). Cell strains from mutant homozygous and hemizygous patients, of course, also show the granules, and an abnormally rapid rate of mucopolysaccharide synthesis as well (Danes and Bearn, 1965, 1966a; Matalon and Dorfman, 1966). However, both the characteristic granules and the elevated rate of synthesis (compared to normal cells in the same medium) disappear if mutant cells are grown in the presence of vitamin A alcohol (Danes and Bearn, 1966b). The absolute rate of mucopolysaccharide synthesis is reduced by vitamin A in cells of all genotypes. The effect is reversible in each case, and the characteristic phenotype slowly reappears in mutant cells after vitamin A has been withdrawn from the medium. These phenomena—the development of a characteristic phenotype in heterozygous cells and the reversal of the mutant phenotype in the presence of vitamin A alcohol—apply equally to the autosomal and sex-linked varieties of the disease. The two loci for Hurler's syndrome, like the locus to be described in the next section, demonstrate the apparently greater lability of cellular phenotype which is sometimes encountered under cultural conditions. The mere stimulus of serial propagation *in vitro* seems sufficient to excite heterozygous cells to disclose the presence of a single dose of the mutant gene, and the addition to the medium of vitamin A alcohol abolishes the phenotypic distinction between mutant cells and normal cells. This latter observation, incidentally, is an argument for using media which are as simple as possible. If vitamin A alcohol were present in all media, the persistence of this interesting abnormality in cultured cells would probably have been overlooked. The effect of vitamin A and the effect of its removal both take considerable time to develop. Hence, it is unlikely that this phenomenon could have been discovered by the classical tissue slice methods of biochemistry.

Orotic Aciduria

Orotic aciduria (Smith *et al.,* 1966) is a rare autosomal recessive disorder. Affected patients are anemic and also have other abnormalities which seem to reflect a disturbance in cellular growth. Most, and perhaps all, of the features of the disease (including the orotic aciduria) respond dramatically to supplementation of the diet with uridine.

Figure 1 summarizes the sequence of reactions whereby normal diploid cell strains synthesize uridine-5′-monophosphate (UMP). Cell strains from affected patients have about one percent of normal activity for the final two enzymes of the pathway (Krooth, 1964; Howell *et al.,* 1967). The earlier enzymes in the pathway do not seem to be affected by the gene.* Heterozygous and mutant homozygous cells develop nearly normal levels of both enzymes affected by the gene if the medium contains certain inhibitors of UMP synthesis (Pinsky and Krooth, 1967a and 1967b). These inhibitors also cause normal cells to develop higher levels of specific activity for the same two enzymes. The effect is readily demon-

* The effect of the gene on the levels of all the enzymes of the pathway, except dihydroorotic acid dehydrogenase (the third enzyme of the sequence), has been studied in erythrocytes (Smith *et al.,* 1961). In diploid cell strains the effect of the gene on the levels of all the enzymes, except the first one (aspartate transcarbamylase), has been studied (Krooth, 1964b; Howell *et al.,* 1967; Wuu and Krooth, 1968a).

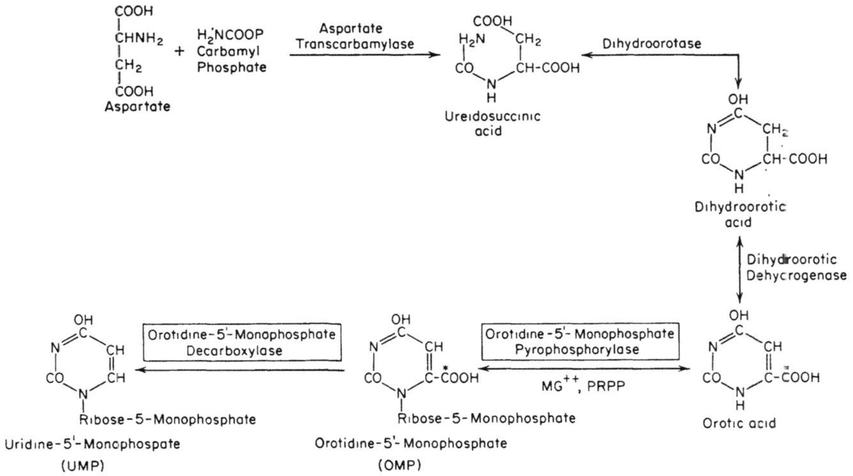

Fig. 1. The catalytic sequence used by human cells for the *de novo* synthesis of uridine-5′-monophosphate.

strable even when the cells are provided with the product of the pathway. An experiment demonstrating the effect of growth in one of the inhibitors on cellular specific activity for the final enzyme of the pathway is shown in Table 3. The present evidence suggest that the inhibitors cause cells of all genotypes to accumulate high levels of intermediates in the pyrimidine pathway, and that one of these intermediates is (at least partly) responsible for the augmentation of enzyme activity (Pinsky and Krooth, 1967b). (However, it has thus far been possible to demonstrate a direct effect of intermediates in the medium on cellular enzyme activity only in the case of mutant homozygous cells.) An experiment showing the effect of growth in one of the intermediates in this pathway on cellular specific activity for the last enzyme of the sequence is given in Table 4. Although growth in the presence of several inhibitors causes an increase in the cellular level of the final two enzymes of the UMP pathway, two of the earlier enzymes (dihydroorotase and dihydroorotic acid dehydrogenase) are unaffected (Pinsky and Krooth, 1967b; Wuu and Krooth, 1968a). The first enzyme of the sequence (aspartate transcarbamylase) has not been studied in this respect.

The chemical sequence by which human cells synthesize uridine-5′-monophosphate appears identical to the one used by other microorganisms in which *de novo* pyrimidine synthesis has been studied. In *Escherichia coli,* all five enzymes of the pathway are repressed by uracil, but the repression is coordinate only in the case of the final four (Beckwith *et al.,* 1962). In yeast, the first enzyme of the pathway is repressed by uracil, while the second is induced by its own substrate, and the final

Table 3: Specific Orotidine-5′-Monophosphate Decarboxylase Activity of Diploid Cell Strains Grown in the Presence and Absence of 1.2×10^{-3}M Barbituric Acid [1]

Cell Strain	Genotype at the Orotic Aciduria Locus [2]	Additive to Reaction Mixture [3]	Additive to Growth Medium	Specific Orotidine-5′-Monophosphate Decarboxylase Activity [4]
RU	RR	0	0	1.088
		Barbituric acid	0	1.052
		0	Barbituric acid	3.191
OFR	Rr*	0	0	0.614
		Barbituric acid	0	0.645
		0	Barbituric acid	1.997
AUC	r*r*	0	0	0.006
		Barbituric acid	0	0.006
		0	Barbituric acid	0.541

[1] The cell strains and the details of the experiment, except for the enzyme assay, are as previously described (Pinsky and Krooth, 1967a). The assay has been modified somewhat, so that activity is now more nearly proportional to protein concentration. (The deviation from proportionality, with the present method, is only about 30% over a 1000-fold range of protein concentrations.) The cells were grown in experimental media for 11 days prior to harvest. All media contained 6×10^{-5}M cytidine.

[2] "R" denotes the normal allele at the orotic aciduria locus, and "r*" its mutant allele.

[3] Barbituric acid either was not added at all, or else was added either to the reaction mixture (at the start of the enzyme assay) or to the growth medium (one day after subculture). The final concentration of added barbituric acid was 1.2×10^{-3}M in either incubation mixture or growth medium.

[4] Millimicromoles orotidine-5′-monophosphate decarboxylated per hour of incubation per milligram cell protein.

three are believed to be induced by an intermediate–dihydroorotic acid (Lacroute, 1966 and 1968).* In the human cell, the final two enzymes seem to change separately from the others (including the one which immediately precedes them), both in response to a Mendelian mutation and to certain inhibitors of UMP synthesis. It may be that during the course of evolution, the cell may have progressively separated, in its biochemical responses, the different enzymes of the pyrimidine sequence.

* The same general scheme applies to leucine biosynthesis in *Neurospora* (Gross, 1965).

Table 4. The Effect of Dihydroorotic Acid on Mutant Homozygous Cells in the Presence of 3-Methylaspartate and Cytidine [1] (AUC Strain, r*r*)

Medium Supplement	Specific Orotidine-5'-Monophosphate Decarboxylase Activity [2]
None	32
Cytidine	13
3-methylaspartate	11
3-methylaspartate plus cytidine	1
3-methylaspartate plus cytidine plus dihydroorotic acid	23

[1] The notation and methods are the same as those for Table 3. In this case, however, the cells were grown in experimental media for only 6 days. The concentration of cytidine was 1.2×10^{-4}M, of 3-methylaspartate 10^{-2}M, and of dihydroorotic acid 1.2×10^{-3}M.

In most experiments (about 90%), growth in cytidine causes about a three-fold reduction in the specific orotidine-5'-monophosphate decarboxylase activity of mutant homozygous cells. (It is not known why the effect is not invariably observed.) 3-methylaspartate, a known inhibitor (Woolley, 1960) of bacterial aspartate transcarbamylase (the first enzyme of the pathway) also causes a three-fold reduction of specific activity in mutant cells. When both these agents are added, their effects appear to be synergistic, and the specific activity of mutant cells is reduced by ten- or twenty-fold. Under these circumstances, about a twenty-fold increase in enzyme activity ascribable to the presence of dihydroorotic acid in the medium, can be observed. Without the use of these compounds, the effect of exogenous dihydroorotic acid is less pronounced (Pinsky and Krooth, 1967b).

[2] $\mu\mu$moles orotidine-5'-monophosphate decarboxylated per milligram cell protein per hour.

The Theory of Intercellular Complementation

The observations just discussed suggest two conclusions. The first is that cultured human cells (at least those of a mutant genotype) increase their catalytic activity for the final enzymes of a synthetic sequence in response to the presence of an intermediate in the sequence. The second conclusion is that as living forms became more complex, the cell may have tended, in its responses, to distinguish increasingly between the several enzymes of UMP synthesis.

Recently, our laboratory has made two further observations. They are both preliminary, in the sense that we must conduct more experiments to feel certain they are correct, but the data thus far obtained are quite striking. Cultured mammalian cells change their specific activity for a number of different enzymes as they grow (cf. review of Pan and Krooth, 1968). This applies, at least under some circumstances, to three of the enzymes we have studied in the UMP sequence. The enzymes are: dihydroorotase, dihydroorotic acid dehydrogenase, and orotidine-5′-monophosphate decarboxylase. These are, respectively, the second, third, and fifth enzymes of the sequence (Fig. 1). At least two enzymes—dihydroorotase and the dehydrogenase—and probably all three change very differently from one another during growth (Kelsey *et al.,* 1968). For example, when the experiments are performed a particular way, specific dihydroorotase activity seems to fall during growth, while dihydroorotic acid dehydrogenase activity rises. This observation suggests that the human cell, unlike *E. coli,* can increase (or decrease) its specific activity for some of the enzymes of the pathway separately from the others. The spontaneous changes in cellular enzyme activity which occur during growth thus support the notion, based on the effect of the gene mutation and of the metabolic inhibitors, that human cells somehow distinguish between individual enzymes of the sequence.

We have also studied the levels of these same three enzymes in a number of intact human tissues. Some tissues, such as whole kidney and liver, seem to have high levels of all the enzymes. In other tissues, however, the pattern looks more interesting. Preliminary experiments suggest that fat and skeletal muscle have high levels of dihydroorotase, one of the early enzymes of the pathway, but very low levels of at least two of the later enzymes—the dehydrogenase and the decarboxylase. It is already known that human red cells have quite significant levels of all the enzymes of the sequence *except* dihydroorotic acid dehydrogenase. Activity for this enzyme is essentially absent from mature erythrocytes (Smith and Baker, 1959; Wuu and Krooth, 1968a). We are presently surveying human organs for the "orotic aciduria tissue": the tissue in normal persons which has the same two deficient enzyme activities as do diploid cell strains cultured from patients with orotic aciduria. It would, of course, be of great interest if such a tissue were found. I have coined the phrase "allotopic mutation" to apply to any mutation which imparts to one tissue a characteristic metabolic feature which is normally found in another tissue. One would suppose that mutations of this class exist, but thus far none has been identified.*

* The definition of an allotopic mutation is intended to express the idea, but it is not a rigorous definition. For example, phenylketonuria is an autosomal recessive disorder. Mutant homozygous individuals have little or no activity for phenylalanine hydroxylase. This enzyme is normally found only in hepatic tissue. Hence, it could be argued that phenylketonuria is due to an allotopic mutation, for it makes liver like all the other tissues of the body. Of course, such an argument is absurd, for it would appear to hold even if the mutation were a simple structural one depriving the enzyme of catalytic activity on account of an amino acid substitution. For a mutation to be allotopic the phenotype which it confers on one kind of cell must be highly characteristic of some other kind of cell. It is difficult to define the phrase "highly

We can summarize all these observations on pyrimidine synthesis in human diploid cells as follows:

1. In human cells, unlike *E. coli,* several of the enzymes of the UMP pathway seem to respond separately from one another. The separate responses can be observed to occur with respect to:

a. The action of a mutant Mendelian gene which affects the last two of the five enzymes of the pathway.

b. The effect of growth of the cells in certain inhibitors of uridine-5'-monophosphate biosynthesis. Growth in these inhibitors is known to cause an increase in cellular specific activity for two of the five enzymes—the same two enzymes which are affected by the gene. On the other hand, the cell does not increase its specific activity for at least two of the remaining three enzymes of the sequence in response to these inhibitors.

c. Spontaneous changes in specific activity which occur during growth. For example, it now seems likely that the second and third enzymes of the sequence change in opposite fashion—one decreasing while the other is increasing.

2. Cells which are mutant homozygous at the orotic aciduria locus, and perhaps normal cells as well, increase their specific activity for the terminal enzymes of the UMP sequence in response to a biosynthetic intermediate.

3. Not all the enzymes of the pathway seem to be represented at appreciable levels in all human tissues. Preliminary experiments suggest that human fat and skeletal muscle have high levels of the second enzyme of the sequence, but very low levels of the third and fifth.

The fact that several enzymes of the pyrimidine pathway respond separately from one another in cultured cells, that they are not all represented amongst the somatic tissues, and that the terminal enzymes of the sequence seem to increase in response to an intermediate suggests an interesting theoretical possibility: perhaps not all the differentiated tissues of a metazoan animal, or even of a chain of yeast cells, engage in the *de novo* synthesis of uridine-5'-monophosphate. Instead, in certain circumstances, some kinds of cells may synthesize an intermediate which, when introduced into other cells, causes an increase in those enzymes necessary to convert the intermediate to product (Pinsky and Krooth,

characteristic" in general terms. In the case of orotic aciduria, one would suspect that the mutation were allotopic if a normal tissue were found which was deficient for both the enzymes that are grossly reduced in mutant homozygous cell strains If the ratio of catalytic activity of the final two enzymes of the pathway to one of the earlier enzymes was approximately the same in the normal tissue as in the cell strains from patients with orotic aciduria, one's suspicion that the mutation was allotopic would be strengthened. Finally, if those agents which increase catalytic activity for the affected enzymes in mutant cells had quantitatively the same effect on the normal tissue from which these enzymes were absent, it would seem even more likely that the mutation truly had an allotopic effect.

1967b). The product might then be delivered, in exchange as it were, to the original donor cells. Hence, the two kinds of cells would be nutritionally dependent on one another (Krooth, 1965).

If this theory, which I call the Theory of Intercellular Complementation, should prove true, and if the mechanism it postulates should prove a general one, it suggests one way whereby eukaryotic cells, especially those in a metazoan organism, might influence and control one another. The complementing "deletions" of catalytic activity due to cellular differentiation might provide one mechanism whereby complicated tissues, containing many cell types, could grow in a coordinated fashion. The nutritional interdependence so created might tend to keep different kinds of cells, or even different organs, in the correct proportion to one another. However, one would guess that if such a mechanism truly operates in mammalian tissues, it is likely to be very ancient, and, specifically, to have antedated endocrine and neural controls over cellular behavior. Hence, even if the theory is correct, it is perhaps more likely to apply to the local relations between cells in a tissue than to the coordination of the growth of anatomically remote organs. Moreover, for reasons to be discussed below, the mechanism the theory postulates is not likely to be the sole system for cellular control even within tissues. However, the theory leads one to conjecture that intermediates and products may have functioned as local agents of intercellular control for a long time. Hence, in some cases contemporary hormones may have evolved from them. For example, thyroxine and the adrenal medullary hormones appear to be tyrosine derivatives, and the adrenocortical hormones derivatives of cholesterol. Therefore, two other places one might look for this kind of nutritional interdependence between cells, or at least for remnants of it, is in tyrosine and cholesterol metabolism.

The fact that this postulated control mechanism, if it truly exists, is probably very ancient has another interesting consequence. The mechanism need not apply to the actual control of cell growth in the tissues of an adult, or even in those of a fairly advanced embryo. In organisms which have differentiated to the point of having a nervous system and specialized endocrine glands, crude nutritional controls of the sort described above may no longer be important. They may have been superseded during the evolution of these more specialized structures for intercellular communication. Moreover, once the organism has access to a rich source of nutrients, through a mouth or placenta, and has begun to distribute them amongst the tissues via a cardiovascular system, the somatic cells of the body may be deluged with the products of all the elementary synthetic pathways. However, it does not follow from these arguments that intercellular complementation is necessarily unimportant in man. For one thing, a control mechanism of this sort might truly function in the human being during the earliest events of embryonic development, prior to the formation of nerves, endocrine glands, or a system for distributing maternal nutrients efficiently. According to this scheme, the persistence of patterns of enzyme complementation in the adult would be merely a remnant of what was once a functional mechanism very early in the development of each individual. It is also possible, of course, that intercellular complementation continues to serve as a controlling mechanism in the mature mammal. Eagle *et al.* (1966) have pointed out that many of the elementary nutrients are present in the blood in concentrations sufficient to satisfy amply the nutritional requirements of growing cultured human cells. Hence, they have very reasonably argued that such nutrients are probably not involved in the control of cellular growth *in vivo*. It is eminently likely that this point is correct. On the other hand, of course, no one knows the nutritional requirements of the highly specialized cells inside the body, nor do we have accurate estimates of the concentration of amino acids or nucleosides, deep within the solid tissues—where many kinds of cells are located. There is unquestionably a dilution of small molecules in the blood at the time the molecules cross the capillary wall into the much larger volume of the intercellular fluid. Moreover, equilibration of the blood with the rest of the extracellular fluid is likely to be quite slow. Consequently, it is conceivable that even compounds such as structural amino acids and nucleosides have something to do with the growth rate of cells, or with their metabolic state, within some of the differentiated solid tissues.

The Theory of Intercellular Complementation, whether right or wrong, has one internal paradox. It postulates that biosynthetic intermediates stimulate cells to develop the catalytic activities which convert the intermediate to product. It also postulates that some cells synthesize only the intermediate (which they then give to other cells) because they lack the catalytic activities necessary to complete the synthesis. Why is it that those cells which make only the intermediate do not generate the remaining catalytic activities in response to the intermediate they themselves synthesize? The answer presumably would be that cells which synthesize the intermediate are differentiated so that the intermediate will not generate the remaining catalytic activities *in them*. Only the recipient cells are differentiated to respond so. However, this statement is extremely vague. The Theory of Intercellular Complementation, of course, makes no postulates about the mechanism of cell differentiation, or about the level of the cellular response to biosynthetic intermediates. For example, either process could conceivably involve transcription, translation, or even protein turn-over (or all three). However, there are two simple mechanisms which could occur in the course of cellular differentiation, and could more specifically resolve the paradox. Both have precedents in the biology of eukaryotic cells. The first is that the donor cells are differentiated so that they are unable to concentrate the intermediates they synthesize. Hence, the intermediate runs out of the cell and cannot stimulate the cell to develop the enzyme activities necessary to complete the synthesis. There is now evidence that at least *cultured* mammalian cells sometimes lose intermediates at rates comparable to the rates at which they synthesize the intermediates (Eagle, 1965, and *vide infra*). This possibility has one hopeful property. It leads to the prediction that it might sometimes be possible to stimulate cells to develop specialized biosynthetic enzymes which they do not spontaneously contain by causing intermediates to accumulate within them. Another possible mechanism for resolving the paradox is to assume that the cell is organized so that the intermediate synthesized by the donor cells never gets to the site within the cell where it could lead to the development of the remaining catalytic activities of the pathway. In other words, as a result of differentiation, the interior organization of the cell excludes, or largely excludes, the intermediate from whatever molecular receptor site is involved in the cellular response to the intermediate. Examples of such compartmentalization of metabolites within eukaryotic cells are known (Davis, 1967).

Perhaps the simplest explicit formulation of the scheme is as follows: Some cells lack (or have very low levels) of activity for the final enzymes of the sequence, but have high levels of activity of the early ones. These cells are called "primary donor cells." They synthesize a biosynthetic intermediate which diffuses (or is secreted) out of the cell. They have low levels of the final enzymes because their intercellular pool of intermediate is low. Other cells lack or have low levels of the early enzyme activities of the sequence, but have high levels of the later activities. These cells are differentiated so that they can concentrate the intermediate. They have high levels of the final enzymes of the sequence because their intracellular pool of the intermediate is high. These cells are called "primary recipient cells." They trap the intermediate, complete the synthesis, and give some of the product back to the primary donor cells. Part of this scheme is outlined in Figure 2.

The primary donor cell is likely to have low levels of intermediate, for the intermediate either diffuses out of the cell or perhaps is actively secreted. The primary recipient cell is likely to have high levels of both intermediate and product. It has high levels of intermediate because it can concentrate intermediate, and high levels of product because it makes product and presumably retains some for its own nutritional requirements. In addition, if there are many primary donor cells for each primary recipent cell, the concentration of intermediate in the primary recipient cell will be increased even further. These arguments are of interest if they are combined with the "sequential induction" scheme (described earlier) which has been found to apply to leucine biosynthesis in *Neurospora crassa* (Gross, 1965) and uridine-5′-monophosphate bio-

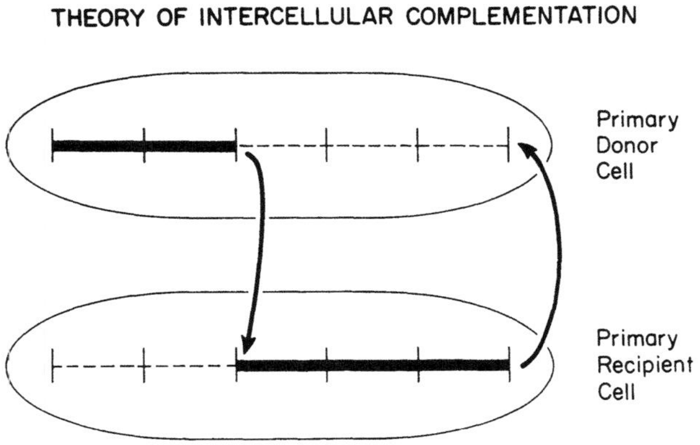

Fig. 2. Diagram of Theory. Each of the two ellipses is supposed to be a cell. The line in each cell represents the uridine-5′-monophosphate biosynthetic pathway, and the five segments of the line represent the five sequentially acting enzymes (Fig. 1). The thickness of each line segment denotes the amount of the corresponding enzyme activity which is supposed to be present in the cell.
The upper cell is a "primary donor cell." It has high levels of the first two enzymes of the sequence and low levels of the final three. It releases dihydroorotic acid—an intermediate in the sequence (Fig. 1). The lower cell is a "primary recipient cell." It has low levels of the first two enzymes of the sequence and high levels of the final three. The primary recipient cell traps intermediate and completes the synthesis. It releases some of the product which is, in turn, trapped by the primary donor cell. The two cells are thus differentiated so that they are nutritionally dependent on one another.

synthesis in *Saccharomyces cerevisiae* (Lacroute, 1966 and 1968). It can be shown that if this scheme applies to human cells, complementary patterns of enzyme activity will develop simply in response to the difference in permeability postulated in the preceding paragraph.

Diploid cell strains from normal individuals have high levels of at least four (and probably all five) of the enzymes of the pyrimidine sequence. Hence, such cells may be differentiated, so that they neither trap nor excrete the biosynthetic intermediate of the pyrimidine sequence. They may synthesize and utilize these compounds autonomously. They may be differentiated so that they do not participate in metabolic complementation. Cell strains from patients with orotic aciduria might be abnormal because they inappropriately excrete one or more of the pyrimidine intermediates. Consequently, they might not accumulate enough intermediate to maintain high levels of the final two enzymes of the sequence.

Some support for the Theory of Intercellular Complementation comes from experiments on the metabolism and synthesis of cystine. All lines of cultured human cells thus far studied have a nutritional requirement for cystine (Eagle, 1959). However, there is evidence suggesting that human heteronuclear cells are able to synthesize this amino acid, despite their nutritional requirement for it. The most plausible explanation for the results thus far obtained is that the heteronuclear cells are ineffective at concentrating the cystine they synthesize; and that, morever, cystine intermediates such as homocystine and cystathionine diffuse (or are secreted) out of the cell into the medium (Eagle and Oyama, 1961; Eagle and Piez, 1962). Normal homonuclear cells seem to possess the early enzymes of the cystine biosynthetic sequence, for they can synthesize homocystine from methionine and cystathionine from homocystine (Eagle *et al.,* 1966). However, such cells cannot grow in cystathionine, in lieu of cystine. Therefore, they may not have appreciable levels of cystathionase or they may be unable to concentrate cystathionine or the cystine which they synthesize (Eagle *et al.,* 1966). The tissue distribution of three enzymes of the cystine sequences has been studied in some detail (Mudd *et al.,* 1965). Relatively high levels of all three enzymes are present in liver, pancreas, and kidney. Skeletal muscle has very low levels of all three enzymes, but especially of the final two. Fat has high levels of the penultimate enzyme of the sequence, but low levels of the first and last enzyme. Testis and heart have appreciable levels of the first enzyme, but not of the final two. The tissue distribution of enzymes in the cystine sequence is comparable to, but not identical with, the distribution of enzymes in the pyrimidine sequence (Wuu and Krooth, 1968b).

The Theory of Intercellular Complementation is not easy to test rigorously. For one thing, some of the mechanisms it postulates are not specified in much detail. There is simply not enough information. Some of the simpler conjectures, such as the idea that orotic aciduric cultured cells are abnormally inefficient at concentrating one of the pyrimidine intermediates which they synthesize, may be feasible to test. However, such predictions are not really critical to the general theory that different kinds of cells within a tissue have complementary patterns of enzyme deficiency. Another problem is that at the present level of our technology it is not easy to measure enzymes in individual cells—especially if it is postulated that intermediates and products run between cells. Nor is it easy to obtain purified populations of a particular kind of cell. Red cells and sperms are available, but other cell types are more difficult to purify. Studies on whole organs, such as those just discussed, of course measure enzyme activities in the chief cell type present. However, the data confound the enzymes of the parenchymatous cells with those of stromal cells, vascular cells, etc. Nor can these cell types, which are infrequent in the tissue, be properly studied in themselves. It should, however, be possible to study the level of activity for enzymes of the pyrimidine pathway in, say, gastrointestinal mucosa, lymphocytes, and sperms. As more pure cell populations are studied, it will be of great interest to determine whether patterns of *complementation* of deficient activity can be pieced together. Finally, it should be feasible to determine whether activity for the final enzymes of the pyrimidine pathway in muscle can be increased by the same agents that increase activity for these enzymes in orotic aciduric cells. Beyond this, however, more work needs to be done on formulating the theory more explicitly, so that truly rigorous and critical predictions can be obtained.

Summary

It is now known that some of the cells of every mammal can grow as microorganisms outside the body. Under appropriate circumstances, these cells will reveal by their biochemistry, nutrition and even appearance, their genetic resemblance to their former host.

Most mammalian cell lines can be classified, on the basis of their chromosomal behavior, into either of two distinctive categories (homonuclear and heteronuclear). Homonuclear cell lines from normal human beings have a diploid chromosome complement and will grow *in vitro* for about 50 cell generations. Homonuclear lines show no more variation from cell to cell in chromosome number than do intact somatic cells. Human homonuclear cells breed true *in vitro* and do not develop abnormalities of either chromosome number or structure, except perhaps at the very end of their *in vitro* life span.

Over 22 mutant Mendelian genes have been shown to confer a distinct phenotype on cultured human diploid cells. These genes are probably at about 17 different loci. Three of the loci are on the X-chromosome. Mutations at each of these sex-linked loci have been used to prove, by microbiological methods, the hereditary dimorphism (with respect to whether the maternal or paternal X-chromosome is functional) of human female somatic cells. In the case of three loci (all autosomal), selective media which disfavor mutant homozygous cells but which will permit the growth of normal or heterozygous cells have been developed. At five loci there are methods for determining the cellular phenotype of individual cells. Two of these five loci are on the X-chromosome.

In two cases, it has been possible to change reversibly the biochemical phenotype of mutant cells by a specific change in the molecular composition of the medium. In one of these instances, the cells respond to a chemical stimulus by acquiring two normal catalytic activities which they are unable to mobilize spontaneously.

Recent studies on diploid somatic cells have suggested the Theory of Intercellular Complementation. The theory states that not all somatic cells engage in the *de novo* synthesis of elementary metabolites, such as certain amino acids and nucleotides. Instead, some cells synthesize only the intermediate in the biosynthetic pathway, which they then donate to other cells. The recipient cells, in response to the intermediate, develop the necessary catalytic activities to convert the intermediate to product. The recipient cells then donate the product to the cells from which they originally obtained the intermediate. The theory predicts that there exist complementary patterns of enzyme deficiencies amongst the cells of intact somatic tissues, and that these deficiencies complement within single biosynthetic sequences. It is postulated that this type of complementation may create nutritional interdependence between different kinds of specialized cells and thereby may tend to keep different types of cells in a fixed proportion during growth. The theory holds that intercellular complementation of this kind was important at the beginning of the evolution of multicellular life and may continue to be important during the early stages of mammalian development. The data thus far collected suggest, but by no means prove, the theory. They also suggest that two compounds likely to be involved in this type of complementation are the amino acid cystine and the nucleotide uridine-5′-monophosphate. Two mammalian tissues in which intercellular complementation involving these compounds may play a role are skeletal muscle and fat.

Acknowledgment

The original investigations described in this paper were supported by Area Grant 1-PO1-GM-15419, National Institutes of Health, United States Public Health Service, Bethesda, Maryland, U.S.A.

References

Baugh, C. L., R. W. Lecher and A. A. Tytell: The effect of pH on the propagation of the diploid cell WI-38 in galactose medium. J. Cell. Physiol. *70*:225, 1967.

Beckwith, J. R., A. B. Pardee, R. Austrian and F. Jacob: Coordination of the synthesis of the enzymes in the pyrimidine pathway of *E. coli*. J. Mol. Biol. *5*:618, 1962.

Danes, B. S. and A. G. Bearn: Hurler's syndrome: Demonstration of an inherited disorder of connective tissue in cell culture. Science *149*:987, 1965.

——— and ———: Hurler's syndrome: A genetic study in cell culture. J. Exp. Med. *123*:1, 1966a.

——— and ———: Hurler's syndrome: Effect of retinol (vitamin A alcohol) on cellular mucopolysaccharides in cultured human skin fibroblasts. J. Exp. Med. *124*:1181, 1966b.

——— and ———: Hurler's syndrome: A genetic study of clones in cell culture with particular reference to the Lyon hypothesis. J. Exp. Med. *126*:509, 1967a.

——— and ———: Cell culture and the Chediak-Higashi syndrome. Lancet *ii*: 65, 1967b.

——— and ———: A genetic cell marker in cystic fibrosis of the pancreas. Lancet *i*:1061, 1968.

Davidson, R. G., H. M. Nitowsky and B. Childs: Demonstration of two populations of cells in the human female heterozygous for glucose-6-phosphate dehydrogenase variants. Proc. Nat. Acad. Sci. *50*:481, 1963.

———, A. M. Glen-Bott and H. Harris: Clonal study of three autosomal loci in heterozygotes. Amer. Soc. Human Genet. Abstracts. (Abstract 45, p. 35: Annual Meeting of 25–27 August, 1965.)

Davis, R.: Channeling in *Neurospora* metabolism. In: Organizational Biosynthesis. H. J. Vogel, J. O. Lampen and V. Bryson, eds. New York: Academic Press Inc., 1967, pp. 303–322.

DeMars, R. and W. E. Nance: Electrophoretic variants of glucose-6-phosphate dehydrogenase and the single active X-theory *in* Defendi, V. (editor): Retention of Functional Differentiation in Cultured Cells. Philadelphia: Wistar Institute Press, 1964, pp. 35–48.

——— and J. G. Leroy: The remarkable cells cultured from a human with Hurler's syndrome: An approach to visual selection for *in vitro* genetic studies. *In Vitro 2*:107, 1966.

Dorfman, A.: Heritable diseases of connective tissues: the Hurler syndrome. In: The Metabolic Basis of Inherited Disease. J. B. Stanbury, J. B. Wyngaarden and D. S. Fredrickson, eds. 2nd ed. New York: McGraw-Hill Book Co., 1966, pp. 963–994.

Eagle, H.: Amino acid metabolism in mammalian cell cultures. Science *130*:432, 1959.

———, K. A. Piez and V. I. Oyama: The biosynthesis of cystine in human cell cultures. J. Biol. Chem. *236*:1425, 1961.

——— and K. Piez: The population-dependent requirement by cultured mammalian cells for metabolites which they can synthesize. J. Exp. Med. *116*:29, 1962.

———: Metabolic control in cultured mammalian cells. Science *148*:42, 1965.

———, C. Washington and S. M. Friedman: The synthesis of homocystine by cultured diploid and heteroploid human cells. Proc. Nat'l. Acad. Sci. *56*:156, 1966.

Gartler, S. M. and D. A. Pious: Genetics of mammalian cell cultures. Humangenetik *2*:83, 1966.

Gitzelman, R.: Hereditary galactokinase deficiency, a newly recognized cause of juvenile cataracts. Pediatrics Res. *1*:14, 1967.

Green, H., B. Goldberg and G. J. Todaro: Differentiated cell types and the regulation of collagen synthesis. Nature *212*:631, 1966.

——— and G. J. Todaro: The mammalian cell as differentiated microorganism. Ann. Rev. Microbiol. *21*:573, 1967.

Gross, S. R.: Proc. Nat. Acad. Sci. *54*:1538, 1965.

Harris, M.: Cell Culture and Somatic Variation. New York: Holt, Rinehart and Winston, 1964.

Howell, R. R., J. R. Klinenberg and R. S. Krooth: Enzyme studies on diploid cell strains developed from patients with hereditary orotic aciduria. Johns Hopkins Med. J. *120*:81, 1967.

Kelley, W. N., F. M. Rosenbloom, J. F. Henderson and J. E. Seegmiller: A specific enzyme defect in gout associated with overproduction of uric acid. Proc. Nat'l. Acad. Sci. *57*:1735, 1967a.

———, F. M. Rosenbloom, J. F. Henderson and J. E. Seegmiller: Xanthine phosphoribosyltransferase in man: Relationship to hypoxanthine-guanine phosphoribosyltransferase. Biochem. Biophys. Res. Comm. *28:*340, 1967b.

Kelsey, W., K. D. Wuu, R. Y. Ma and R. S. Krooth: Unpublished observations, 1968.

Krooth, R. S. and A. N. Weinberg: Studies on cell lines developed from the tissues of patients with galactosemia. J. Exp. Med. *113:*1155, 1961.

———: Characterization of genetically marked human diploid cell strains. Proceedings of the Symposium on the Characterization and Uses of Diploid Cell Strains, Opatija 1963. Zagreb: International Association of Microbiological Societies, 1963, pp. 79–95.

———, ed.: Somatic Cell Genetics. (Fourth Macy Conference on Genetics.) Ann Arbor: University of Michigan Press, 1964a.

———: Properties of diploid cell strains developed from patients with an inherited abnormality of uridine biosynthesis. Cold Spring Harbor Symp. Quant. Biol. *29:*189, 1964b.

———: Study of galactosemia, acatalasemia, and other human metabolic mutants in cell culture. In: Somatic Cell Genetics. (Fourth Macy Conference on Genetics.) R. S. Krooth, ed. Ann Arbor: University of Michigan Press, 1964c, pp. 167–194.

———: The future of mammalian cell genetics. Birth Defects (Original Article Series) *1*(2):21, 1965.

———: Some properties of diploid cell strains developed from the tissues of patients with inherited biochemical disorders. *In Vitro 2:*82, 1966.

———, G. A. Darlington and A. Velazquez: The genetics of cultured mammalian cells. Ann. Rev. Genetics *2* (in press), 1968.

Lacroute, F.: Thèses: Regulation de la Chaine de Biosynthèse de l'Uracile chez *Saccharomyces cerevisiae.* Faculté des Sciences, Université de Paris, Juin, 1966.

———: Regulation of pyrimidine biosynthesis in *Saccharomyces cerevisiae.* J. Bacteriol. *95:*824, 1968.

Levintow, L. and H. Eagle: Biochemistry of cultured mammalian cells. Ann. Rev. Biochemistry *30:*605, 1961.

Lyon, M.: Chromosomal and subchromosomal inactivation. Ann. Rev. Genetics *2:* (in press), 1968.

Matalon, R. and A. Dorfman: Hurler's syndrome: Biosynthesis of acid mucopolysaccharides in tissue culture. Proc. Nat'l. Acad. Sci. *56:*1310, 1966.

Migeon, B. R., V. M. Der Kaloustian, W. L. Nyhan, W. J. Young and B. Childs: X-linked hypoxanthine-guanine phosphoribosyltransferase deficiency: Heterozygote has two clonal populations. Science *160:*425, 1968.

Mudd, S. H., J. F. Finkelstain, F. Irreverre and L. Laster: Transsulfuration in mammals. J. Biol. Chem. *240:*4382, 1965.

Nyhan, W. L., J. Pesek, L. Sweetman, D. G. Carpenter and C. H. Carter: Genetics of an X-linked disorder of uric acid metabolism and cerebral function. Pediat. Res. *1:*5, 1967.

Pan, Y. and R. S. Krooth: The influence of progressive growth on specific catalase activity. I. Effect of cellular genotype: Homozygous strains. J. Cell. Physiol. *71:*151, 1968.

Pinsky, L. and R. S. Krooth: Studies on the genetic control of pyrimidine biosynthesis in human diploid cell strains. I. Effect of 6-azauridine on cellular phenotype. Proc. Nat. Acad. Sci. *57*:925, 1967a.

——— and ———: Studies on the genetic control of pyrimidine biosynthesis in human diploid cell strains: II. Effects of 5-azaorotic acid and pyrimidine precursors on cellular phenotype. Proc. Nat. Acad. Sci. *57*:1267, 1967b.

Rosenbloom, F. M., W. N. Kelley, J. F. Henderson and J. E. Seegmiller: Lyon hypothesis and X-linked disease (Lesch-Nyhan Syndrome). Lancet *ii*:305, 1967.

Seegmiller, J. E., F. M. Rosenbloom and W. N. Kelley: Enzyme defect associated with a sex-linked human neurological disorder and excessive purine synthesis. Science *155*:1682, 1967.

Sigman, B. and S. M. Gartler: The absence of inactivation at two autosomal loci. Humangenetik *2*:372, 1966.

Smith, L. H. and F. A. Barker: Pyrimidine metabolism in man. I. Biosynthesis of orotic acid. J. Clin. Investig. *38*:798, 1959.

———, Jr., M. Sullivan and C. M. Huguley, Jr.: Pyrimidine metabolism in man. IV. Enzymatic defect of orotic aciduria. J. Clin. Investig. *40*:656, 1961.

———, C. M. Huguley and J. A. Bain: Hereditary orotic aciduria. Chap. 33 in The Metabolic Basis of Inherited Disease. J. B. Stanbury, J. B. Wyngaarden and D. S. Fredrickson, eds. 2nd ed. New York: McGraw-Hill Book Co., Inc., 1966, pp. 739–758.

Tedesco, T. A. and W. J. Mellman: Arginosuccinate synthetase activity and citrulline metabolism in cells cultured from a citrullinemic subject. Proc. Nat. Acad. Sci. *57*:829, 1967.

Uhlendorf, B. W. and S. H. Mudd: Cystathionine synthetase in tissue cultures derived from human skin: Enzyme defect in homocystinuria. Science *160*:1007, 1968.

Wajntal, A. and R. DeMars: A tetrazolium method for distinguishing between cultured human fibroblasts having either normal or deficient levels of glucose-6-phosphate dehydrogenase. Biochemical Genetics *1*:65, 1967.

Woolley, D. W.: 3-methylaspartic acid as a potent antimetabolite of aspartic acid in pyrimidine biosynthesis. J. Biol. Chem. *235*:3238, 1960.

Wuu, K. D. and R. S. Krooth: Dihydroorotic acid dehydrogenase activity of human diploid cell strains. Science *160*:539, 1968a.

——— and ———: Unpublished observations, 1968b.

CHROMOSOME MOSAICS AS MARKERS IN EMBRYOLOGY

H. G. Schwarzacher
Department of Anatomy, Justus Liebig University, Giessen, Germany

In many of the reported cases of human chromosome mosaics, the different types of cells present have been found in more than one tissue of the body. It can be postulated that in such cases the event leading to a mosaic condition happens early in development. Otherwise, a higher percentage of aberrant cells could only arise by repetitions of the same event, or by an advantage in proliferation of the aberrant cells. However, in a mosaic that arises before differentiation of the main tissues, a certain spatial distribution of marker cells should be expected if the assumption is made that different regions of the early embryo have certain prospective developmental significance. Detailed analyses of the distribution of the cell types in mosaics may help clarify this problem. In practically all cases of human mosaics only a few tissues could be investigated. Klinger and Schwarzacher (1962) reported one case of a human embryo whose chromosomally aberrant cells could be identified because they were sex chromatin positive, in contrast to the remaining cells which were sex chromatin negative. In the present paper some further investigations of this case are reported.

Distribution of Sex Chromatin Positive Cells in Various Tissues of a Human XY/XXY Mosaic Embryo

A full description of this 60 mm CRL 46,XY/47,XXY mosaic embryo was given by Klinger and Schwarzacher (1962). In tissue culture 5.3–17.9% of the cells of five different tissues of the embryo and 1.3% of cultured amnion cells had a 47/XXY chromosome complement, the rest were 46/XY. The frequency of sex chromatin positive cells ranged from 5–18%

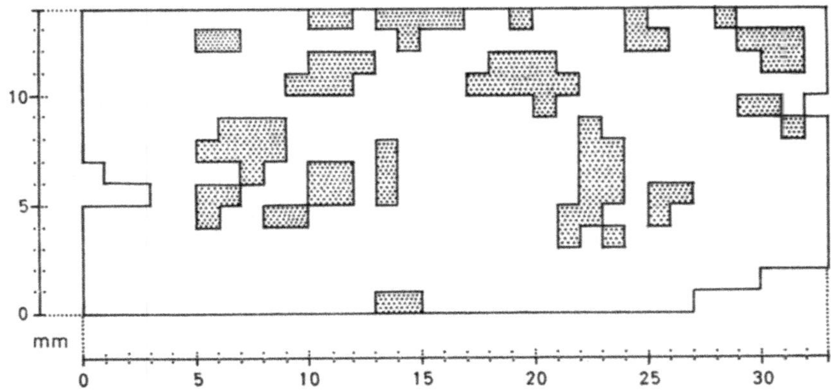

Fig. 1. Distribution of sex chromatin positive cells in a 4.48 cm² field of the amnion epithelium. Areas of 1 mm² containing more than 40% sex chromatin positive cell nuclei are marked.

in cultured cells and from 6.7–15% in different tissues representing all three primary germ layers.

The mosaic pattern is apparent, in that sex chromatin positive nuclei are collected in more or less circumscribed clusters. In less systematically organized tissues, such as the loose connective tissue, clusters of marker cells are less distinct, whereas in well organized tissues, such as the epithelial layers, the clusters of sex chromatin positive cells are well defined. In internal organs such sex chromatin positive patches are not bigger than about 100μ in diameter (Fig. 4 in Klinger and Schwarzacher, 1962). Further investigations of different epithelial layers revealed that in the epidermis such patches occasionally reach a diameter of 400μ. In the amnion epithelium, the patches of sex chromatin positive cells are still larger, as has been described in the previous paper. A detailed study of an area measuring 9.4 cm² (2 fields of 4.5 and of 4.9 cm²) revealed 24 patches, irregular in shape and ranging in size from 1 to 28 mm² (Fig. 1). Marker cells were also found in the embryonic parts of the placenta and in the chorion. In the chorion epithelium (trophoblast) irregular small patches of sex chromatin positive cells occur.

Possible Origin of the Mosaic

The possible ways in which this mosaic condition could have arisen have been discussed in the previous paper (c.f.). The simplest assumption would be that a somatic non-disjunction of a 46/XY cell led to 47/XXY and 45/Y cells, the latter cell type is presumably not viable. Another possibility is that the embryo is a chimera of XY and XXY zygotes. Other possibilities require more than one aberrant event or the repeated occurrence of somatic non-disjunctions.

Embryologic Consequences

If the assumption that one aberrant event led to the origin of marker cells is accepted, the following conclusions may be drawn:

1. The event must have occurred very early, before any differentiation took place, since marker cells are found in all germ layers of the embryo proper, the placenta, and the extra-embryonic membranes. Disregarding differential growth rates of the two cell types, one would assume that the event took place around the 8-cell stage. A non-disjunctional event at the third somatic division would lead to 14.3% of marker cells (see Fig. 8 in Klinger and Schwarzacher, 1962).

2. The cells must have undergone considerable intermingling after the chromosomal aberrant event, since marker cells are present in all parts of the embryo.

It can be assumed that the mixture of cells already occurs in the morula stage, which would account for the inclusion of aberrant cells in the trophoblast as well as in the embryoblast. Cell mixture is also likely to occur when the morula develops into the blastocyst. In this process the outer cell layer is pressed against the zona pellucida and the inner cell group containing the stem cells for the embryo proper is formed. Thus, after the formation of the blastocyst all regions would consist of a mosaic of cells. Within each region further mixture of cells may continue, up to the stage when formation of clones may occur.

Since intermixture of cells must occur at very early stages of development, the assumption may be made that determination has not yet been attained by the cells of the early morula. At the blastocyst stage trophoblast and embryoblast may be differentiated. Within these two types of tissues cell mixture may continue. Thus, whenever at a later stage a layer or group of cells differentiates, it consists of a mixture of cells.

It is conceivable that within the embryo proper extensive intermingling of cells occurs due to the cell movements connected with gastrulation and organ formation. In some tissues the cells may never cease to intermingle, as for instance in connective tissue. Clones of marker cells may develop only in tissues which grow without such cell mixture, as for instance in the epithelial layers. This is the case in the epidermis in which little cell mixture may occur after general body form has been established. The cells of the amnion epithelium probably cease to intermingle at a relatively early stage and, therefore, large clones may develop. In the case of our mosaic embryo, 24 patches of marker cells were found in an area of about 10 cm^2. The total area of the amnion in an embryo of that size is on the average 70 cm^2. Therefore, the total number of patches would be 168. Assuming 14% of marker cells were present also in earlier stages, the amnion epithelium must have consisted of about

1000 cells when clonal growth started. An embryo with an amnion of about 1000 cells is comparable to the Edwards-Jones-Brewer ovum (Brewer, 1937), in which the primitive streak just begins to appear.

Consequences for the Lyon Hypothesis

According to the hypothesis of Lyon (1962) an inactivation of one of the two X chromosomes takes place in females early in development. This inactivation may be related in time with the appearance of sex chromatin bodies. In man, sex chromatin bodies have been reported to appear in trophoblast and chorionic mesoderm at the 12–14th day of gestation, and in embryonic tissues at the 18th day (Park, 1957). In the amnion epithelium cell mixture ceases probably not long after the formation of sex chromatin bodies. Therefore, relatively large clones of cells with either the paternal or the maternal X chromosome inactivated may be formed. In tissues in which cell mixture continues for longer periods, the mosaic of these cells may be rather fine grained. The observation of Linder and Gartler (1965), that in women heterozygous for electrophoretic variants of G6PD even the smallest sample (about 1 mm^3) of normal uterine muscle tissue shows both G6PD variants, is therefore not surprising and cannot be taken as an argument against the Lyon hypothesis, as has been done by Grüneberg (1967).

The observations on sex linked ocular albinism are also in agreement with a cell mixture in early embryonic stages. From the figures given by Fall (1951) and Francois and Deweer (1953) the maximal size of pigment clusters in the ocular fundus of heterozygous females can be estimated to be about 0.1–0.2 mm in diameter. This would require a considerable mixture of cells which form the outer layer of the retina. It was impossible to check on the size of sex chromatin positive patches in our mosaic embryo, since the retinal pigment cell nuclei contain several chromocenters in both sexes.

In the epidermis of our embryo, the patches of sex chromatin positive cells are about 0.01–0.2 mm^2. Allowing for a 1000-fold increase in epidermal area during growth, this would give rise to 0.1–2.0 cm^2 patches in the adult. Gartler and Linder (1964) had similar results in their study of adult women heterozygous for a G6PD deficiency mutation. They estimated the size of mosaic patches in the epidermis to be 0.1–0.2 cm^2, and from this they concluded that the epidermis consisted of 5×10^5 cells at the time when the patches arise. Since X inactivation presumably occurs at an earlier stage when much fewer epidermal cells are present, they considered the possibility of cell mixture continuing until a stage of essential two-dimensional "balloon like" epidermal growth is reached. In the mouse, assuming similar growth conditions as in man, mosaic patches of the epidermis of about 0.1–1 mm^2 would be expected. In the cases of X linked genes which affect coat color and produce a mosaic phenotype when heterozygous in the female mouse, the mosaic areas are usually larger (e.g., Fraser *et al.,* 1953; Dickie, 1954; Lyon, 1960). In man, a case of 46,XX/46,XY mosaicism was reported in which rather large and differently pigmented skin areas were found (Corey *et al.,* 1967). In these cases migration and distribution of melanoblasts are as yet poorly understood. It has been shown, however, that in the mouse embryo the melanoblasts arrive at their destination in the skin at about the 12th day of gestation (Rawles, 1948). In man, melanoblasts have been demonstrated in the dermis at the 10th week (Zimmerman and Becker, 1959). At that time they are presumably still few in number and widely separated. They may, therefore, be able to form quite large clones.

Summary

The distribution of marker cells in a XY/XXY human embryo described by Klinger and Schwarzacher (1962) leads to the conclusion that a considerable mixture of cells must take place during the earliest stages of development. Therefore, single cells or cell groups of the early morula are not yet determined and do not have prospective significance. The mixture of cells seems to continue within different regions and tissues of the embryo until later stages at a different degree.

Some consequences for the Lyon hypothesis result: in diffuse growing tissues (e.g., connective and muscle tissues) of adult females, the mosaic of cells with one or the other X chromosome inactivated is very fine grained. Larger clones may develop only in tissues growing relatively undisturbed such as epithelial layers. The relatively large patches of coat color in female mice heterozygous for sex-linked color genes may be explained by the fact that melanoblast cells migrate into skin late in development and may therefore originate large clones.

References

Brewer, J. I.: A normal human ovum in a stage preceding the primitive streak. (The Edwards-Jones-Brewer Ovum.) Amer. J. Anat. *61:*429, 1937.

Corey, M. J., J. R. Miller, J. R. MacLean and B. Chown: A case of XX/XY mosaicism: Amer. J. Hum. Genet. *19:*378, 1967.

Dickie, M. M.: The tortoise-shell house mouse. J. Hered. *145:*158, 1954.

Fraser, A. S., S. Sobey and C. C. Spicer: Mottled, a sex-modified lethal in the house mouse. J. Genet. *51:*217, 1953.

Gartler, S. M. and D. Linder: Selection in mammalian mosaic cell populations. Cold Spring Harbor Symp. Quant. Biol. *29:*253, 1964.

Grüneberg, H.: Sex-linked genes in man and the Lyon hypothesis. Ann. Hum. Genet., Lond., *30:*239, 1967.

Klinger, H. P. and H. G. Schwarzacher: XY/XXY and sex chromatin positive cell distribution in a 60 mm human fetus. Cytogenetics *1*:266, 1962.

Linder, D. and S. M. Gartler: Glucose-6-phosphate dehydrogenase mosaicism: utilization as a cell marker in the study of leiomyomas. Science *150*:67, 1965.

Lyon, M. F.: A further mutation of the mottled type in the house mouse. J. Hered. *51*:116, 1960.

———: Sex chromatin and gene action in the mammalian X-chromosome. Amer. J. Human Genet. *14*:135, 1962.

Park, W. W.: The occurrence of sex chromatin in early human and macaque embryos. J. Anat. *91*:369, 1957.

Rawles, M. E.: Origin of melanophores and their role in development of color patterns in vertebrates. Physiol. Rev. *28*:383, 1948.

Zimmermann, A. A. and S. W. Becker, Jr.: Precursors of epidermal melanocytes in the negro fetus. Pigment Cell Biol., New York, 1959, pp. 159–170.

MULTIPOLAR MITOSIS AND SOMATIC SEGREGATION IN CELL CULTURES OF MICROTUS AGRESTIS

H. G. SCHWARZACHER AND F. PERA
Department of Anatomy, Justus Liebig University, Giessen, Germany

In tissue cultures derived from epithelial kidney cells of *Microtus agrestis* multinucleated cells and multipolar mitoses are present to a relatively high degree. In a previous study of living cells, cultured in a Sykes-Moore chamber, the formation and division of multinucleated cells was described (Pera and Schwarzacher, 1968). In the course of this study, a binucleated cell was followed through mitosis to the formation of daughter cells.

The two nuclei of the cell started mitosis almost synchronously, and the chromosomes of both nuclei formed one common metaphase plate. A tripolar spindle developed, and the cell divided into 2 cells, one with 2 nuclei and the other with 1 nucleus. DNA measurements of these nuclei (Feulgen staining, histophotometry with a Barr and Stroud microdensitometer) showed values that suggested the presence of a triploid nucleus in the mononucleated cell, and a haploid and a tetraploid nucleus in the binucleated cell. Also, the observed number of chromocenters per nucleus was in agreement with the distribution of the chromosomes to the daughter nuclei in this way. In dipolid cells of both sexes in *Microtus agrestis* the 2 very large sex chromosomes may form 2 large chromocenters in interphase (Wolf *et al.,* 1965; Schmid *et al.,* 1965). Since one chromocenter is expected to occur per haploid set, it should be possible to estimate the ploidy of cells by the number of chromocenters. The formation of chromocenters is, however, not a constant characteristic of all cells (Schmid *et al.,* 1965). Therefore, estimations of the ploidy of a cell nucleus by the number of chromocenters alone is not reliable. On the other hand, a concurrence of the number of chromocenters with the ploidy grade found by photometric DNA measurement in a nucleus may be taken as strong supporting evidence that this nucleus indeed contains the full chromosome set in question. In the present case, in the nuclei of the daughter cells of the tripolar mitosis there was agreement between the number of chromocenters and the nuclear DNA content: the nucleus with a triploid DNA value contained 3, the nucleus with a haploid DNA value contained 1, and the nucleus with a tetraploid value contained 4 chromocenters. Therefore, little doubt is left that in this case the 2 diploid nuclei of a binucleated cell fused to constitute a common metaphase plate with a tetraploid number of chromosomes, and that 3 daughter nuclei were formed by a tripolar mitosis, the daughter nuclei being triploid, haploid and tetraploid (Fig. 1).

Multipolar mitoses constitute about 3% of all mitoses in *fixed* preparations. Tripolar mitoses occur much more frequently than mitotic figures with more than 3 poles. In a few cases of tripolar mitoses, the arrangement of chromosomes, particularly the large sex chromosomes, can be seen. The pattern of distribution of the chromosomes to the daughter cells in these cases indicates that tetraploid cells divide into 2 triploid and 1 diploid daughter cells. This is illustrated quite clearly in the case of Figure 2; a female cell in tripolar anaphase is shown; 8 large sex chromatids are seen that are distributed to the prospective daughter nuclei in a ratio of 3:3:2. Also the bulks of the other chromosomes give

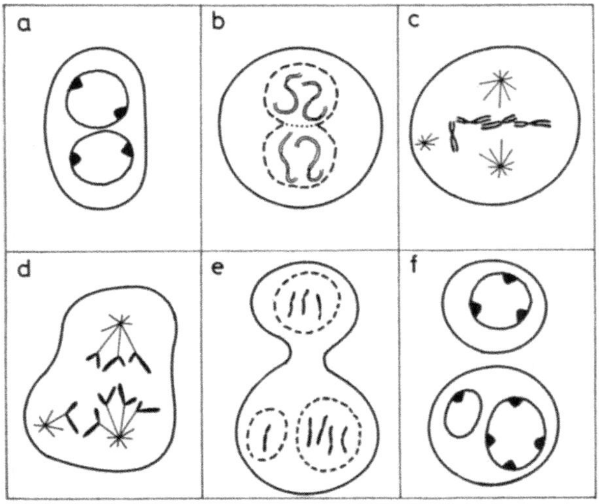

Fig. 1. Schematic representation of the division of a binucleated cell. The 2 diploid nuclei (a) form a common metaphase (c). A tripolar spindle leads to the formation of 3 nuclei, one haploid, one triploid, and one tetraploid (d, e). Cytoplasmic cleavage results in one mononucleated and one binucleated cells (e, f). For original photographs of this observation see Pera and Schwarzacher (1968).

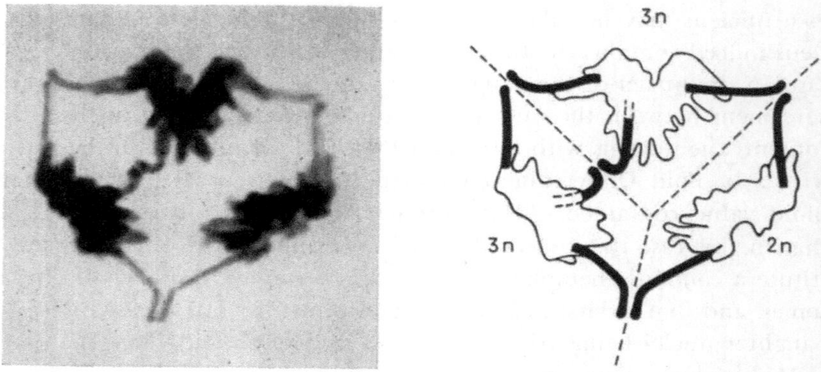

Fig. 2. Tripolar anaphase of a tetraploid mother cell with formation of 2 triploid and 1 diploid daughter cells. Left: microphotography, Feulgen stain, magnification 1,000. Right: schematic representation; the sex chromosomes are drawn in black.

the impression of this distribution, DNA measurement gave values in agreement with it.

Discussion and Conclusions

The formation of daughter nuclei with different ploidies as a result of multipolar mitosis furnishes evidence for the assortment of chromosomes into haploid (or eventually into diploid and triploid) sets. If the chromosomes would not be sorted out into complete sets, a random distribution of chromosomes, perhaps approximating but never giving exactly euploid numbers, would result. Our observations on multipolar mitosis support a distribution of complete sets of chromosomes to the daughter cells. This could be established by a selective alignment of the sets on the metaphase plate as has been discussed by Ohno (1966) for tetrapolar mitosis in tetraploid cells. Likewise, the same mechanism for somatic segregation, proposed by Ohno (1966) for the 4 diploid daughter cells of a tetrapolar mitosis, can be applied to a tripolar mitosis leading to 1 diploid and 2 triploid daughter cells (Fig. 3).

In our material, tetrapolar and more-than-tetrapolar mitoses are very rare compared with tripolar mitoses. Tripolar mitosis seems not to be a very uncommon form of division of a tetraploid cell. This may also be true for other tissues and other species. In this connection we wish to point out that our material consisted of untreated short term cultures, basically of diploid cells.

The factors causing assortment and selective alignment of the chromosomes in multipolar mitosis are not yet understood. A possibility would be that each centriole controls a complete haploid (or diploid, or tri-

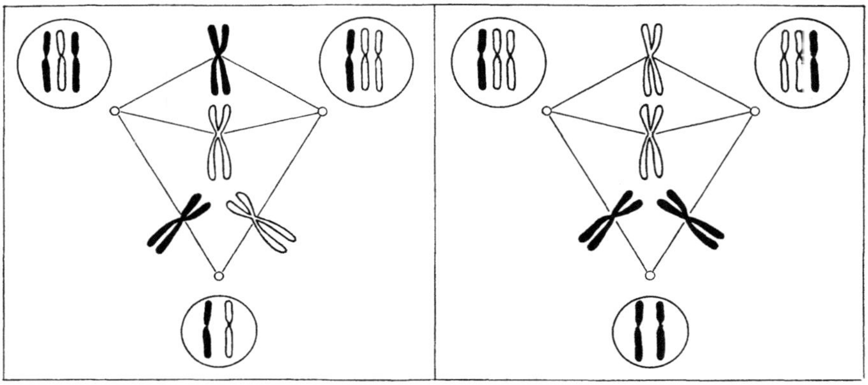

Fig. 3. Possibility of somatic segregation in a tripolar mitosis of a tetraploid mother cell leading to 2 triploid and 1 diploid daughter cells. The 2 homologues of one parent are drawn in black, those of the other parent are left white. The combination at right results in a diploid segregant daughter cell.

ploid) set. Also, causes for the formation of multipolar spindles are not exactly known. Their formation may be related to a surplus of centrioles. No information, however, on the number and spatial distribution of centrioles in multipolar mitoses is available.

Summary

In short term tissue cultures from *Microtus agrestis*, multipolar mitoses constitute about 3% of all mitoses. The observation of a multipolar dividing living cell, as well as analysis and DNA measurement of multipolar mitotic figures and their daughter nuclei in fixed preparations, revealed that, in general, multipolar spindles may develop mostly in polyploid cells and that they lead to euploid daughter nuclei. Among multipolar mitoses, tripolars in tetraploid cells dividing into 2 triploid and 1 diploid daughters are most frequent. An assortment of chromosomes into complete haploid (or diploid, or triploid) sets must be postulated. This includes also the possibility of somatic segregation (Ohno, 1966).

Acknowledgment

This work was supported by the Deutsche Forschungsgemeinschaft.

References

Ohno, S.: Cytologic and genetic evidence of somatic segregation in mammals, birds, and fishes. In: Phenotypic Expression *in vitro* 2:46, Tissue Culture Assoc. Baltimore: Williams and Wilkins, 1966.

Pera, F. and H. G. Schwarzacher: Formation and division of binucleated cells in kidney cell cultures of *Microtus agrestis*. Humangenetik *6*:158, 1968.

Schmid, W., D. W. Smith and K. Theiler: Chromatinmuster in verschiedenen Zelltypen und Lokalisation von Heterochromatin auf Metaphasechromosomen bei *Microtus agrestis, Mesocricetus auratus, Cavia cobaya* und beim Menschen. Arch. Klaus-Stift. Vererb. Forsch. *40*:35, 1965.

Wolf, U., G. Flinspach, R. Böhm and S. Ohno: DNS-Reduplikationsmuster bei den Riesen-Geschlechtschromosomen von *Microtus agrestis*. Chromosoma (Berl.) *16*:609, 1965.

CYTOGENETICS OF MARSUPIALS

D. L. HAYMAN AND P. G. MARTIN
Departments of Genetics and Botany,
University of Adelaide, South Australia

Introduction

The *Marsupialia* are one of the orders of the *Mammalia*. They are of interest for several reasons. They diverged from the general line of mammalian evolution at an early stage and may therefore give us some insight into the early progenitors of present day eutherian mammals. The entry of marsupials into Australia and South America was accompanied by an explosive radiation of adaptations to occupy a diversity of ecological niches. Living members of the group are found in their greatest diversity and abundance in Australia and New Guinea where there are some 160 species (Darlington, 1957). The remainder of the order, about 60 species (Cabrera, 1957), are found in Southern and Central America, and one species extends into North America.

The chromosomes of the Australian marsupials have been studied in some detail, and presently about one half of the Australian species have been examined (McIntosh and Sharman, 1953; Sharman, 1961; Martin and Hayman, 1967). Consequently, relatively more is known of the chromosomes of this group than of any other group of mammals. Studies of the chromosome number and morphology of the American marsupials are beginning (e.g., Biggers *et al.*, 1965).

Cytogenetic information can be used in a number of ways. It provides additional information of a unique kind to complement taxonomic studies and it increases our knowledge of aspects of cell biology which otherwise would not be revealed. Chromosome number and morphology still provide the basic information but increasing use is being made of autoradiography and of measurements of the DNA content of the nucleus to allow additional criteria to be applied.

The purpose of this review is to report on the current status of marsupial cytogenetics with an emphasis on those points of particular interest which have been revealed, even where the analysis of the problem is incomplete.

Chromosome Numbers in the Marsupialia

General

The chromosome numbers known in the *Marsupialia* are set out in Table 1. In general, the marsupials have the smallest number of chromo-

Table 1: Chromosome Numbers of Marsupials *

Species †	2n	Sex Chromosomes ‡	Reference §
Superfamily *Didelphoidea* Gray Family *Didelphidae* Gray			
Didelphis marsupialis (Linnaeus) (Syn. D. aurita, D. virginiana)	22	XY/XX	2
Didelphis azarae (Temminck) (Syn. D. paraguayensis)	22	XY/XX	2
Lutreolina crassicaudata (Desmarest)	22	XY/XX	2
Philander opossum (Linnaeus)	22	XY/XX	7
Caluromys derbianus (Waterhouse)	14	XY/XX	7
Marmosa mexicana (Merriam)	14	XY/XX	7
Superfamily *Dasyuroidea* Goldfuss Family *Dasyuridae* Goldfuss			
Antechinus flavipes (Waterhouse)	14	XY/XX	2
Sminthopsis crassicaudata (Gould)	14	XY/XX	2
Sminthopsis macroura (Gould)	14	XY/XX	2
Sminthopsis frogatti (Ramsay)	14	XY/XX	1
Antechinomys spenceri (Thomas)	14	XY/XX	1
Dasycercus cristicauda (Krefft)	14	XY/XX	3
Dasyuroides byrnei (Spencer)	14	XY/XX	4
Dasyurus geoffroii (Gould)	14	XY/XX	3
Dasyurus hallucatus (Gould) (Syn. Satanellus hallucatus)	14	XY/XX	2
Dasyurus viverrinus (Shaw) (Syn. D. quoll)	14	XY/XX	2
Dasyurus maculatus (Kerr)	14	XY/XX	2
Sarcophilus harrisii (Boitard)	14	XY/XX	2
Myrmecobius fasciatus (Waterhouse)	14	XY/XX	2

Table 1 (Continued)

Species †	2n	Sex Chromosomes ‡	Reference §
Superfamily *Perameloidea* Waterhouse			
Family *Peramelidae* Waterhouse			
Isoodon macrourus (Gould)	14	XY/XX	2
Isoodon obesulus (Shaw)	14	XY/XX	2
Perameles nasuta (Geoffroy)	14	XY/XX	2
Perameles gunnii (Gray)	14	XY/XX	2
Perameles bougainville (Quoy and Gaimard)	14	XY/XX	2
Macrotis lagotis (Reid)	19♂ 18♀	XY_1Y_2/XX	3
Superfamily *Vombatoidea* Iredale and Troughton			
Family *Vombatidae* Iredale and Troughton			
Lasiorhinus latifrons (Owen)	14	XY/XX	2
Vombatus ursinus (Shaw) (Syn. Phascolomis mitchelli ref. 3.) (Syn. Phascolomis ursinus ref. 2.)	14	XY/XX	2
Family *Phascolarctidae* Owen			
Phascolarctos cinereus (Goldfuss)	16	XY/XX	2
Superfamily *Phalangeroidea* Thomas			
Family *Phalangeridae* Thomas			
Phalanger sp.	14	XY/–	1
Trichosurus vulpecula (Kerr)	20	XY/XX	2
Trichosurus caninus (Ogilby)	20	XY/XX	2
Trichosurus arhemensis	20	–/XX	1
Family *Petauridae* Gill			
Petaurus breviceps (Waterhouse)	22	XY/XX	2
Pseudocheirus peregrinus (Broddaert)	20	XY/XX	2
Pseudocheirus archeri (Collett)	16	–/XX	1
Hemibelideus lemuroides (Collett)	20	–/XX	1
Schoinobates volans (Kerr)	22	XY/XX	2
	+2–6 supernumeraries		4
Family *Burramyidae* Broom			
Burramys parvus (Broom)	14	XY/–	5
Acrobates pygmaeus (Shaw)	14	XY/XX	5
Cercartetus concinnus (Gould)	14	XY/XX	2

Table 1 (Continued)

Species †	2n	Sex Chromosomes ‡	Reference §
Cercartetus lepidus (Thomas)	14	XY/XX	2
Cercartetus nanus (Desmarest) (Cercartetus syn. Cercaërtus)	14	XY/XX	3
Family *Macropodidae* Gray			
Hypsiprymnodon moschatus (Ramsay)	22	XY/XX	2
Potorous apicalis (Kerr) (Syn. P. tridactylus)	13♂ 12♀	XY_1Y_2/XX	2
Bettongia gaimardi (Desmarest) (Syn. B. cuniculus)	22	XY/XX	2
Bettongia lesueur (Quoy and Gaimard)	22	XY/XX	2
Bettongia penicillata (Gray)	22	XY/XX	2
Aepyprymnus rufescens (Gray)	32	XY/XX	1
Lagorchestes hirsutus (Gould)	22	XY/XX	2
Lagorchestes conspicillatus (Gould)	15♂ 16♀ (also 14♂)	X_1X_2Y/ $X_1X_1X_2X_2$	6
Lagostrophus fasciatus (Péron)	24	XY/XX	2
Onychogalea unguifera (Gould)	20	XY/XX	2
Petrogale penicillata pearsoni (Thomas)	22	XY/–	2
Petrogale penicillata hacketti (Thomas)	20	XY/–	1
Petrogale xanthopus (Gray)	22	XY/XX	6
Petrogale rothschildi (Thomas)	22	XY/XX	2
Petrogale penicillata signata (Thomas)	18	–/XX	1
Petrogale penicillata inornata (Gould)	18	XY/–	1
Dendrolagus lumholtzi (Collett)	14	XY/–	1
Dendrolagus goodfellowi (Thomas)	14	XY/–	1
Dendrolagus matschiei (Förster and Rothschild)	14	–/XX	1
Dendrolagus ursinus (Müller)	12	XY/–	1
Thylogale stigmatica (Gould)	22	XY/XX	2
Thylogale thetis (Lesson)	22	XY/XX	2
Thylogale billardieri (Desmarest)	22	XY/XX	2
Setonix brachyurus (Quoy and Gaimard)	22	XY/XX	2
Dorcopsis sp.	22	XY/–	1
Wallabia bicolor (Desmarest) (Syn. Protemnodon bicolor)	11♂ 10♀	XY_1Y_2/XX	2
Macropus (For following species Protemnodon or Wallabia may have been used synonymously)			

Table 1 (Continued)

Species †	2n	Sex Chromosomes ‡	Reference §
M. parryi (Bennett)	16	XY/XX	2
M. eugenii (Desmarest)	16	XY/XX	2
M. dorsalis (Gray)	16	XY/XX	2
M. parma (Waterhouse)	16	XY/XX	1
M. agilis (Gould)	16	XY/XX	2
M. rufogrisea (Desmarest)	16	XY/XX	2
M. irma (Jourdan)	16	XY/XX	2
M. robustus (Gould)	16	XY/XX	2
M. fuliginosus (Desmarest) (Syn. M. major varieties fuluginosus, melanops and ocydromus in ref. 2.)	16	XY/XX	2
M. giganteus (Shaw) (Syn. M. major varieties major and tasmaniensis in ref. 2.)	16	XY/XX	2
Megaleia rufa (Desmarest) (Syn. Macropus rufus)	20	XY/XX	2
Superfamily *Tarsipedoidea* Gill			
Family *Tarsipedidae* Gill			
Tarsipes spencerae (Gray)	24	–/XX	1

* North American Marsupials according to J. S. Miller and R. Kellogg, *Bulletin U.S. Nat. Mus. 205:* (1955). South American Marsupials according to A. Cabrera. *Rev. Mus. Argentino Ciencias Barnardino,* Rivadavia *4* (1): (1957). Australian Marsupials according to W. D. L. Ride. A guide to the native mammals of Australia, Oxford University Press, Melbourne (in press).

† Synonyms are only given when we know that recent cytological papers have used these names. They need not have taxonomic validity. Sub-species are only listed when chromosome numbers differ.

‡ If only one sex is known to have been studied, the other sex is not recorded.

§ The last list similar to this appears in reference 2. All previous references are referred to in that paper. Where the animals have been available (90% species) all numbers reported by Sharman (1961) have been verified by us. 1. First recorded in this paper. 2. Sharman, 1961. 3. Martin and Hayman, 1967. 4. Hayman and Martin, 1965. 5. Gunson, Thompson and Sharman, 1968. 6. Martin and Hayman, 1966. 7. Biggers *et al.,* 1965.

somes and the largest chromosomes of the mammals. The species with the largest number is *Aepyprymnus rufescens* (2n = 32; Fig. 6C), the species with the smallest *Wallabia bicolor* (2n = 10 ♀, 11 ♂).

The relationships between chromosome number and taxonomic grouping have been pointed out by Sharman (1961) and extended with further observations by Martin and Hayman (1967). The distribution of chromosome numbers shows a pronounced bimodality—the modes being 2n = 20 or 2n = 22 and 2n = 14 (Fig. 1). This bimodality is distributed amongst the families in the manner summarized in Table 2.

Because three families (one American, two Australian) themselves show the same bimodality it seems probable that it has evolved inde-

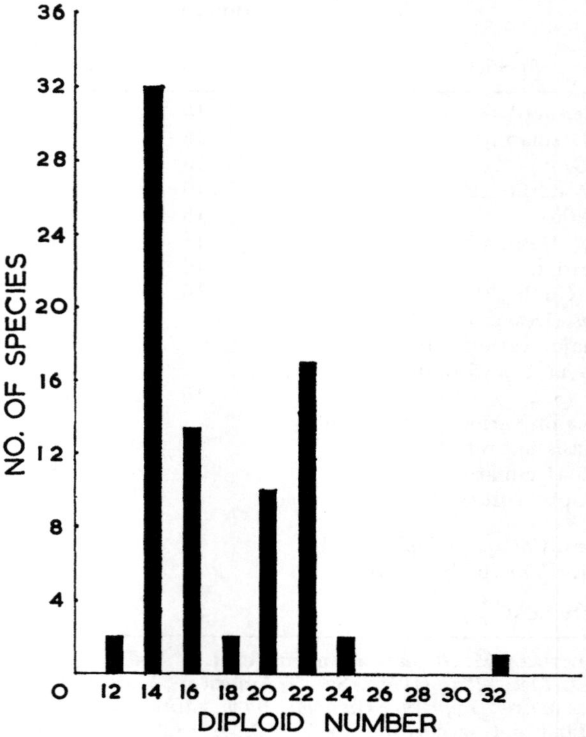

Fig. 1. Histogram of the distribution of the chromosome numbers in the *Marsupialia* (data from Table 1).

pendently a number of times. The three most obvious hypotheses are (1) that the ancestral chromosome number was 2n = 14, evolution to 2n = 20 or 22 having occurred independently a number of times; (2) that the ancestral chromosome number was 2n = 22, evolution to 2n = 20 and/or 2n = 14 having occurred independently a number of times; (3) the ancestral number was neither 2n = 14 nor 2n = 22, but that evolution towards 2n = 14 and 2n = 20 and 2n = 22 has occurred several times. A decision between (1) and (2) could be made if it were shown that the 2n = 14 karyotypes in different families were very similar, while the 2n = 20 or 22 were different or vice versa. This is the approach followed in this review. Since it involves comparisons of karyotypes our methods are briefly outlined in the next paragraph. However, these methods are very time consuming and they have not yet been applied to all the comparisons which will be discussed.

Two comparisons are commonly made between the chromosomes of different species in order to determine what relationship may exist between them. The first of these is to compare chromosome arms—the "nombre fondamental" (Matthey, 1949). We have found this difficult to apply to the marsupials because of the different size of the small arms of more or less acrocentric chromosomes. This tended to make decisions on whether or not they should be called arms, and counted, arbitrary. Sharman (1961) has had the same problem. The concept, however, has proved useful in certain comparisons. The second is to determine the mean percent length of each chromosome or arm in terms of the total complement (Rothfels and Siminovitch, 1958). Where the DNA contents of the nucleus are the same, comparisons between species based on this criterion would be quite valid. However, in order to facilitate the comparisons which can be made between the chromosomes of different species it was suggested (Martin and Hayman, 1965) that if the measurements of chromosome length are corrected for the total DNA content of the nucleus, then a quantitative method of comparing the chromosomes could be developed. The assumption is that in an interspecific comparison the ratio of total lengths will be the same as the ratio of amounts of nuclear DNA, i.e., that the DNA is distributed uniformly along the

Table 2: Genera of Marsupials Grouped According to Chromosome Number

	20 or Over	*14*	*Intermediate*
Didelphoidea			
Didelphidae	3 genera 4 species	2 genera 2 species	
Dasyuroidea			
Dasyuridae		8 genera 13 species	
Perameloidea			
Peramelidae	1 genus 1 species	2 genera 5 species	
Vombatoidea			
Vombatidae		2 genera 2 species	
Phascolarctidae			1 genus 1 species
Phalangeroidea			
Phalangeridae	2 genera 3 species	1 genus 1 species	
Petauridae	4 genera 4 species		1 genus 1 species
Burramyidae		3 genera 5 species	
Macropodidae	11 genera 18 species	2 genera 5 species	3 genera 13 species
Tarsipedoidea			
Tarsipedidae	1 genus 1 species		

compacted metaphase chromosome. Critical evidence for this assumption is still not available, although the evidence that is available supports it (Carlson *et al.*, 1963; Radley, 1966).

Marsupials with 2n = 14

Using this comparative technique the chromosome morphology in representatives of the fourteen chromosome species of three of the superfamilies of Australian marsupials have been examined in detail (Martin and Hayman, 1967).

The five species of *Dasyuridae* studied showed very similar autosomal karyotypes with only few variations. Ignoring these variations a "standard" karyotype could be constructed on the basis of the common chromosomes. From this "standard" it was possible to construct the exceptions by postulating interchanges between arms of the "standard" karyotype. While the constancy of chromosome number has been maintained within the whole superfamily there has been slight modification of the chromosome morphology during evolutionary diversification.

Three species of *Peramelidae* with the germ line chromosome number of 2n = 14 have been analyzed and all show an essentially similar autosomal complement with small but significant differences between many of the chromosome arms and it was not possible to construct a "standard" karyotype. The same situation applies to the two fourteen chromosome species of the *Vombatoidea*. Within each group there were found to be more chromosomes with similar lengths than would be expected by chance alone.

Comparative DNA measurements are not available for the American species in which 2n = 14 has been found, and so a comparison of the form followed in the previous study cannot be made. However, a visual comparison can be made between representatives of the 2n = 14 species of the marsupials from all living families. In Figures 2 and 3 the chromosomes of *Sminthopsis crassicaudata* (Dasyuridae), *Caluromys derbianus* (Didelphidae), *Acrobates pygmaeus* (Burramyidae), the 2n = 13 complement of *Perameles nasuta* (Peramelidae), *Vombatus ursinus* (Vombatidae), *Phalanger sp.* (Phalangeridae), *Dendrolagus goodfellowi* (Macropodidae) and *Potorous apicalis* (Macropodidae) are arranged. Considering the chromosomes in Figure 2 first there is a striking constancy of chromosome form in the karyotype; each species has three large pairs of autosomes, 1 intermediate sized pair, and 2 small pairs. In addition, in three of the species the satellited chromosomes—presumably those involved in the nucleolar organization—are very similar in morphology (see arrows in Fig. 2). The failure to recognize these satellited chromosomes in the other complements is probably due to the quality of the preparations. We are sure that in *Cercartetus nanus*—a species with a very similar karyotype to that of *Acrobates pygmaeus* (Martin and Hayman, 1967)

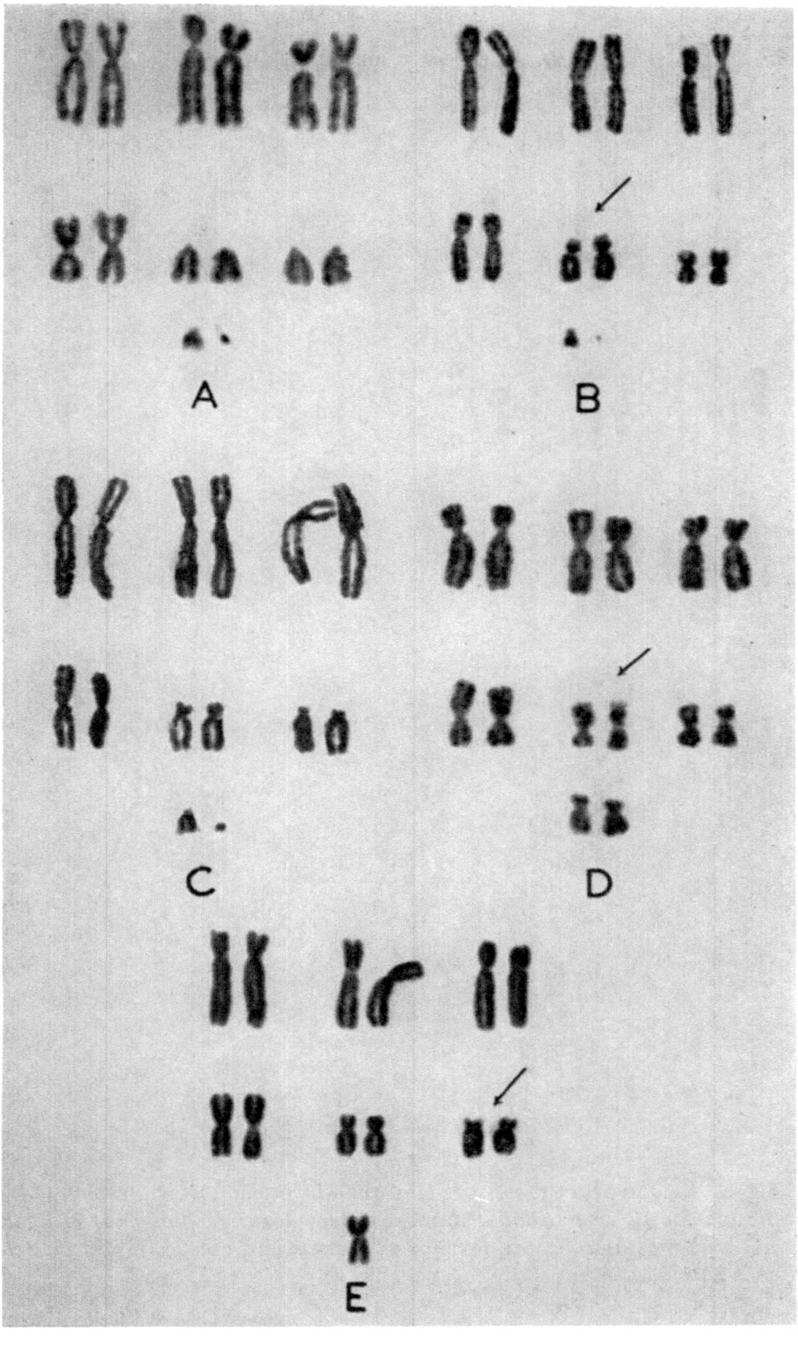

Fig. 2. Karyotypes of typical representatives of those families of the *Marsupialia* which have 2n = 14 and have a similar form. A. *Acrobates pygmaeus;* B. *Sminthopsis crassicaudata;* C. *Caluromys derbianus;* D. *Vombatus ursinus;* E. *Perameles nasuta* (2n = 13 tissue). The arrows indicate the satellited chromosomes. See the text for discussion.

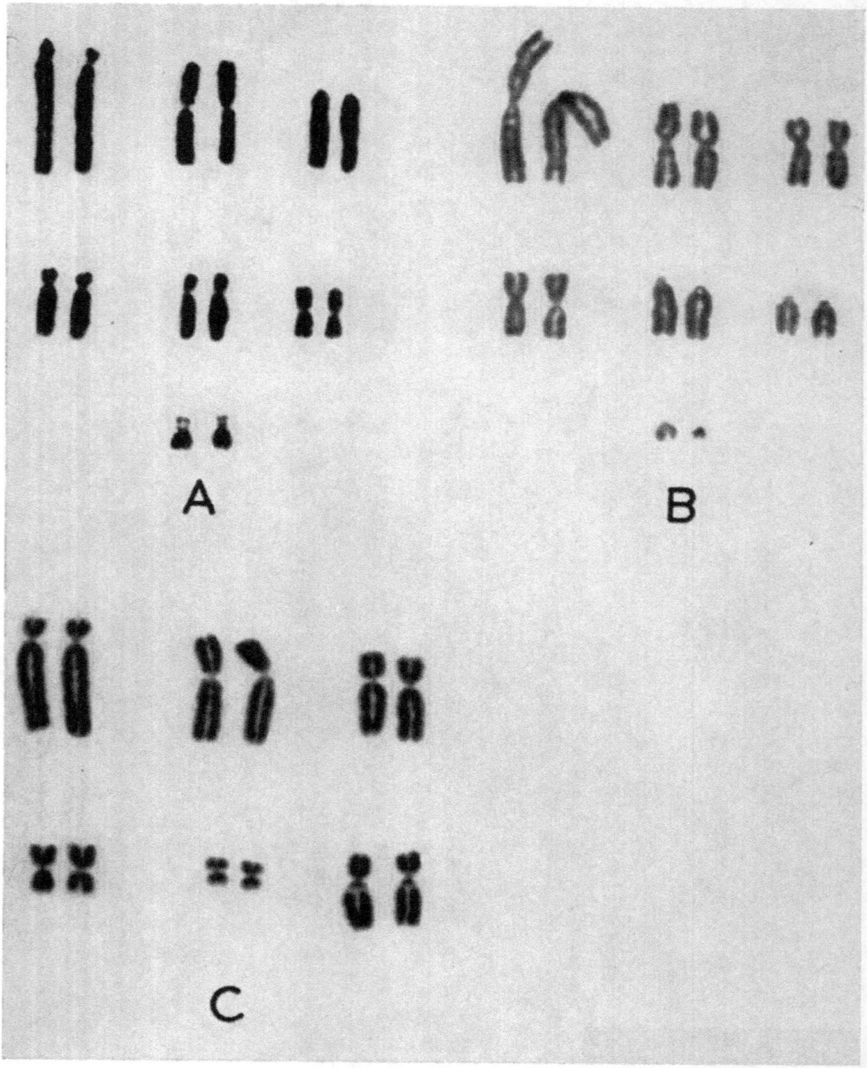

Fig. 3. Karyotypes of those species of the *Marsupialia* which have 2n = 14 and do not obviously have a similar form. A. *Dendrolagus goodfellowii;* B. *Phalanger* sp.; C. *Potorous apicalis.* See text for explanation of the derivation of *Potorous* from a 2n = 14 species.

and one which is in the same family—the satellited chromosomes are one of the two small autosomal pairs.

Considering the range of morphologies possible with a chromosome complement of 2n = 14 it is clear that these karyotypes—representative of quite distinct lines of marsupial evolution—show a similarity of chromosomal form in their complement which cannot be a matter of chance. These common features, form, number, and the location of the satellites have been achieved in spite of at least a 30% difference in DNA content. Allotting *Sminthopsis* the arbitrary value of 100 for its DNA content, the DNA content of *Perameles* is 130.

Three other 2n = 14 complements are shown in Figure 3. One of them is from a species of *Phalanger* from a family in which all the other species so far counted have 2n = 22; the others are from *Macropodidae*. In this first family (*Phalangeridae*) there is the same bimodal distribution within the family as exists over the *Marsupialia* as a whole, although the species counted are still only a portion of those extant. The 2n = 14 complement does not show the same basic form as do other 2n = 14 species, although they do possess a general degree of similarity. When further species of *Phalanger* are studied a better test can be made of the presence of this basic form in this family. At present no decision can be made. The 2n = 14 representatives from the *Macropodidae* both differ from the general 2n = 14 form. One of them, *Potorous apicalis*, has 2n = 12 but must have been derived from a 2n = 14 species in which the additional chromosomal pair were acrocentrics of the size of the long arm of the X-chromosome (Sharman and Barber, 1952). In neither case is the similarity to the general 2n = 14 form immediately obvious. In these three species it would be possible to derive the general form by appropriate exchanges. Possibly with additional information on other species in these groups a decision can be made; at present we interpret the situation in the *Macropodidae* as we indicate in the section dealing with that group, and the interpretation of the *Phalangeridae* is unresolved.

Marsupials with Chromosome Numbers Other Than 2n = 14

The other diploid chromosome numbers which occur most frequently in the marsupials are 2n = 20 and 2n = 22. Apart from *Macrotis lagotis* (Peramelidae) which must have been derived from a 2n = 20 species to give its multiple sex chromosome system, these numbers are confined to some American species and to some of the Australian species of the *Phalangeroidea*. Comparative DNA measurements are not yet available for sufficient of these species and so comparisons are confined to gross morphological features. In the American species only 2n = 22 has been recorded and all the chromosomes are acrocentric (Biggers *et al.*, 1965). This same chromosome number is found in the *Petauridae* where all chromosomes are metacentric or have obvious short arms. *Hemibelideus lemuroides* is probably derived by centric fusion from a chromosome complement like that of *Schoinobates volans* and we presently incline to the view that *Pseudocheirus archeri* also arose by chromosome reduction from 2n = 22. In the *Phalangeridae* the chromosome number found is 2n = 20 and all the chromosomes are acrocentric. Representative chromosome complements are shown in Figure 4. Within each family the chromosome complements are very similar, although they show great differences between the families.

The *Macropodidae* possess the greatest range of diploid chromosome numbers and in this respect are quite different from the other families of marsupials. They present, therefore, a number of problems in relating them to each other and to the general distribution of chromosome numbers shown in other groups. As indicated previously, the relationship between macropods with 2n = 14 and those with higher numbers is still uncertain and this discussion omits them.

On the argument that *Hypsiprymnodon* retains a greater number of primitive characters than any other macropod and that its autosomal chromosome complement (2n = 22) is like that of four other genera,

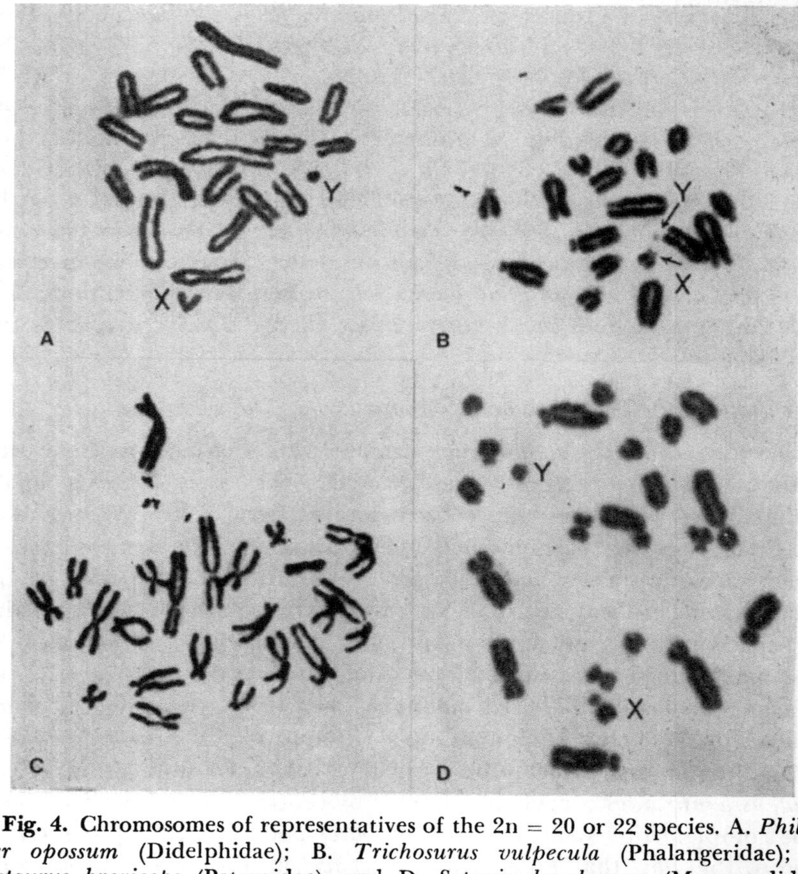

Fig. 4. Chromosomes of representatives of the 2n = 20 or 22 species. A. *Philander opossum* (Didelphidae); B. *Trichosurus vulpecula* (Phalangeridae); C. *Petaurus breviceps* (Petauridae), and D. *Setonix brachyurus* (Macropodidae). See Fig. 5 for other relevant karyotypes of the *Macropodidae*.

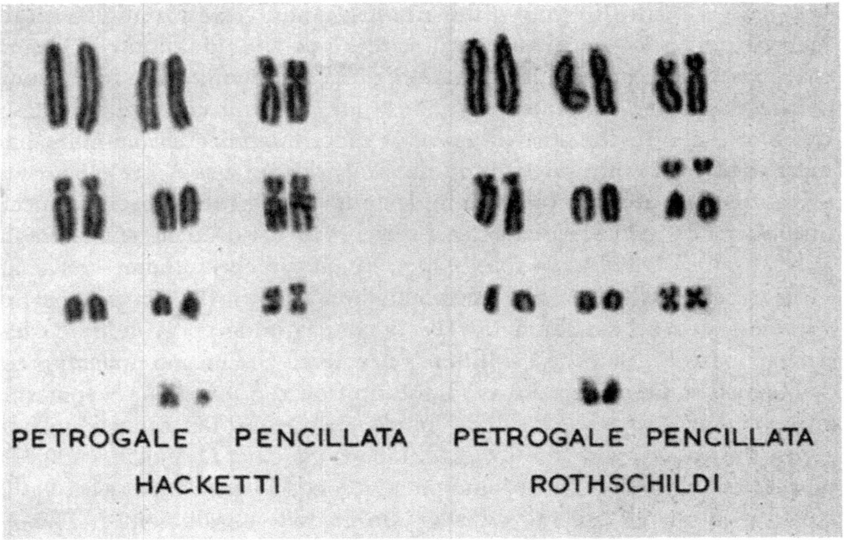

Fig. 5. The somatic chromosomes of *Petrogale penicillata hacketti* (2n = 20) and *Petrogale penicillata rothschildi* (2n = 22) showing the evidence for the presumed fusion of acrocentric elements which gave rise to the reduction in chromosome number.

Petrogale, Lagorchestes, Dorcopsis and *Thylogale,* there is a strong suggestion that 2n = 22 is the basic number of this group. Consequently, a reduction in chromosome number must have occurred in order to give rise to the majority of the present species in this family. Evidence for this reduction can be seen in the following observations. The chromosomal difference between *Petrogale penicillata hacketti* (2n = 20) and *Petrogale rothschildi* (2n = 22) with which it forms a viable but sterile hybrid appears to be a centric fusion (Fig. 5). The chromosome complements of the two species of *Petrogale* with 2n = 18 are different from one another, suggesting two separate instances of chromosome reduction from 2n = 22. One genus, *Macropus,* is a group of species with 2n = 16 which show some cytological similarities to each other. Two of these hybridize with *Megaleia rufa* (2n = 20), forming sterile progeny. At present we cannot say, on cytological grounds, what relationships exist between the 2n = 22 groups and *Megaleia rufa* so that the possibility that the 16 chromosome species gave rise to this 2n = 20 species cannot be excluded. Some of the peculiarities of the 2n = 16 species with respect to their sex chromosome construction are discussed in the section relating to sex chromosomes.

The problems posed by species with numbers greater than 2n = 22 are familiar ones to cytogeneticists and not solved by what is known of the cytology of the marsupials. On the argument of a basic diploid number of 2n = 22 in this group, these numbers must have resulted from an increase, and such an increase requires the possible donation of chromosome ends and certainly the donation of centromeres. The only source of these is other chromosomes. These may occur as mosaic patches of trisomic tissue in the germ line or as supernumerary chromosomes and must then be incorporated into selectively advantageous exchanges with the autosomes of the 2n = 22 complement. While the theoretical mechanism for such an increase is clear, there is as yet no demonstration that it has occurred. We have found supernumerary chromosomes in a marsupial species with 2n = 22 (Hayman and Martin, 1965c) so they can apparently occur in mammals. We do not consider polyploidy to have been responsible for the evolutionary increase in chromosome numbers.

There is some evidence for a limitation on the length of chromosome arms in the *Macropodidae*. The longest arm measured occurs in *W. bicolor,* the species with the lowest number, and in *M. rufa* 2n = 20. The postulated reduction in chromosome number is associated with an increase in the length of the arms of intermediate chromosomes. This can be interpreted by supposing that the physical size of the spindle and the cell in which division occurs imposes a limit to the absolute length to which a chromosome arm may "grow" by translocations.

The increase and reduction in chromosome number postulated to occur in the *Macropodidae* is not associated with a steady gain or loss of DNA. In those species for which determinations have been made it appears that the 2n = 22 species have DNA values within the range 80–96 (on an arbitrary scale) and species with higher chromosome numbers fall within this range. The greatest increase in DNA content occurs in the 2n = 16 group where values as high as 124 occur. *Wallabia bicolor* (2n = 10 ♂ 11 ♀) has a value of about 100. Some of this variation can be accounted for in terms of unusual sex chromosomes (see later). However, there is still a considerable range in DNA content in this group and over the marsupials as a whole. This range is greater than that described by Atkin *et al.* (1965) for an extensive range of eutherian mammals and suggests that it would be unwise to generalize on the uniformity of DNA values in mammals.

The superfamily *Tarsipedoidea* is known from its only representative. This species has 2n= 24 and all the chromosomes are acrocentric.

Conclusions

In summary, there are five marsupial families in which 2n = 14 has been found and in which there is a strong case for considering the basic chromosome form to be the same. It is possible that the family *Phalangeridae* may belong here as well. If it does, then three of these families (*Didelphidae, Peramelidae* and *Phalangeridae*) also include species with 2n = 20 or 22. Two families, *Petauridae* and possibly *Macropodidae,* are presently thought to be restricted to species in which 2n = 20 or 22 is the basic number. It is characteristic of the 2n = 20 or 22 species that each family has to its own individual morphological pattern of chromosomes while the 2n = 14 complement is much more conservative. This arrangement of the families is based on current taxonomic groupings.

A point of considerable importance in interpreting these observations is the fact that both the American species of *Didelphidae* and the Australian fauna show this same bimodality and both share the same 2n = 14 form. It is currently estimated that the separation of the precursors of these two faunas occurred about the Eocene—some fifty million years ago (Ride, 1957). We favor the following interpretation. Both the Australian species and the *Didelphoidea* were derived from a 2n = 14 stock and the chromosome form of this stock was very like that which we find today. Subsequently and independently 2n = 22 or 20 forms arose and some of these have given rise to present-day families, while others are present in conjunction with 2n = 14 in the same family. It is less likely that bimodality was present in the progenitors since this would require it to have been maintained in at least two and possibly three present-day families.

It is apparent that an acrocentric complement of both 2n = 20 or 2n = 22 could have evolved from a 2n = 14 form, given the necessary addition of telomeres and centromeres. Although there is no good evidence to show that this has happened, chromosomal fission may also be a mechanism. While there is no evidence to suggest why a bimodal distribution should evolve, we are examining the size of the spindle and the arrangement of the chromosomes upon it and studying the distribution and frequency of chiasmata to see if they throw any light on the matter.

A number of species remain to be studied. We are engaged upon this and upon a more detailed study of those species for which we have some information already. Our present conclusions, therefore, should be regarded as tentative. A particular deficiency is the absence of karyotypes from the *Caenolestoidea* the other surviving American superfamily. Information from this group will allow another test of the generality of this distribution and may contribute to our knowledge of the entry into Australia of the precursors of our marsupial fauna.

Autosomal Chromosomal Polymorphism

The majority of chromosome determinations made in the *Marsupialia* have been of only a few members of each species. This is generally the result of their scarcity. Consequently, the existence of polymorphism for the autosomal chromosomes has only been demonstrated in two species and may well prove to be more frequent.

In the greater glider *Schoinobates volans* five different chromosome numbers were found among seven individuals examined from one population. The variation in chromosome numbers was independent of sex. It was suggested that the basic chromosome number was 2n = 22 (XX/XY) and that the additional small metacentric chromosomes were supernumerary chromosomes (Hayman and Martin, 1965). This is the only report of such chromosomes in a mammal (see Fig. 6B). There is as yet no evidence for any adaptive role they play in this species. The high frequency found in this population, from two to six, contrasts with the usual numbers of supernumerary chromosomes found in plant populations.

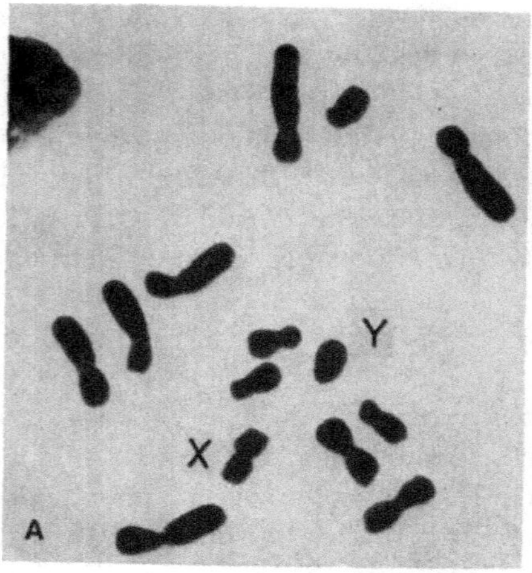

Fig. 6A. Chromosomes from the corneal epithelium of *Perameles nasuta* male.

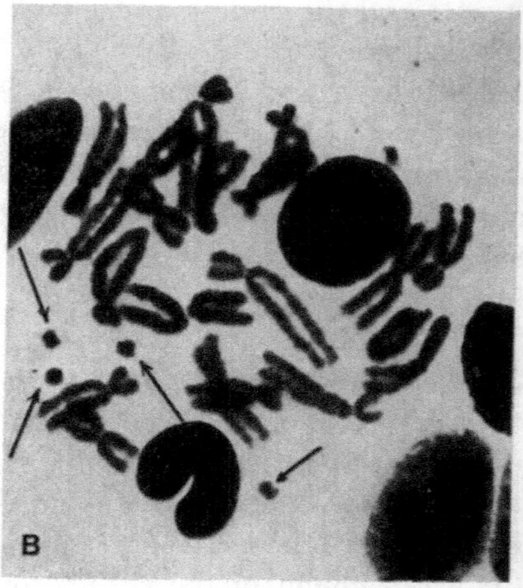

Fig. 6B. Chromosomes from bone marrow of *Schoinabates volans,* arrows indicating supernumerary chromosomes.

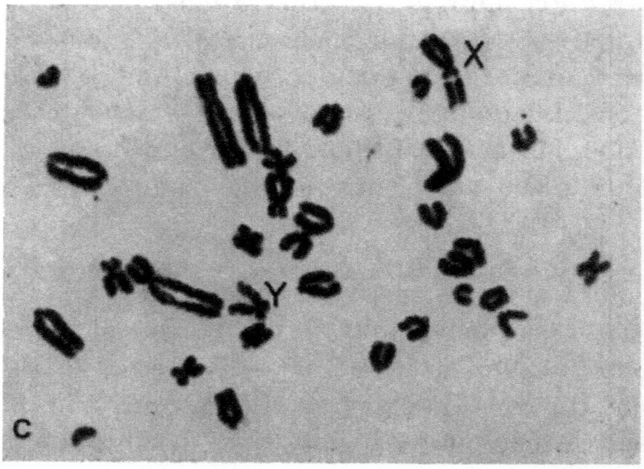

Fig. 6C. Chromosomes from leucocyte culture of *Aepyprymnus rufescens*—the species with the largest number known in the *Marsupialia.*

Associated with the complex sex-chromosome system observed in the hare-wallaby *Lagorchestes conspicillatus* is a polymorphism for a centric fusion involving two non-homologous autosomal chromosomes (Martin and Hayman, 1966). A point of theoretical interest is shown in the behavior of the resulting trivalent at male meiosis. Although Anaphase I disjunction is such that a high proportion of unbalanced Metaphase II spermatocytes would be expected, these are not found. This suggests that, in this instance at least, unbalanced secondary spermatocytes do not survive to enter the second division of meiosis. Insufficient numbers of available animals prevent any further consideration of this system.

Sex Chromosomes in the Marsupials

The study of the variations in sex chromosome complement and their associated phenotypes, and the behavior of the sex chromosomes in terms of their cycle of DNA synthesis during cell division, has been responsible for a considerable advance in interpreting their role in the determination of sex and the expression of sex-linked genes in mammals (Lyon, 1962; Ohno, 1967). Information on these and on other aspects of the role of sex chromosomes is available from what is known of the sex chromosomes in the marsupials.

Sex Determination in the Marsupials

An examination of a range of sex chromosome abnormalities in man has established that the Y-chromosome is male determining (Jacobs, 1966). The same situation apparently occurs in the mouse (Russell, 1961) and the cat (Thuline and Norby, 1961). It is thought that this may be the general rule in mammals. Sharman *et al.* (in press) have identified both an XXY and an X0 tammar (*M. eugenii*), and describe these animals as follows: "The XXY individual had a male phallus and a completely male reproductive system being in these respects like XXY individuals of eutherian species. It differed from XXY eutherians in having undescended testes and in being a female body weight. The X0 tammar was phenotypically female and the gonads were intermediate between testis and ovary in structure. The accessory reproductive structures were typically female indicating that in the absence of the Y-chromosome female characteristics appeared. Since in this individual the chromosomes were studied from only one of the body tissues—the gonads—the possibility cannot be excluded that the individual was a sex chromosome mosaic." The authors conclude that the marsupial Y-chromosome, like that of the eutherian mammals examined, is strongly male determining.

Multiple Sex Chromosome Systems

In three species of marsupials (see Table 1) an XX/XY_1Y_2 sex chromosome system has been identified. Such a system results from an X-autosome fusion and the subsequent loss of an acentric fragment. In another species an $X_1X_1X_2X_2/X_1X_2Y$ system has been described and it is suggested that this involved both an X-autosome and a Y-autosome interchange (Martin and Hayman, 1966).

Interchanges between the X-chromosomes and autosomes postulated to account for the construction of multiple sex chromosome systems pose special situations for the "inactive-X" hypothesis. Dosage compensation requires inactivation of only that portion of the X-chromosome which is unequally represented in both males and females. One would, therefore, expect that the compound X-chromosome present in multiple sex chromosome systems would show a specific pattern of DNA synthesis in the female, i.e., the original X-chromosome regions should show asynchrony of DNA synthesis, the original autosomal regions need not. This expectation has been tested and the data confirm the hypothesis (Hayman and Martin, 1965a; Martin and Hayman, 1966).

X-autosome interchanges have been described in other mammals and here they are associated with instability of the genetic expression of the adjacent autosomal loci (Cattanach, 1961). In order for a multiple sex chromosome mechanism to evolve following an X-autosome interchange not only has the appropriate disjunctional orientation to be preferred but there must also be rapid stabilization of the inactivation mechanism. Genetic factors for such stabilization have been shown to exist (Cattanach and Isaacson, 1965). The special features necessary for the successful survival of a multiple sex-chromosome system may explain their infrequent occurrence.

The XX/XY Sex-chromosome Systems of Marsupials

In an examination of the XX/XY systems a number of interesting observations have been made (Hayman and Martin, in preparation). The most complete information comes from the *Macropodidae* and this discussion is confined to them. The X-chromosomes of all the species in this family possess a characteristic achromatic region which we believe to be the nucleolar organizer (Hsu *et al.*, 1967). In one species, *Lagostrophus fasciatus,* this region has not been identified positively. However, its presence in the X-chromosome of every other species, the variation in size of the region which is found between species, and the poorer quality of the only preparations available make us confident that it is also present in this species.

In Figure 7 the X-chromosomes of a representative series of species of macropods are shown. The size is relative and allows comparisons between the species. Among those species with an XX/XY sex-chromosome system there are two distinct groups with respect to the size of the X-chromosomes. The members of the smaller group (A–H) show some variation in size which is significant between the extremes. However, the

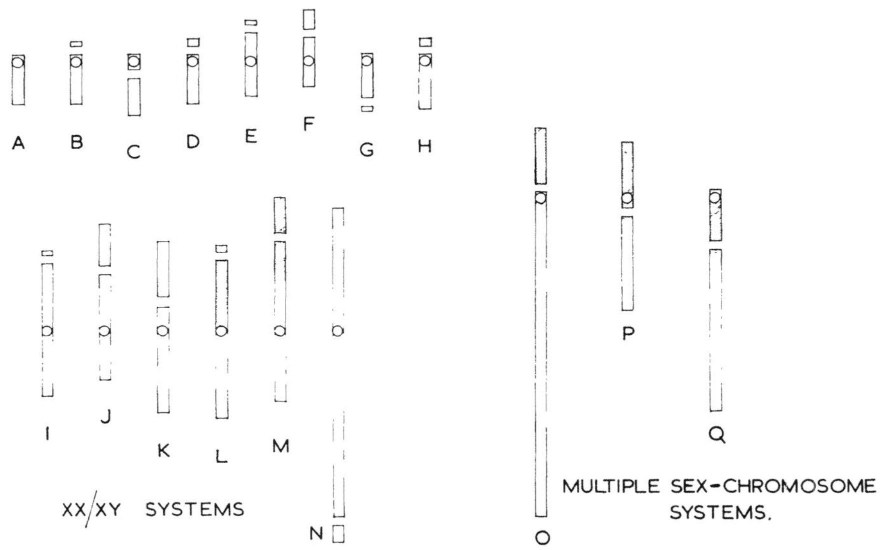

Fig. 7. Diagrammatic representation of a sample of the X-chromosomes of the *Macropodidae*. The relative size of the chromosomes is determined by the method outlined in the text. The species represented are (A) *Lagostrophus* Sp., (B and C) *Petrogale* Sp., (D) *Dendrolagus* Sp., (E) *Dorsalis* Sp., (F and G) *Thylogale* Sp., (H) *Macropus giganteus,* (I) *Macropus irma,* (J) *Onychogalea* Sp., (K) *Aepyprymnus rufescens,* (L) *Megaleia rufa,* (M) *Macropus rufogriseus,* (N) *Macropus parryi,* (O) *Wallabia bicolor,* (P) *Potorous apicalis,* (Q) *Lagorchestes conspicillatus.* The shading is explained in the text.

most pronounced variation is that involving the position of the centromere and the nucleolar organizer. It is clear that these X-chromosomes have been subjected to a great deal of structural change, presumably inversions or shifts, in the differentiation of each of the species. Each such change must survive in the population initially as a heterozygote prior to fixation in the homozygous form. These changes involve considerable regions of the X-chromosome and would affect crossing-over in the female and possibly the X- and Y-chromosome pairing and disjunction in the male. The bases on which this selection has operated are unknown. The factors involved in the variation in size are considered in the discussion of the larger X-chromosomes.

The groups of larger X-chromosomes associated with XX/XY sex-chromosome systems show a variation in size ranging from chromosomes which are approximately twice as big as the average smaller X-chromosome group (Fig. 7I–L), to three times as big (Fig. 7M), to nearly four times as big (Fig. 7N). Again, there is variation in the position of the nucleolar organizer in these chromosomes.

In our opinion the smaller group of X-chromosomes constitute the primitive X-chromosome or basic X-chromosome. This is because we have not found smaller X-chromosomes than these in any of the marsupials. Further, the size of these smaller chromosomes is of the same order as that of the postulated basic X-chromosome in the mammals—although somewhat smaller. Finally, when a study is made of the accumulation of tritiated thymidine, as an index of DNA synthesis, these chromosomes behave uniformly over their length in the time of initiation and completion of their synthesis (Marshall-Graves, 1967; Hayman and Martin, unpublished).

The size of the larger X-chromosomes is due to the presence of additional chromosomal material which could be either of autosomal or X-chromosomal origin. Those situations in which it is known that autosomal material has been translocated on to the X-chromosome are illustrated in Figure 7 as multiple sex-chromosome systems. Studies of the synchrony of DNA synthesis over the X-chromosomes of females shows that the asynchronous regions are confined to the shaded portions of the X-chromosomes in Figure 7. The other regions of autosomal origin replicate synchronously and earlier (see section on multiple sex chromosomes). The size of these shaded regions obviously corresponds to the basic X-chromosome size and includes the nucleolar organizer and the centromere. Consequently, the problem of dosage compensation is coped with adequately.

The pattern of the terminal stages of DNA synthesis has been studied in two of those species with large X-chromosomes. In both these species (*M. rufa* and *M. rufogriseus*) the main features of behavior show a consistent pattern. In the male the X-chromosome does not replicate uniformly, one portion replicating much later than the other, and later than the majority of the autosomal complement but not later than the Y-chromosome (see Fig. 8). The size of the early replicating portion is indicated by the shading in Figure 7. In the female one X-chromosome is uniformly late replicating while the other behaves as does the single X-chromosome in the male. In *M. rufa* some cells very late in synthesis are found in which only the non nucleolar-organizer arm of one chromosome was labelled.

These observations are similar to those of other workers, in particular those of Ohno (1967). The problem arises in their interpretation. In terms of dosage compensation, equating late replication with inactivation, the two larger X-chromosomes in the female contain between them only one active region—and as can be seen from the diagram this corresponds to the basic X-chromosome in size. However, the origin of the extra chromosomal material on these larger X-chromosomes is not so easily explained. If it is multiplication of sex chromosomal material then how do such repeats survive without difficulties at meiosis? All other such chromosomes prefer to pair and exchange internally rather than frater-

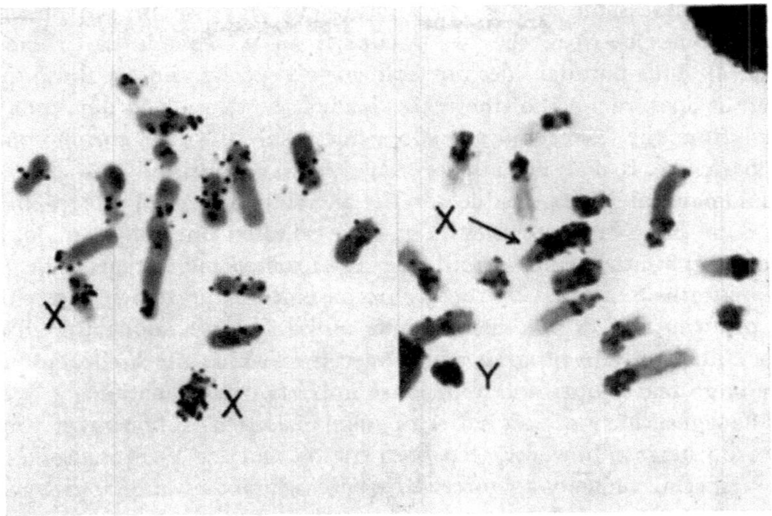

Fig. 8. Autoradiographs of the final stages in DNA synthesis of the chromosomes of a male and female *M. rufogriseus*. The sex chromosomes are indicated.

nally. Further, the only way to suppose dosage compensation is occurring in these species is to imagine three quarters of the X-chromosomes to be inactive in all tissues. The X-chromosomes are subject to structural changes as are other chromosomes and many opportunities must have occurred for extensive inert regions such as these to be lost. We feel that since these regions are present they must be doing something that it is selectively advantageous to retain. The observation in *M. rufa* that there is asynchrony between the late replicating sections of the X-chromosome suggests that there may be differences between these regions in their function. The evidence of Hill and Yunis (1967) is relevant here but comparable observations are not available for marsupials.

If the material was autosomal in origin then, unless this same material was present on the Y-chromosome, there would be a genetic imbalance which would be lethal. Figure 9 shows the size of the Y-chromosomes in

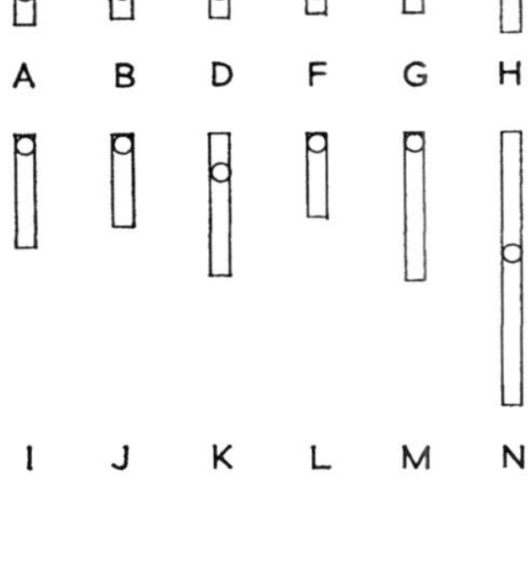

Fig. 9. Diagrammatic representation of a sample of Y-chromosomes from the *Macropodidae;* the letters correspond to the species in Fig. 7. The chromosomes are drawn to the same scale.

most of these same species. Paralleling the increase in size of the X-chromosomes between the two groups is an increase in Y-chromosome material. This parallel does not hold for all species and for those species where it does, removal of the equivalent of the extra Y-chromosomal material from the X-chromosome does not result in an X-chromosome of the basic size. It does not appear that there is a consistent case that autosomal material is responsible for the increase in size of the X-chromosomes or the Y-chromosomes. If it were so then one ought to find the labelling pattern found in multiple sex-chromosome systems. The labelling patterns exhibited by the Y-chromosome in the two species studied are uniform and do not demonstrate any differences along the chromosome. The study in progress of *M. parryii* in which the Y-chromosome is very large may contribute additional important information.

The significance of this series of observations is unknown at present. They do suggest, however, that both the X- and the Y-chromosomes possess a special capacity to increase in size which is not shared by autosomes. Since the sex-chromosomes are already specialized in other aspects, such as their cycle of condensation and time of DNA synthesis, which make them different to the autosomes, then these features may all be interrelated. It is relevant that while the similarity of the form of the $2n = 14$ species has been emphasized already (*vide* Fig. 2) this applies to the autosomes and not to the sex-chromosomes which show great variation. A further instance of the special features shared by sex-chromosomes is provided by the mosaicism discussed in the next section.

Sex Chromosome Mosaicism

Among the five peramelid species in which both somatic and reproductive tissue has been examined, four of them were found to possess a remarkable mosaicism for the sex-chromosome complement (Hayman and Martin, 1965b). In adult animals not all the somatic cells contain the normal diploid complement, although this is always found in the germ line cells (see Table 3). The chromosomes missing are the Y-chromosome from the male and the X-chromosome from the female. In most tissues in each individual more than 200 cells have been counted so that the possibility that mosaicism exists for the normal complement has been excluded at a high level of probability.

It was postulated that this mosaicism was the result of elimination during the early stages of embryogenesis of an X-chromosome from the female and a Y-chromosome from the male. The most reasonable hypothesis is to relate the elimination of the X-chromosome to the X-inactivation mechanism, and it was suggested that the delay in initiating and completing DNA synthesis associated with this inactivation could be the means whereby elimination occurred. Could it be shown that different X-chromosomes were randomly eliminated by identifying phe-

Table 3: The Chromosome Number and Sex Chromosome Complements in Different Tissues of Adults of *Isoodon spp.* and *Perameles spp.*

Species	Sex	Tissue	2n	Sex Chromosome
Isoodon macrourus / *Isoodon obesulus*	♂ and ♀	bone marrow	13	XO
		leucocytes	13	XO
		spleen	13	XO
		corneal epithelium	13	XO
		intestinal epithelium	13 *	XO
	♂	testes	14	XY
	♀	ovarian tissue	14	XX
Perameles nasuta / *Perameles gunni*	♂ and ♀	corneal epithelium	14	XX and XY
		intestinal epithelium	14 †	XX and XY
		bone marrow	13	XO
		leucocytes	13	XO
		spleen	13	XO
	♂	testes	14	XY
	♀	ovarian tissue	14	XX

* Only *I. obesulus* examined.
† Only *P. nasuta* examined.

notypic mosaicism in a female heterozygous for a sex-linked gene this would be important confirmatory evidence. Tests for electrophoretic variants of glucose-6-phosphate dehydrogenase have been unsuccessful. The same argument was applied to the elimination of the Y-chromosome since in the somatic tissue of all mammalian species so far examined, the Y-chromosome also appears to be late replicating. The attractiveness of this hypothesis lies in the fact that the presence or absence of the X- or the Y-chromosome is identical in the same tissue in both sexes of any one species so that identical features of the X- and the Y-chromosome are responsible for their elimination and this elimination probably evolved simultaneously in both sexes.

This explanation, while it may still prove to be basically correct, requires modification in view of the following observations. The mosaicism is not the same in all species—the complete diploid complement has not been found in any of the somatic tissue of *Isoodon* spp. examined, but it has been found in tissue cultures derived from skin (Ellem and Jackson, personal communication) and in the divisions of the corneal epithelium, and in the lining of the intestine of *Perameles nasuta* and *Perameles gunnii* (see Fig. 6A). It would seem that elimination must alter at different stages of development in these two genera. Using the technique of Schmid (1967) the cells at the base of the hair follicle have been examined for the presence of the sex-chromatin body in male and female *Perameles nasuta*. There is a difference between the cells from the two sexes in the heterochromatin present at interphase. This would suggest that the sex-chromosomes are behaving in these tissues, as they do in man. Hence, their presumed "inertness" now is not affecting their survival in these cells. The Y-chromosome in the corneal epithelium of the male *Perameles nasuta* is late replicating in that it replicates as late as the last autosomal regions. Comparable information is not yet available for the X-chromosomes in the female.

The difference in the mosaicism between the two genera affects tissue of ectodermal and endodermal origin. A more complete understanding of the mosaicism will come from an examination of the chromosomes in tissue from early embryos. Such material has not been easy to obtain as the species have proved difficult to breed in captivity.

It may be of significance that the X-chromosomes of *Perameles spp.* are larger than those of *Isoodon spp.*

There seems no complete parallel to this situation in other mammals. Bianchi and Contreras (1967) describe what may be the initiation of a similar situation in *Akadon azarae* but not all members of the population show the elimination they describe.

Marsupials as Animals for Genetic Research

Certain of the marsupials because of their small size, reproductive capacity, and low chromosome number have attracted attention as possible laboratory animals. Martin (1965) has described the successful breeding of *Sminthopsis crassicaudata;* Hope and Godfrey (1968) report the genetic analysis of transferrin variation in another laboratory culture of this species. Their data suggest that the variation is due to three allelic genes.

Population genetic studies of variation in marsupial species are limited to the few species in which sufficient numbers of animals can be sought in the wild. Hope (personal communication) has found considerable variation between disjunct populations of *Trichosurus vulpecula* in the frequency of three allelic genes determining transferrin types. There is also considerable variation between these populations in the frequency of high and low potassium content of the red cells. Both these situations show evidence of genetic polymorphism.

Kirsch and Poole (1967) described genetically determined polymorphism for transferrins in *Macropus giganteus* (Shaw); animals from Eastern Australia have only one transferrin type while animals from western populations have two types. The absence of natural hybrids between these two forms, the difficulty in obtaining hybrids between them in captivity, and the difference in their gestation time and coat color has resulted in them being considered separate species.

Genetically determined variation in transferrins has also been described in *Megaleia rufa* (Cooper and Sharman, 1964).

Acknowledgments

We would like to express our appreciation to the many colleagues whose cooperation, particularly in helping us obtain specimens of rare species, has enabled us to make our survey as complete as it is. We are particularly grateful to Professor H. Waring, Professor A. R. Main, Dr. John Shield and Dr. John Kirsch of the University of Western Australia; to Dr. David Ride of the Museum of Western Australia; to Dr. John Thompson of the University of Melbourne; to Mr. R. Strahan of the Taronga Park Zoo, Sydney; to Professor G. B. Sharman of the University of New South Wales; to Mr. John Calaby and Mr. W. Poole of the C.S.I.R.O. Wildlife Research Division; to Mr. Eric Worrell of Gosford, N.S.W.; and to the Royal Zoological Society of South Australia. We acknowledge the generosity of Dr. Ira Fritz who allowed us to borrow the slides of *Caluromys derbianus* and *Philander opossum*. We are grateful for the financial support we have had, initially from the C.S.I.R.O. and lately from the Australian Research Grants Commission.

We have been ably assisted by Mrs. Rosalie Trezise, Miss Christine Smith and Mr. Peter Waller.

References

Atkin, N. B., Gail Mattison, W. Beçak and S. Ohno: The comparative DNA content of 19 species of placental mammals, reptiles and birds. Chromosoma *17*:1, 1965.

Bianchi, N. O. and J. R. Contreras: The chromosomes of the field mouse *Akadon azarae* (Cricetidae, Rodentia) with special reference to sex chromosome anomalies. Cytogenetics *6*:306, 1967.

Biggers, J. D., H. I. Fritz, W. C. D. Hare and R. A. McFeely: Chromosomes of American marsupials. Science *148*:1602, 1965.

Cabrera, A.: Catalogo de los Mamiferos de America del sur. Pt. 1 Rev. Mus. Argent. Cianc. Nat. No. 4, 1957.

Cattanach, B. M.: A chemically-induced variegated-type position effect in the mouse. Z. Vererbungsl. *92*:165, 1961.

——— and J. H. Isaacson: Genetic control over the inactivation of autosomal genes attached to the X-chromosome. Z. Vererbungsl. *96*:313, 1965.

Cooper, D. and G. B. Sharman: Transferrin variation in kangaroos. Nature *203*:1094, 1964.

Darlington, P. J., Jr.: Zoogeography: The Geographical Distribution of Animals. New York: John Wiley and Sons, Inc., 1957.

Graham, M. A. and M. L. Barr: Sex chromatin in the opossum *Didelphis virginiana*. Arch. Anat. Micro. Morph. Exp. *48*:111, 1959.

Gunson, Mary M., G. B. Sharman and J. A. Thompson: The affinities of *Burramys* (Marsupiala: Phalangeroidea) as revealed by a study of its chromosomes. Aust. J. Science (in press), 1968.

Hayman, D. L. and P. G. Martin: An autoradiographic study of DNA synthesis in the sex chromosomes of two marsupials with an XX/XY_1Y_2 sex chromosome mechanism. Cytogenetics *4*:209, 1965a.

——— and ———: Sex chromosome mosaicism in the marsupial genera *Isoodon* and *Perameles*. Genetics *52*:1201, 1965b.

——— and ———: Supernumerary chromosomes in the marsupial *Schoinabates volans* (Kerr). Aust. J. Biol. Sci. *18*:1081, 1965c.

Hill, R. N. and J. J. Yunis: Mammalian X chromosomes: Changes in patterns of DNA replication during embryogenesis. Science *155*:1120, 1967.

Hope, R. M. and G. K. Godfrey: Transferrin polymorphism in the Australian marsupial mouse *Sminthopsis crassicaudata* (Gould). Aust. J. Biol. Sci. *21*:587, 1968.

Hsu, T. C., B. R. Brinkley and F. E. Arrighi: The structure and behavior of the nucleolus organisers in mammalian cells. Chromosoma *23*:137, 1967.

Jacobs, P. A.: Abnormalities of the sex chromosomes in man. In: Advances in Reproductive Physiology Vol. I. Anne McLaren, ed. London: Logos Press, 1966.

Kirsch, J. A. W. and W. E. Poole: Serological evidence for speciation in the grey kangaroo, *Macropus giganteus* (Shaw 1790) (Marsupialia: Macropodidae). Nature *215*:1097, 1967.

Lyon, M. E.: Sex chromatin and gene action in the mammalian X-chromosome. Amer. Jour. Human Genet. *14*:135, 1962.

Marshall-Graves, Jennifer A.: DNA synthesis in chromosomes of cultured leucocytes from two marsupial species. Exp. Cell Res. *46*:37, 1967.

Martin, P. G.: The potentialities of the fat-tailed marsupial mouse, *Sminthopsis crassicaudata* (Gould) as a laboratory animal. Aust. J. Biol. Sci. *13*:559, 1965.

——— and D. L. Hayman: A quantitative method for comparing the karyotypes of related species. Evolution *19*:157, 1965.

——— and ———: A complex sex-chromosome system in the hare-wallaby *Lagorchestes conspicillatus* (Gould). Chromosoma *19*:159, 1966.

——— and ———: Quantitive comparisons between the karyotypes of Australian marsupials from three different superfamilies. Chromosoma *20*:290, 1967.

Matthey, R.: Les Chromosomes des Vertébrés. F. Rouge, Lausanne, 1949.

Muntzing, A.: A new category of chromosomes. Proc. 10th Int. Cong. Genet. Montreal *1*:453, 1958.

Ohno, S.: Sex Chromosomes and Sex-linked Genes. Monographs on Endocrinology Vol. 1. New York: Springer-Verlag Inc., 1967.

Radley, J. M.: DNA content of individual chromosomes of *Protemnodon bicolor*. Exp. Cell. Res. *41*:217, 1966.

Ride, W. D. L.: On the evolution of Australian marsupials. In: The Evolution of Living Organisms. G. W. Leeper, ed. Melbourne, Australia: Melbourne University Press, 1962.

Rothfels, K. H. and L. Siminovitch: The chromosomes of the rhesus monkey (*Macaca mulatta*) determined in kidney cells cultivated *in vitro*. Chromosoma *9*:163, 1958.

Russell, L. B.: Genetics of mammalian sex determination. Science *133*:1795, 1961.

Schmid, W.: Sex chromatin in hair roots. Cytogenetics *6*:342, 1967.

Sharman, G. B.: The mitotic chromosomes of marsupials and their bearing on taxonomy and phylogeny. Aust. J. Zool. *9*:38, 1961.

——— and H. N. Barker: Multiple sex chromosomes in the marsupial, *Potorous*. Heredity *6*:345, 1952.

———, E. S. Robinson, S. M. Walton and P. J. Berger: The sex chromosomes and reproductive anatomy of some intersexual marsupials. In press, 1968.

Thuline, H. C. and D. E. Norby: Spontaneous occurrence of chromosome abnormalities in cats. Science *134*:554, 1961.

INSECTIVORA CYTOGENETICS

D. S. Borgaonkar

Division of Medical Genetics, Department of Medicine, The Johns Hopkins University School of Medicine, Baltimore, Maryland

Cytogenetic analyses among the Insectivora Bowdich are available for seven families and a total of 61 species have been investigated. Painter (1925) was the first to report on chromosome counts for any Insectivore and Makino's (1951) list of chromosome numbers in animals gave information on only 8 species comprising altogether 10 reports. However, during the last decade, a number of investigators have concerned themselves with systematic studies of certain groups of Insectivores and these can be expected to contribute to a more adequate understanding of speciation and evolution in the insectivores. Cytogenetic analyses of "species hybrids" per se are not available, although a little information is available on the reproductive mechanisms of these animals. However, recently from studies on their physiology, breeding behavior, and ecology, information necessary for maintaining the animals in captivity has been made available.

At the outset it should be unequivocally stated that the karyotype of a species is a morphological attribute of that animal and when phylogenetic interpretations are made, evidences from other disciplines are also to be considered.

In this review I shall report on the chromosome studies of insectivores and consider, among other questions, viewpoints on the basic chromosome number, i.e., *x*, evolution of the karyotype, sex-determination and sex-differentiation, phylogenetic trends among the families including intra- and inter-relationships, possible role of the karyotype in speciation and evolution of taxa.

When we report the chromosome number of a species, we shall use this information as reported by the concerned writer. The species problem has been amply discussed elsewhere in this symposium (Mayr, 1968) and we shall use that term as a *concept:* a taxonomic entity which is dynamic and consists of a group of individuals which are reproductively isolated from other groups. (Species have been described on a single isolated museum specimen of insectivores. We are not sure whether these "species" are valid or whether after a biosystematic evaluation they will have to be reclassified.)

As Dr. Gropp is reviewing material on the Talpidae and Erinaceidae, I shall concern myself to studies on the Macroselidae, Soricidae, Tenrecidae, Solenodontidae, and Tupaiidae. Significant findings in this group of interesting mammals are: chromosomal polymorphism within the species of *Sorex araneus* giving us an insight into the intraspecific variation present in this interesting European insectivore, evidence that chromosomal rearrangements may play a causative and determining role in the speciation process along with the long recognized processes of geographic, allopatric, sympatric and isolation phenomena, behavior of chromosomes during meiosis in both testicular and ovular material, sex-determination, and evidence that the sex chromosomes translocate (interchange) segments of chromosomes with the autosomes and change the sex chromosome constitution.

As early as 1925 Painter asked: "Have the changes which have occurred in mammals been due entirely to mutations within the genes, or have extensive and gross changes in the chromatic material played a role, such as the botanists are apparently finding in the case of many plants?" And now the stasipatric model of speciation (White, 1968), based on chromosomal rearrangement, seems to have played a role in insectivores (Borgaonkar and Gould, 1968; Meylan, 1965, 1966), in rodents (Matthey, 1966) and probably in other mammals. Even though one species population indicates a group of individuals, changes of the type which cytogeneticists would find must necessarily arise in single individuals at a given time. Hence the spread of this new mutant type would have to be viewed in a different magnitude than one would consider geographic and allopatric models of speciation which are population phenomena.

Macroscelidae

The information on chromosome numbers in this family is presented in Table 1. Brenner (1946) from South Africa was the first investigator to report on the chromosome complement of *Elephantulus myurus jamesoni* from a study of five male and four female animals. He reported that this species has a diploid number of 14 chromosomes and thought that the low chromosome number indicates a possible relationship with the Marsupials. Subsequently the eminent mammalian cytogeneticist Robert Matthey (1954) evaluating the paper of Brenner observed that Ellerman, Morrison-Scott and Hayman (1953) include *E. myurus* as a sub-species of *E. rupestris*. Proper identification of the specimens studied

Table 1: List of Families and Species with Detailed Information Includ

Taxon.	No. of Animals Studied		Tissue Studied	Chr. No. 2n	Nombre Fonda-mental
	Males	Females			
Family Tupaiidae					
Urogale everetti				26c	
		2		44	
Tupaia glis	1	1		60	
		1	Skin	60	70
	2		B.M., Skin	62	
		2	B.M., Skin	62	72
				60	
		2	Kidney	62	72
	2		Kidney	62	72
T. chinensis	2		Blood, Skin	62	72
		3	Skin, Spleen	62	72
				62	
T. longpipes				60	
T. minor				66	72
				66	
T. montana				68	
T. palowanensis				52	
Family Macroscelidae					
Elephantulus myurus	5		Testis	14	25
jamesoni		4	Ovary	14	
E. myurus			Testis	30	
E. rufescens	1, other unkn.				
	Total 2			34	42
Macroscelides rozeti	1		Testis	28	51 *
Family Talpidae					
Talpa europaea	x		Testis	38	
Talpa europaea	6		Testis	34	68
Talpa europaea	1		Bone	34	63
		1	Marrow	34	64
T. europaea cinerea	2			34	68
Talpa caeca caeca	2			36	70

* Calculated from the figures of Matthey (1954) on page 672.

ing Chromosome Number, Chromosome Arms, and Sex Chromosomes

Meiosis	Sex Chr.	Authors	Remarks
		Dodson	
	XX	Chu & Bender, 1962	
	XY	Chu & Bender, 1962	
	XX	Chu & Bender, 1962	
	XX	Hsu & Johnson, 1963	
	XY	Klinger, 1963	(X-Meta, Y acro)
	XX	Klinger, 1963	
		Arrighi (pers. comm.)	
	XX	Egozcue *et al.*, 1968	
31 II	XY	Egozcue *et al.*, 1968	
	XX	Borgaonkar	
	XY	Borgaonkar	
		Arrighi	
		Arrighi	
		Elliot & Lisco	
		Arrighi	
		Arrighi	
		Arrighi	
	XY	Brenner, 1946	
	XX	Brenner, 1946	
		Ford & Hamerton, 1956	$n = 15$
		Chu & Bender, 1962	2 SAT pairs
14 II	XY	Matthey, R., 1954	
x	XY	Koller, 1936	Probably a mistake in reporting 38. Ferret (*Putorius fume*) has 38 and mole 34 chromosomes.
17 II	XY	Bovey, 1949	
	XY	Gropp, Geisler & Citoler, 1966	
	XX		
		Meylan, 1966	
		Meylan, 1966	

Table 1

Taxon.	No. of Animals Studied Males	Females	Tissue Studied	Chr. No. 2n	Nombre Fonda-mental
Family Talpidae (cont.)					
Condylura cristata				34	
Mogera insularis				32	58
Family Erinaceidae					
Erinaceus europaeus	x		Testis	48	
Erinaceus europaeus	10		Testis	48	90
E. europaeus		2	Bone		
europaeus	2		Marrow	48	
E. europaeus	10		Bone	48	92
europaeus		10	Marrow	48	92
Erinaceus roumanicus	6		Testis	48	
E. e. roumanicus	1			48	
		1		48	
E. concolor	9		Bone	48	92
roumanicus		6	Marrow	48	92
E. amurensis				44	
Family Soricidae					
Sorex araneus	2		Testis	23	44
Sorex araneus	2		Testis	23	44
Sorex araneus	4		Bone		
		2	Marrow	22–25	40
Sorex araneus			Testis	22–27	
				22–25	
S. araneus I	2	3		28–31	
II	3	6		28–31	
Sorex araneus					
Type A	x		Testis	23	42
	x		+	23	42
		x	Blood	22	42
Type B			Testis, Blood	21–31	40
		x		20–32	40
	x			21–33	
Sorex caecutiens	1		Testis	42	67
Sorex cinereus				66	
Sorex fumeus				66	

(Continued)

Meiosis	Sex Chr.	Authors	Remarks
		Meylan, 1968	
16 II	XY	Tateishi, 1938	
x		Painter 1925	First count of any insectivore
23 II, 2 I	XY	Bovey, 1949	
	XX	Geisler & Gropp, 1967	
	XY	Geisler & Gropp, 1967	
	XY	Kral, 1967	
	XX	Kral, 1967	
	XY	Jordan, 1960	
	XY	Geisler & Gropp, 1967	Karyotype similar to Bovey (1949)
	XX	Geisler & Gropp, 1967	
	XY	Kral, 1967	
	XX	Kral, 1967	
		Kang and Kim, 1963	
10 II, 1 III	XXaY	Bovey (1949)	
	X–Xa–Y	Bovey (1949)	Sex trivalent
1 III	XY_1Y_2	Sharman, 1956	X-autosome trans.
	XX	Sharman, 1956	No. of autosome arms 36
	XY_1Y_2	Ford, Hamerton, Sharman, 1957	No. of autosome arms 36
	XX		
	XY_1Y_2	Matthey & Meylan, 1961	
	XY_1Y_2		
	XY_1Y_2	Meylan, 1964	No. of autosome arms 38
	XY_1Y_2	Meylan, 1965	
	XX	Meylan, 1965	
		Meylan, 1964	No. of autosome arms 36
	XX	Meylan, 1965	
	XY_1Y_2	Meylan, 1965	
21 II	XY	Skaren & Halkka, 1966	
		Meylan, 1968	
		Meylan, 1968	

Table 1

Taxon.	No. of Animals Studied Males	Females	Tissue Studied	Chr. No. 2n	Nombre Fonda-mental
Family Soricidae (cont.)					
Sorex unguiculatus	2		Testis	41	71±±
	3		Lung & Kidney	42	62
		2	Lung & Kidney	42	62
S. shinto saevus	3		Lung & Kidney	42	62
		3	Lung & Kidney	42	62
Sorex minutus	5		Testis	42	56
		1	Ovary	42	56
Crocidura murina			Testis	40	46
C. occidentalis kivu	2			50	64
		1		50	64
Crocidula suaveoleus mimula	1		Embryo	40	50
C. leucodon leucodon	1	3		28	56
Crocidura russula russula	10		Testis	42	52 *
C.r. pulchra				42	52
Neomys fodiens	2		Testis	52	92
N. anomalus milleri Motta		1	Ovary	52	94
Suncus murinus			Bone Marrow	40	50
	8	6	Spleen, Testis	40	
Blarina brevicauda	11			50	52
		5		50	52
	4			49	52
		1		49	52
Family Tenrecidae					
Centetes ecaudatus		4	Blood	38	52
Schreber	4		Spleen, Skin	38	51
Hemicentetes nigriceps Gunther		1	Spleen	38	52
H. semispinosus		3	Blood	38	52
Cuvier	4		Blood, Spleen	38	51
Echinops telfairi		2	Blood, Spleen	40	58
Martin	3		Blood, Spleen	40	57
	2		Blood, Spleen	41	58
Setifer setosus		5	Blood, Spleen	40	54
Froriep	5		Blood, Spleen, Testes	40	53

* Calculated from the karyotype.

(Continued)

Meiosis	Sex Chr.	Authors	Remarks
21 II	X0	Halkka & Skaren, 1964	
	XY	Takagi & Fujimaki, 1966	
	XX	Takagi & Fujimaki, 1966	
	XY XX	Takagi & Fujimaki, 1966	X & Y telocentric, Y heteropycnotic
21 II	XY	Meylan, 1965	
	XX	Meylan, 1965	
20 II	XY	Tateishi, 1937	
	XY	Meylan, 1967	
	XX	Meylan, 1967	
		Meylan, 1964	
		Meylan, 1964	
21 II	XY	Bovey, R., 1949	Pre-reduction of XY bivalent
		Meylan, 1966	
26 II	XY	Bovey, 1949	
		Meylan, 1966	
		Tateishi, 1937	
	XX/XY	Raychaudhari *et al.*, 1968	
25 II	XY	Meylan, 1967	
	XX	Meylan, 1967	
	XY	Meylan, 1967	
	XX	Meylan, 1967	
	XX XY	Borgaonkar & Gould, 1965 and Borgaonkar	
	XX	Borgaonkar	
	XX	Borgaonkar	
	XY	Borgaonkar	
	XX	Borgaonkar	
	XY	Borgaonkar	
	XY_1Y_2	Borgaonkar	
	XX	Borgaonkar	
20 II	XY	Borgaonkar	

Table 1

Taxon.	No. of Animals Studied Males	Females	Tissue Studied	Chr. No. 2n	Nombre Fonda-mental
Family Tenrecidae (cont.)					
Microgale dobsoni		5	Blood, Spleen	30	52
O. Thomas	1		Blood, Spleen	30	51
M. talazaci		3	Blood, Spleen	30	52
F. Major	7		Blood, Spleen & Testes	30	51
M. cowanii O. Thomas	1		Spleen, Testes	54	55
Family Solenodontidae					
Solenodon paradoxus		1	Blood, Skin	34	

is absolutely critical. Matthey counted 28 chromosomes in somatic metaphase cells of *Macroscelides rozeti* and pointed out that Brenner (1946) misinterpreted his data and *E. myurus* also has 28 chromosomes. Matthey was investigating the question of existence of polyploidy and concluded that in fact both these genera have the same chromosome number. Unfortunately, these two are the only papers which furnish karyotypes and the two other papers of Chu and Bender (1962) and Ford and Hamerton (1956) only mentioned having studied the insectivore species. However, Chu and Bender (1962) reported that *Elephantulus rufescens* has 2 satellited chromosome pairs, 5 pairs of metacentrics and 11 pairs of subtelocentrics or acrocentrics. It appears, though, that structural rearrangement of chromosomes may have caused variation in chromosome numbers of 28, 30 and 34 in this family.

The important observation in this family is that the sex determining mechanism appears to be XX female and XY male (Brenner, 1946; Matthey, 1954).

Talpidae

Information on this family will be reviewed by Gropp (1968). Table 1 lists chromosome numbers for 3 genera and five species. Three taxa have 34 chromosomes. The other two species have 32 and 36 chromosomes. However, they seem to differ in the number of chromosome arms. Sex chromosomes are XX female and XY male.

Bovey (1949) studied meiosis in *Talpa europaea* and found 17 bivalents and noted that there is a pre-reduction of XY bivalents in his material. These observations led White (1954) to state that there is a pre-reduction of sex bivalents in the insectivores!

(Continued)

Meiosis	Sex Chr.	Authors	Remarks
	XX	Borgaonkar & Gould, 1968	
15 II	XY	Borgaonkar & Gould, 1968	
	XX	Borgaonkar & Gould, 1968	
15 II	XY	Borgaonkar & Gould, 1968	
	XY	Borgaonkar	
	XX	Borgaonkar	

Erinaceidae

Data on this family also is being reviewed by Gropp (1968). Chromosome number for only one genus and four of its species is known (Table 1). All species except one have diploid number of 48 chromosomes. Sex chromosomes are XX female and XY male. The number of chromosome arms present appear to be 92. Kral (1967) found differences in karyotypes of two species of European hedgehogs, *Erinaceus europaeus* and *Erinaceus concolor*, on the basis of size and shape of four chromosome pairs. Geisler and Gropp (1967) reported chromosomal polymorphism in *E. e. europaeus* and *E. e. roumanicus* which may have resulted from a pericentric inversion and translocation resulting in variation of three pairs of chromosomes. Both the taxa are known to occur in a transitional zone but no information as to the fertility of F_1 hybrids is available (Herter, 1938). Hence, the role of karyotype variation is unclear, although it appears that these two taxa may represent two distinct species.

Soricidae

This insectivore family is one of the most extensively and widely studied and illustrates some interesting cytogenetic mechanisms. Chromosome numbers for five genera, sixteen species, and five sub-species have been determined (Table 1). Chromosomal polymorphism, sex-chromosome trivalents, structural rearrangement of chromosomes including Robertsonian fusion within a species complex are but a few of the findings in this family. Bovey (1949) in an excellent paper reviewed the Insectivora cytogenetics and described three sex chromosomes for *Sorex araneus,* X-Xa-Y for two male specimens. Sharman (1956) following this work examined 4 male and 2 female specimens and found two sex chromosomes, XX, in the females and concluded that the compound X-chromosome has arisen by a reciprocal translocation between an ancestral X-chromosome and one member of a pair of acrocentric autosomes. He also found variation in chromosome number and explained this as having occurred due to centric fusion (Robertsonian fusion of centromeres) as all the animals had the same number of chromosome (autosome) arms, 36. Ford, Hamerton, and Sharman (1957) in the following year extended this study to fifty more shrews and found the chromosome number to vary from 22 to 27 in males and 22 to 25 in females. The number of autosomal arms was constantly 36, as before. The chromosome number was found to be constant within individuals in different tissues. Whether the cytological types consisting of homozygotes and heterozygotes had any selective advantage over one another was unclear.

Cytological studies of Meylan (1960, 1964, 1965, etc.) on the European specimens of *S. araneus* showed two distinct types: A and B. The number of chromosome arms was 42 in type A and chromosome number was 23 in males and 22 in females, whereas in type B the number of chromosome arms was 40 and the chromosome number varied from 21 to 33 in the males and 20 to 32 in females. Structural rearrangement of the Robertsonian type was found in 6 autosome pairs (3 to 8) in type B. The difference between the types A and B was explained as due to separation of an ancestral population during the last glaciation, an example of allopatric speciation whereas the Robertsonian polymorphism was considered as a prelude to sympatric speciation. White (1968) considers the variation in karyotypes of these European shrews as possible examples of stasipatric speciation. Here again, as no data are available on the hybrids of these two types of *Sorex araneus,* the question of them being genetically isolated remains unanswered. I would like to question the conclusion of Meylan (1964): "In the lack of any morphological differences, the important chromosome modifications which have been observed for *S. araneus* confirm the independence between chromosome and organic evolution." I do not think it is necessary to draw this conclusion. The chromosomal variation and speciation phenomena (organic evolution) could be closely related, in fact, could go hand in hand, and variation may go undetected by the present techniques available. Mere absence of morphological variation is not sufficient to warrant this opinion. Instead we need information on what these arrangements can do to interracial and interspecific hybrids. Till then the role of karyotype variation in speciation will remain unanswered. That this may be reason is because in plants and other animals where hybrids have been obtained and studied, variation and non-homology of chromosomes has been decidedly known to have affected sterility of the hybrids (Benirschke, 1966) which plays a determining role in speciation.

Another important finding in the *Sorex araneus* material is the presence of three sex chromosomes in males (Bovey, 1949; Sharman, 1956; Ford *et al.,* 1957; Meylan, 1965, 1966). It does not seem to affect the reproductive mechanism in the males and the XX female, XY male, XY_1Y_2 males all seem to coexist. One interesting point about the sex chromosomes is that the X chromosome is acrocentric in these species.

Chromosome number determinations have been made for other species of *Sorex* and the number goes up to 42 as in *S. unguiculatus* (Table 1). The question of species identification and taxonomy has been discussed by Skaren and Halkka (1966) and Siivonen (1965) with regard to *S. unguiculatus.* The work of Takagi and Fujimaki (1966) on 2 species of *Sorex, S. shinto saevus* and *S. unguiculatus* shows the karyotypes to be identical. *Sorex unguiculatus* was found to have 2n = 41 and 2n = 42 chromosomes respectively by the Finnish and Japanese workers. I suspect Halkka and Skaren (1964) are in error in concluding that *S. unguiculatus* has a diploid chromosome number of 41. Their findings are based on studies from testicular tissue where good spreads are difficult to obtain; however, they did count 21 formations at meiotic metaphase and if these are all bivalents, including the X-Y pair, then this should give a count of 42 chromosomes. Besides, their hypothesis of maleness on account of X0 constitution is not sound due to paucity of good data.

Chromosomal evolution in the genus *Sorex* appears to be extremely interesting with diploid numbers ranging from 20 to 66 and the number of chromosome arms from 42 to 71 or more. Skaren and Halkka (1966) considered the possibility of a series of pericentric inversions having altered the karyotype in different species. According to them, Dr. R. Matthey also holds similar views. Here again information on hybrids will be critical.

Data on other genera are presented in Table 1. The difficulty in arriving at a conclusion with regard to a basic diploid chromosome number for the Soricidae and number of basic chromosome arms has been pointed out by previous workers and I would agree that the data are insufficient at the present time, especially in view of the large variation one finds in this family.

Recently, Meylan (1967) has studied 21 specimens of *Blarina brevicauda* and found that the two chromosome types with 50 and 49 chromosomes differ in a single chromosome pair, one submetacentric chromosome in the 49 type. If this is due to centric fusion of 2 acrocentrics from the 50 chromosome type, the possibility exists that the 49 chromosome type may become, or might have resulted, homozygous with a 48 chromosome type consisting of one pair of submetacentric chromosomes. The X chromosome is metacentric in this genus.

Tenrecidae

This primitive insectivore family is endemic to Madagascar and consists of 10 genera and about 40 species. Because of its restricted distribution and optimum number of taxa described, we believe that there exists an ideal opportunity to document and compare the karyotype information for the entire family. Table 1 lists the diploid chromosome counts,

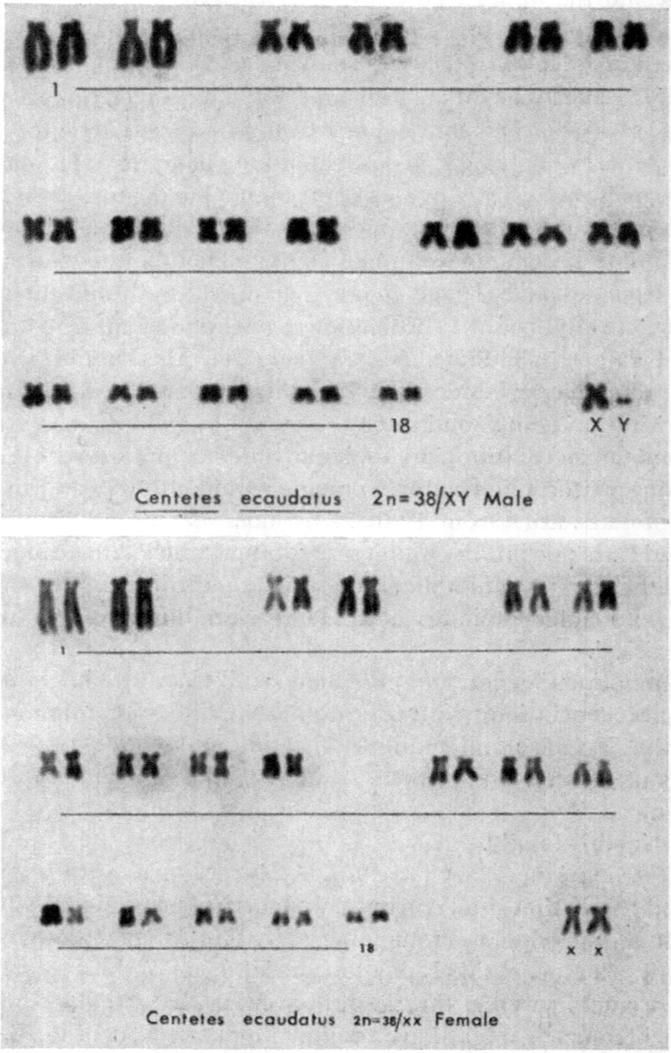

Fig. 1. Karyotypes of a male and female *Centetes ecaudatus*.

sex chromosomes, and the number of chromosome arms for the species investigated. The range of diploid chromosome numbers is 38 to 54 but the number of chromosome arms do not seem to vary as much, 51 to 58.

Three species and two genera have the same number of chromosomes, 38, and number of chromosome arms, 51. Interestingly, one of them, *Hemicentetes semispinosus*, is considered to be a neotenic form of another *Centetes ecaudatus* (Gould and Eisenberg, 1966). Be that as it may, these two genera do not appear to have any karyotypic differences (Figs. 1, 2 and 3). The sex determining mechanism appears to be XX female and XY males. The geographic distribution of *H. nigriceps* and *H. semispinosus* is such that they do not seem to overlap, i.e., they are allopatric, and no natural hybrids are known to occur. Both these species are known to breed in captivity. It will be of interest to explore the possibility of species hybridization in captivity and to see if hybrids could be obtained and, if so, whether there is any possibility of free gene exchange. The species differ considerably in morphology, natural habitat and behavior (Gould and Eisenberg, 1966). They are highly specialized in their feeding habits. If chromosome morphology is any indication of potentialities of further evolution, these three species have the lowest number of acrocentric chromosome pairs within the sub-family Tenrecinae. I have been unable to find any satellited chromosome pair so far in this family and

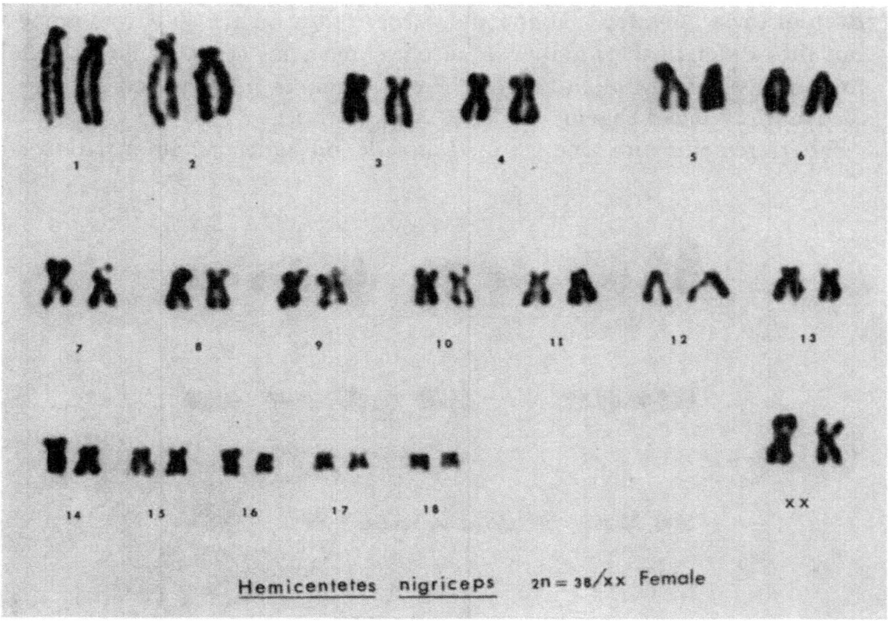

Fig. 2. Karyotype of a female *Hemicentetes nigriceps.*

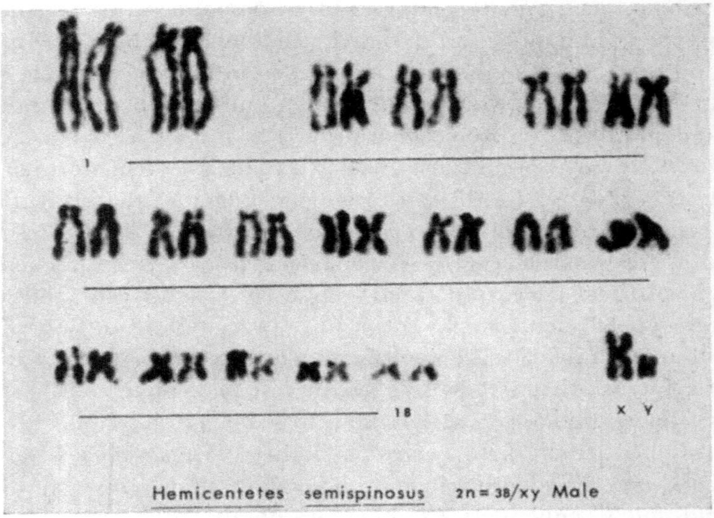

Fig. 3. Karyotype of a male *Hemicentetes semispinosus.*

I have not come up with any indication why this is so. The X chromosome is metacentric and the Y appears to be submetacentric or acrocentric and is the smallest chromosome in the genome.

Two other species in the sub-family Tenrecinae have been studied. Both of these, *Echinops telfairi* and *Setifer setosus,* have 40 chromosomes but they differ in the number of chromosome arms (Figs. 4 and 5). The former has 58 and the latter 54. These two genera differ in morphology, distribution, and behavior.

There remains only one genus, *Dasogale,* on which no information is

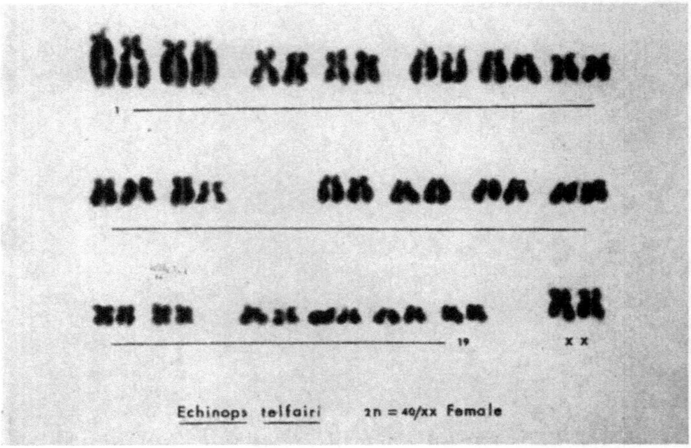

Fig. 4A. Karyotype of a female *Echinops telfairi.*

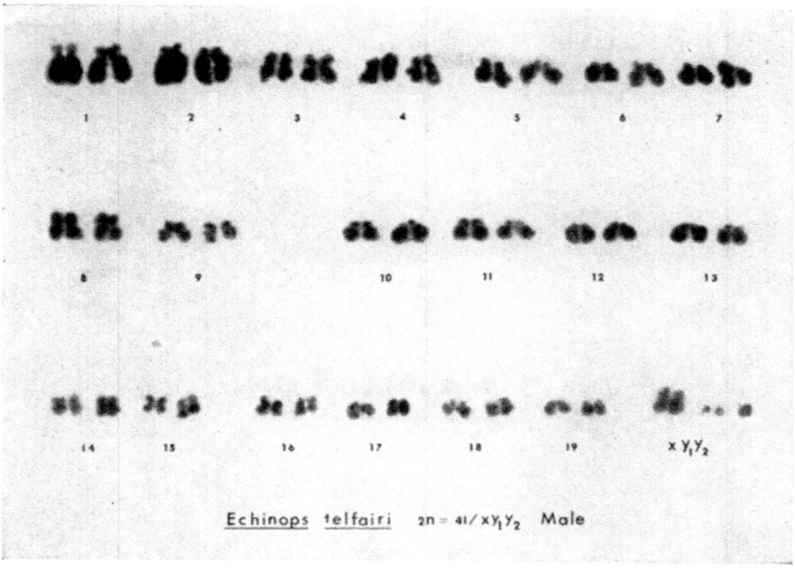

Fig. 4B. Karyotype of a male *Echinops telfairi* with XY_1Y_2 chromosome.

available. This taxon was described on the basis of a museum specimen in 1928 and was considered by Grandidier (1928) to have close affinities with both the sub-families.

Two male specimens of *E. telfairi* were found to have 41 chromosomes and the extra chromosome is a small acrocentric chromosome. This

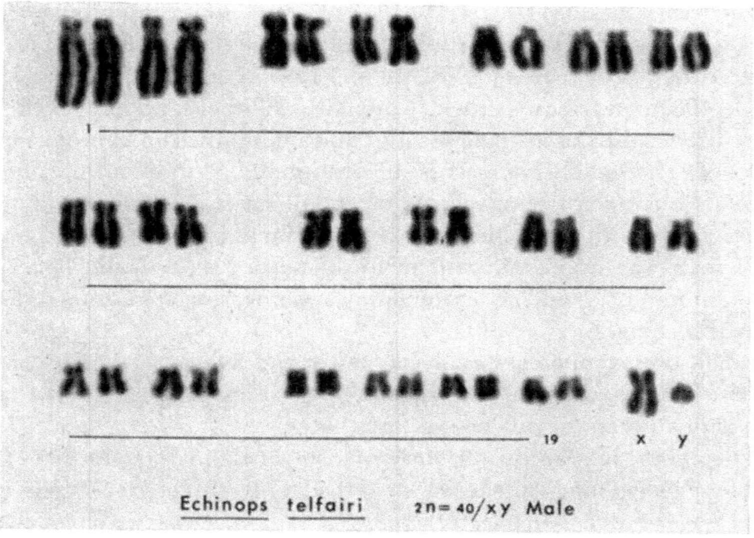

Fig. 4C. Karyotype of a male *Echinops telfairi* with an XY chromosome.

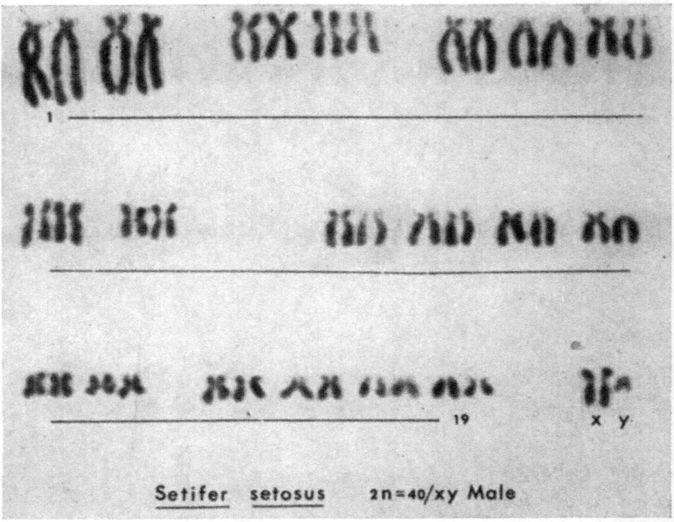

Fig. 5. Karyotype of a male *Setifer setosus.*

chromosome was interpreted as another Y chromosome since it is similar to the Y chromosome found in these animals. Unfortunately, meiosis could not be studied in these animals, but the chromosome was found in both blood and spleen cultures. In many cells, 7/14 in one animal and 14/22 in another, this extra chromosome was not found. However, absence or inability to detect the Y chromosome in cells of tenrecs is not uncommon (Borgaonkar and Gould, 1965). This is a minute and lightly stained chromosome and it may be that with our technique we do not see this element. Two Y chromosomes have been described for other insectivores, *Sorex araneus* (Sharman, 1956; Ford *et al.,* 1957; Meylan, 1964, 1965) and many other mammals. All the specimen of *Echinops* were collected from a single locality and maintained in captivity for over 2 years. The exact life span of these animals is not known. These two animals did not have any gross morphological abnormalities but there exists a possibility that this extra chromosome is an autosome.

Meiosis was studied in two males of *Setifer setosus* and 20 bivalents were counted. X and Y chromosomes were associated into a bivalent formation (Fig. 6).

When one compares the karyotypes of the five species, it becomes apparent that the largest chromosome pair is essentially acrocentric and this is of interest as will be discussed later.

The other sub-family Oryzorictinae is inadequately studied. Out of the five genera and 30 species described only three have been studied. Two species of *Microgale, M. dobsoni* and *M. talazaci,* have a diploid

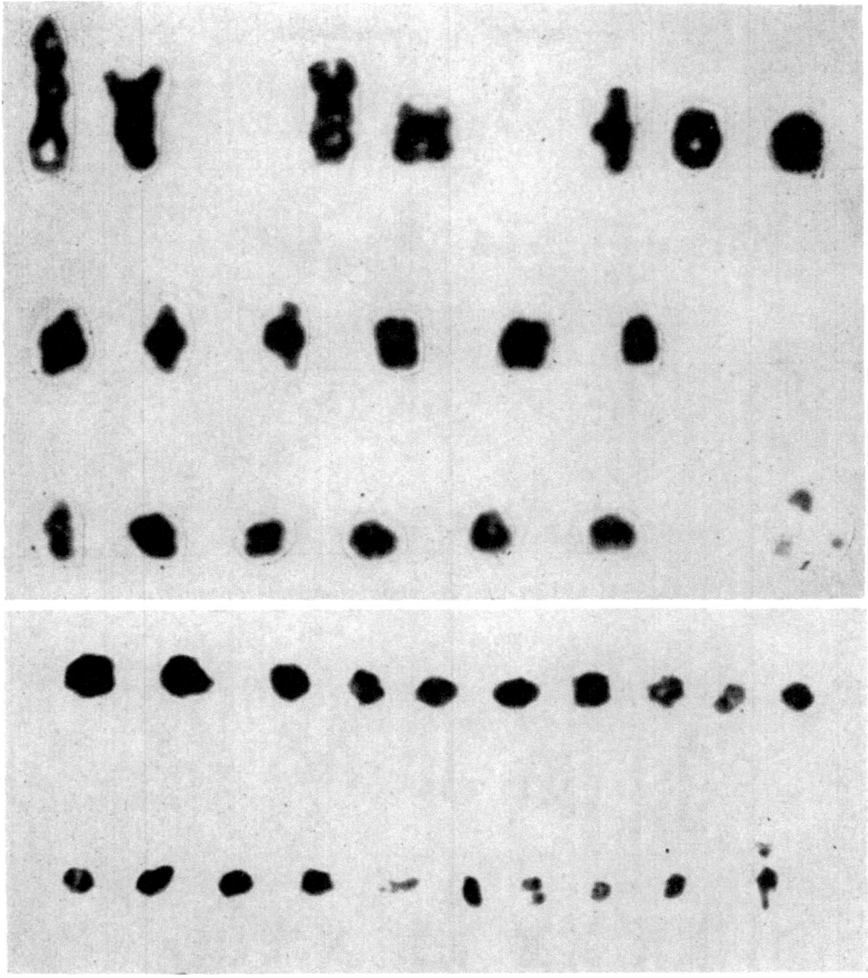

Fig. 6. First meiotic metaphase configurations from testicular preparations of *Setifer setosus*. The last bivalent is the sex chromosome pair.

chromosome count of 30 chromosomes (Figs. 7 and 8). But they have the same number of chromosome arms and seem to differ karyotypically from each other in two chromosome pairs. We have hypothesized (Borgaonkar and Gould, 1968) that this arose due to reciprocal translocation of chromosome segments which later became homozygous in one of the two species. We have further maintained that this chromosomal rearrangement may have played a causative and determining role in causing speciation between these two species. Recently, my colleagues at the National Zoological Park in Washington have been successful in getting

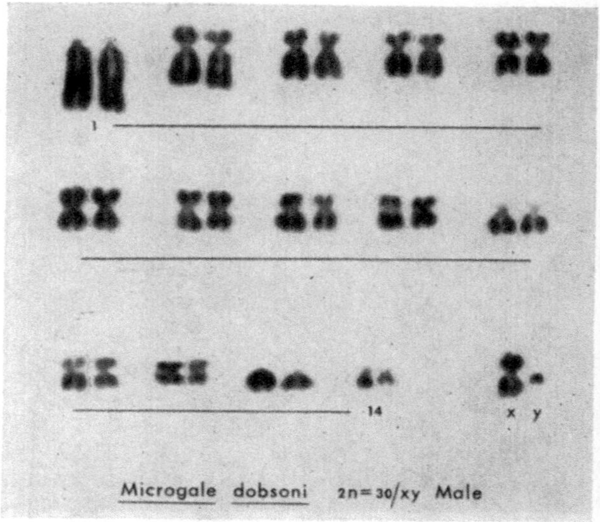

Fig. 7A. Karyotype of a male *Microgale dobsoni.*

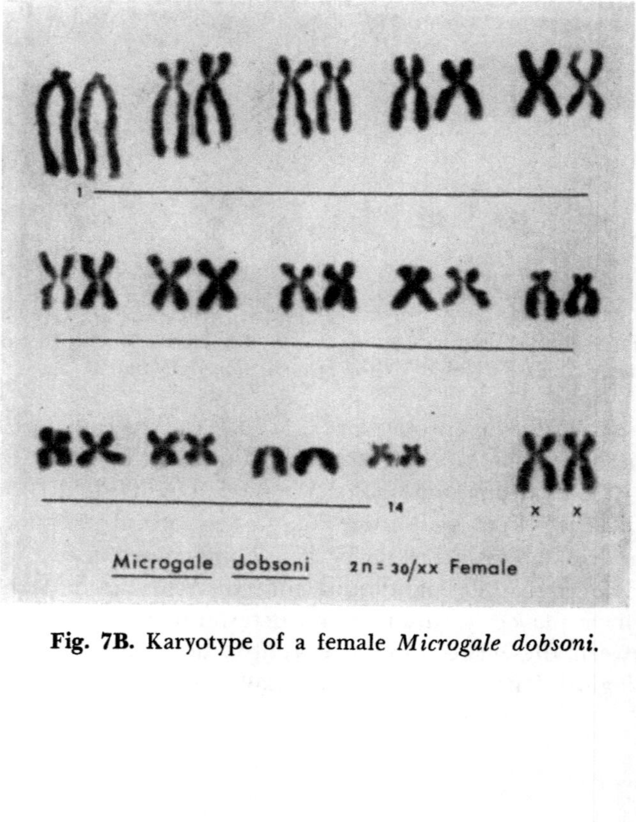

Fig. 7B. Karyotype of a female *Microgale dobsoni.*

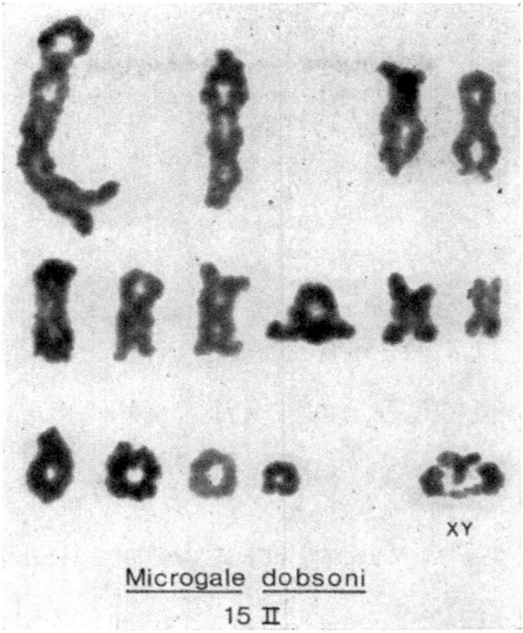

Fig. 7C. First meiotic metaphase photograph showing 15 bivalents from a male *Microgale dobsoni*.

these animals to breed in captivity. Undoubtedly, hybridization will be attempted between these two species and the hybrids will be extensively studied. From the distribution of these taxa (Fig. 9) it can be seen that such hybridization was possible; however, presently no hybrids are known to occur in nature (personal observations). If one is able to study meiosis in hybrids of these two species, a quadrivalent association of 4 chromosomes involved in this chromosomal rearrangement should be visible. These two species are distinguishable on morphological and anatomical grounds and now on the basis of karyotypes also.

Only one male specimen of *Microgale cowanii* caught near Lakota was available for study. This animal has 54 chromosomes and only one metacentric chromosome is present, and this is probably the X chromosome (Fig. 10). All the other chromosomes are acrocentrics or have very short short arms. The number of chromosome arms is 55. Ironically enough, this group of species, according to Grandidier and Petit (1930), "Les *Geogale* et les *Microgale,* du groupe *cowanii,* viennent y rappeler par leur allure et sans doute aussi, leurs moeurs, les Soricides, qui du reste, a diverse titres, se rapprochent des Centetidae," is thought to be the most primitive of all tenrecs. I am excited by this finding as there is excellent corroboration of evidence from two separate sets of data collected independently of one another.

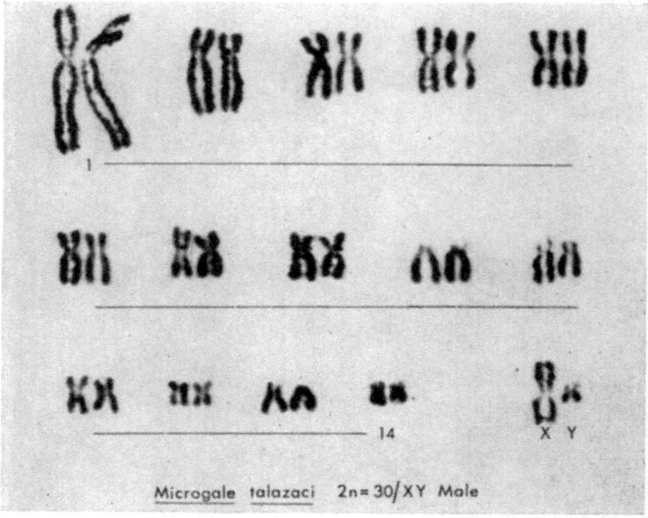

Fig. 8A. Karyotype of male *Microgale talazaci.*

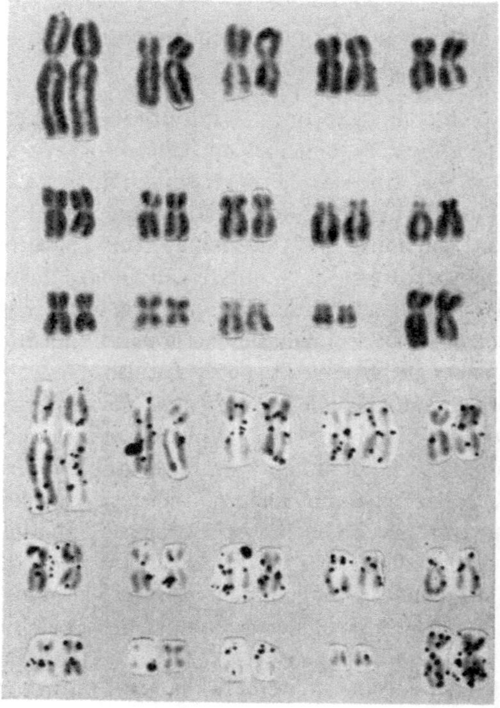

Fig. 8B. Karyotype of female *Microgale talazaci* with its tritiated thymidine labelled counterpart below.

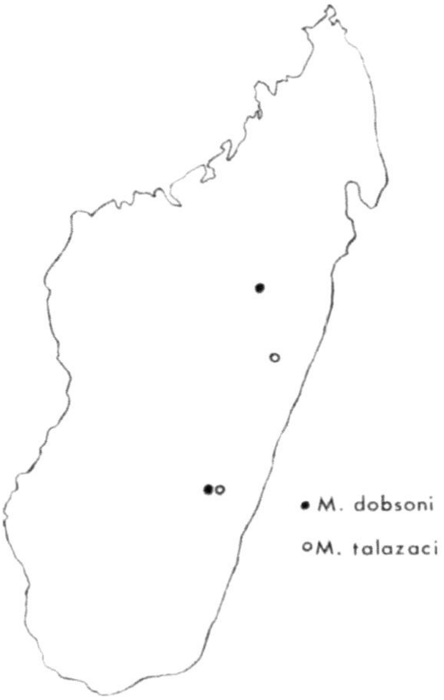

Fig. 9. Geographic distribution of the localities where the two species of *Microgale* are found.

When one compares the karyotypes of all the tenrec species (Fig. 11) it becomes apparent that only *M. talazaci* has the largest chromosome pair different from others. This may be an indication that it evolved from *M. dobsoni*. Is it possible that when studying closely related species, such as we are attempting, we will be able to trace the evolutionary pathways? We are able to draw support for the idea that *M. talazaci* has

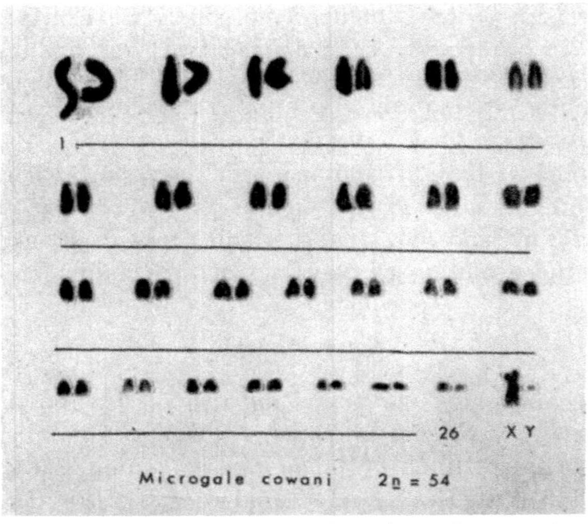

Fig. 10. Karyotype of a male *Microgale cowani*.

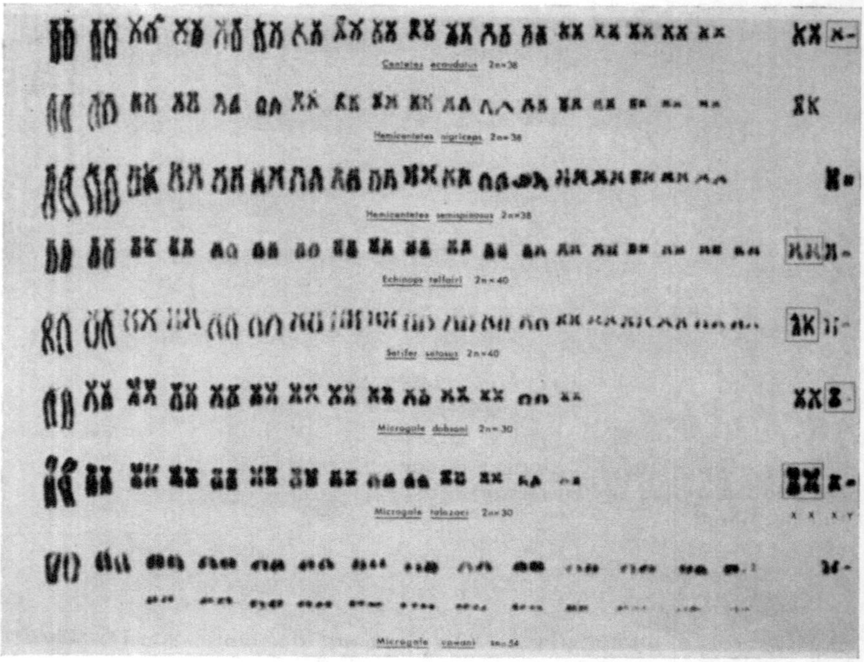

Fig. 11. Composite karyotype of all tenrec species.

evolved from *M. dobsoni* from the similarity of chromosome No. 1 of *M. cowani* and *M. dobsoni* and the sub-family Tenrecinae where it is always acrocentric. Thomas (1918) erected a new genus, *Nesogale* to accommodate *dobsoni* and *talazaci*. Grandidier and Petit (1931), describing a new tenrec genus *Paramicrogale,* state "Ce genre *Paramicrogale* apparaitra bien comme derive des *Microgale,* et s'acheminant plutôt vers les Centetinae, comme le *Nesogale* du reste, mais selon un processus different." These observations also support our observation of the evolutionary pathways in the Tenrecidae (Fig. 12).

I believe that we have arrived at a very interesting stage of this study and karyotyping of other genera such as *Limnogale,* the aquatic tenrec, *Geogale, Dasogale,* and *Paramicrogale* will probably be of great help in ascertaining the evolutionary pathways of this family.

Solenodontidae

The family consists of one genus and two species and they are found in the Caribbean countries of Haiti, Cuba and Dominican Republic. Only one female specimen of *Solenodon paradoxus* Cabrera was available for study and the chromosome number was 34 (Fig. 13). This family is thought to be somewhat closely related to the Tenrecidae (Simpson,

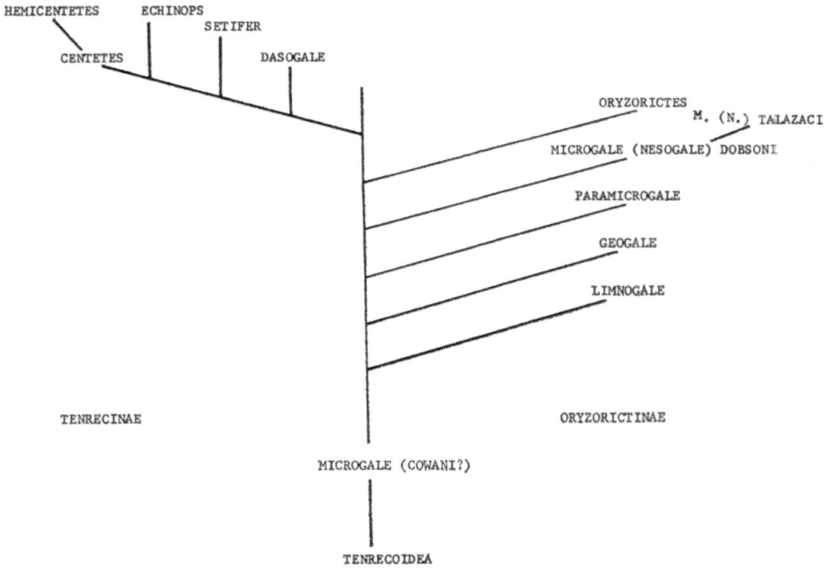

Fig. 12. Evolutionary pathways in the Tenrecidae.

1945; Eisenberg and Gould, 1966) and is included by him in the super-family Tenrecoidea. Because of the meagre amount of data available we are unable to draw any conclusions as regards the relationship of these two families.

Tupaiidae

This family of tree-shrews, Tupaiidae, is included by many in the order of Primates and by some in the order Insectivora. Van Valen (1965) has recently concluded that the better course is to refer the Tupaiidae to the Insectivora without taxonomically expressing a special relationship to any of the other groups of Insectivores. We have included Tupaiidae in this survey by mutual agreement between myself, Benirschke, Egozcue and Hsu, and I presume without any involvement in the controversy of placing the tree shrews in either groups. I have summarized the available information on the chromosomes of this group in Table 1. *Tupaia* seems to be a very interesting genus with a range of chromosome numbers from 52 to 68. Two species of this genus, *T. glis* and *T. chinensis*, are distinguishable, among other features, by the number of mammae present, i.e., *T. glis* has four and *T. chinensis* has six (Lyon, 1913; Elliot, Wong and Borgaonkar, 1968). Chu and Bender (1962), Hsu and Johnson (1963), and Arrighi (1968) have reported chromosome number of 60 for *T. glis* and the number of chromosome arms is 72. Whereas Klinger

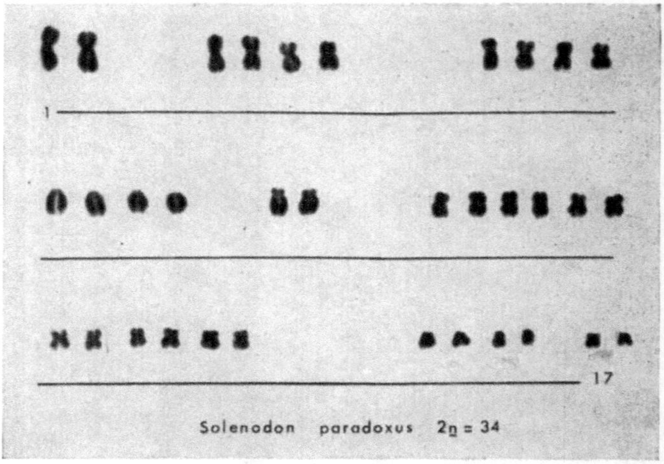

Fig. 13. Karyotype of a female *Solenodon paradoxus.*

(1963) and Egozcue *et al.* (1968) report chromosome number of 62 for their specimen but the number of chromosome arms is the same 72. The specimen of Klinger was identified by competent primatologists and he favors the view that the chromosomal variation found is evidence for speciation. Our specimen of *T. chinensis* and the report of Arrighi on this species indicates a count of 62 chromosomes for this species and the karyotype of these two reports agree with one another and also with the karyotypes reported for *T. glis* by Klinger (1963) and Egozcue *et al.* (1968). If there is no mistake in the identification of the specimen, which is not unlikely, here we have a case of one species, *T. glis,* with two chromosome numbers and karyotypes, one of which is similar to another species, *T. chinensis.* But if there is a mistake in identification, then it points to the importance of proper identification. Only Hsu and Johnson (1963) document the locality of their specimen of *T. glis* as from Malaya. Our specimen were collected in Thailand. Maybe if we know the localities of specimen we will be able to synthesize this information in a much more fruitful way. Elliot and Lisco (1968) report chromosome number of 66 for *T. minor* and the number of chromosome arms is 72 and the same diploid number is found by Arrighi (1968) in the specimen studied by her.

From the description of the 44 chromosomes of *Urogale everetti* by Chu and Bender (1962) I calculate that the number of chromosome arms for this species is 84. This seems to be quite different from the karyotype of *Tupaia.* However, chromosome numbers of other species of *Tupaia* give an indication that they might be closely related.

If centric fusion of Robertsonian type, or centromere fission has caused this variation of 52, 60, 62, 66 and 68 chromosome numbers in the genus

Tupaia then this could be taken as an excellent example of the usefulness of cytogenetic data in the understanding of speciation. Only when information on hybrids and/or potentialities of gene exchange between these taxa is known, one will be able to conclude as to what happened first—chromosomal rearrangement or speciation. As chromosomal changes are known to disrupt genetic imbalance and gene effects it is more likely that the rearrangement gave rise to morphological variation leading to speciation. This would give additional support to White's concept of stasipatric speciation.

Comments and Summary

In this review I have made an attempt to summarize all the available information on Insectivora cytogenetics and hope that this is as complete and exhaustive as possible. Interesting findings are those of structural changes involving only a few pairs of chromosomes, Robertsonian fusion type, centromere fission type (or even may be cross-over in the centromere region), and reciprocal translocations between two chromosome pairs in closely related species, sub-species, ecotypes, or just groups of animals within a species. That these are incidental and have had no or little role to play in causing morphological variation and/or reproductive isolation is too naive a conclusion, but I am sure there are some who hold this view. I believe that such karyotype changes, at least in some instances are a prelude to speciation. Only further intensive workup, by a large group of mammalian cytogeneticists, will satisfy the critics of cytotaxonomy.

The insectivores are a small but highly specialized group and studies of certain families in this order undoubtedly will be rewarding in tracing the phylogenetic pathways. One of the puzzling findings is that the X chromosome in some species is an acrocentric chromosome whereas in most species of insectivores and mammals it is submetacentric or metacentric. It will be interesting to test these species for X-linked genes such as that for G6PD known to occur in animals of rather widely separated taxa.

Multiple sex chromosomes have been found in this group including XY_1Y_2 sex chromosome constitution for males. The hypothesis of translocation between Y and an autosome was invoked to explain the phenomena. Only recently cases of X chromosome/autosome translocation have been found in man. What effects such chromosomal interchanges between sex chromosomes and autosomes will have on the evolution of sex chromosomes only time can tell.

Cytogenetic studies of hybrids, of any sort, are notably missing in this review. Whether hybridization is possible, and even if fertilization is successful whether such hybrids will live to maturity is entirely unpredictable from the present state of knowledge. The importance of meiotic studies of such hybrids is a matter on which, I am confident, we all will agree.

Acknowledgment

Part of the original work reviewed in this article was supported by grant GB-4978 from the National Science Foundation and grant #4314 from the Penrose Fund of the American Philosophical Society.

References

Arrighi, F. E.: Personal communication, 1968.

Benirschke, K.: Sterility and fertility of interspecific mammalian hybrids. In: Comparative Aspects of Reproductive Failure. K. Benirschke, ed. New York: Springer-Verlag Inc., 218, 1966.

Borgaonkar, D. S. and E. Gould: Chromosome complement of Tenrecs, *Centetes ecaudatus* (Order Insectivora, Class Mammalia). Experientia *21*:631, 1965.

——— and ———: Homozygous reciprocal translocation as a mode of speciation in *Microgale* Thomas 1883 (Tenrecidae Insectivora). Experientia *24*:506, 1968.

Bovey, R.: Un type nouveau d'heterochromosomes chez un Mammifère: le trivalent sexuel de *Sorex araneus*. Arch. Jul. Kl-St. Vere. *23*:506, 1948.

———: La formule chromosomique de quelques Insectivores indigenes. Experientia *5*:72, 1949.

———: Les chromosomes des chiropteres et des Insectivores. Rev. Suisse Zool. *56*:371, 1949.

Brenner, S.: The chromosome complement of *Elephantulus*. So. Afri. J. Med. Sci. (Biol. Suppl.) *11*:71, 1946.

Chu, E. H. Y. and M. A. Bender: Cytogenetics and evolution of primates. Ann. N.Y. Acad. Sci. *102*:253, 1962.

Dodson, E. D., cited by M. A. Bender and C. H. Y. Chu. The chromosomes of primates. In: Evolutionary and Genetic Biology of Primates. *1*:261, 1963.

Egozcue, J., B. Chiarelli, M. Sarti-Chiarelli and F. Hagemenas: Chromosome polymorphism in the tree shrew (*Tupaia glis*). Folia Primat. *8*:150, 1968.

Eisenberg, J. F. and E. Gould: The behavior of *Solenodon paradoxus* in captivity with comments on the behavior of other Insectivora. Zoologica *51*:49, 1966.

Ellerman, J. R., T. C. S. Morrison-Scott and R. W. Hayman: Southern African mammals 1758 to 1951; a reclassification. London: Brit. Mus. Nat. Hist.

Elliot, O. S., M. Wong and D. S. Borgaonkar: Karyological study of *Tupaia* from Thailand. In manuscript.

——— and H. Lisco: Personal communication, 1968.

Ford, C. E. and J. L. Hamerton: Chromosomes of five rodent species. Nature *177*:140, 1956.

Ford, C. E., J. L. Hamerton and G. B. Sharman: Chromosome polymorphism in the common shrew. Nature *180*:392, 1957.

Geisler, M. and A. Gropp: Chromosome polymorphism in the European hedgehog *Erinaceus europaeus* (Insectivora). Nature *214*:396, 1967.

Gould, E. and J. F. Eisenberg: Notes on the biology of the Tenrecidae. J. Mammal. *47*:660, 1966.

Grandidier, G.: Un nouveau type de Mammifère insectivore de Madagascar. Bull. Acad. Malgache N.S. *11*:85, 1928.

——— and G. Petit: Etude d'un Mammifère Insectivore Malgache le *Geogale aurita* A.M.E. et A.G. Faune des Colonies Fran. *4*:441, 1930.

——— and ———: Un type nouveau de Centetidae Malgache (*Paramicrogale occidentalis*). Bull. Soc. Zool. France *56*:126, 1931.

Gropp, A., M. Geisler and P. Citoler: The chromosomes of the hedgehog (*Erinaceus europ.*) and the mole (*Talpa europ.*). Mamm. Chr. Newsletter *22*: 181, 1966.

———: These proceedings.

Halkka, O. and U. Skaren: Evolution chromosomique chez genre *Sorex:* nouvelle information. Experientia *20*:314, 1964.

Herter, K.: Die Biologie der Europäisehen Igel. Monograph. Leipzig, 1938.

Hsu, T. C. and M. L. Johnson: Karyotypes of two mammals from Malaya. Amer. Naturalist *97*:127, 1963.

Jordon, M.: Formules chromosomiques de quelques Insectivores de Bialowieza. I. *Erinaceus roumanicus*. Folia Biologica (Krakow) *8*:151, 1966.

Kang, Y. S. and Y. J. Kim: Studies on chromosomes of the Korean mammals. I. Karyotypes of squirrel and others. Zoologica (Seoul) *2*:1, 1963.

Klinger, H. P.: The somatic chromosomes of some primates (*Tupaia glis, Nycticebus coucang, Tarsius bancanus, Cercocebus aterrimus, Symphalangus syndactylus*). Cytogenetics *2*:140, 1963.

Koller, P. C.: Chromosome behaviour in the male ferret and mole during anoestrus. Proc. Roy. Soc. Lond. Ser. B. *121*:192, 1936.

Kral, B.: Karyological analysis of two European species of the genus *Erinaceus*. Zool. Listy *16*:239, 1967.

Lyon, M. W. Jr.: Tree shrews: An account of the family Tupaiidae. Proc. U.S. Nat. Mus. *45*:1, 1913.

Makino, S.: An Atlas of the Chromosome Numbers in Animals. Ames, Iowa: The Iowa State College Press, 1951.

Matthey, R. and A. Meylan: Le polymorphisme chromosomique de *Sorex araneus* L. Etude de deux portées de 5 et 9 petits. Rev. Suisse Zool. *68*:223, 1961.

———: Le polymorphisme chromosomique des *Mus* africains du sous-genre *Leggada*. Revision generale portant sur l'analyse de 213 individus. Rev. Suisse Zool. *73*:585, 1966.

———: Les chromosomes des *Macroscelides rozeti* Duvernoy (Mammalia-Insectivora). Existet-il une serie polyploide chez les Macroscelidae? Rev. Suisse Zool. *61*:669, 1954.

Mayr. F.: These proceedings.

Meylan, A.: Contribution a l'étude du polymorphisme chromosomique chez *Sorex araneus* L. Rev. Suisse Zool. *67*:258, 1960.

———: Le polymorphisme chromosomique de *Sorex araneus* L. Rev. Suisse Zool. *71*:903, 1964.

———: Le formule chromosomique de *Sorex minutus* L. Experientia *21*:268, 1965.

———: Repartition geographique des races chromosomiques de *Sorex araneus* L. en Europe (Mamm.-Insectivora). Rev. Suisse Zool. *72:*636, 1965.

———: Données nouveues sur les chromosomes des Insectivores européens (Mamm.). Rev. Suisse Zool. *73:*548, 1966.

———: Formules chromosomiques et polymorphisme Robertsonien chez *Blarina brevicauda* (Say) (Mammalia-Insectivora). Canad. J. Zool. *45:*1119, 1967.

———: La formule chromosomique de *Crocidura occidentalis kivu* Osgood (Mammalia-Insectivora). Rev. Suisse Zool. *74:*685, 1967.

———: In: Chromosome Numbers of Eutherian Mammals. 6th supplement. R. Matthey, ed. January 6, 1968. Mamm. Chr. Newsletter *9:*72, 1968.

Painter, T. S.: A comparative study of the chromosomes of Mammals. Amer. Naturalist *59:*385, 1925.

———: Chromosome numbers in mammals. Science *61:*423, 1925.

Raychaudhari, S. P., P. V. Ranjini and T. Sharma: Karyological studies of 16 species of Indian mammals. M.C.N. *9:*82, 1968.

Sharman, G. B.: Chromosomes of the common shrew. Nature *177:*941, 1956.

Siivonen, L.: *Sorex isodon* Turov (1924) and *S. unguiculatus* Dobson (1890) as independent shrew species. Aquilo Ser. Zool. *4:*1, 1965.

Simpson, G. G.: The principles of classification and a classification of mammals. Bull. Amer. Mus. Nat. Hist. *85:*1, 1945.

Skaren, U. and O. Halkka: The karyotype of *Sorex caecutiens* Laxmann. Hereditas *54:*376, 1966.

Takagi, N. and Y. Fujimaki: Chromosomes of *Sorex shinto saevus* Thomas and *Sorex unguiculatus* Dobson. Jap. J. Genet. *41:*109, 1966.

Tateishi, S.: On the chromosomes of the two species of Insectivora. A preliminary note. Jap. J. Genet. *13:*211, 1937.

———: The chromosomes of two species of Insectivora. Annot. Zool. Jap. *17:* 515, 1938.

Thomas, O.: On the arrangement of the small Tenrecidae hitherto referred to *Oryzorictes* and *Microgale*. Ann-Mag. Nat. Hist. *1* (Ser. 9): 302, 1918.

Van Valen, L.: Treeshrews, primates, and fossils. Evolution *19:*137, 1965.

White, M. J. D.: Animal Cytology and Evolution, 2nd ed. Cambridge: University Press, 1954.

———: Models of Speciation. Science *159:*1065, 1968.

CYTOLOGIC MECHANISMS OF KARYOTYPE EVOLUTION IN INSECTIVORES

A. GROPP
Abteilung für Cytopathologie und Cytogenetik des Pathologischen Instituts der Universität Bonn, Germany

It seems safe to accept the generalization that a species can only be defined by a whole array of morphological, physiological and behavioral features and, of course, of those features pertinent to the notion of a reproductive community. Is the karyotype only one of these criteria, or does it represent a superposed principle? The latter assumption seems to be more acceptable, mainly with regard to the karyotype as the particular structural mould of the genome on the cytological level, modeled by the forces of evolution, and constantly remodeled as a prerequisite of speciation. But some facts, difficult to understand in mammalian cytogenetics, should be explained by a theory based on this assumption: the most important of these difficulties is the existence on the one hand of

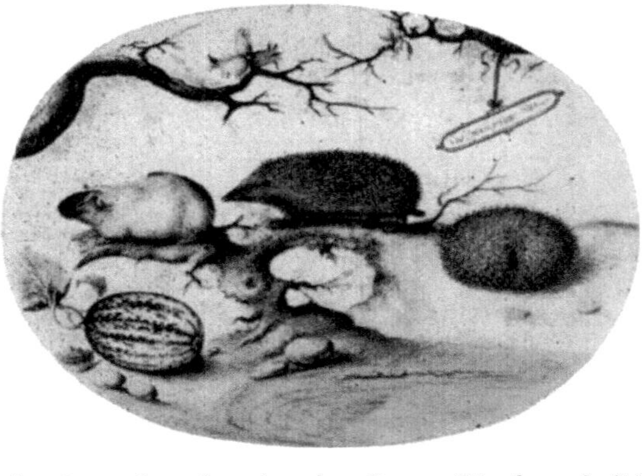

Fig. 1. Reproduction of a drawing by Georg Hoefnagel (1542–1600), from *Terra: Animalia Quadrupedia et Reptilia,* Four sketchbooks, dedicated to Emperor Rudolf II. Printed by permission of Mrs. Edith G. Rosenwald, Alverthrope Gallery, Jenkintown, Pennsylvania.

families with a very extensive karyotype variation between closely related phenotypes, and on the other hand of families displaying a considerable similarity of karyotype patterns. The extensive karyotype variations found in ground squirrels of Western North America by Nadler (1968) and in certain African subgenera of *Mus* by Matthey (1963a, 1964) are examples of the first type. The other type is represented by the karyotype uniformity in most *Felidae,* in the *Ursidae* and in other families. Taylor *et al.* (1968), reviewing their own results and those of Benirschke (1967) on the *Camelidae,* discussed the divergent evolution of the members of this family in spite of the near-identity of their karyotypes in terms of being accomplished by single gene variation or minor chromosome rearrangements. This antagonistic situation in the occurrence of karyotype variations within different taxonomic groups can be explained in the way Matthey (1964) does. He distinguishes taxonomic groups with prevailing major chromosomal mutations as *primum movens* of the evolutionary steps, and others in which the genetic changes occur most probably at the gene level.

It is the aim of this study, not only to present some data on a limited number of mainly European species of the order *Insectivora,* but to analyze also more thoroughly the chromosome polymorphism in two European sibling species of the genus *Erinaceus.* The results should help to reconcile the apparent discrepancy existing between the concept of a genetic variation at the gene level or at the level of minor rearrangements

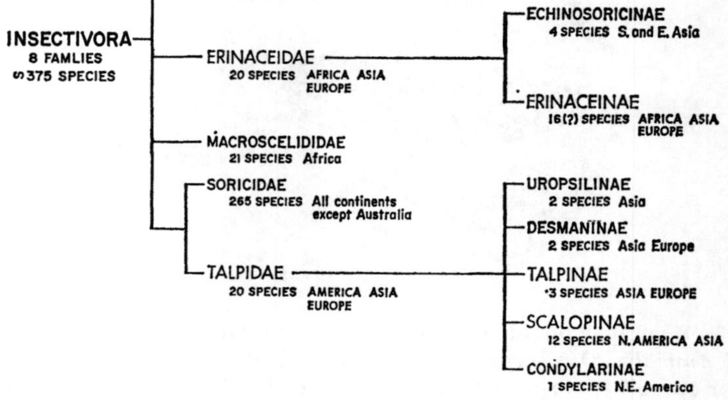

Fig. 2A. Families and subfamilies of the order *Insectivora.*

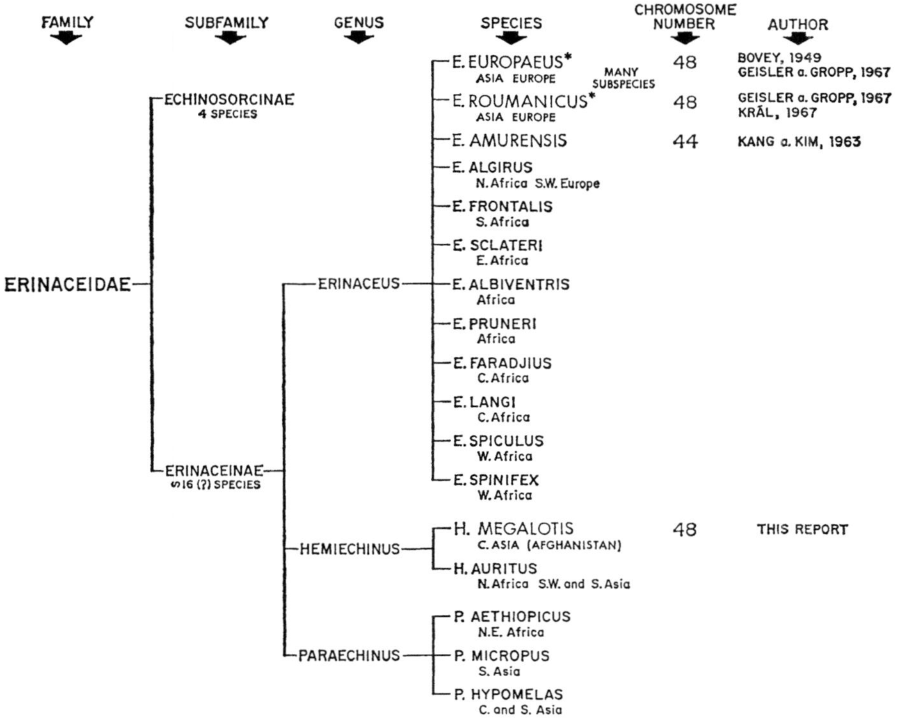

Fig. 2B. Family *Erinaceidae* (after Herter, 1938; v. Wettstein, 1941; Ellerman and Morrison-Scott, 1951). Only the subfamily *Erinaceinae* is represented in detail. *Separate list of subspecies and of questionable further east Asiatic Species/Subspecies in Fig. 3.

and of that of a chromosomal variation becoming evident cytologically. The findings in the European *Erinaceidae* may contribute also to understanding the importance of *cytologic isolation* as an evolutionary mechanism of similar significance as geographic isolation. The latter aspect shall be further strengthened by cytogenetic observations in a newly recognized European rodent species of the genus *Mus*.

The members of the 8 families of the order Insectivora (Fig. 2A) are obviously endowed with a high degree of karyotype variation (Borgaonkar, 1965–1967). Cytogenetic analysis revealed distinct types of chromosome variation in association to certain taxonomic subgroups. Robertsonian fusion changes are apparently of great importance in some European and North American shrews (Soricidae; Ford *et al.*, 1957; Meylan, 1960, 1964, 1965, 1967). Although only few members of the *Talpidae* (Fig. 2A) and *Erinaceidae* (Fig. 2B) have yet been investigated cytogenetically, there is a good possibility that a Robertsonian mechanism occupies no major role in the evolution of the recent species. Instead, they display structural chromosome rearrangements best explained by pericentric inversions and translocations. In the following two sections of this report the findings in these families are reviewed.

Talpidae

The taxonomic position of this family is represented in Fig. 2A, partly using the check-list of Palaearctic and Indian mammals by Ellerman and Morrison-Scott (1951). In only three species karyotype analyses have been performed. These are the European moles *Talpa europaea* with two subspecies (Bovey, 1949; own observations on 7 males and 8 females from Sechtem and Bad Godesberg near Bonn) and *T. caeca (*Meylan, 1966) and the North American hairy-tailed mole *Parascalops breweri* (Subfamily *Scalopinae*). Chromosome preparations of the latter have been made available through the courtesy of R. Low, Dartmouth Medical School, Hanover, N.H. (2 males from Plainfield, N.H.). A synopsis of the karyotypes of *T. europaea europaea* and *P. breweri* in Figure 4 shows some striking common features which should be considered with regard to the wide geographic separation of both species and to their taxonomic position in two different genera: both species have 34 chromosomes with nearly identical sex chromosomes. The common marker, a pair of metacentric chromosomes with a paracentromeric gap in the short arms, is very remarkable. The arrangement of both karyotypes, nevertheless, underlines the existence of a chromosome diversity of both species characterized by a shift between the metacentrics and acrocentrics or subacrocentrics, respectively. Pericentric inversions must have occurred at some time or other. But the presence of two small acrocentrics in *Para-*

	CHROMOSOME NUMBER	GEOGRAPHIC ORIGIN OF INVESTIGATED SPECIMENS
E. EUROPAEUS		
E. EUR. EUROPAEUS +	48	DIFFERENT SITES IN WESTERN GERMANY, SWITZERLAND
E. EUR. ITALICUS +	48	MONTECATINI
E. EUR. CONSOLEI		
E. EUR. HISPANICUS		
E. EUR. NESIOTES		
E. EUR. RHODIUS		
E. EUR. CENTRAL ROSSICUS		
E. EUR. CENTRAL ROSSICUS PALLIDUS		
E. EUR. CONCOLOR		
E. (EUR.?). TRANSCAUCASICUS (SATUNIN, 1905)		
E. ROUMANICUS		
E. ROUM. ROUMANICUS +	48	LOWER AUSTRIA TATRA MOUNTAINS (ČSSR)
E. ROUM. DROZDOVSKY		
E. (EUR.?) AMURENSIS +	44	
E. (EUR.?) DEALBATUS		
E. (EUR.?) MIODON		
E. DAUURICUS		

Fig. 3. Table of the Eurasian species and subspecies of hedgehogs (***Erinaceus***). For bibliography, see Fig. 2B and text. For *E.* (*eur.?*) ***transcaucasicus,*** compare Fig. 6. The classification of the East-Asiatic species (and subspecies) remains doubtful. + = karyotyped species or phenotypes.

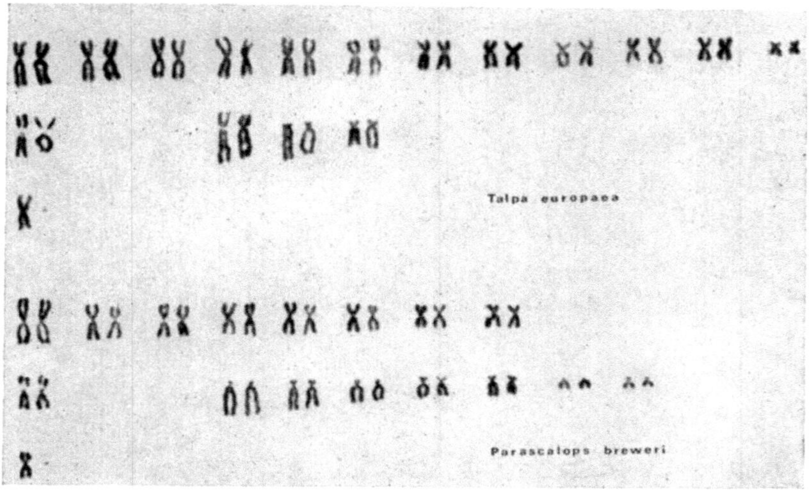

Fig. 4. Synoptic presentation of karyotypes of *T. europaea* (*Talpinae*) and *P. breweri* (*Scalopinae*). First row, *metacentrics;* second row, metacentric marker chromosome–acrocentrics and subacrocentrics; third row, XY-chromosomes.

scalops proves that also translocations were probably of some importance in producing this karyotype variation.

Meylan (1966) described the karyotype of a European mole subspecies identified as *T. europaea cinerea*. Careful analyses of the karyotype published by this author, in comparison of moles of the area of Bonn, Germany (identified as *T. europaea europaea*), revealed slight structural differences, particularly in the occurrence of a large subacrocentric in *T. europaea europaea*. Figure 5 shows the karyotype of three moles trapped near Bonn. While the karyotypes in the middle and below represent a standard male and female karyotype found in our material, there was a structural heterozygosity in one single male exhibiting an aberrant phenotype with a light-greyish pelage (karyotype in top of Fig. 4). The chromosomal variation was most distinct in the second pair of the first and second rows of the karyotype. This finding can, indeed, be best explained by the assumption of a "Europaea-Cinerea"-Hybrid. Again, pericentric inversion may be responsible for the structural changes in two closely related phenotypes of the European mole.

Meylan's (1966) findings of 36 chromosomes in *T. caeca,* due to the presence of two extremely small metacentrics in addition to the other chromosomes of similar structure and shape as in *T. europaea cinerea,* is difficult to explain convincingly. More knowledge on mole karyotypes is necessary to discuss the origin of these "minutes."

Conclusion Cytogenetic investigations in the three species (4 phenotypes), covering two genera of the *Talpidae* revealed a general frame of

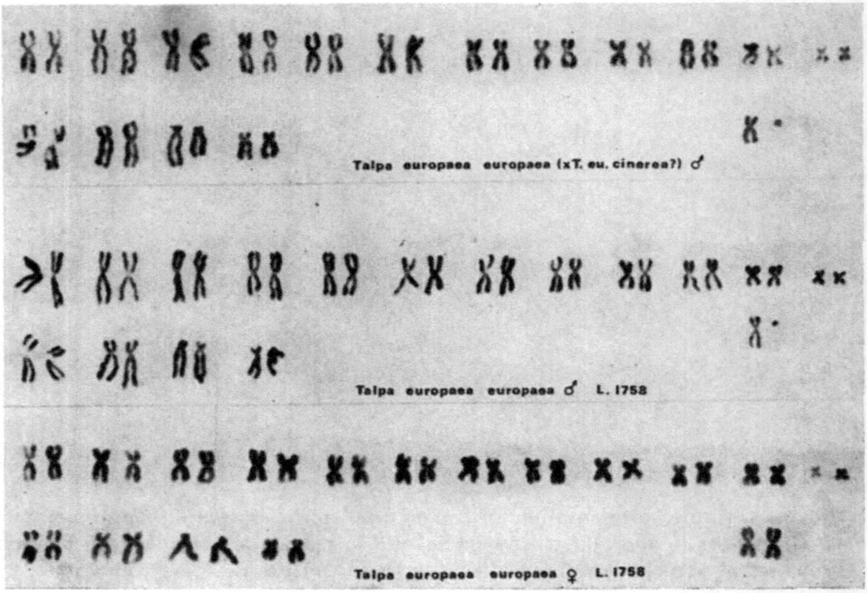

Fig. 5. Synopsis of karyotypes of a male (middle) and a female (below) European mole (*T. europaeus europaeus* L. 1758). Presumably heterozygote karyotype of a male (top). First row, metacentrics; second row, marker chromosome--subacrocentrics and acrocentrics—Sex chromosomes.

striking uniformity of the basic numbers of chromosomes (except the "minutes" of *T. caeca*) and of the occurrence of a marker chromosome. Within this frame, a remarkable structural variation occurs which is best explained as being produced by processes of pericentric inversion and reciprocal translocation.

Erinaceidae

Pelage, cranial and ecological characteristics of the European hedgehog are very well studied in many local phenotypes. The pedigree of Figure 2B marking the position of the European hedgehogs within the *Erinaceidae,* and Figure 3 tabulating the species and subspecies of the genus *Erinaceus* are based on the somewhat controversial classifications proposed by Herter (1938), v. Wettstein (1941) and Ellerman and Morrison-Scott (1951). Most of the diverging opinions of these authors have to do with the debatable determination of the species or subspecies character of the European and Asiatic phenotypes of *Erinaceus* (*europaeus*) as indicated in Figure 3.

There is general agreement on the extension of the respective dispersal areas of the Western (*E. europaeus europaeus*) and the Eastern

E. roumanicus roumanicus) European hedgehogs (Fig. 6). Their habitats adjoin in Central Europe forming a relatively sharp border-line from the Baltic Sea, following the Oder river, crossing Western Silesia, the Bohemian-Moravian plane, Upper Austria, and ending at the Istrian peninsula (Herter, 1938; Kratochvil, 1966). Herter (1938) described intermediate phenotypes in this border region, whereas Kratochvil (1966) denied the occurrence and even the possibility of hybrid types from his studies in Bohemia-Moravia. Herter, however, succeeded in cross-breeding *E. europaeus* and *E. roumanicus* in an outdoor experiment on an island previously free from hedgehogs but, unfortunately, he did not investigate the reproductive capacity of the F_1-generation. Both authors agree in recognizing the two European phenotypes as separate nominate species, each of them with several subspecies (Fig. 3). Following Herter (1938), their speciation is understandable by the assumption of an isolating mechanism exerted by a vast Central European diluvial glacial barrier. Nevertheless, the same author finds considerable difficulties in explaining the fact that they retained a fairly sharp separation after the post-diluvial dispersal into their neighboring recent habitats, in spite of the apparently proven capability to interbreed.

v. Wettstein (1941), having studied extensively the Mediterranean hedgehog phenotypes, came to a very controversial conclusion. He considered all European and Pontico-Mediterranean phenotypes as belonging to a single species *E. europaeus* with at least 12 subspecies. This

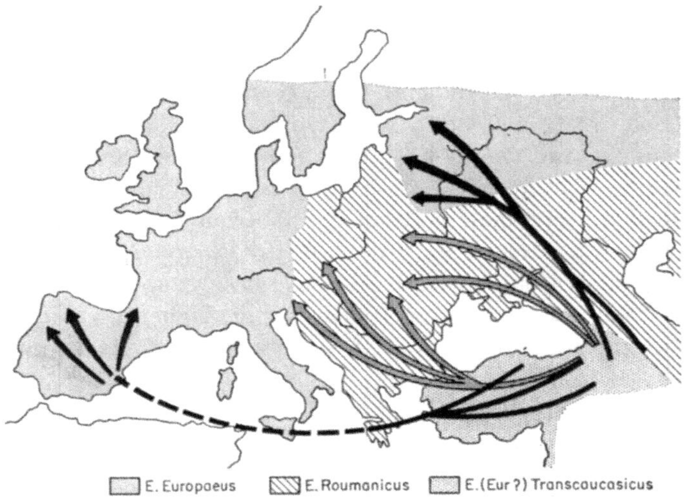

Fig. 6. Habitats of *E. europaeus* and *E. roumanicus* (adapted from Herter, 1938). The hypothetical migrations of Pontic origin (v. Wettstein, 1941) are indicated by arrows (black indicates older migration wave; shaded indicates younger migration).

opinion results, in part, from his morphologic findings which suggest a common Pontic origin of all Near-East and European phenotypes. The Pontic *E. europaeus transcaucasicus* (Figs. 3 and 6) is in this theory an ancestor type which had given rise to different vagrant pushes. An older migratory wave of the *europaeus*-type is assumed by v. Wettstein to have touched in one extension the Mediterranean islands, Italy and Spain, and North-Western Europe in another, whereas the *roumanicus*-type should correspond to a younger race group of similar Pontic origin. These hypothetical migrations are represented by arrows in Figure 6.

This situation in the European hedgehog phenotypes demands categorically cytogenetic investigations in the genus *Erinaceus* and also in other related genera (Figs. 2B and 3). Bovey (1949), in a study on the mitotic and meiotic chromosomes of *E. europaeus,* found a chromosome number of 48. However, comparative studies using improved cytologic techniques were only recently undertaken in *E. europaeus* and *E. roumanicus* by Geisler and Gropp (1967) and Král (1967). They confirmed the already established chromosome number of 48 for both species. The main result of these two investigations was the finding of a very distinct structural chromosome variation concerning at least three larger subacrocentrics or submetacentrics and an acrocentric (of medium size in *E. europaeus,* and of a very small size in *E. roumanicus*). Further specimens of *E. europaeus* and *roumanicus* as well as a male ear-hedgehog, *Hemiechinus megalotis* (see Fig. 3B), have been investigated since our first report (Geisler and Gropp, 1967), partly with the use of ^{3}H-thymidine autoradiography for the study of the late labelling pattern of the chromosomes. Karyotype analysis has not yet been performed in the genus *Hemiechinus.*

Four hedgehogs (one male and one female from Wolkersdorf/Lower Austria; one male and one female from outside Vienna) identified as *E. roumanicus,* and twenty hedgehogs (10 males and 8 females from the Siebengebirge/Rhineland, Paderborn/Westphalia, and the East-Friesian island Borkum; one male and one female from Montecatini/Italy) with pelage and cranial characteristics of *E. europaeus europaeus* or *E. eur. italicus* (animals from Montecatini), have been studied. The male specimen of *H. megalotis* was trapped near Kabul/Afghanistan; it was kindly offered by the helpful courtesy of Dr. Klockenhoff, Zoological Project, University of Kabul/Afghanistan.

Chromosome preparations were obtained from bone marrow, short term lymphocyte cultures or kidney epithelium-cultures. ^{3}H-thymidine autoradiography was performed in lymphocyte cultures of one male and female *E. europaeus* and in kidney cell cultures of one male *E. roumanicus,* and of the male *H. megalotis.* Meiotic preparations were made from minced testicular tissue by hypotonic pretreatment, followed by fixation in alcohol: acetic acid and airdrying on thoroughly cleaned slides (description of this latter material in Gropp and Citoler, Patterns of autosomal heterochromatin, this volume).

Figure 7 represents synoptically the karyotypes of males of the two European phenotypes *E. europaeus* and *E. roumanicus* compared with that of an *H. megalotis male.* The chromosomes are arranged according to their centromeric index and their size: first row, large subacrocentrics; second row, large to small meta- and submetacentrics; third row, acrocentrics, metacentric "minutes," XY-chromosomes.

Several very remarkable common characteristics and diverging features as well can be ascertained in analyzing these species which belong to two different genera. (1) Their common chromosome number is 48. (2) The presence of two pairs of metacentric "minutes" is a common feature of the three karyotypes. They are, moreover, very similar to the smallest pair of chromosomes in *Talpa caeca.* (3) The presumable X-chromosome is of somewhat similar size in the three species, whereas the size of the small Y-chromosome differs slightly. Compared with these concordant points, the next points stress the principal divergences. (4) The respective numbers of subacrocentrics and metacentrics differ markedly. (5) *E. europaeus* and *E. roumanicus* both have a pair of acrocentrics, of medium sized in *E. europaeus* and of very small size in *E. roumanicus.* Acrocentrics are absent in *H. megalotis.*

Altogether, the chromosomal characteristics emanating from the comparative synopsis in Figure 7 demonstrate a certain resemblance of the three karyotypes and thus underline the relationship of these species.

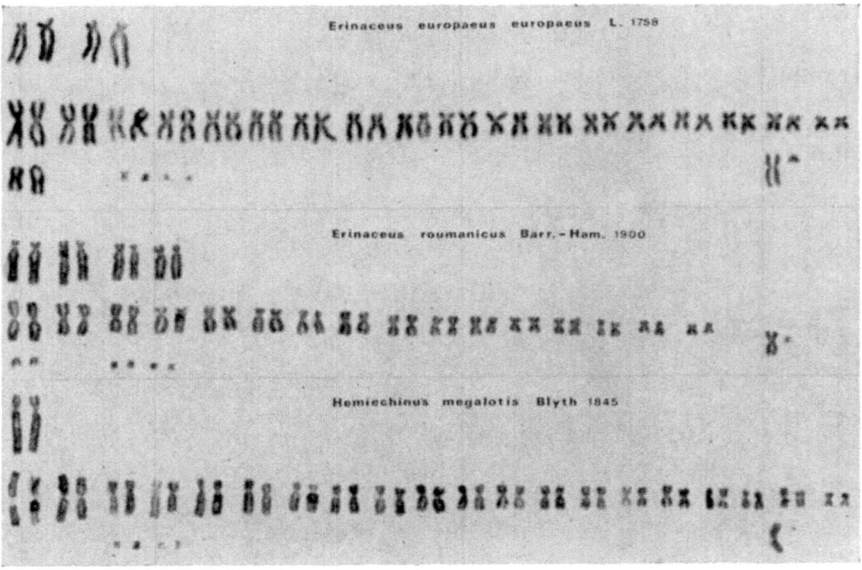

Fig. 7. Synopsis of the karyotypes of three hedgehog species.

The variation of the karyotypes is most probably brought about by the mechanism of pericentric inversions and translocations. But the comparison of the karyotypes alone cannot give any further suggestions as to the closer mutual taxonomic relationships of the three species. The gross karyotype variation seems to be of approximately the same extent between the closely related sibling species of *Erinaceus* and between these and the kin of a different genus.

Facing this general problem in mammalian cytogenetics, a considerable further elucidation was provided in the case of *Erinaceus* by the analysis of the autosomal labelling pattern in ^{3}H-thymidine autoradiography. Figure 8A compares the two karyotypes of *E. europaeus* (female) and *E. roumanicus* (male). The same features as discussed before are recognizable.

Careful observation reveals, in addition, signs of chromatid apposition in the long arms of some of the larger subacrocentrics and in one of the X-chromosomes of *E. europaeus* (see also Fig. 7, with the same phenomenon). Schmid (1967) alluded to this sign as an attribute of heterochromatic chromosomal parts.

In Figure 8B the autoradiographs of the same karyotype as in Figure 8A corresponding to a late labelling pattern are represented. The most outstanding feature is the presence of a large amount of late labelling autosomal material in both species differing in that respect distinctly

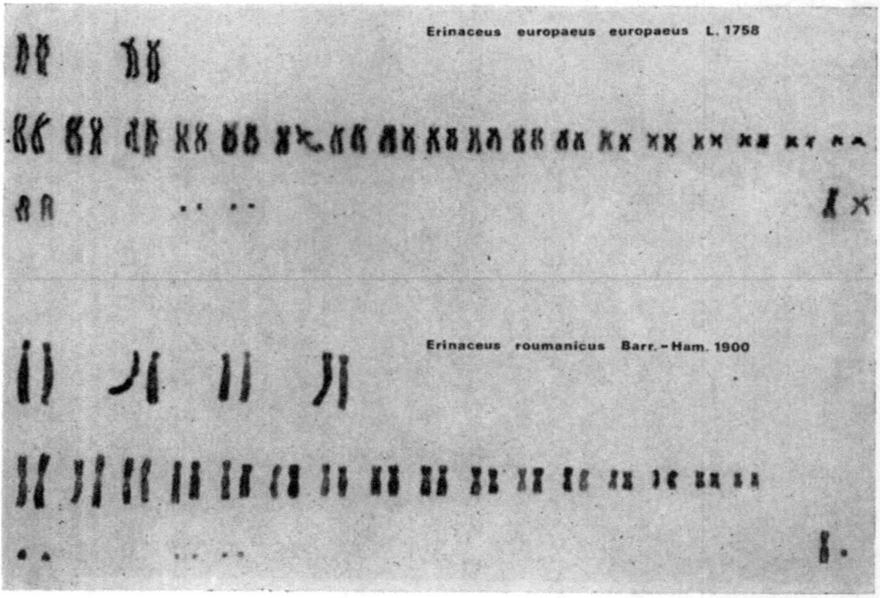

Fig. 8A. Comparative presentation of the karyotypes of two European hedgehogs.

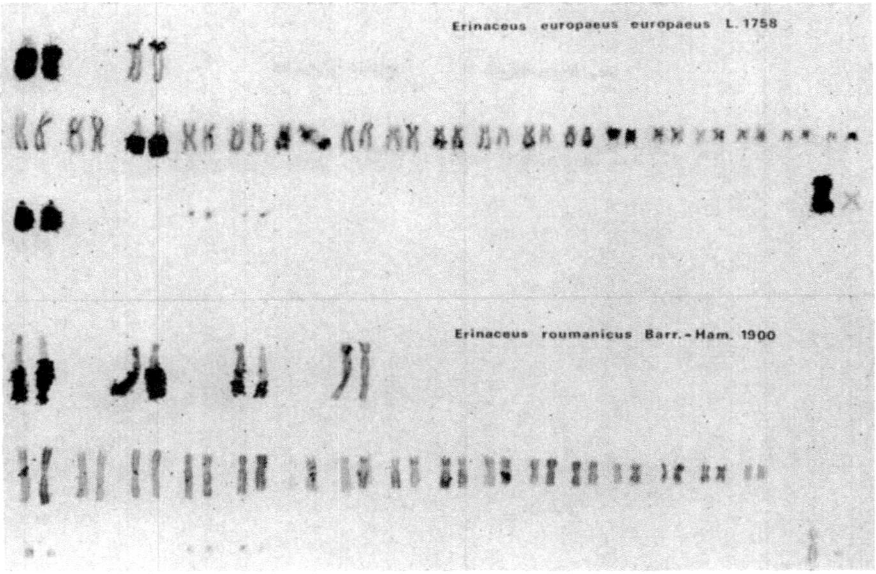

Fig. 8B. ^{3}H-thymidine-autoradiographs of the same karyotypes as in Fig. 8A, corresponding to a late replication, exhibiting large coherent blocks of labelling.

from a more diffuse pattern in *H. megalotis* (Fig. 9). The significance of this finding with regard to the heterochromatic nature of the late labelling autosomal regions and to genetic inactivation shall be briefly discussed in a separate contribution to this volume (Gropp and Citoler). The total amount of late labelling chromosomal portions seems to be equal in both *Erinaceus* species, but probably different from that of *H. megalotis* (Fig. 9). A comparison of the distribution of the most prominent blocks of late labelling on the autosomes of *E. europaeus* and *roumanicus* shows that the differences of their localization could be explained convincingly by a transfer or translocation of this material from one chromosome to another. Such changes can certainly not account for all particular variations between both karyotypes. Yet, they make understandable the most characteristic divergences in the size of the acrocentric chromosome and the appearance of a further large subacrocentric with great segments of late labelling in *E. roumanicus*. Thus, careful analysis of the late labelling acrocentric in *E. europaeus* showed that a small pericentromeric part, most accurately corresponding to the size of the acrocentric in *E. roumanicus*, remains free of grains. It is suggested that the larger late labelling segment of this chromosome in *E. europaeus* is transferred, presumably joining the late labelling material of one of the metacentrics, thus resulting in one of the large subacrocentrics of the *E. roumanicus* karyotype.

Assuming the heterochromatin nature of the late labelling segments

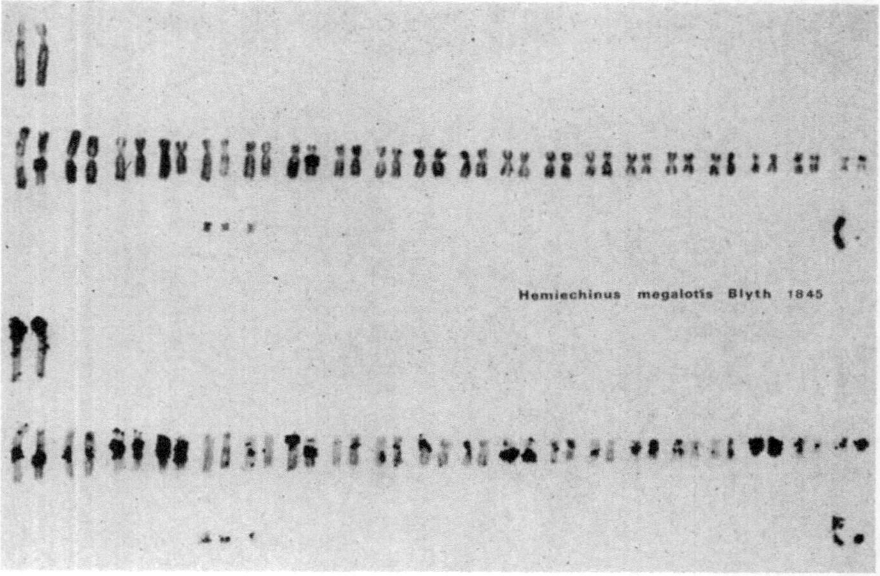

Fig. 9. Karyotype and ^{3}H-thymidine-autoradiograph of the same karyotype of the ear-hedgehog, *H. megalotis*. (Note the more diffuse type of late labelling.)

of the autosomes (Gropp and Citoler, this volume) it seems obvious that at least some of the most prominent variations of the chromosomes in *E. europaeus* and *roumanicus* concern heterochromatin and probably genetically inactivated material. Thus, from a point of view of euchromatin only, the apparent gross variations of both karyotypes would become substantially reduced to minor structural variations. This conception can be elucidated further by looking at both karyotypes under the assumption that the late labelling autosomal and heterochromatic blocks remain out of consideration. The comparison with the late labelling pattern of *H. megalotis* (Fig. 9) reveals that, on the contrary, the difference between both *Erinaceus*-species and *H. megalotis* are much deeper than shown by the karyotypes only.

Conclusion The karyotype differences in both European hedgehog phenotypes, the Western and the Eastern hedgehogs, prove definitively their different taxonomic position which seems to be best characterized by the species level. The nomination as *E. europaeus* and *E. roumanicus* used in this review takes into account the cytogenetic findings and seems at the same time to be corroborated by the classical taxonomic point of view expressed by Herter (1938) and mainly by Kratochvil (1966). It may be even more appropriate to consider *E. europaeus* and *roumanicus* as phenotypes being underway in the realization of their speciation (Matthey, 1964). The disputable occurrence of intermediate hybrid types in the border zone of the respective habitats is compatible with the close relationship of both sibling species. Cytologically, the hybridization which was performed experimentally by Herter (1938) is not difficult to understand. The decisive assay of the reproductive capacities of the F_1, however, either by meiotic studies in the males or by the breeding experiment, was not done. Our own attempts to crossbreed *E. europaeus* and *E. roumanicus* in the laboratory have, after two years, not yet brought any results.

Both *Erinaceus* species possess large amounts of late labelling autosomal blocks which, in analogy to the late labelling X-chromosome and for other reasons (Gropp and Citoler, this volume), are equivalent to heterochromatin and probably to genetically inactivated or inert chromosomal material. The most convincing hypothesis for both species is to assume a chromosomal difference of an ambiguous character: the gross karyotype variations are mainly changes on a cytologic level and result predominantly from translocations and probably also pericentric inversions of heterochromatin. The genetically effective changes, on the other hand, are mainly caused by minor rearrangements. It seems safe to assume that the presence of sizeable and coherent blocks of autosomal heterochromatin is an old property of the two species, *E. europaeus* and *E. roumanicus* and probably of their common ancestor type. In *Hemiechinus megalotis*, on the contrary, there is also a fair amount of late labelling in autosomes but its distribution pattern is fundamentally different.

It is not possible to draw a picture of the origin of all the phenotypes mentioned in Figure 3 by the isolated findings in few species only. Likewise, the theory of different migratory waves taking rise from a Pontic ancestor type (v. Wettstein, 1941) or the other theory of a primordial separation by a diluvial glacial barrier (Herter, 1938) can only be confirmed or rejected when the karyotypes of more and mainly of the Pontico-Mediterranean phenotypes are known.

Yet, the concept of ambiguity of gross cytological changes versus possible or probable minor rearrangements to be assumed for both recent European hedgehog species bears several consequences for the problem of speciation, not only in the genus *Erinaceus* but also in a more general respect. Heterochromatinization in the context of evolution and speciation has presumably the meaning of enhancing gross chromosomal rearrangements in avoiding at the same time to interfere to a similar extent with the genetically important chromosomal parts. However, the polymorphic condition, for instance in both European *Erinaceus*-species, could be incompatible with a normal prophasic pairing or diakinetic separation of homologous chromosomes in a F_1 for cytologic reasons. This implies that not mainly a genetic, but a predominantly cytologic mechanism builds up a barrier against reciprocal reproduction.

The theory founded on the importance of heterochromatin in karyotype rearrangements, and relying on the possibility of an enhancement of major chromosomal changes due to heterochromatinization, is a new concept of karyotype variation and karyotype evolution in mammals. It could be interpreted as being a mechanism of equal importance as the other known principles of karyotype rearrangements. It is conceivable that the concept of *heterochromatinization in karyotype evolution* implies the possibility of a more or less frequent repetition of steps of similar or identical chromosome rearrangements, even simultaneously and in a smaller isolated or quasi-isolated population. This could facilitate an understanding of the spread of a chromosomal change in a given population. Furthermore, the application of this principle must not necessarily be limited to cases revealing as large blocks of late labelling heterochromatin as in both investigated *Erinaceus*-species. It could prove to be true as well on a smaller scale, perhaps on a level beyond distinct microscopic visibility.

Comparative Presentation of a Further Case of Robertsonian Changes in the Genus *Mus*

Matthey (1963–1965) first described a sympatric chromosomal polymorphism of the Robertsonian type in Pigmy-mice (Subgenus *Leggada* of *Mus*) from different localities in Africa. A similar condition has not yet been known in other members of the genus *Mus*.

Professor E. v. Lehmann (Museum Alexander Koenig, Bonn, Germany), studying the habitat and the morphological characteristics of a dark colored wild-type mouse, previously described as *M. poschiavinus* Fatio, 1869, discovered a remarkably reduced fertility of the F_1-generation, which had resulted from the crossbreeding of the *poschiavinus*-type with the common *M. musculus* (E. v. Lehmann, 42. Hauptvers. Dtsch. Ges. f. Säugetierkde, Bern, Sept. 1968). *Poschiavinus*-mice are only found at a single locality in a Swiss high-altitude valley, the Val poschiavo (Fig. 10), where they coexist with the common *Mus musculus*. The fluctuation of both populations up from their first recognition to this time will be reviewed by v. Lehmann (l.c.). Cytogenetic investigations done in our laboratory in collaboration with v. Lehmann on specimens of the *poschiavinus*-type, on F_1 from crosses with the laboratory mouse, and from backcrosses and F_2, gave the following results: the *poschiavinus*-mice have 26 chromosomes (Fig. 11A), 14 of them large to medium-sized metacentrics, 12 acrocentrics (Nombre Fondamental—Matthey: 40). The F_1 from *poschiavinus* $\times$ *M. musculus* ($2n = 40$, all acrocentrics) showed always a somatic chromosome set of 33 (Fig. 11B), 7 of them metacentric and 26 acrocentric (N.F. = 40).

Meiotic figures of diakinesis (Fig. 12) and metaphase I of F_1-males

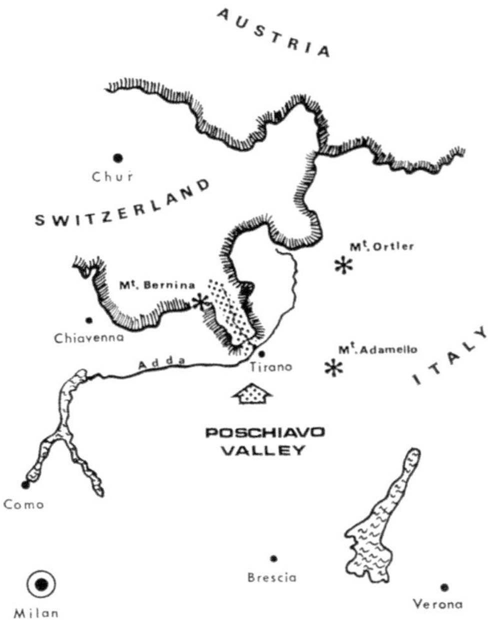

Fig. 10. Geographic localization of the Poschiavo valley in southeastern Switzerland.

revealed the presence of 13 elements: 7 trivalents, 5 bivalents and a characteristic end-to-end associated XY-bivalent. The backcross- and F_2-mice could be bred only at a very limited rate. Some of them displayed marked malformations. The results of the cytogenetic studies in these mice shall be discussed elsewhere.

Conclusion The *poschiavinus*-mice and *M. musculus* display an ideal Robertsonian variation. The karyotype differences, the difficulties to interbreed the F_1, even under the condition of force-breeding in the laboratory, and the sympatric coexistence of both types allude to a taxonomic classification of the *poschiavinus* mice as a separate species, *M. poschiavinus.*

The evolution of this variant species is probably sympatric, even if conditions of quasi-isolation in a high-altitude valley may have enhanced the establishment and spread of the different karyotype. The conditions of the fusion mechanism may in itself be more complicated than by simple fusion-translocation (Matthey, 1963b). This mechanism represents, however, in an antagonistically different way a rearrangement comparable to that in *Erinaceus* in bringing about extensive cytologic variation and reducing the interference with the global genetically active material to a minimum degree. The gross cytologic variation builds up at the same time an effective barrier securing the existence of the species.

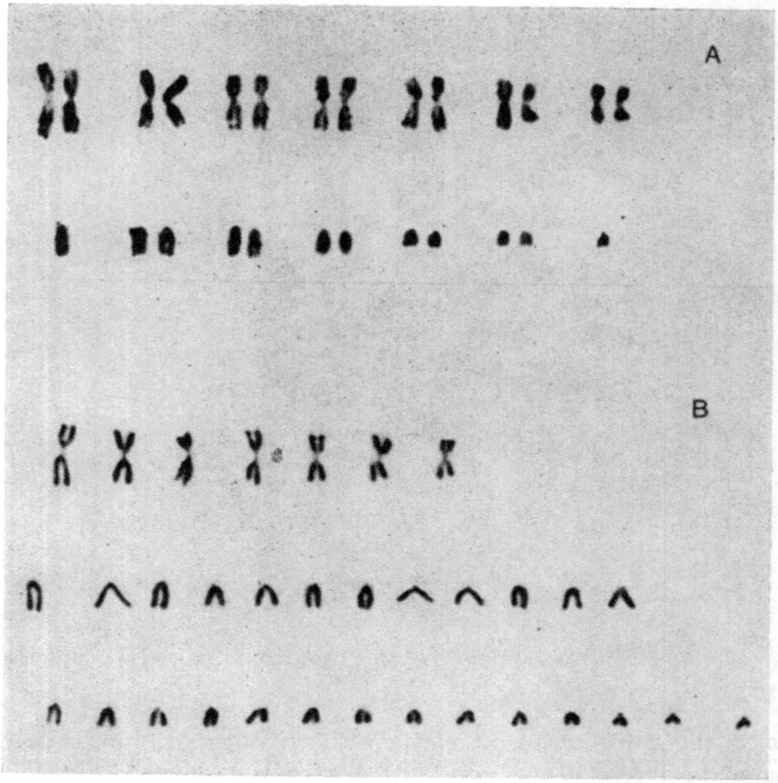

Fig. 11. A. Karyotype from spermatogonial metaphase of a male *M. poschiavinus*. Chromosome number 26; N. F. = 40. B. Karyotype from bone marrow of a male F_1 (*M. poschiavinus* ♂ × *M. musculus* ♀). Chromosome number 33; N. F. = 40.

General Concluding Remarks

Two main conclusions seem to follow from the observations in some *Erinaceidae* and from the findings in *M. poschiavinus*. The first is that it must not be assumed that a real discrepancy exists between evolutionary steps caused by changes at the gene level or by minor chromosomal rearrangements and those brought about by gross karyotype variations, provided that the ambiguous character of cytologic events and genetic mutations is considered. It becomes conceivable that gross structural rearrangements of the chromosomal set, fusions or inversions and translocations of heterochromatinized chromosomal regions do not necessarily change fundamentally the whole content of genetically active material. Thus, even in the presence of those rearrangements, the genetically effective changes probably occur on the level of the gene or of minor structural modifications, possibly further influenced by position effects or similar phenomena. The most plausible explanation for the gross karyotype variations in the cases studied in this review is that they build up a very powerful cytologic isolation mechanism for a new species or, also, for a population underway in speciation. A characteristic peculiarity of the mechanisms discussed here may be seen in an intrinsic possibility to repeat identical or similar changes, because the structural variations represent cytologic phenomena, which are conditioned by factors subject to cellular biology. Thus, fusions of acrocentrics are dependent on the structure and function of the centromere (Matthey, 1963b), and the remodeling of heterochromatin is, at least in part, due to its allocyclic behavior, to the particular biochemical and biophysical status, and to its specific irritability. This principle of karyotype variation is basically different from the classic point of view of genetic mutation. It makes at least understandable that structural homozygosity can take place easier or even faster than by the notion of random chromosomal mutations.

The second conclusion to be drawn from the cytogenetic observations presented here is that a cytologic isolation of an incipient species can arise within, peripheral or outside the range of an already existing species (or proto-species). The notion of cytologic isolation does not necessarily compel to accept a more or less dogmatic position of an exclusive sympatric or allopatric origin of a species. In the case of *Erinaceus*, a

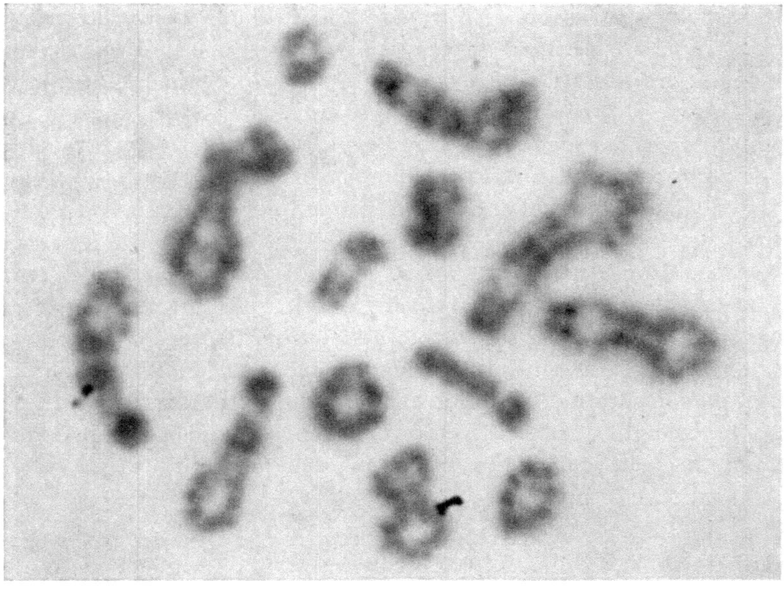

Fig. 12. Male F_1 (*M. poschiavinus* ♂ × *M. musculus* ♀). Testis. Diakinesis: 7 tri- (tetra-) valents, 5 bivalents and XY-bivalent.

sympatric speciation is possible, yet not proven. However, in *M. poschiavinus* a primary karyotypic variation must have occurred, most likely at the beginning of a sympatric process of speciation. The allopatric model of speciation, emphatically supported by Mayr (1963), cannot be the only universal mechanism of the evolution of mammalian species. White (1968), in a recent study of models of speciation, based on findings in *Morabinae* grasshoppers, critically discussed the exclusive acceptance of an allopatric origin of speciation. He introduced the notion of a stasipatric model involving the spread of a chromosomal rearrangement throughout a substantial part of the range of an already existing species. This model fits exactly the possibility of the development of cytogenetic isolating mechanisms within a given and already existing species. Both geographic and cytologic isolation are apparently essential mechanisms of speciation. It should not even be excluded that more and different mechanisms may exist bringing forth isolating barriers and conditioning the origin of a species. Geographic and cytologic mechanisms of isolation represent two different extremes, but they must not exclude each other. Most probably they can also overlap to a variable extent, or they may, in one case or the other, be linked together in bringing about conditions which render possible their mutual effectiveness.

Summary

The karyotype variation in *Talpidae* and *Erinaceidae* (as far as cytogenetic analyses are performed) seems to be caused mainly by pericentric inversions or translocations. In two European sibling species, *Erinaceus europaeus* and *E. roumanicus,* further studies using ^{3}H-thymidine-autoradiography revealed suggestions of a new concept of karyotype variation based on heterochromatinization. Heterochromatic and probably genetically inactivated autosomal regions are prone to enhance gross structural variation.

Further data on a newly recognized mouse species related to *M. musculus* by an ideal Robertsonian fusion mechanism, together with the findings in *Erinaceus,* lend support to a view considering cytologic mechanisms of karyotype evolution as isolating factor of similar importance as geographic separation. Isolation, apparently, is the basic principle in speciation, brought forth either by prevailing cytologic or prevailing geographic mechanisms, or possibly even other mechanisms.

Acknowledgments

I thank Drs. P. Citoler, M. Geisler (*Erinaceidae*) and U. Tettenborn (*M. poschiavinus*) for their collaboration and great help. Much valuable help in preparing the figures and the manuscript was offered by Miss F. Klee and Mrs. R. v. Hahn.

These investigations were partly supported by the Deutsche Forschungsgemeinschaft und Landesamt für Forschung, Nordrhein/Westf.

References

Benirschke, K.: Personal communication, 1967.

Borgaonkar, D. S.: A list of chromosome numbers in the Order Insectivora. J. Hered. *56*:266, 1965.

———: Addendum—Chromosome numbers in the Order Insectivora. J. Hered. *57*:28, 1966.

———: Additions to the lists of chromosome numbers in the insectivores and primates. J. Hered. *58*:211, 1967.

Bovey, R.: Les chromosomes des chiroptères et des insectivores. Rév. suisse de Zool. *56*:371, 1949.

Citoler, P. and A. Gropp: DNS-Replikation von autosomalem Heterochromatin. Exptl. Cell. Res. (in press).

Ellerman, J. R. and T. C. S. Morrison-Scott: Checklist of Palaearctic and Indian Mammals 1758–1946. London (Brit. Mus.), 1951.

Ford, C. E., J. L. Hamerton and G. B. Sharman: Chromosome polymorphism in the common shrew. Nature (Lond.) *180*:392, 1957.

Geisler, M. and A. Gropp: Chromosome polymorphism in the European hedgehog *Erinaceus europaeus* (Insectivora). Nature (Lond.) *214*:396, 1967.

Herter, K.: Die Biologie der europäischen Igel. Leipzig: Verlag Dr. Paul Schöps, 1938.

Kang, Y. S. and J. K. Yung: Studies on chromosomes of the Korean mammals. I. Karyotypes of squirrel and others. Zoologica (Seoul) *2*:1, 1963.

Král, B.: Karyological analysis of two European species of the genus *Erinaceus*. Zool. Listy *16*:239, 1967.

Kratochvil, J.: Zur Frage der Verbreitung des Igels (*Erinaceus*) in der CSSR. Zool. Listy *15*:291, 1966.

Matthey, R.: Cytologie comparée et polymorphisme chromosomique chez des *Mus* africains appartenants aux groupes *bufo-triton* et *minutoides* (Mammalia-Rodentia). Cytogenetics *2*:290, 1963a.

———: Polymorphisme chromosomique intraspecifique et intra-individuel chez *Acomys minous* Bate. Etude des hybrides *A. minous* ♂ × *A. cahirinus* ♀ (Rodentia–Murinae). Le méchanisme des fusions centriques. Chromosoma (Berl.) *14*:468 1963b.

———: Evolution chromosomique et spéciation chez les Mus du sous-genre *Leggada* Gray 1837. Experientia *20*:657, 1964.

———: Une inversion péricentrique à l'origine d'un polymorphisme chromosomique non-Robertsonian dans une population de *Mastomys* Rodentia-Murinae). Chromosoma (Berl.) *18*:188, 1966.

Mayr, E.: Animal Species and Evolution. Cambridge: Harvard University Press, 1963.

Meylan, A.: Contribution à l'étude du polymorphisme chromosomique chez *Sorex araneus* L. (Mamm.–Insectivora). Note préliminaire. Rév. suisse Zool. *67*:258, 1960.

Meylan, A.: Le polymorphisme chromosomique de *Sorex araneus* L. (Mamm.–Insectivora). Rév. suisse Zool. *71*:903, 1964.

———: Répartition géographique des races chromosomiques de *Sorex araneus* L. en Europe (Mamm.–Insectivora). Rév. suisse Zool. *72*:636, 1965.

———: Données nouvelles sur les chromosomes des Insectivores européens (Mamm.). Rév. suisse Zool. *73*:548, 1966.

———: Formules chromosomiques et polymorphisme Robertsonien chez *Blarina brevicauda* (SAY) Mamm.: Insectivora. Canad. J. Zool. *45*:1119, 1967.

Nadler, C. F.: The chromosomes of *Spermophilus townsendi* (Rodentia: Sciuridae) and report of a new subspecies. Cytogenetics *7*:144, 1968.

Schmid, W.: Heterochromatin in mammals. Arch. J. Klaus-Stift. Vererb.-Forsch. *42*:1, 1967.

Taylor, K. M., D. A. Hungerford, R. L. Snyder and F. A. Ulmer, Jr.: Uniformity of karyotypes in the Camelidae. Cytogenetics *7*:8, 1968.

Wettstein, v. O.: Die Säugerwelt der Ägäis, nebst einer Revision des Rassenkreises von *Erinaceus europaeus.* Ann. Naturhist. Mus. Wien *52*:245, 1941.

White, M. J. D.: Models of speciation. Science *159*:1065, 1968.

Wurster, D. H.: Cytogenetic and phylogenetic studies in carnivora. This volume.

PATTERNS OF AUTOSOMAL HETEROCHROMATIN

A. GROPP AND P. CITOLER
Abteilung für Cytopathologie und Cytogenetik des Pathologischen Instituts der Universität Bonn, Germany

The relationship of chromocenters of interphase nuclei to the heterochromatin of metaphase chromosomes is well studied for the sex chromosomes of man and mammals (Ohno *et al.*, 1959; Schmid, 1967; Tiepolo *et al.*, 1967; Zakharov and Egolina, 1938). One of the most important implications of this relationship is the inactive-X hypothesis expressed by Lyon (1961). Genetic inactivation and heterochromatinization can be assumed to be characteristic features of one of the X-chromosomes in the female. Asynchronous late replication of the DNA is a further remarkable phenomenon shown by the X-heterochromatin.

However, many mammalian species exhibit non-sex-chromosomal chromocenters of the interphase nuclei. The elucidation of their variability and of their origin seems to be important. The main problems to be faced are those of possible relations between chromocenter formation, autosomal heterochromatinization and genetic inactivation. Yet, an appropriate exploration depends on the size and the total amount of interphase nuclear chromocenters and of heterochromatin of autosomes in metaphase preparations.

An object presenting ideal conditions for this study was found in the European hedgehog species *E. europaeus* and *E. roumanicus* (Insectivores). The findings pertinent to the problems of karyotype variation and of speciation of these animals are discussed in the contribution to this volume entitled "Cytologic Mechanisms of Karyotype Evolution in Insectivores." The observation, however, of the presence of large chromocenters of interphase nuclei (Fig. 1) and of important blocks of late autosomal DNA-synthesis in cells of male and female hedgehogs, together with results on male meiosis, gave rise to a study which endeavors to answer the following questions:

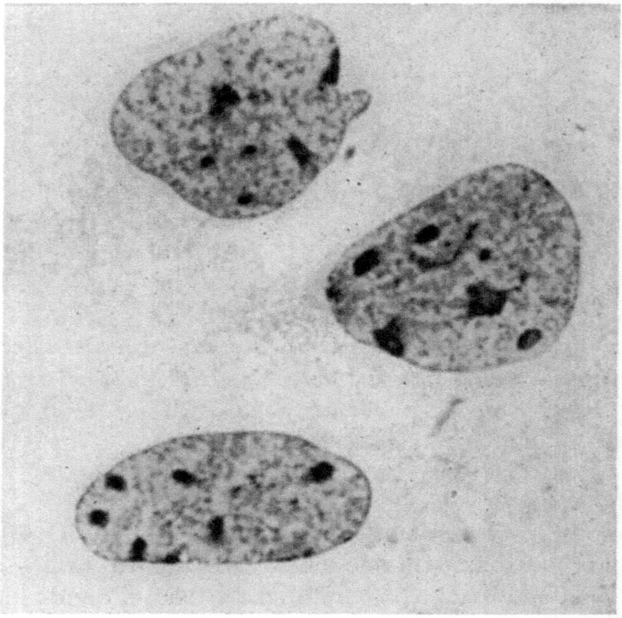

Fig. 1. Interphase nuclei of a fibroblast culture from a male fetus of *Erinaceus europaeus.* 5–7 large chromocenters. (Thionin stain.)

1. Are the chromocenters of the interphase nuclei equivalent to particular autosomal segments?
2. Does autosomal heterochromatin show conditions of DNA-replication similar to the allocyclic sex chromosomes and to sex-chromatin?
3. Which indications exist pointing to a genetic inactivation of autosomal heterochromatin analogous to the presumable inactivation of allocyclic sex chromosomes?

Material and Methods

1. Fibroblast cultures on inserted coverslips of European hedgehogs (*E. europaeus* and *E. roumanicus;* see Gropp, this volume, l.c.) were prepared from skin and periostal tissue explants of two *male* fetuses of 3 cms crown-rump length. The primary cultures were pulse labelled by the addition of ^{3}H-thymidine 5 minutes prior to fixation in ethanol (5 and 18 μC/ml of culture fluid. Spec. activity 15,000 mC/mMol). The cells adherent to the coverslips were stained with orcein and then exposed for autoradiography (Emulsion Gevaert 7,15; exposure 7 days).

2. Metaphase plates were obtained from lymphocyte cultures (with phytohemagglutinin) of adult male and female hedgehogs and from kidney cell cultures of an adult female. ^{3}H-thymidine (1,2 μC/ml of cul-

ture fluid) and colchicine were added to the cultures 6 hours, and 3 hours, respectively, before harvesting the cells. The hypotonic treatment, fixation in ethanol-acetic acid and air drying on cleaned slides was done as usual, and followed by orcein staining and exposure for autoradiography (Emulsion Ilford L4, 3–14 days).

Results

1. *Chromocenters in interphase nuclei in relation to the cycle of DNS-synthesis.* The pulse labelling with ^{3}H-thymidine used in this study, permits to distinguish nuclei in active DNA-synthesis (S-phase-labelled nuclei) and those being in a pre- or post-synthetic period (G_1- or G_2-phase-unlabelled nuclei). All interphase nuclei contain several chromocenters and mostly one nucleolus (Figs. 1–4). Nuclei of male origin have been used in order to avoid interference by sex-chromatin. Since it is difficult to identify with certainty the nucleolus in stained cells, particularly if superposed by the photographic grains of an autoradiograph, all larger dense structures of the nuclei were counted in the phase-contrast microscope. The evaluation of 50 labelled (S-phase) and 100 unlabelled (G_1- and G_2-phase) nuclei of *E. europaeus* displayed an identical frequency distribution: S-phase nuclei contained on the average 5.28 ($s = \pm 1{,}358$) contrasted dense structures while the combined G_1- and G_2-nuclei averaged 5.18 ($s = \pm 1{,}328$).

Conclusion The findings of González and Nardone (1968) have shown that an equal occurrence of nucleoli in S-, G_1- and G_2-nuclei cannot be assumed: their number in the G_1-phase may exceed that in the S-phase slightly, and is somewhat less in the G_2-phase. In an actively growing cell population these differences are most probably compensated and may be neglected. Thus, the actual average number of large chromocenters in a nucleus results from the number of all dense structures by deduction of the mean number of nucleoli, that is, approximately 5 minus 1 = 4.

2. *Type of ^{3}H-thymidine-labelling of interphase nuclear chromocenters.* Non-selected labelled interphase nuclei in areas of active growth of pulse labelled fibroblast cultures of *E. europaeus* (male) were evaluated according to the density and distribution of photographic grains. Altogether about 300 labelled (S-phase) nuclei could be subdivided into the following groups:

(1) Lightly labelled nuclei with grains mainly localized at the euchromatic nuclear areas, whereas the chromocenters were not or only very little labelled: *12%*.

(2) More heavily labelled nuclei with equal distribution at chromocenters and euchromatin (Fig. 2A): *64%*.

(3) Nuclei with heavy labelling of the chromocenters, whereas euchromatin is only slightly or not labelled (Figs. 2B and 3): *24%*.

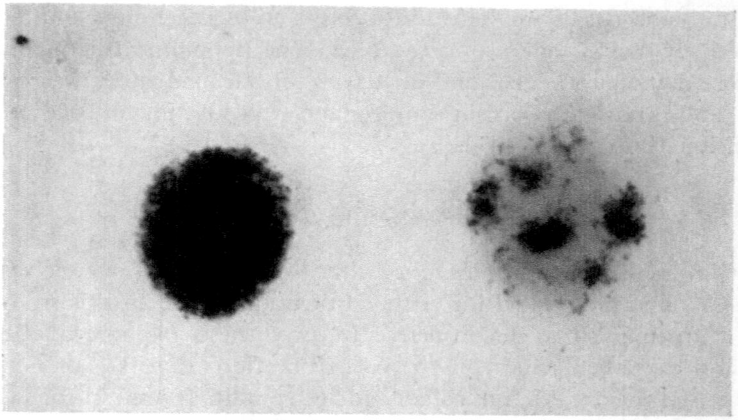

Fig. 2. ^{3}H-thymidine-autoradiograph of nuclei from a kidney cell culture of a female *E. europaeus*. Left: heavy labelling of equal distribution. Right: exclusive labelling of chromocenters. (Orcein stain.)

Nuclei of intermediate behavior occurred, but those of the groups 2 and 3, nevertheless, could be well distinguished. The S-phase nuclei with prevailing or exclusive labelling of chromocenters show mostly one, rarely more, unlabelled dense structures (Fig. 3). It appears more distinctly by the observation in the phase-contrast-microscope (Fig. 4) and corresponds to the nucleolus.

Conclusion Fibroblasts of the hedgehog contain a surprisingly high percentage of S-phase nuclei with an unequal pattern of labelling (24%

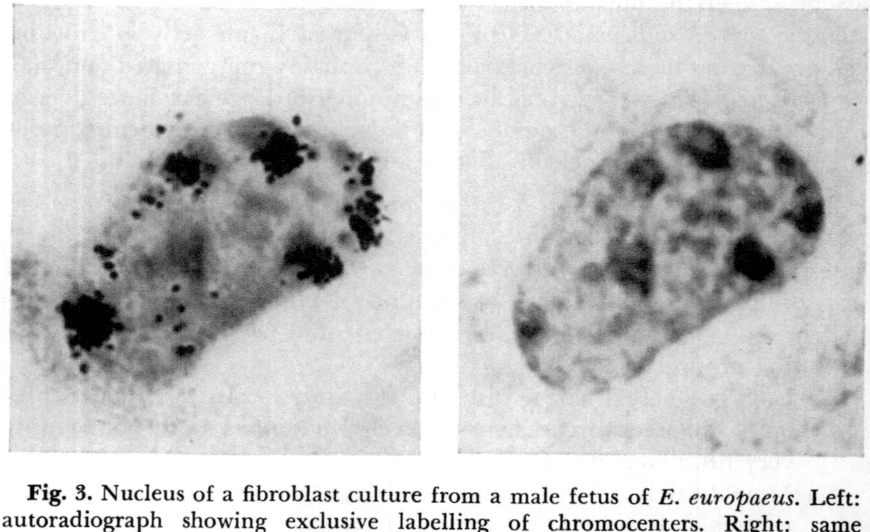

Fig. 3. Nucleus of a fibroblast culture from a male fetus of *E. europaeus*. Left: autoradiograph showing exclusive labelling of chromocenters. Right: same nucleus after removal of the photographic grains. The dense unlabelled structure in the center corresponds to the nucleolus.

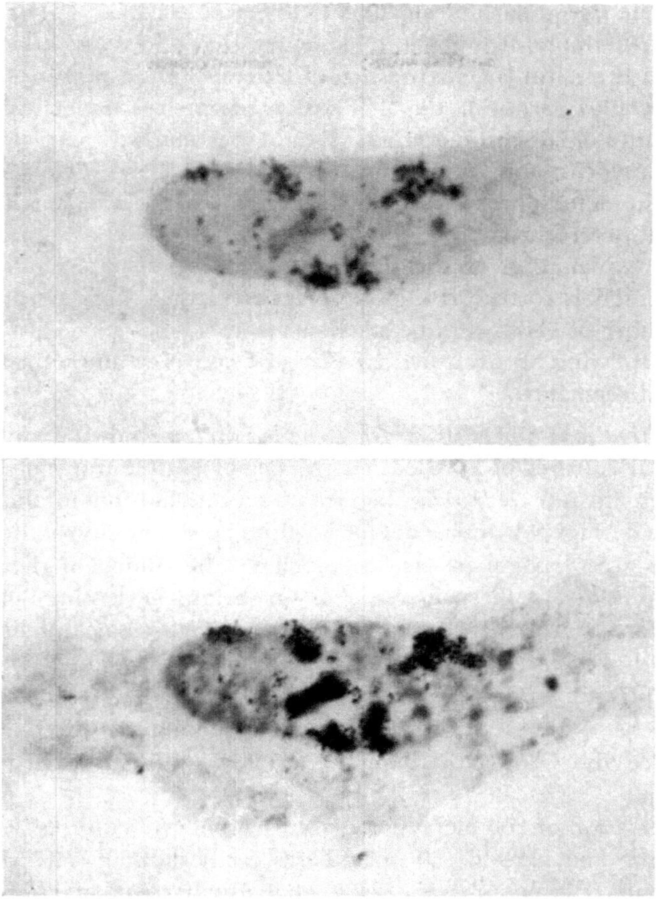

Fig. 4. ³H-thymidine-autoradiograph of a nucleus of a fibroblast culture from a male fetus of *E. europaeus*. Top: appearance in the light microscope. 5 labelled chromocenters. Below: same nucleus in the phase contrast microscope. The unlabelled nucleolus appears distinctly.

of all labelled nuclei). Careful analysis reveals that the labelling pattern of these nuclei corresponds to chromocenters which presumably are not of sex-chromosomal, but autosomal origin. It seems to be safe to assume the equivalence of chromocenters (interphase nuclear heterochromatin) and heterochromatic segments of metaphase chromosomes. The predominant labelling of chromocenters in interphase nuclei of the (male) hedgehog fibroblasts is analogous to the behavior of sex-heterochromatin, either the heterochromatic original-type of the X-chromosome or the "accessory gonosomal heterochromatin" (Schmid, 1967) of the composite-type sex-chromosomes. For these latter heterochromatic structures of the interphase nucleus it was shown that synthesis of DNA can occur in a condensed state (Grumbach *et al.*, 1963; Klinger *et al.*, 1967; Pera and Wolf, 1967). Furthermore, it is generally accepted that an exclusive labelling of sex-heterochromatin is an expression of late replication within the S-phase (Comings, 1967; Schmid, 1967). Thus, predominant or exclusive labelling of autosomal chromocenters should be similarly explained as a result of asynchronous replication in the late S-phase. This assumption implies the actual process of DNA-synthesis in the condensed state of autosomal heterochromatin.

If this explanation of the prevalent labelling of interphase nuclear chromocenters is correct, it should be expected that there are among the total number of ^{3}H-thymidine-labelled metaphases a good deal of metaphases displaying an exclusive labelling of only few autosomes or large autosomal segments.

3. *Observations on ^{3}H-thymidine-labelled metaphases.* Karyotypes from kidney cell cultures of labelled metaphases of both European species of hedgehogs are represented in the separate contribution to this volume (Gropp, l.c., Figs 8A and B). The labelling pattern shown there corresponds to a replication in the late S-phase. The timing of the addition of ^{3}H-thymidine to the cultures (6 hours before harvesting) is, indeed, suitable for an observation of a higher percentage of metaphases with a late labelling pattern (approximate length of the G_2-phase 4 hours). The most surprising finding was the presence of a labelling pattern with large blocks of late replication distributed among only few autosomes (Fig. 5; see also Gropp, this volume, l.c., Fig. 8, and the pertinent discussion).

An evaluation of 200 metaphases from lymphocyte cultures of *E. europaeus* (male and female, 100 metaphases each) showed a ^{3}H-thymidine-labelling in 186 metaphases = *93%*. The qualitative analysis of these 186 metaphases gave the following results: *39%* of metaphases contained more than 7 labelled chromosomes or they displayed a diffuse labelling pattern. *45%* showed a labelling of only 5 to 7 chromosomes (Fig. 5), *16%* a labelling of less than 5 chromosomes.

Conclusion The high percentage of metaphases with a ^{3}H-thymidine labelling of sizable blocks of only few autosomes (61% of all labelled metaphases contained up to 7 labelled chromosomes) is in surprising agreement with the similarly high percentage of interphase nuclei displaying labelling of chromocenters (24%). It can be postulated that the real percentage even exceeds this latter value, provided that the capacity of absorption of dense cellular structures for β-rays (Maurer and Primbsch, 1964) is taken into consideration.

The parallelism manifested by this behavior and the near-conformity of the numbers of chromocenters in interphase nuclei (mean value in the male: 4) and the numbers of late replicating autosomal blocks (most

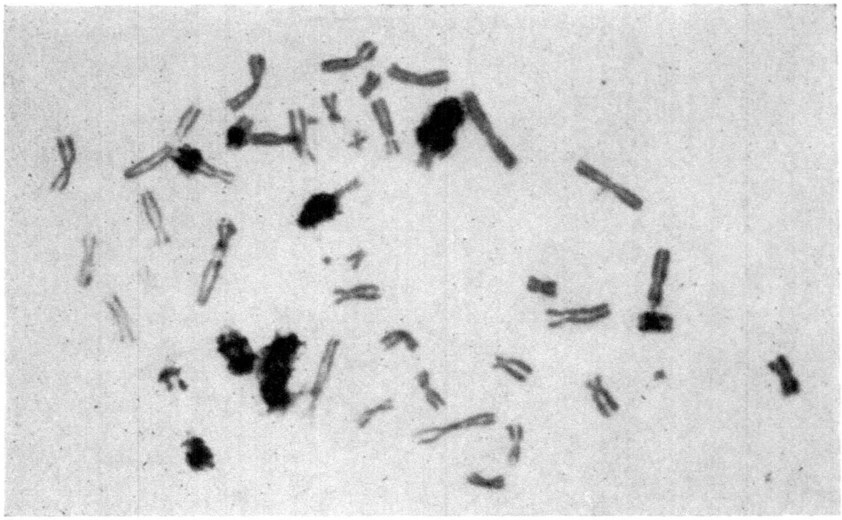

Fig. 5. ^{3}H-thymidine-autoradiograph from kidney cell culture of a female *E. europaeus*. 7 large blocks of late replication, among these one X-chromosome, the others representing autosomal segments.

metaphases less than 5 up to 7, minus one X-chromosome in female) are strong arguments in favor of an equivalence of chromocenters and autosomal heterochromatin. Both are distinguished by the property of late replication, analogous to the allocyclic X-chromosome and to the sex-heterochromatin in the interphase nucleus.

4. *Autosomal heterochromatin in meiosis.* In meiotic prophase (Fig. 6A) of *E. europaeus* non-sex-chromosomal segments with positive heteropycnosis occur. These segments are noticeable in the pachytene figures of Figure 6A by their denser appearance. In concordance with this phenomenon, diakinesis and metaphase I (Figs. 6B,C) show three bivalents displaying large achiasmatic segments. These can be distinguished from the other bivalents by their "open" configuration. An identical phenomenon could be observed in the diakinesis (Fig. 7A) and metaphase I of *E. roumanicus*. This is to be expected because this species exhibits similar features as to chromocenters and chromosomal heterochromatin. The peculiarity of this finding is demonstrated by a comparison with Figure 7B of a diakinesis figure of *Hemiechinus megalotis*. This animal shows mainly closed ringshaped bivalents and chiasma formation throughout.

Conclusion There are mainly three bivalents with large achiasmatic segments in both *Erinaceus*-species. Thus, their number and size is in agreement with the number and size of autosomes showing large blocks of late replication (4 to 6). The conclusion can be drawn that both phenomena are different expressions of the heterochromatic nature of iden-

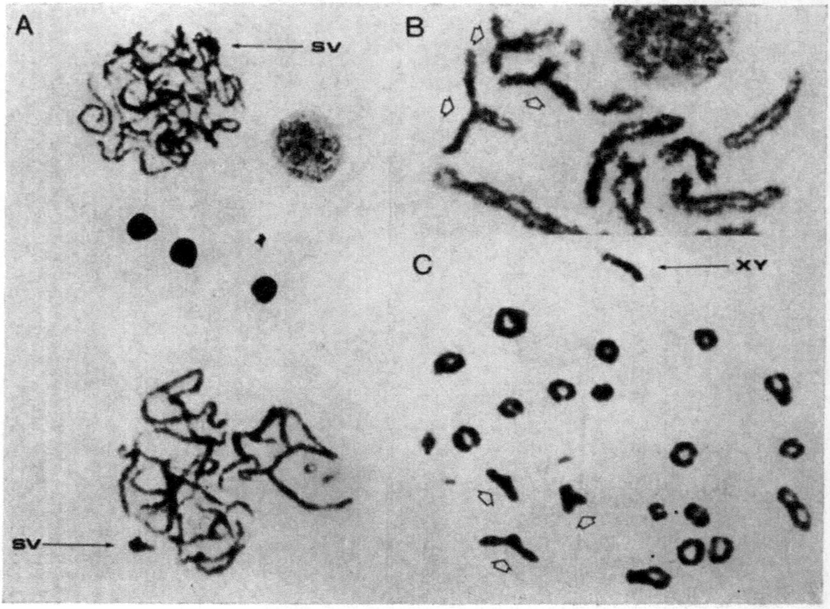

Fig. 6. Meiotic preparation. Testis, *E. europaeus*. A: Pachytene figures with non-sexchromosomal dense segments. SV = sex vesicle. B: Diakinesis (part) with three bivalents showing achiasmatic segment (arrows). C: Metaphase I with three bivalents (arrows) corresponding to those in B.

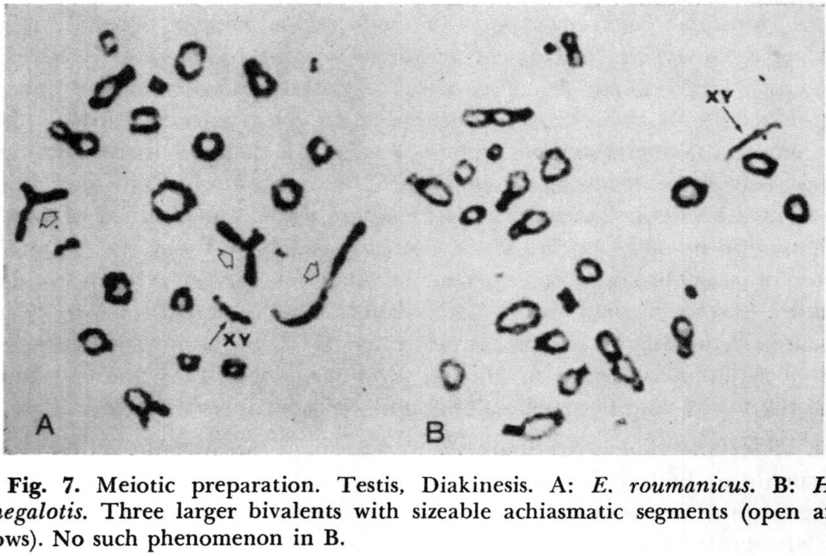

Fig. 7. Meiotic preparation. Testis, Diakinesis. A: *E. roumanicus*. B: *H. megalotis*. Three larger bivalents with sizeable achiasmatic segments (open arrows). No such phenomenon in B.

tical chromosomal segments. This assumption implies an achiasmatic behavior of large heterochromatic autosomal segments.

The peculiarity of this observation in both *Erinaceus*-species and the relation to the particular distribution of heterochromatin is underlined by the finding of different patterns of diakinesis and of chromosomal DNA-replication in *Hemiechinus megalotis*. Although a certain amount of late labelling in ^{3}H-thymidine-autoradiography is present in this species, there are, nevertheless, no such large coherent blocks of autosomal heterochromatin as in *Erinaceus*.

General Concluding Remarks and Summary

The results presented are likely to give answers to the questions put at the outset of this study. In nuclei of hedgehog fibroblasts, the chromocenters are most probably equivalent to autosomal heterochromatin in the interphase nucleus and at metaphase chromosomes. They display features of DNA-replication identical to that of allocyclic sex-chromosomes and of sex-chromatin. Yet, the final question concerning the problem of genetic inactivation of autosomal heterochromatin cannot be answered directly. There exist, however, strong arguments indicative of an inactivation of autosomal heterochromatic segments. One of these arguments is the striking analogy of the cytologic behavior of autosomal heterochromatin and of sex-heterochromatin. Genetic inactivation of the latter (Ohno *et al.*, 1959; Lyon, 1961) seems to be well established. Further evidence of an exemption of heterochromatic segments from the pool of genetically active material is to be seen in the behavior of these segments in meiosis. It seems safe to assume at least a certain degree of genetic insignificance of a chromosomal segment which is exempt from chiasma formation. A final argument for reduced or missing genetic activity is discussed in a separate contribution to this volume (Gropp, l.c.). There, the occurrence of extensive karyotype changes in the closely related species *E. europaeus* and *roumanicus* was found to be connected to the presence of late labelling and apparently heterochromatic autosomal segments. The most appropriate explanation for this finding is to assume a genetic inactivation of the chromosomal segments involved. Otherwise the great differences of the karyotypes of both species are difficult to understand.

Thus, there is reason to support genetic inactivation of autosomal heterochromatin. This concept may be of significance not only in the species discussed in this study. It seems to be of importance to investigate other species exhibiting similar favorable conditions for the establishment of the cytologic principles of the behavior and patterns of heterochromatin.

Acknowledgment

This work was supported in part by a grant of the Ministry of Cultural Affairs, Nordrhein/Westf., for Cancer Research.

References

Comings, D. E.: The duration of replication of the inactive X-chromosome in humans based on the persistence of the heterochromatic sex chromatin body during DNA synthesis. Cytogenetics *6:*20, 1967.

González, Sister Paula and R. M. Nardone: Cyclic nucleolar changes during the cell cycle. I. Variations in number, size, morphology and position. Exptl. Cell Res. *50:*599, 1968.

Grumbach, M. M., A. Morishima and J. H. Taylor: Human sex chromosome abnormalities in relation to DNA replication and heterochromatinization. Proc. Nat. Acad. Sci., Wash. *49:*581, 1963.

Klinger, H. P., H. G. Schwarzacher and J. Weiss: DNA content and size of sex chromatin positive female nuclei during the cell cycle. Cytogenetics *6:*1, 1967.

Lyon, M. F.: Gene action in the X-chromosome of the mouse (*Mus musculus* L.). Nature (Lond.) *190:*372, 1961.

Maurer, W. and E. Primbsch: Größe der ß-Selbstabsorption bei der ^{3}H-Autoradiographie. Exptl. Cell Res. *33:*8, 1964.

Ohno, S., W. D. Kaplan and R. Kinosita: Formation of the sex chromatin by a single X-chromosome in liver cells of *Rattus norvegicus*. Exptl. Cell Res. *18:* 415, 1959.

Pera, F. and U. Wolf: DNS-Replikation und Morphologie der X-Chromosomen während der Syntheseperiode bei *Microtus agrestis*. Chromosoma (Berl.) *22:* 378, 1967.

Schmid, W.: Heterochromatin in mammals. Arch. Klaus-Stift. Vererb. Forsch. *42:* 1, 1967.

Tiepolo, L., M. Fraccaro, M. Hultén, J. Lindsten, A. Mannini and P. M. L. Ming: Timing of sex chromosome replication in somatic and germ-line cells of the mouse and the rat. Cytogenetics *6:*51, 1967.

Zakharov, A. F. and N. A. Egolina: Asynchrony of DNA replication and mitotic spiralization along heterochromatic portions of Chinese hamster chromosomes. Chromosoma (Berl.) *23:*365, 1968.

CHROMOSOMAL EVOLUTION IN RODENTS

CHARLES F. NADLER
Department of Medicine, Northwestern University Medical School, Chicago, Illinois

Introduction

The order Rodentia, composed of 1687 recent species belonging to 43 families and 354 genera, occupies a wide range of habitats throughout the world (Anderson and Jones, 1967). Rodents have been classified within three suborders: the Sciuromorpha, Myomorpha, and the Hystricomorpha, although the validity of suborders has been questioned (Wood, 1958). The various gross morphological features characterizing these groups, as well as the families and genera of rodents, are presented in several recent publications (Walker, 1964; Anderson and Jones, 1967). An invaluable taxonomic reference that described species and subspecies of North American rodents is the publication by Hall and Kelson (1959).

Rodents are readily available from different ecological niches and are easily collected and maintained in the laboratory, so it is not surprising that chromosomes have been studied more extensively in rodents than any other mammals. Among the most important early investigators in rodent cytogenetics are Professor Robert Matthey and Professor Sajiro Makino, who between them described the chromosome numbers of nearly all the 171 species analyzed up to 1962 (Matthey, 1958; Matthey, 1962). With the increased interest in mammalian cytogenetics stimulated by newer techniques developed in the 1950's, rodents have now been studied even more intensively. These recent investigations have made important contributions to our understanding of chromosome evolution in mammals.

The purpose of this paper is to review the recent studies of rodent chromosomes and evaluate them within an evolutionary context. This review will consider first the cytological mechanisms responsible for karyotypic change; next, the succeeding stages of karyotype evolution dominated by chromosome polymorphism; and finally, the events leading to fixation of a "new" karyotype within an evolving species by either allopatric (Mayr, 1963) or stasipatric (White, 1968) speciation. Evolution of sex chromosome mechanisms, hybridization and applications of cytogenetics to rodent taxonomy will also be considered. Enumeration of the chromosome complements of all rodents studied to date is beyond the scope of this work; these data have been published by Matthey (1938, 1962) and brought up to date by him since 1962 in annual supplements (1964–1968) appearing in the *Mammalian Chromosomes Newsletter*.

Chromosome Rearrangements

Changes in the chromosome complement accompanying speciation are brought about by events that are remarkably similar in all animals (White, 1954; Matthey, 1958). This is fortunate because the mechanisms responsible for chromosomal rearrangements are much more easily studied in insects (White, 1954; 1968), especially *Drosophila* (Patterson and Stone, 1952), than in mammals. High diploid numbers and lack of landmarks within the chromosomes themselves make mammalian rearrangements more difficult to analyze. From insect studies the pathways of karyotypic change in higher forms can be postulated.

Occasionally confirmation of interspecific differences suggested by comparisons of mitotic chromosomes in mammals can be obtained by analysis of meiotic chromosomes from hybrids or animals heterozygotic for a chromosomal rearrangement. Translocations (Fig. 1) have been confirmed by

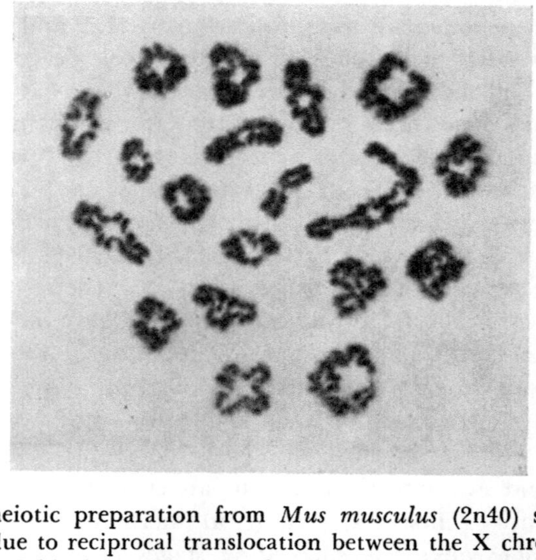

Fig. 1. A meiotic preparation from *Mus musculus* (2n40) showing a chain quadrivalent due to reciprocal translocation between the X chromosome and an autosome, and 18 bivalents. (From Ford and Evans, in Cytogenetics, 1964.)

meiotic studies in *Mus musculus* (Ford and Evans, 1964) and *Mus* (*Leggada*) *minutoides* (Matthey, 1963a).

The initiating event in karyotype evolution is a chromosome rearrangement in the germ cells of an individual animal which should not produce a deleterious effect in the offspring but ideally confer some selective advantage. Unfortunately this advantage may not be identified easily. However, White (1958) has been able to show the positive effect of chromosomal alterations in the evolution of the grasshopper *Moraba scurra;* the alterations in chiasmata frequency that presumably conferred a selective advantage were the result of centric fusions and fissions.

Among the more easily identifiable chromosome rearrangements are pericentric inversions, which may shift the position of the centromere, and reciprocal translocations, also called centric fusions, which may alter both the morphology and diploid chromosome number. More difficult to detect are paracentric inversions, tandem fusions, duplications, deletions, isochromosome formation and ring chromosome formation.

Examples of pericentric inversions in rodents are found in *Mus* (*Leggada*) *minutoides* (Matthey, 1964b), the rat *Mastomys natalensis* (Matthey, 1966a), and the rat *Rattus rattus (Yosida et al.,* 1965a). Other situations where pericentric inversions are postulated to have occurred are in the evolution of chipmunks of the genus *Eutamias* (Nadler, 1964a), the prairie dog *Cynomys ludovicianus* (Nadler and Harris, 1967), jirds of the genus *Meriones* (Nadler and Lay, 1967), and in mice of the genus *Peromyscus* (Ohno *et al.,* 1966; Hsu and Arrighi, 1966).

The mechanism of centric fusion is perhaps the commonest cytological rearrangement observed in rodents and it accounts for evolution from a high to low diploid number (2n). Fusion requires two breaks, one occurring proximally and one distally adjacent to the centromeres of two, nonhomologous acrocentric chromosomes. Three consequences of centric fusion include the formation of a new bi-armed chromosome, formation of a small centric fragment that is lost in subsequent cell divisions, and a reduction in chromosome number by one. Computation of the Fundamental Number or FN (Matthey, 1945) by counting bi-armed chromosomes as two arms and acrocentric chromosomes as one arm provides a means for detecting this mechanism. Thus, if two related species have the same number of chromosome arms (FN) but different 2n's, it frequently can be shown that their karyotypic differences are due to a series of centric fusions in which two acrocentrics have combined to produce a metacentric. This situation is termed a Robertsonian system (Robertson, 1916). Among rodents, Robertsonian centric fusion has been observed in *Mus* (*Leggada*) *minutoides* (Matthey, 1966b), *Acomys* (Matthey, 1963b), *Gerbillus* and *Tatera* (Matthey, 1958), *Meriones* (Matthey, 1957b; Nadler and Lay, 1967), ground squirrels of the genus *Spermo-*

philus (Nadler, 1966a), and various species and genera of Microtinae (Matthey, 1957a).

Opposed to centric fusion is centric fission or dissociation, conceptually important as a means for explaining evolution from lower to higher diploid numbers. In this process, common in plants (Marks, 1957) and among some Lepidoptera (White, 1957), fragmentation of metacentric chromosomes occurs to produce two acrocentrics and thereby increase the chromosome number. Marks (1957) suggests, with particular reference to plants, that fragmentation may arise from breakage transversely through the centromere. In contrast, White (1957) believes that fission usually requires the presence of a supernumerary chromosome to provide an additional centromere, and that fission thus behaves as a reciprocal translocation between a small supernumerary and a metacentric.

Controversy presently exists regarding the relative frequency of fission and fusion in mammalian karyotype evolution. This issue merits careful consideration because the Robertsonian system theoretically can characterize both fission and fusion. Thus, postulates concerning directions of evolution and identification of ancestral and specialized forms within a taxon are derived from a system that may proceed in opposite directions.

Traditionally, centric fusion is accepted as more common than fission, and this view is based upon observations on many kinds of animals where the predominant direction of evolution seems to be from a higher 2n to a lower 2n (White, 1957; Matthey, 1958). White (1957) reviewed evidence from a variety of different insects where both fusion and fission occurred together and found fusion usually more common. An exception was noted in the butterfly genus *Lysandra* where only one species showed the usual (n) of 23 or 24 displayed by other members of the family; the other species had higher chromosome numbers ranging from 45 to 190–191 that could have arisen only by fission.

With respect to fusion in eutherian mammals, Matthey (1958; 1968) analyzed chromosomes from a total of 668 species and subspecies. He found the probability that a species from a given taxon would display a 2n ranging between 40–56 was greater than 50% (95% confidence limits). The primitive modal chromosome number was 48; Matthey (1958) therefore postulated that mammals evolved from a pool of species possessing 40–56 chromosomes. Species with a 2n above or below the 40–56 range were randomly distributed and thus do not invalidate his hypothesis. Some of the genera Matthey (1958) studied were *Meriones* (2n72–42), *Gerbillus* (2n66–30), *Tatera* (2n72–42), *Microtus* (2n72–17), and *Acomys* (2n66–38) and in these cases a Robertsonian fusion system seemed clearly present.

Direct evidence of centric fusion has been seen in humans, where reciprocal translocations between acrocentric autosomes are observed fairly commonly (Hamerton *et al.*, 1963). Fission of metacentrics has not been

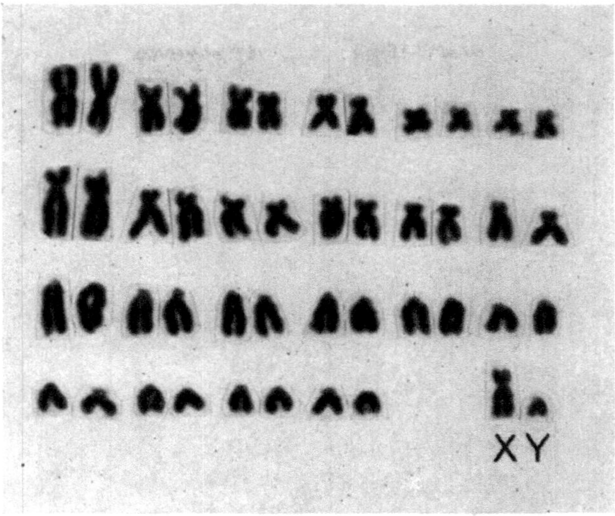

Fig. 2. Karyotype of *Spermophilus townsendi canus,* 2n46. (From Nadler, in Cytogenetics, 1968b.)

observed directly in man. Fusion cannot be differentiated from fission by meiotic analysis because in both cases a trivalent, formed from a metacentric and its two acrocentric homologues, will be observed.

Correlation of other lines of evidence with chromosome data suggest that fusion rather than fission occurred in the ground squirrel *Spermophilus townsendi* (Nadler, 1968b). In that species chromosomally homogeneous subspecies (possibly sibling species) exhibit 2n's of 46, 38 or 36 (Figs. 2–4). Central populations show the 2n of 46; however, geographi-

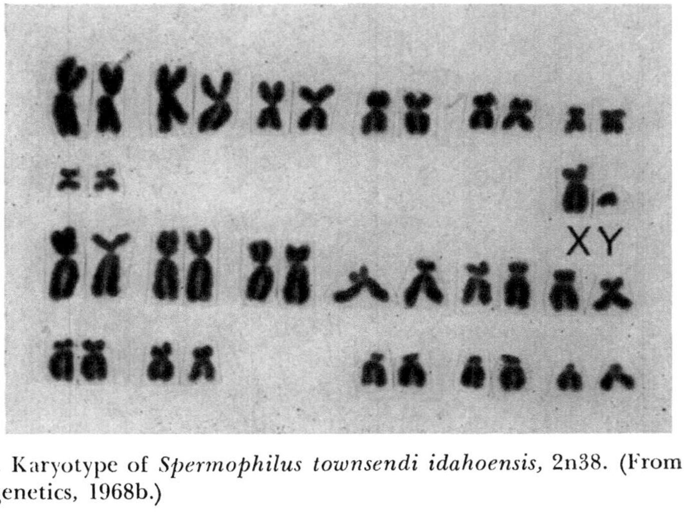

Fig. 3. Karyotype of *Spermophilus townsendi idahoensis,* 2n38. (From Nadler, in Cytogenetics, 1968b.)

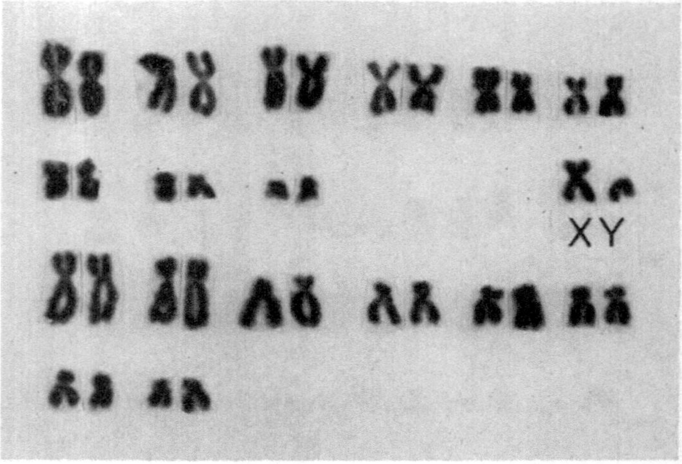

Fig. 4. Karyotype of *Spermophilus townsendi townsendi,* 2n36. (From Nadler, in Cytogenetics, 1968b.)

cally peripheral populations isolated by river barriers have a 2n of 36 or 38 (Fig. 5). This distributional pattern favors evolution from higher to lower 2n's and suggests that the 2n46 pattern is the ancestral one.

Recently, Todd (1967) and Staton (1967) stimulated interest in the fission mechanism by suggesting that the pre-eutherian 2n was 14; if so, evolution in a variety of carnivores, perissodactyla and primates must be explained by a fissioning process. Although the evidence for fission in certain insects with diffuse centromeres or supernumerary chromosomes is good, White (1957) suggests that the isochromosomes formed by transverse breaks through the centromere may result in genic excesses or deficiencies and also that interarm pairing might lead to abnormalities in meiosis.

Despite lack of direct evidence for and the theoretical evidence against fission as an evolutionary mechanism, several studies in rodents lend support to this explanation. The prairie dog *Cynomys ludovicianus,* a recently evolved and highly specialized terrestrial squirrel (Bryant, 1945), has a 2n of 50, a karyotype lacking acrocentrics and a FN of 96 (Nadler and Harris, 1967). The taxon considered most closely related to *Cynomys* is the ground squirrel subgenus *Spermophilus* (Bryant, 1945; Black, 1963), and among its species *S. townsendi* (2n46, FN 66) most nearly resembles *Cynomys* from a chromosomal standpoint.

Evolution of the *Cynomys* karyotype (2n50 without acrocentrics) from an ancestral *Spermophilus*-like pattern (2n46 with 22 acrocentrics) can best be explained by centric fission with subsequent conversion of acrocentrics to metacentrics by means of a series of pericentric inversions.

A study of chromosomes from North American marmots, which is discussed in the section on taxonomy, also suggests that fission and fusion may have both occurred within the same genus (Hoffmann and Nadler, 1968).

Although the issue is not yet settled it would appear on the basis of both theoretical arguments and indirect observations that centric fusion rather than fission is the more frequent mechanism accounting for the Robertsonian system in rodents. In certain cases such as those cited above fission seems to be operative and cannot be easily discounted. Perhaps analysis of large samples from mammalian species whose normal karyotype is known in detail will provide cytological proof of fission by demonstrating two acrocentrics derived from a metacentric. Until that time, careful correlation of chromosomal patterns with phylogenies based on other kinds of taxonomic evidence offers the best means for establishing the fission hypothesis.

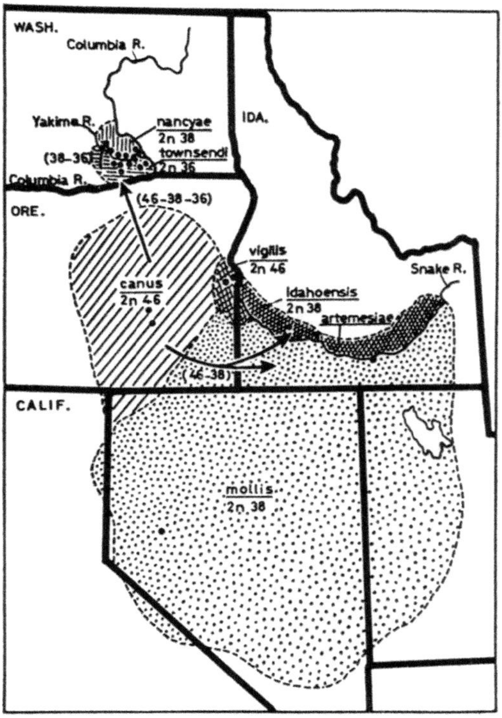

Fig. 5. Map illustrating the geographical distribution of subspecies of *Spermophilus townsendi*. Localities of specimens examined cytologically are illustrated by heavy dots. Arrows indicate the presumed direction of spread from the more ancestral population; the diploid numbers adjacent to the arrows represent the probable 2n's of animals at the time they spread to new habitat. (From Nadler, in Cytogenetics, 1968b.)

Chromosome Polymorphism

After the initial chromosomal aberration appears it may become established within a local population through interbreeding between normal animals and a chromosomal heterozygote. Subsequently, interbreeding between heterozygotes may produce a chromosome polymorphism characterized by: (1) the presence of animals with the original homozygous karyotype; (2) heterozygous animals with one original and one aberrant chromosome; and (3) animals that display a new homozygous karyotype with two aberrant chromosomes. Other heterozygous karyotypes may theoretically occur due to differences in meiotic segregation and they have been discussed by Drs. Ford and Hirschhorn. This kind of chromosome polymorphism, existing within a panmictic population, is the forerunner of the karyotype evolution accompanying speciation. It should be distinguished from examples where allopatric, chromosomally distinct populations appear within a species because the latter occurrence represents a much later stage of speciation.

One of the most important and carefully studied examples of chromosome polymorphism occurs in the mouse *Mus* (*Leggada*) *minutoides* and related forms (Matthey, 1963a; 1964b; 1966b), and these data are summarized in Figures 6–8.

Early studies described diploid numbers of 34, 33, 32 and 31 from specimens of *Mus* (*Leggada*) *minutoides* from the Congo and all specimens had a FN of 36 (Matthey, 1963a). The karyotype of animals with 2n34 contained 32 acrocentric autosomes and large submetacentric X and Y chromosomes. Evolution of karyotypes with lower diploid numbers appeared to result from Robertsonian centric fusions between acrocentric autosomes of unequal size. Confirmation of a Robertsonian relationship was derived from the observation of trivalents in meiotic preparations and Matthey invoked fusion as the responsible mechanism. I question whether the differentiation of fusion from fission can be made by the observation of trivalents alone. A number of additional studies expanded the early observations, and Matthey (1966b) summarized the data on 213 animals from a variety of different locations (Fig. 8). He concluded that the ancestral karyotype of this taxonomic group contained 36 acrocentric chromosomes, including the sex chromosomes, and that *M. minutoides*, *M. minutoides indutus*, *M. bufo*, *M. tenellus* and *M. setulosus* were characterized by the primitive chromosome pattern.

Several other intraspecific polymorphisms were discovered in addition to the one originally reported (Matthey, 1966b). In the first, a population of *M. m. minutoides* displayed chromosome patterns of 2n18–19, FN 36 and 2n18, FN 35, the latter FN being the result of a presumed pericentric inversion. Within the *M. triton* group, one population was composed of

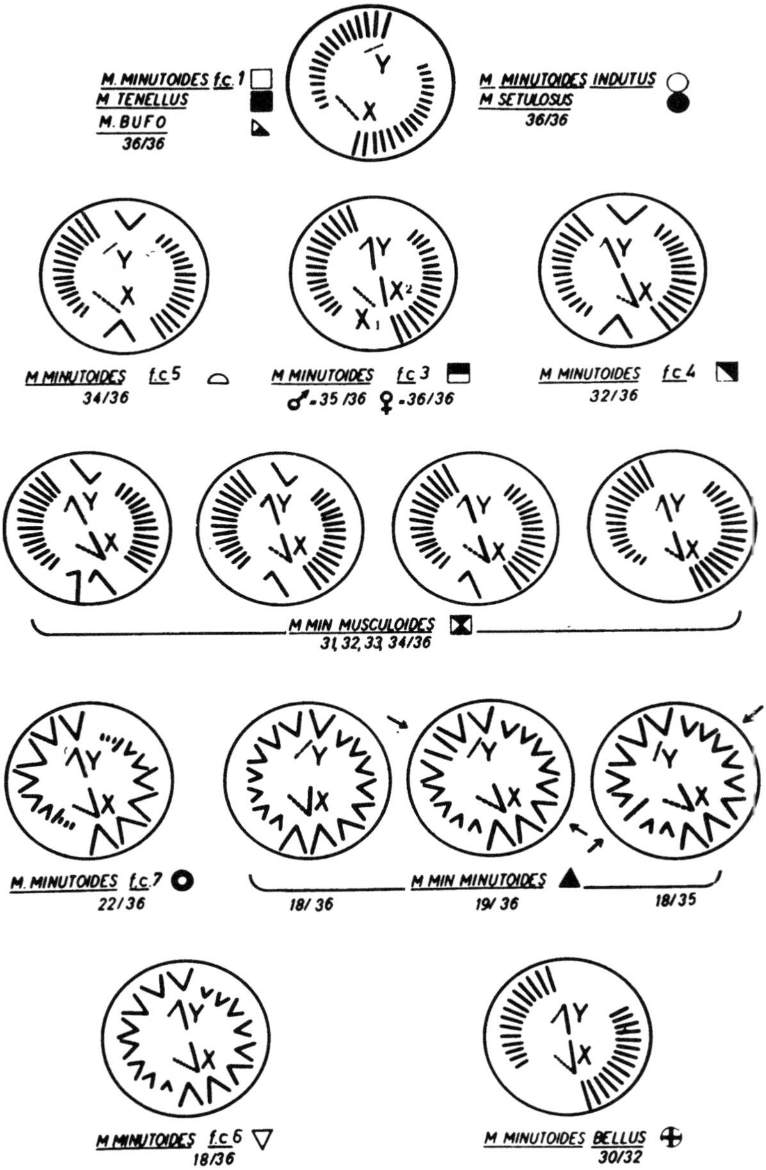

Fig. 6. A diagrammatic representation of the karyotypes observed in the *Mus* (*Leggada*) *minutoides* complex. The symbols adjacent to each taxon refer to the localities depicted in Fig. 8. In the numbers below the karyotypes, the numerator refers to the 2n, while the denominator indicates the FN. (From Matthey, in Rev. Suisse Zool., 1966b.)

animals having a 2n of 32 and FN's of 34 and 33, the latter FN occurring in females that apparently underwent a deletion of one arm of an X chromosome. Another population of *M. triton* was composed of animals having a 2n of 20 or 21 and FN 32 while a third population of *M. Triton* exhibited 2n's of 20, 21 and 22 with FN of 34 uniformly present. Changes in the sex chromosome mechanism and morphology from the primitive acrocentric complement are discussed in a later section but they too represent Robertsonian fusions sometimes associated with a deletion.

The importance of the chromosome studies in the *Leggada* group lies in the strong support provided for Robertsonian fusion, especially the evidence of fusions converting the primitive acrocentric sex chromosomes to bi-armed chromosomes. Matthey (1964b, 1966b) also believes the data are indicative of sympatric speciation originating within the ancestral polymorphic population. He suggests that the first step in speciation might have been an intersterility between mice with sex chromosomes of primitive type (acrocentric) and individuals having undergone the translocation between the sex chromosomes and autosomes.

Chromosome polymorphism within a population was also noted in a study of the genus *Gerbillus* (Wahrman and Zahavi, 1955). Specimens of *G. pyramidum* from the same locality in the Negev desert of Israel were characterized by either 33 or 31 autosomal elements in the first meiotic metaphase.

Another example of intraspecific chromosome polymorphism was found in the spiny mouse *Acomys minous* which also displays chromosomal variation among cells of individual specimens (Matthey, 1963b).

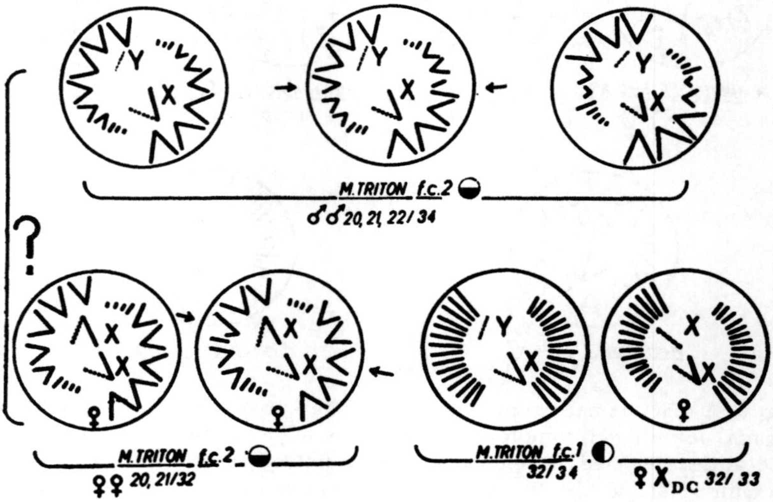

Fig. 7. A diagrammatic representation of the karyotypes of *Mus triton* from the *Leggada* complex. (From Matthey, in Rev. Suisse Zool., 1966b.)

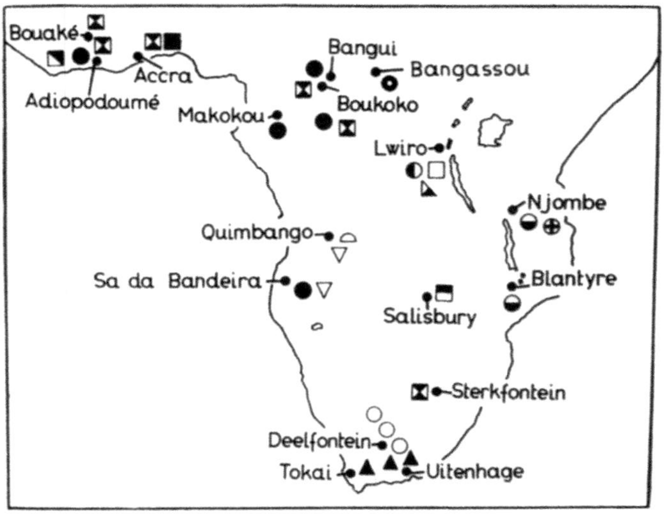

Fig. 8. The collecting localities of specimens from the *Mus* (*Leggada*) *minutoides* complex whose karyotypes are illustrated in Figs. 6 and 7. (From Matthey, in Rev. Suisse Zool., 1966b.)

In the rat *Mastomys natalensis,* a non-Robertsonian polymorphism due to pericentric inversion, has been described (Matthey, 1966a). Four specimens from a population had 8 small acrocentric autosomes, the other chromosomes being bi-armed; two animals had 6 acrocentrics; and two animals constituting the heterozygotes for the inversion had 7 acrocentric autosomes (Fig. 9).

Yosida *et al.* (1965a) noted polymorphism in the rat *Rattus rattus* from Japan. In normal karyotypes of this species, the largest chromosome pair is telocentric. Heteromorphism for this pair was found in 18.4% and 40% of the two populations studied, the aberrant chromosome being a subtelocentric that probably resulted from a pericentric inversion. Homozygous karyotypes characterized by a pair of subtelocentrics were not observed.

Laboratory and wild stock of *Rattus norvegicus* have provided another example of polymorphism involving the third largest pair of chromosomes, pair 3 (Yosida and Amano, 1965b). In this polymorphism, pair 3 appeared either as a homozygous pair of telocentrics or subtelocentrics or as a heterozygous pair composed of one telocentric and one subtelocentric chromosome.

Laboratory colonies of the guinea pig *Cavia porcellus* (2n64) (Cohen and Pinsky, 1966) were examined, and two types of heteromorphism were recognized with respect to the largest pair of autosomes. Normally these chromosomes are subtelocentric. The first heteromorphism was con-

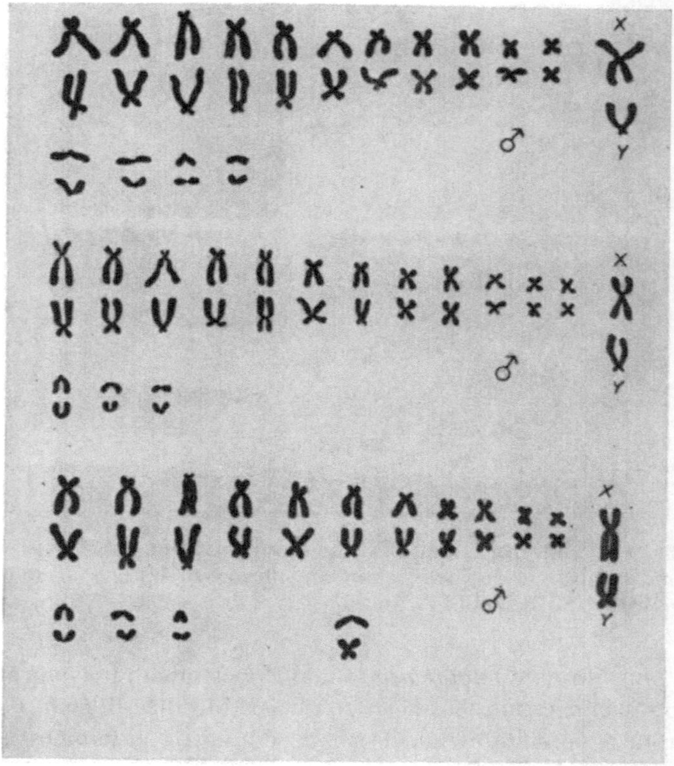

Fig. 9. Chromosome polymorphism in *Mastomys natalensis* due to a pericentric inversion. Top: a homozygous male with 22-bi-armed and 8 acrocentric autosomes. Middle: a second homozygous karyotype with 24 bi-armed and 6 acrocentric autosomes. Bottom: a heterozygous karyotype composed of 23 bi-armed and 7 acrocentric autosomes. (From Matthey, in Chromosoma, 1966a.)

sidered the result of a duplication or translocation involving the short arm while the second probably resulted from a deletion of the short arm. Manna and Talukdar (1964) also described an autosomal polymorphism in *Cavia porcellus* that involved a different pair of chromosomes.

Finally, an autosomal polymorphism due to pericentric inversion was found in a laboratory colony of deer mice, *Peromyscus maniculatus* (2n48) (Ohno *et al.,* 1966). Two subspecies, *P. m. bairdii* and *P. m. hollisterii,* exhibited polymorphism of a different pair of autosomes, but in each case the heteromorphic pair of autosomes was of acrocentric and submetacentric morphology. In *P. m. bairdii,* a second pair of autosomes was also polymorphic, and in the heteromorphic state they consisted of a small mediocentric and a small acrocentric chromosome. Sparkes and Arakaki (1966) found intrasubspecific and intersubspecific chromosomal polymorphism in *Peromyscus maniculatus rubidus* (2n48), *P. m. gracilis*

(2n48) and *P. m. bairdii* (2n48). The diploid numbers remained constant while individual animals exhibited variation in the number of acrocentric and metacentric chromosomes, a change again attributed to pericentric inversion.

Mammalian chromosome polymorphism may exist in a presumably balanced state as was suggested by data in the shrew *Sorex araneus* (Ford *et al.,* 1957) and perhaps some populations of the *Leggada* group of mice examined by Matthey. In a balanced polymorphism, heterozygotes are presumed to have a degree of selective superiority. Another more transient type of polymorphism may occur in association with mammalian speciation, and in this type of transient polymorphism, selection may favor chromosomal homozygotes rather than heterozygotes (White, 1968).

Speciation and Chromosomal Diversity

A discussion of present views of speciation is now required to arrive at an understanding of how evolving species develop new and distinctive chromosome complements. The chromosomal prerequisites, including the initiating rearrangement and subsequent chromosome polymorphism, have already been discussed. Among the current theories of speciation, the allopatric model, based upon geographic isolation of a peripheral population, is most widely accepted (Mayr, 1963). Other speciation models, including sympatric speciation, are convincingly shown by Mayr (1963) to be less tenable. Recently, another model, based on cytogenetic observations at the population level, has been proposed by White (1968) who terms it stasipatric speciation. These models and the possible methods whereby new chromosome rearrangements spread through a population are illustrated in Figure 10.

In the allopatric model (Fig. 10A), the major event is geographic isolation of a population, usually peripherally located, for a sufficient length of time to establish reproductive isolation (Mayr, 1963). While isolating mechanisms are developing, it seems probable that selective factors or perhaps chance alone influence both the chromosomally polymorphic ancestral population and the newly-isolated population. As a consequence, in time each population may attain a different homozygous karyotype originating from the initial chromosome polymorphism. At this final stage, should the two populations become sympatric, each will have developed a state of reproductive isolation and distinct chromosome complements. Alternatively, the isolated population may develop a chromosome polymorphism independently *after* it becomes geographically separated from the ancestral population. In either case, chromosomes may not only differentiate the two species but perhaps contribute to their reproductive isolation as well.

The stage of speciation at which chromosome polymorphism disap-

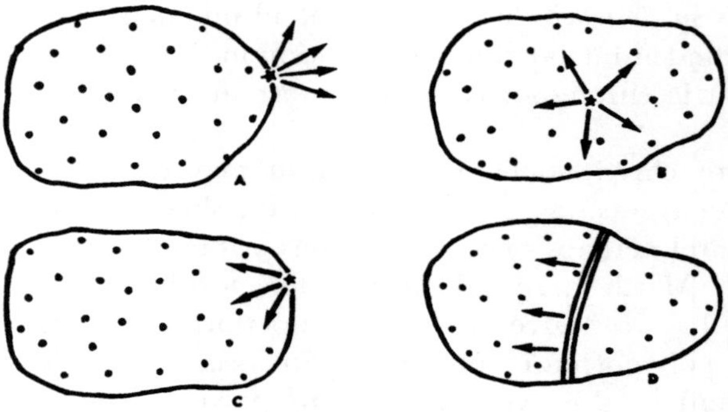

Fig. 10. Allopatric and stasipatric models of speciation. A. The chromosomal rearrangement becomes established in an isolated peripheral population and spreads into territory previously unoccupied by the species (allopatric model). B. The rearrangement arises in a nonperipheral area and spreads through the range of the species (stasipatric model). C. The rearrangement establishes itself in a peripheral population and then spreads throughout the species range (modified form of stasipatric model). D. A narrow zone of hybridization (non-heterotic) occurs between cytologically distinct, contiguous taxa. (From White, in Science, 1968.)

pears leaving chromosomally distinct populations is difficult to establish because the process is probably very slow, taking place over thousands of years (Nadler, 1968b). This situation is, however, of practical interest because a number of cases have been reported where populations classified as subspecies display different, apparently homogeneous karyotypes, and these findings suggest that the groups in question are perhaps behaving as true species. For example, in the ground squirrel *Spermophilus richardsoni,* the geographically isolated subspecies *S. r. elegans* and *S. r. nevadensis* display a 2n of 34, FN 64 in contrast to the presumably more ancestral subspecies *S. r. richardsoni* with 2n36, FN 64 (Nadler, 1964b). *S. r. aureus* (2n34, FN 64) is contiguous with *S. r. richardsoni* in Montana, and the two taxa are similar in gross morphology and are said to intergrade, a condition not yet confirmed cytologically (Nadler, 1968a). Here the chromosomal patterns may denote a late stage of speciation or perhaps the presence of morphologically similar sibling species.

A similar finding was noted in the ground squirrel *Spermophilus townsendi* where chromosomally distinct populations with 2n46, 2n38 and 2n36 are seen (Nadler, 1968b). Figure 5 illustrates the geographic distribution of the chromosomally distinct forms comprising these subspecies (or perhaps sibling species) and indicates the original polymorphisms from which the present taxa probably evolved.

Wahrman and Zahavi (1955) described yet another example of incipient speciation where three morphologically indistingiushable allopatric populations of *Gerbillus pyramidum* each displayed homozygous chromosome complements with 2n40, 52 and 66. The lack of gross morphological differentiation indicated that speciation was not complete in these three forms although it is possible that they represent sibling species.

Finally, Matthey (1963b) has discovered two races of *Acomys minous* characterized by 2n38, FN 66 and 2n40, FN 68 respectively. These forms were also not distinguished as different species.

An alternative mechanism to geographic or allopatric speciation has been proposed as a result of cytogenetic investigations in morabine grasshoppers of the *viatica* group. In stasipatric speciation, as described by White (1968), there was cytological and zoogeographical evidence for a direct conversion of an essentially continuous population into a number of contiguous taxa with different karyotypes. Low vagility of these grasshoppers with consequent restriction of gene flow between populations are probable prerequisites for this model. Because of these conditions, a chromosomal rearrangement arising in a single animal may become established and spread successfully throughout a small local population. This concept that speciation results from a cytological point mutation differs from the usual views of speciation where mutations functioning as a population phenomenon are the major factors. Support for the role of chromosomes as primary genetic isolating mechanisms between incipient species of morabine grasshoppers was derived in part from the observation of narrow zones of overlap (hybridization). Thus, a transient type of non-heterotic polymorphism with strong selection for homozygous karyotypes and rapid elimination of heterozygotes appeared operative. Once the new rearrangement becomes established, it may then spread from a geographically isolated peripheral (Fig. 10C) or central (Fig. 10B) population throughout the originally continuous species, and where cytologically different populations meet, the zones of hybridization will be characteristically narrow (Fig. 10D). It is not inconceivable that a similar mechanism might account for the presence of the chromosomally distinct contiguous populations seen in ground squirrels (Nadler, 1964b, 1968a, 1968b), chipmunks (Sutton and Nadler, 1968), gerbils (Wahrman and Zahavi, 1955) and pocket gophers (Patton and Dingman, 1968).

Ultimate clarification of taxonomic status and origin of the chromosomally different populations now classified as subspecies is dependent upon the investigation of specimens in zones of supposed intergradation. There the presence or absence of intermediate or heterozygous karyotypes will confirm or deny intergradation in a more precise manner than can be achieved using the minimal and usually only relative differences in gross morphology and pelage that are found among subspecies or sibling species. Study of zones of contact will also characterize the type of polymorphism present, heterotic or otherwise, and through the study of hybrids, if they exist, shed light on the contribution of karyological differences to reproductive isolation. By these means, it may be possible to show whether a stasipatric type of speciation occurs in mammals as well as insects or whether the allopatric process is predominant.

Hybridization

Mayr (1963) classified isolating mechanisms which are necessary for maintenance of species integrity into two categories termed premating and postmating mechanisms. The former category includes seasonal and habitat, ethological, and mechanical isolation where transfer of sperm does not occur. Premating mechanisms are often best studied under natural conditions. Postmating mechanisms include gametic and zygote mortality, hybrid inviability and sterility or partial sterility of hybrids.

Among the postmating mechanisms of greatest interest to cytogeneticists is hybrid sterility. A number of different cytological conditions may exist at meiosis in hybrids (White, 1954): (1) complete pairing of the two sets of chromosomes, spindle morphology and anaphase normal; (2) incomplete pairing, meiotic divisions otherwise normal; (3) complete pairing, but physiology of meiosis abnormal; (4) incomplete pairing, spindle mechanism abnormal; (5) no pairing, meiosis otherwise normal; (6) no pairing, spindle mechanism abnormal. Hybrids of groups (1) and (2) are considered potentially fertile whereas hybrids in groups (3)–(6) are usually sterile.

Some of the better studied rodent hybrids are derived from interspecific crosses among jirds of the genus *Meriones*. Zahavi and Wahrman (1957) reported the following crosses: (1) *M. crassus* ♀ (2n60) × *M. tristrami* ♂ (2n72); (2) *M. tristrami* ♀ × *M. crassus* ♂; (3) *M. sacramenti* ♀ (2n46) × *M. shawi* ♂ (2n44); (4) *M. tristrami* ♀ × *M. sacramenti* ♂; (5) *M. crassus* ♀ × *M. sacramenti* ♂. Disturbed meiosis was noted in F_1 males of mating (1) and infertility was suspected in F_1 females of the same cross when backcrossed to *M. crassus* males.

Matthey (1957b) examined hybrids derived from a cross between a *Meriones libycus* ♀ (2n44) × *M. tristrami* ♂ (2n72) and reported a diploid number of 58 in the hybrids. The male hybrid showed no spermatogenesis or spermatozoa.

Most recently, Lay and Nadler (1968) studied hybridization between two partially sympatric species, *M. shawi* ♀ (2n44) × *M. libycus* ♂ (2n44). The *M. shawi* karyotype (Fig. 11A) contains 20 metacentric, 12 submetacentric and 10 acrocentric autosomes, a large submetacentric X and a smaller submetacentric Y chromosome. In contrast, *M. libycus* (Fig. 11B) displays 20 metacentric, 10 submetacentric and 12 acrocentric autosomes, a large acrocentric X and a submetacentric Y, indistinguishable from the Y of *M. shawi*. Hybrids (Fig. 11C) display the anticipated haploid contribution from each parent species; the X chromosomes, the largest pair of submetacentrics and the largest pair of acrocentric autosomes can be identified as to species origin. Thus, three chromosome rearrangements in addition to a fourth required to account for the different numbers of acrocentric and submetacentric autosomes of each species must be postulated during the evolutionary divergence of *M. shawi* and *M. libycus*.

Breeding data show that F_1 females are fully fertile whereas F_1 males are infertile; the testes show either absent or markedly diminished spermatogenesis with an arrest of meiosis in prophase 1. Insufficient meiotic material was obtained from the male hybrids to confirm the three pericentric inversions and possible translocation of X chromosome material that best explain the karyotypic differences observed between the two species.

The discovery of apparently normal fertility in F_1 females backcrossed to *M. shawi* contrasts with the decreased fertility, small litter size and absence of male offspring seen when F_1 females are backcrossed with *M. libycus* males. It is suggested that while premating mechanisms may exist, postmating mechanisms in the form of F_1 male infertility and reduced fertility in one backcross are sufficient to maintain effective reproductive isolation between the two species. Zygote mortality or hybrid inviability may be operative as additional isolating mechanisms to prevent introgression back into the *M. libycus* gene pool. Analysis of the reciprocal cross, *M. libycus* ♀ × *M. Shawi* ♂, is in progress.

Successful hybridization between *M. shawi* and *M. libycus* might be considered evidence favoring a close phylogenetic relationship between the two species. However, many gross morphological features suggest that these two partially sympatric species are not particularly close relatives (Lay, unpublished data). In this context, it is well to remember the admonition of White (1954) that the strength or extent of the isolating mechanism is not necessarily an accurate measure of the degree of phyletic relationship; some closely related species are separated by strong isolating mechanisms whereas less closely related species may have weaker but no less effective isolating mechanisms under natural conditions.

Hybridization between *Acomys cahirinus* ♀ (2n38, FN 70) × *A. minous* ♂ (2n40, FN 68) produced sterile hybrids with abnormal meiosis and karyotypes showing a haploid chromosome set from each parent (Matthey, 1963b). Intraindividual chromosomal variation was also described and discussed in this publication.

Among natural populations of rodents, an interesting example of hybridization was reported in the pocket mice *Perognathus penicillatus* (2n46) × *P. pernix* (2n52) (Patton and Soule, 1967). Thirty-five specimens were collected from a limited area of sympatry, and two female hybrids with 2n49 were obtained. Placental scars were noted in both females but because the scars were less than the normal number the possibility of partial sterility of the F_1 hybrids was suggested. Backcross specimens were

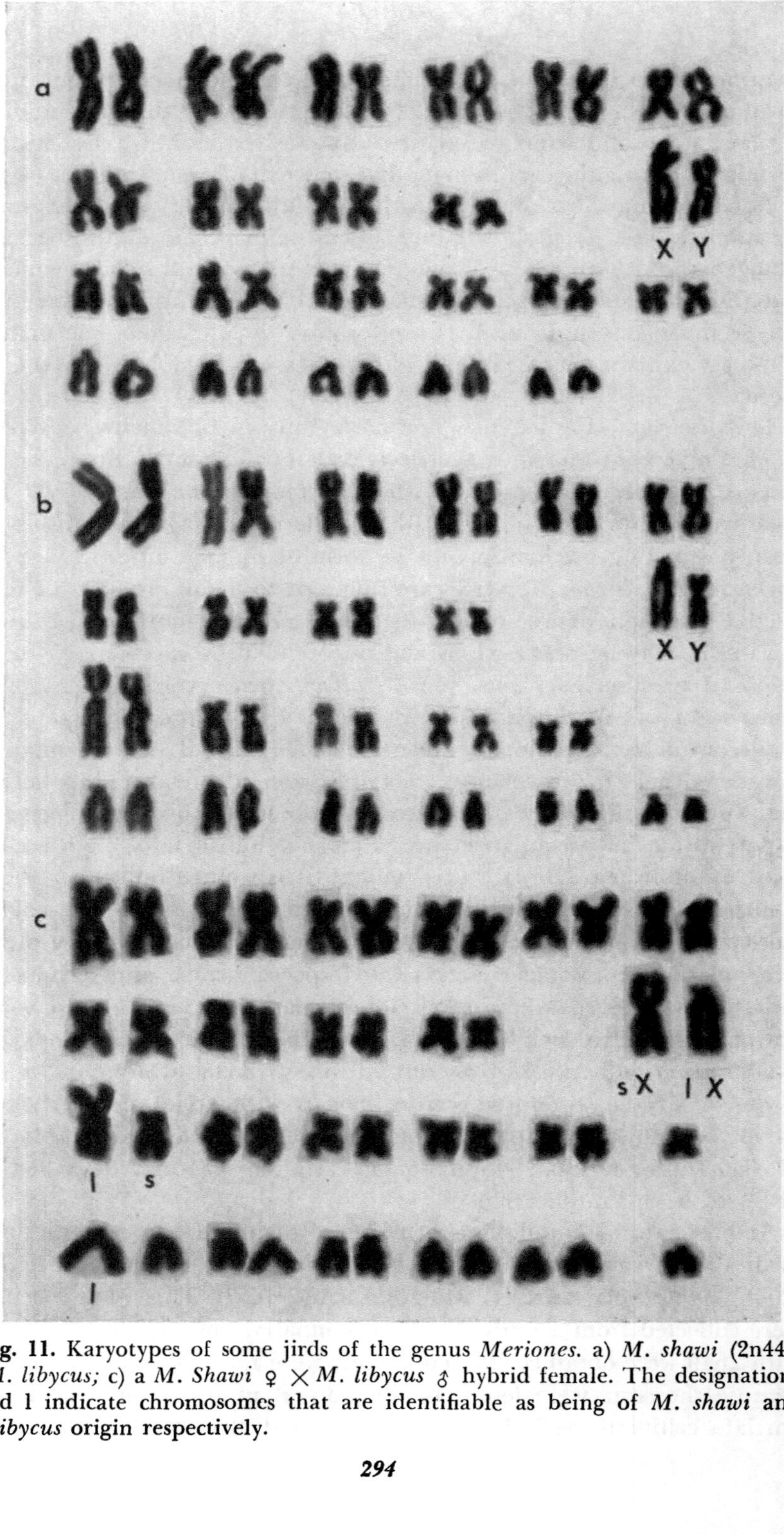

Fig. 11. Karyotypes of some jirds of the genus *Meriones.* a) *M. shawi* (2n44); b) *M. libycus;* c) a *M. Shawi* ♀ × *M. libycus* ♂ hybrid female. The designations s and l indicate chromosomes that are identifiable as being of *M. shawi* and *M. libycus* origin respectively.

not found which also indicated that introgression was slight, if it occurred at all.

Naturally occurring hybridization is also recorded in a narrow zone of sympatry involving the pocket gophers *Thomomys bottae* (2n76) and *T. umbrinus* (2n78) (Patton and Dingman, 1968). A single F_1 female with a 2n of 77 was pregnant with three nearly full term embryos. Thus, fertility appears normal in this hybrid and the presence of embryos indicates probable backcrossing against one of the two parental species. Three additional specimens displayed 2n's of 76, 76 and 77 respectively and the karyotypes were not exactly intermediate between the parental types, an observation that suggested further backcrossing following hybridization. The data raise interesting questions concerning the subspecific versus specific status of *T. bottae* and *T. umbrinus* that are not yet answerable with the present information.

These examples serve to illustrate the usefulness of hybridization as a tool for retrospective appraisal of chromosome evolution, and as a model for investigating the role of chromosomes as isolating mechanisms in mammals. Study of natural hybridization merits special attention because premating mechanisms are more easily analyzed than under laboratory conditions.

Sex Chromosomes

A review of chromosome evolution is incomplete without mention of the many unusual sex chromosome mechanisms exhibited by rodent species. Detailed discussions of the sex chromosomes are available elsewhere (White, 1954, 1957; Ohno, 1965, 1967; Mittwoch, 1967). Ohno (1965) has proposed that mammalian sex chromosomes evolved from a pair of autosomes originally present in the ancestral protoinsectivore karyotype. He also suggested that the primitive X chromosome has persisted in its entirety in all mammals, the original autosomal genes remaining as X-linked genes. The Y chromosome achieved greater specialization and lost most of its original autosomal genes. This disparity between X and Y gene content required some form of dosage compensation, and inactivation of one of the two female X chromosomes resulted.

From a mechanical viewpoint, morphological changes in the sex chromosomes most commonly result from translocation of autosomal material onto one or both of the primitive sex chromosomes to produce neo-X and neo-Y chromosomes (White, 1954, 1957).

Of great interest are the studies of African mice comprising the *Mus* (*Leggada*) group (Matthey, 1966b, 1967). The primitive sex chromosomes are represented by a pair of acrocentrics (Fig. 12). As evolution proceeded by centric fusion, certain species developed a large metacentric X and a large submetacentric Y due to translocation of a pair of acrocentric auto-

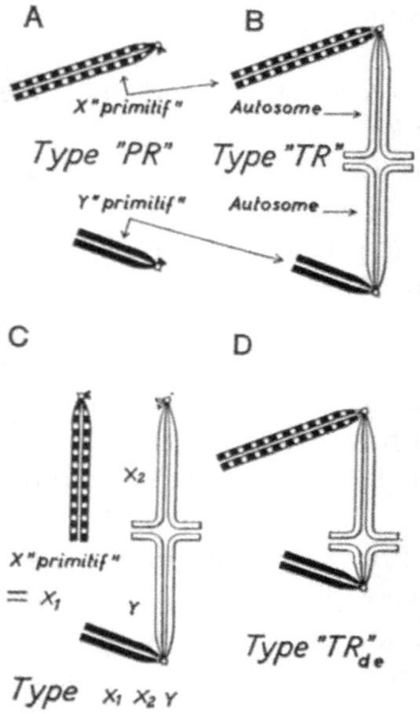

Fig. 12. Diagrammatic representation of the sex chromosome mechanisms in the *Mus (Leggada) minutoides* complex. A and B represent the primitive and translocation type of sex chromosomes while C and D illustrate the postulated mechanisms responsible for the X_1X_2Y and X deletion patterns. (From Matthey, in Rev. Suisse Zool., 1966b.)

somes onto the primitive sex chromosomes (Matthey, 1967). During meiosis the pairing segments derived from the autosomes exhibit chiasma formation.

Another sex chromosome mechanism occurs in *Mus minutoides* from Rhodesia where X_1X_2/Y and X_1X_2/X_1X_2 patterns were seen in males and females with 35 and 36 chromosomes respectively (Matthey, 1965). At meiosis the translocation type Y chromosome formed a trivalent with the primitive (acrocentric) X_1 and the homologue of the autosome that was translocated onto the Y chromosome.

A fourth mechanism characterized by a deletion involving the X chromosome is exhibited by *Mus minutoides musculoides* and *Mus triton* (Matthey, 1967). Males of the former species have a normal (translocation type) XY pattern; females have one normal metacentric X chromosome while the other, Xde, is submetacentric due to deletion of a major part of one arm. In *Mus triton* the normal X is submetacentric and the Y is a short acrocentric. Within one population, however, 11 of 21 females showed only one typical X while the other X was an acrocentric equal in length to the long arm of a normal X; the deleted (Xde) chromosome was not seen in any of the 18 males examined.

Within the genus *Gerbillus*, an XY_1Y_2/XX mechanism was present in *G. gerbillus gerbillus* which had 2n's of 43 and 42 in males and females respectively (Matthey, 1954; Wahrman and Zahavi, 1955). A trivalent was seen in meiosis. Another gerbil, *G. pyramidum* (2n40, Algeria), seemed to display a multiple sex chromosome mechanism (Matthey,

1952), but later studies by Wahrman and Zahavi (1955) failed to confirm this finding.

Pairing of the X and Y chromosomes during meiosis in most mammals occurs in an end-to-end fashion which suggests that the homologous segments are short. However, when the Y chromosome is large, side-to-side pairing between the X and Y has been recognized, as in the rat *Rattus natalensis* (Hamerton, 1958) and among certain cricetene rodents (Matthey, 1961). In the Chinese hamster, *Cricetus griseus* (2n22), pairing of isopycnotic segments of the X-Y bivalent was seen (Ohno and Weiler, 1962). The explanation for this behavior is thought to be a translocation of a pair of homologous autosomes onto the primitive X and Y chromosomes (Matthey, 1961).

The unusually large size of the sex chromosomes of certain rodents has stimulated investigation into their origin. Ohno, Beçak and Beçak (1964), using two dimensional measurements, showed that the original or ancestral type of X chromosome averaged 5% of the homogametic haploid (autosome + X) set. When the X chromosome was larger than the original type it occurred as a duplicate, triplicate or quadruplicate type X that comprised 10, 15 or 20% of the haploid set. Thus, the golden hamster *Mesocricetus auratus* (2n44) Ohno, Beçak and Beçak (1964), the chinchilla *Chinchilla lanigera* (2n64) (Nes, 1963), and the porcupine *Erethizon dorsatum* (2n42) (Benirschke, 1968) have a duplicate type X. A triplicate type X is seen in *Microtus oregoni* (2n17/18) (Ohno, Beçak and Beçak, 1964), and a quadruplicate type X constituting 20% of the haploid set is observed in *Microtus agrestis* (2n50) (Wolf *et al.*, 1965). Another vole, *Microtus chrotorrhinus* (2n60), has a very large submetacentric X and a large acrocentric Y chromosome resembling the sex chromosomes of *M. agrestis* (Meylan, 1967).

An investigation of South American rodents from the suborder Hystricomorpha has demonstrated the occurrence of large X chromosomes in *Cuniculus paca* (2n74), *Dasyprocta aguti* (2n64) and *Myoprocta acouchy* (2n62) (Fredga, 1966). A fourth species, *Myocastor coypus* (2n42) had medium-sized X chromosomes that are probably not of duplicate type. The Y chromosome of all these species was small.

Radioautographic techniques (Wolf *et al.*, 1965; Galton *et al.*, 1965) have been used to investigate the origin and behavior of these large sex chromosomes, and in each species with a large X all but a segment equivalent to the original type X (5% of the haploid complement) undergoes late replication. The origin of the large X and Y chromosomes is obscure, but translocation of autosomal heterochromatin onto the X chromosome or duplication of the original type X chromosome are possible explanations (Wurster *et al.*, 1968). Radioautographic investigation of the sex chromosome mechanisms in the *Mus* (*Leggada*) complex where the origin of the sex chromosomes is well understood might lead to a broader interpretation of the findings in other rodent species.

Additional rodents with unusual attributes pertaining to the sex chromosomes are two voles with diploid numbers of 17. In the case of *Microtus oregoni* a 2n of 17 was found in the germ cells of the male and somatic cells of the female, while male somatic cells displayed 18 chromosomes (Ohno *et al.*, 1963). This species may behave as a gonosomic mosaic: ♂ somatic cells (2n18, XY), ♂ spermatogonia (2n17, OY); ♀ somatic cells (2n17, XO), ♀ oogonia (2n18, XX) (Ohno, 1967). The other vole, *Ellobius lutescens*, also has 17 chromosomes in the somatic cells of both sexes and the spermatogonia of the male (Matthey, 1964). During the second meiotic division some cells display 8 while others contain 9 chromosomes. It was proposed that the unpaired element was a fused X-Y in the male and X-X in the female and that viable zygotes were formed only when a gamete with 9 chromosomes was fertilized by a gamete with 8 chromosomes. Thus, an egg with 9 fertilized by a sperm with 8 chromosomes would produce a female; the reverse would produce a male. Inviable offspring would result from gamete combinations having both 8 or 9 chromosomes.

Another instance of unusual behavior of the sex chromosomes occurs among species of mice belonging to the genus *Apodemus*. The sex chromosomes are of the XY, XX type. However, unlike most mammals whose separation of the X and Y complex occurs at anaphase of the first meiotic division (pre-reductionally), most species of *Apodemus* undergo a post-reductional separation of the X and Y chromosomes (Matthey, 1947; Makino, 1951).

Several kinds of sex chromosome aberrations have been reported in addition to the X chromosome deletions found in the *Mus* (*Leggada*) group (Matthey, 1967). In the mouse, *Mus musculus* (2n40), animals with 39 chromosomes and an X0 complement were described (Welshons and Russell, 1959); these animals were phenotypically normal, fertile females. Other specimens of *Mus musculus* with 41 chromosomes and XXY sex chromosomes were studied by Russell and Chu (1961) and found to be sterile but otherwise phenotypically normal males.

Among wild populations, Nadler and Hughes (1966a) reported an aberration in female specimens of *Spermophilus beldingi* (2n30) consisting of deletion of both arms of a chromosome morphologically resembling an X chromosome. Familial transmission of the aberration was demonstrated by the recognition of the same aberration in a phenotypically normal, lactating adult and a juvenile presumed to be mother and daughter. In an effort to implicate the X as the aberrant chromosome, kidney cells grown in culture and histological sections of the genital tract were examined for the presence, or possible absence, of nuclear sex dimorphism in normal and aberrant females (Nadler, unpublished data).

Unfortunately, sex chromatin bodies are not identifiable in *S. beldingi*, a condition frequently reported in other rodents (Moore, 1966). Despite some uncertainty, it seems likely that the X chromosome is aberrant and that the aberrant females have attributes similar to the X0 mice.

Future studies of the various sex chromosome mechanisms in rodents will be of value as models for investigating not only sex chromosome function and behavior but phylogenetic relationships as well. Already the elucidation of the sex chromosome mechanisms has been of prime importance in establishing the direction of evolution in the *Mus* (*Leggada*) group (Matthey, 1964).

Cytotaxonomy

Proper interpretation of cytogenetic data demands identification of the animals under investigation, a requirement all too frequently overlooked by cytogeneticists. Specimens can be most reliably identified when the skin, skull and collecting locality are available for comparison with taxonomic descriptions in the literature or with correctly identified museum specimens. Preservation of properly labelled skin and skull preparations (Hall and Kelson, 1959) from specimens utilized in cytogenetic studies is encouraged, and it is mandatory when taxonomic revisions are proposed.

Chromosomes are utilized in rodent taxonomy as characters for diagnosis and as a means for establishing phylogenetic relationships. In specimen identification chromosomes may vary in their applicability at different taxonomic levels, and the level of applicability must be ascertained for each taxon. For example, chromosomes are diagnostic at the generic level in the chipmunk genera *Tamias* and *Eutamias* (Nadler, 1964a; Sutton and Nadler, 1968) and in tree and flying squirrels of the genera *Sciurus, Tamiasciurus* and *Glaucomys* (Nadler and Sutton, 1967). More frequently, however, chromosomes are distinctive at the species level and many examples are cited throughout this paper.

At times, even populations and subspecies may be differentiated by their chromosomes. Good examples exist in the *Mus* (*Leggada*) complex (Matthey, 1966b), the ground squirrel *Spermophilus richardsoni* (Nadler, 1964b), the ground squirrel *Spermophilus spilosoma* (Nadler and Hughes, 1966b), and in *Spermophilus townsendi* (Nadler, 1968b). A new, chromosomally homogeneous subspecies, *S. townsendi nancyae* (2n38), was recognized in the latter species because it differed from an adjacent, chromosomally homogeneous population *S. townsendi townsendi* (2n36). Shellhammer (1967) also found karyotypic differences between subspecies of harvest mice, *Reithrodontomys raviventris raviventris* (2n38) and *R. r. halicoetes* (2n38), and he presented evidence that these forms were in a terminal stage of speciation—if in fact they were not already acting as full species.

An equally important application of chromosomes in taxonomy is their use as indicators of phylogenetic relationship. Here it is mandatory that all available lines of evidence be considered before drawing final conclusions because chromosomes may vary in their validity at different taxonomic levels, as already discussed. In this context, chromosomes may not parallel the divergence shown by other characters; conversely, chromosomes may show great diversity at population or species levels while gross morphological divergence is minimal. Another limitation of cytogenetic evidence by itself is the difficulty ascertaining direction of karyotypic change. A good example of this problem occurs in *Marmota* (Hoffmann and Nadler, 1968); chromosomes considered alone suggest that evolution progressed from the *M. caligata–M. flaviventris* complex (2n42, FN 62) → *M. monax* (2n38, FN 62) → *M. broweri* (2n36, FN 62) by Robertsonian centric fusion. Careful evaluation of gross morphological characters, ecology and zoogeography of Holarctic species of *Marmota* suggests instead that *M. monax* is the least specialized and most ancestral species of the group, and that *M. broweri,* which is a specialized late Pleistocene invader of North America, evolved by centric fusion, and the specialized *M. caligata–M. flaviventris* complex evolved by fission.

A phylogeny based on a massive amount of gross morphological data from recent and fossil specimens of Sciuridae (Black, 1963) is illustrated in Figure 13, and the available chromosomal data are listed for each genus.

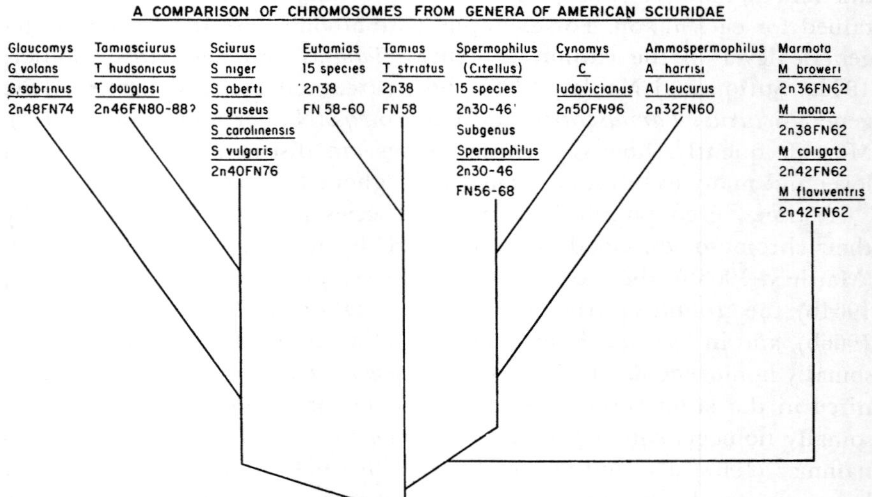

Fig. 13. A phylogeny of North American Sciuridae illustrating the diploid chromosome number and FN of each genus. (After Black, 1963.)

The great length of time that has elapsed since the divergence of most genera of Sciuridae (Black, 1963) makes it difficult to determine chromosomal relationships at generic levels. Among the tree and flying squirrels, chromosomes favor as close a relationship between *Sciurus* and *Glaucomys* (2n40, FN 76 *vs.* 2n48, FN 78) as between *Sciurus* and *Tamiasciurus* (2n46, FN 80–88) (Nadler and Sutton, 1967). The karyotypes confirm the view that *Sciurus* and *Glaucomys,* except for the gliding membrane, are morphologically quite similar and perhaps evolved from a common ancestral stock (Black, 1963).

The two chipmunk genera *Eutamias* and *Tamias,* which occupy an intermediate phylogenetic position between the tree squirrels and ground squirrels (Black, 1963), are very closely related both chromosomally and morphologically to each other (Nadler, 1964a). Although there is no obvious relationship between the chipmunks and either the tree or terrestrial squirrels, the chipmunk chromosomes resemble most the chromosomes of ground squirrels of the genus *Spermophilus* (Fig. 13).

Chromosomes of the genus *Spermophilus* bear some resemblance to those of *Marmota,* whose intrageneric relationships were discussed previously; both taxa have 2n's and FN's in the same range (Fig. 13). *Cynomys* with its high 2n of 50 and extremely high FN of 96 stands at an extreme position in the family, and there is little chromosomal support for its supposed proximity to the subgenus *Spermophilus* (Bryant, 1945; Black, 1963).

Controversy exists concerning the taxonomic status of *Ammospermophilus.* One recent investigator regards it as a subgenus of *Spermophilus* (Howell, 1938), while others consider its morphological features distinctive enough to warrant generic rank (Bryant, 1945; Black, 1963). The chromosomes of *A. harrisi* and *A. leucurus* are indistinguishable with 2n32, FN 60 and a karyotype composed of 30 bi-armed autosomes, a submetacentric X and acrocentric Y chromosome (Nadler and Sutton, 1962). There is a close similarity between the chromosomes of these species and those comprising the subgenera *Ictidomys* (2n32–34, FN 60–64) and *Spermophilus* (2n30–46, FN 56–66) of the genus *Spermophilus.* Chromosome studies thus do not support the generic status of *Ammospermophilus* and instead favor its reclassification as a subgenus of *Spermophilus.* The karyotypes also suggest that *Ammospermophilus* is more closely related to *Spermophilus* than is *Cynomys,* a conclusion in disagreement with the morphologically oriented views of Bryant (1945) and Black (1963).

At infrageneric levels, where taxa are less divergent, chromosomes are more revealing, particularly in the genus *Spermophilus.* There, the subgenera *Ictidomys, Poliocitellus,* and *Spermophilus* can be shown to share an equally close chromosomal relationship to one another while the subgenera *Callospermophilus, Otospermophilus* and *Xerospermophilus* seem more distantly related to the first three (Nadler, 1966b).

The subgenus *Ictidomys* has provided a particularly interesting model for testing the validity of chromosomal characters as a tool for classification. It is composed of two species, *Spermophilus tridecemlineatus* and *S. mexicanus,* that are similar in many pelage and skeletal features, and a third species, *S. spilosoma,* that is morphologically more divergent (Bryant, 1945). The fact that *S. tridecemlineatus* and *S. mexicanus* (both 2n34, FN 64 with similar karyotypes) differ from *S. spilosoma* (2n32, FN 60) supports the gross morphological data and substantiates the present taxonomy of this group (Nadler and Hughes, 1966b).

The chromosomal rearrangements that best account for karyotype evolution among species of the subgenus *Spermophilus* are outlined in Figure 14. A close cytological relationship exists between the *S. townsendi* complex and *S. washingtoni,* and also between *S. richardsoni, S. armatus* and *S. undulatus;* yet chromosomal affinities between *S. beldingi, S.*

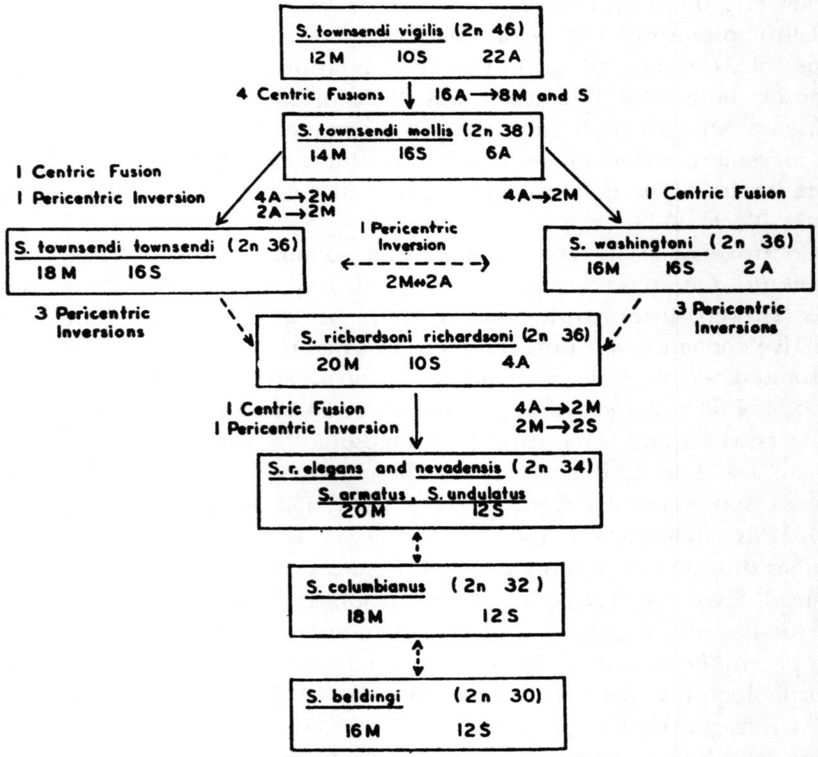

Fig. 14. Karyotype evolution in the subgenus *Spermophilus.* M, S, and A represent metacentric, submetacentric and acrocentric respectively. Solid arrows indicate conversions based upon reasonable evidence, whereas broken arrows designate more speculative rearrangements. (From Nadler, in J. Mammalogy, 1966a.)

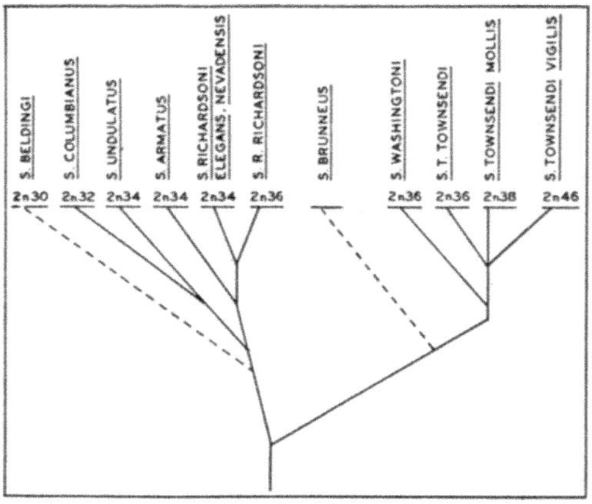

Fig. 15. Phylogenetic relationships among species of the subgenus *Spermophilus* based on data from chromosomes, gross morphology, ecology, zoogeography, and ectoparasite studies. (From Nadler, in J. Mammalogy, 1966a.)

columbianus and the other species are not clear (Nadler, 1966a). When information from gross morphology, ecology, comparative biochemistry, zoogeography and parasitology are correlated with the chromosome data, a more valid appraisal of interspecific relationships can be made (Fig. 15). The combined evidence (Nadler, 1966a) strongly suggests there are two species groups in the subgenus, the *S. townsendi–S. washingtoni* group and the *S. richardsoni–S. armatus–S. undulatus–S. columbianus–S. beldingi* group. In the latter group, the first three species are closely related despite the widely separated distribution of *S. undulatus.* Finally, *S. beldingi* and *S. columbianus* can be shown by serum protein, parasite, and zoogeographical data to belong to the same group as *S. richardsoni, S. armatus* and *S. undulatus,* despite divergent chromosome patterns.

Investigations of infraspecific relationships within the subgenus *Spermophilus* have been discussed elsewhere in this chapter. In both *S. richardsoni* (Nadler, 1964b) and *S. townsendi* (Nadler, 1968b) (Figs. 2–5) chromosomes characterize populations presently classified as subspecies although their possible status as sibling species has been suggested. Here cytogenetic studies offer promise as a means for detecting the presence or absence of intergradation in zones of contact, thus clarifying the taxonomy of these forms.

These data taken from the Sciuridae illustrate the point that as the closeness of relationship increases from suprageneric to specific and finally infraspecific levels, the reliability and usefulness of the cytological data increases proportionally.

The subfamily Microtinae has also been studied extensively, and Matthey (1957a) has constructed a phylogeny based upon chromosomal and other evidence (Fig. 16). Robertsonian centric fusions have played a major role in the evolution of this group with less frequent occurrences of pericentric inversions and fragmentations or fissions.

At the generic level, Patton (1967) successfully used chromosomal characters to evaluate the taxonomic status of pocket mice belonging to the genus *Perognathus,* subgenus *Chaetodipus*. He found evidence for four species groups based on the diploid and fundamental numbers, and his findings confirm the classification based on gross morphological characters. Although the 2n's ranged from 38–54 in the *intermedius* group the FN's clustered in the range of 54–58 and suggested that centric fusions were the major rearrangements operative in this karyotype evolution. Evidence for sibling species status was obtained for *P. artus* (2n54) and *P. goldmani* (2n52).

These are but a few examples in which chromosome studies have contributed to taxonomic evaluations of rodents. Many important investigations will be made in the future, particularly at specific and infraspecific levels. It is at these levels that chromosome analysis contributes most as one of the "new methods" in systematics.

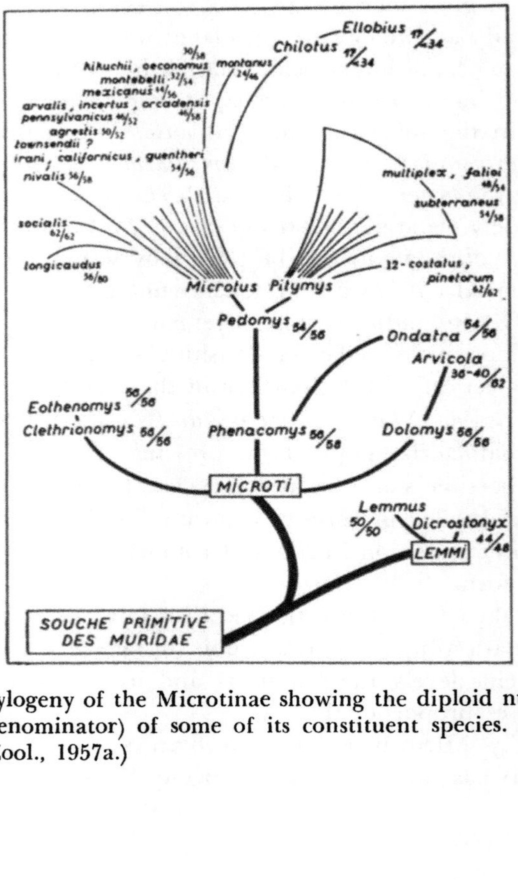

Fig. 16. A phylogeny of the Microtinae showing the diploid number (numerator) and FN (denominator) of some of its constituent species. (From Matthey, in Rev. Suisse Zool., 1957a.)

Summary

Surveys of the chromosomes from various taxa of the Rodentia have provided the basis for much of our understanding of chromosomal evolution in mammals as a whole. Although much descriptive work still remains to be done in the Rodentia, the trend of recent investigation has been directed toward specific aspects of chromosomal function and behavior. This review has considered certain areas where investigations to date indicate further work will be fruitful. These areas include: (1) the mechanisms responsible for chromosome evolution which may be confirmed by analysis of meiotic pairing from heterozygotes or hybrids; (2) the extent, stability and role of chromosome polymorphism in speciation; (3) the selective advantages of certain karyotypes within a polymorphic population either in natural or laboratory conditions; (4) evolution of sex chromosome systems in rodents, including the behavioral aspects of multiple sex chromosomes and the "duplicate" or larger size sex chromosomes; and (5) the role of chromosomes as isolating mechanisms. A greater understanding of these phenomena based upon examination of larger samples using population concepts will provide new and important insight into taxonomic relationships among rodents, particularly at specific and infraspecific levels.

Acknowledgment

Financial support was received from National Science Foundation Grants GB-3251 and GB-5676X.

References

Anderson, S. and J. K. Jones, Jr.: Recent Mammals of the World—A Synopsis of Families. New York: Ronald Press Co., 1967.

Benirschke, K.: The chromosome complement and meiosis of the North American porcupine. J. Hered. *59*:71, 1968.

Black, C. C.: A review of the North American Tertiary Sciuridae. Bull. Mus. Comp. Zool., Harvard Univ. *130*:109, 1963.

Bryant, M. D.: Phylogeny of Nearctic Sciuridae. Amer. Midl. Nat. *33*:257, 1945.

Cohen, M. M. and L. Pinsky: Autosomal polymorphism via a translocation in the guinea pig, *Cavia porcellus*. L. Cytogenetics *5*:120, 1966.

Ford, C. E., J. L. Hamerton and G. B. Sharman: Chromosome polymorphism in the common shrew. Nature *180*:392, 1957.

——— and E. P. Evans: A reciprocal translocation in the mouse between the X chromosome and a short autosome. Cytogenetics *3*:295, 1964.

Fredga, K.: Chromosome studies in five species of South American rodents (suborder Histricomorpha). Mammal. Chrom. Newsl. *20*:45 (April), 1966.

Galton, M., K. Benirschke and S. Ohno: Sex chromosomes of the chinchilla: allocycly and duplication sequence in somatic cells and behavior in meiosis. Chromosoma *16*:668, 1965.

Hall, E. R. and K. R. Kelson: The Mammals of North America. New York: Ronald Press Co., 1959.

Hamerton, J. L.: Mammalian sex chromosomes. In: Symposium on Nuclear Sex. D. R. Smith and W. M. Davidson, eds. London: Heinemann, 1958.

———, F. Giannelli and C. O. Carter: A family showing transmission of a D/D reciprocal translocation and a case of regular Downs syndrome. Cytogenetics *2*:194, 1963.

Hoffmann, R. S. and C. F. Nadler: Chromosomes and systematics of some North American species of the genus *Marmota* (Rodentia: Sciuridae). Experientia *24*:740, 1968.

Howell, A. H.: Revision of North American ground squirrels, with a classification of the North American Sciuridae. N. Amer. Fauna *56*:1, 1938.

Hsu, T. C. and F. E. Arrighi: Chromosomal evolution in the genus *Peromyscus* (Cricetidae, Rodentia). Cytogenetics *5*:355, 1966.

Lay, D. M. and C. F. Nadler: Hybridization in the rodent genus *Meriones*: I. Breeding and cytological analyses of *Meriones shawi* ♀ × *Meriones libycus* ♂ hybrids. Cytogenetics *8*:35, 1969.

Makino, S.: Studies on the murine chromosomes. V. A study of the chromosomes in *Apodemus*, especially with reference to the sex chromosomes in meiosis. J. Morph. *88*:93, 1951.

Manna, G. K. and M. Talukdar: Chromosomal polymorphism in the guinea pig, *Cavia porcellus*. Experientia *20*:324, 1964.

Marks, G. E.: Telocentric chromosomes. Amer. Naturalist *91*:223, 1957.

Matthey, R.: L'évolution de la formule chromosomiale chez les Vértébres. Experientia *1*:50, 78, 1945.

———: Encore les hétérochromosomes des *Apodemus*. Arch. Julius Klaus Stift. *22*:85, 1947.

———: Chromosomes sexuels multiples chez un Rougeur (*Gerbillus pyramidum* Geoffry). Arch. Julius Klaus Stift. *27*:163, 1952.

———: Un cas nouveau de chromosomes sexuels multiples dans le genre *Gerbillus* (Rodentia-Muridae-Gerbillinae). Experientia *10*:464, 1954.

———: Cytologie comparée systématique et phylogenie des microtinae (Rodentia-Muridae). Rev. Suisse Zool. *64*:39, 1957a.

———: Cytologie et taxonomie du genre *Meriones*, Illiger (Rodentia-Muridae-Gerbillinae). Säugetierk. Mitt. *5*:145, 1957b.

———: Les chromosomes des mammifères euthériens. Liste critique et essai sur l'évolution chromosomique. Arch. J. Klaus Stift. *33*:253, 1958.

———: Cytologie comparée des Cricetinae paléarctiques et americains. Rev. Suisse Zool. *68*:41, 1961.

———: Les nombres diploides des Mammiféres euthériens. Liste critique. Mammal. Chrom. Newsl. *8*:17 (August), 1962.

———: Polymorphisme chromosomique intraspécifique chez un mammifère *Leggada minutoides* Smith (Rodentia-Muridae). Rev. Suisse Zool. *70*:173, 1963a.

———: Polymorphisme chromosomique intraspecifique et intraindividuel chez *Acomys minous* Bate (Mammalia-Rodentia-Muridae). Etude cytologique des hybrides *Acomys minous* ♂ × *Acomys cahirinus* ♀. Le mecanisme des fusions centriques. Chromosoma *14*:468, 1963b.

———: Etudes sur les chromosomes d'*Ellobius lutescens* (Mammalia-Muridae-Microtinae). II. Informations complémentaires sur les divisions meiotiques. Rev. Suisse Zool. *71*:401, 1964a.

———: Evolution chromosomique et spéciation chez les *Mus* du sous-genre *Leggada* Gray 1837. Experientia *20*:1, 1964b.

———: Un type nouveau de chromosomes sexuels multiples chez une Souris africaine du groupe *Mus* (*Leggada*) *minutoides* (Mammalia-Rodentia). Male: X_1X_2/Y. Female: X_1X_2/X_1X_2. Chromosoma *16*:351, 1965.

———: Une inversion péricentrique á l'origine d'un polymorphisme chromosomique non-Robertsonian dans une population de *Mastomys*. (Rodentia-Murinae). Chromosoma *18*:188, 1966a.

———: Le polymorphisme chromosomique des *Mus* africains du sous-genre *Leggada*. Révision générale portant sur l'analyse de 213 individus. Rev. Suisse Zool. *73*:585, 1966b.

———: Etude de deux femmelles hetérozygotes pour une delétion partielle portant sur un bras du chromosome X chez *Mus* (*Leggada*) *minutoides musculoides* Temm. Cytogenetics *6*:168, 1967.

———: General survey of the chromosome numbers of eutherian mammals. Mammal. Chrom. Newsl. *9*:17, 1968.

Mayr, E.: Animal Species and Evolution. Cambridge: Harvard University Press, 1963.

Meylan, A.: *Microtus chrotorrhinus* (Miller) another species with giant sex chromosomes. Mammal. Chrom. Newsl. *8*:280 (October), 1967.

Mittwoch, U.: Sex Chromosomes. New York: Academic Press, 1967.

Moore, K. L.: Sex chromatin patterns in various animals. In: The Sex Chromatin. K. L. Moore, ed. Philadelphia: W. B. Saunders Co., 1966.

Nadler, C. F.: Contributions of chromosomal analysis to the systematics of North American chipmunks. Amer. Midland Nat. *72*:298, 1964a.

———: Chromosomes and evolution of the ground squirrel *Spermophilus richardsoni*. Chromosoma *15*:289, 1964b.

———: Chromosomes and systematics of American ground squirrels of the subgenus *Spermophilus*. J. Mammalogy *47*:579, 1966a.

———: Chromosomes of *Spermophilus franklini* and taxonomy of the ground squirrel genus *Spermophilus*. Syst. Zool. *15*:199, 1966b.

———: Chromosomes of the ground squirrel *Spermophilus richardsoni aureus* (Davis). J. Mammalogy *49*:312, 1968a.

———: The chromosomes of *Spermophilus townsendi* (Rodentia: Sciuridae) and report of a new subspecies. Cytogenetics *7*:144, 1968b.

——— and K. E. Harris: Chromosomes of the North American prairie dog *Cynomys ludovicianus*. Experientia *23*:41, 1967.

——— and C. E. Hughes: Chromosomal aberrations in a population of ground squirrels. Science *151*:579, 1966a.

——— and ———: Chromosomes and taxonomy of the ground squirrel subgenus *Ictidomys*. J. Mammalogy *47*:46, 1966b.

——— and D. M. Lay: Chromosomes of some species of *Meriones* (Mammalia: Rodentia). Zeit. F. Säugetierk. *32*:285, 1967.

——— and D. A. Sutton: Mitotic chromosomes of some North American Sciuridae. Proc. Soc. Exper. Biol. Med. *110*:36, 1962.

——— and ———: Chromosomes of some squirrels (Mammalia-Sciuridae) from the genera *Sciurus* and *Glaucomys*. Experientia *23*:249, 1967.

Nes, N.: The chromosomes of *Chinchilla lanigera*. Acta vet. Scand. *4*:128, 1963.

Ohno, S. and C. Weiler: Relationship between large Y-chromosome and side-by-side pairing of the XY-bivalent observed in the Chinese hamster, *Cricetulus griseus*. Chromosoma *13*:106, 1962.

———, J. Jainchill and C. Stenius: The creeping vole (*Microtus oregoni*) as a gonosomic mosaic. I. The OY/XY constitution in the male. Cytogenetics *2*:232, 1963.

———, W. Beçak and M. L. Beçak: X-autosome ratio and the behavior pattern of individual X-chromosmes in placental mammals. Chromosoma *15*:14, 1964.

———: A phylogenetic view of the X-chromosome in man. Ann. Génét. *8*:3, 1965.

———, C. Weiler, J. Poole, L. Christian and C. Stenius: Autosomal polymorphism due to pericentric inversions in the deer mouse (*Peromyscus maniculatus*) and some evidence of somatic segregation. Chromosoma *18*:177, 1966.

———: Sex Chromosomes and Sex-linked Genes. New York: Springer-Verlag, 1967.

Patterson, J. T. and W. S. Stone: Evolution in the Genus *Drosophila*. New York: Macmillan Co., 1952.

Patton, J. L.: Chromosome studies of certain pocket mice, genus *Perognathus* (Rodentia: Heteromyidae). J. Mammalogy *48*:27, 1967.

——— and O. H. Soule: Natural hybridization in pocket mice, genus *Perognathus*. Mammal. Chrom. Newsl. *8*:263 (October), 1967.

——— and R. E. Dingman: Chromosome studies of pocket gophers, genus *Thomomys*. I. The specific status of *Thomomys umbrinus* (Richardson) in Arizona. J. Mammalogy *49*:1, 1968.

Robertson, W. R. B.: Chromosome studies. I. Taxonomic relationships shown in the chromosomes of Tettigidae and Acrididae: V-shaped chromosomes and their significance in Acrididae, Locustidae, and Gryllidae: Chromosomes and variation. J. Morphology *27*:179, 1916.

Russell, L. B. and E. H. Y. Chu: An XXY male in the mouse. Proc. Natl. Acad. Sci. U.S. *47*:571, 1961.

Shellhammer, H. S.: Cytotaxonomic studies of the harvest mice of the San Francisco Bay region. J. Mammalogy *48*:549, 1967.

Sparkes, R. S. and D. T. Arakaki: Intrasubspecific and intersubspecific chromosomal polymorphism in *Peromyscus maniculatus* (Deer Mouse). Cytogenetics *5*:411, 1966.

Staton, D.: Karyotypes and phylogeny of New World primates. Mammal. Chrom. Newsl. *8*:203 (July), 1967.

Sutton, D. A. and C. F. Nadler: Chromosomes of the North American chipmunk genus *Eutamias*. J. Mammalogy (in press), 1968.

Todd, N. B.: A theory of karyotypic fissioning, genetic potentiation and eutherian evolution. Mammal. Chrom. Newsl. *8*:268 (October), 1967.

Wahrman, J. and A. Zahavi: Cytological contributions to the phylogeny and classification of the rodent genus *Gerbillus*. Nature *175*:600, 1955.
Walker, E. P.: Mammals of the World. Baltimore, Md.: Johns Hopkins Press, 1964.
Welshons, W. J. and L. B. Russell: The Y-chromosome as the bearer of the male determining factors in the mouse. Proc. Natl. Acad. Sci. U.S. *45*:560, 1959.
White, M. J. D.: Animal Cytology and Evolution. Cambridge, England: Cambridge University Press, 1954.
———: Some general problems of chromosomal evolution and speciation in animals. Surv. Biol. Prog. *3*:109, 1957.
———: Restrictions on recombination in grasshopper populations and species. Cold Spring Harbor Symp. Quant. Biol. *23*:307, 1958.
———: Models of speciation. Science *159*:1065, 1968.
Wolf, U., G. Flinspach, R. Bohm and S. Ohno: DNS-Reduplikations-muster bei den Riesen-Geschlechtschromosomen von *Microtus agrestis*. Chromosoma *16*: 609, 1965.
Wood, A. E.: Are there rodent suborders? Syst. Zool. *7*:169, 1958.
Wurster, D. H., K. Benirschke and H. Noelke: Unusually large sex chromosomes in the Sitatunga (*Tragelaphus spekei*) and the Blackbuck (*Antilope cervicapra*). Chromosoma *23*:317, 1968.
Yosida, T. H., A. Nakamura and T. Fukaya: Chromosomal polymorphism in *Rattus rattus* (L.) collected in Kusudomari and Misima. Chromosoma *16*:70, 1965a.
——— and K. Amano: Autosomal polymorphism in laboratory bred and wild Norway rats, *Rattus norvegicus*, found in Misima. Chromosoma *16*:658, 1965b.
Zahavi, A. and J. Wahrman: The cytotaxonomy ecology and evolution of the gerbils and jirds of Israel (Rodentia: Gerbillinae). Mammalia *21*:341 1957.

CYTOGENETIC AND PHYLOGENETIC STUDIES IN CARNIVORA

DORIS H. WURSTER
Department of Pathology, Dartmouth Medical School, Hanover, New Hampshire

Let us look at the carnivores from the point of view of their presently accepted taxonomy and phylogeny and see how the study of their chromosome constitution fits with the accepted scheme, or where it may give us something new to consider. The earliest Carnivora were the creodonts which appeared in the Paleocene about 75 million years ago. By the Eocene, 60 million years ago, a branch called the Miacidae was flourishing. The creodonts subsequently became extinct, but toward the end of the Eocene the Miacidae diverged into three major branches of the Carnivora (Fig. 1*): these are first the Feloidea, which include the cats, the hyenas, and the mongooses and civets; the second the Canoidea, which include the raccoons, bears, dogs, wolves and foxes, and the weasels, minks, martens, badgers, skunks and otters; and the third the Pinnipedia or marine carnivores, which include the walruses, seals and sea lions. The Pinnipedia are, today, generally not included with the Carnivora but are placed in a separate Order. They did, however, arise from the Canoidea and still retain close serological (Leone and Wiens, 1956; Pauly and Wolfe, 1957) and chromosomal relationships. The latter can be seen in the figure by comparing the diploid number, fundamental number and marker chromosomes with those of the Canoidea.

In our laboratory we have studied the chromosomes of about 65 species of the Fissipedia, or land-living carnivores, and altogether at least 105 species of the approximately 250 species in the Order have been investigated here or elsewhere. This is a large Order within which there is much diversity of chromosome number and morphology. The diploid numbers range from a low of 30 in the mink to a high of 78 in the dogs and wolves; the fundamental number for the whole Order

* Figs. 1–8 and 11 appeared in Chromosoma *24*:336, 1968.

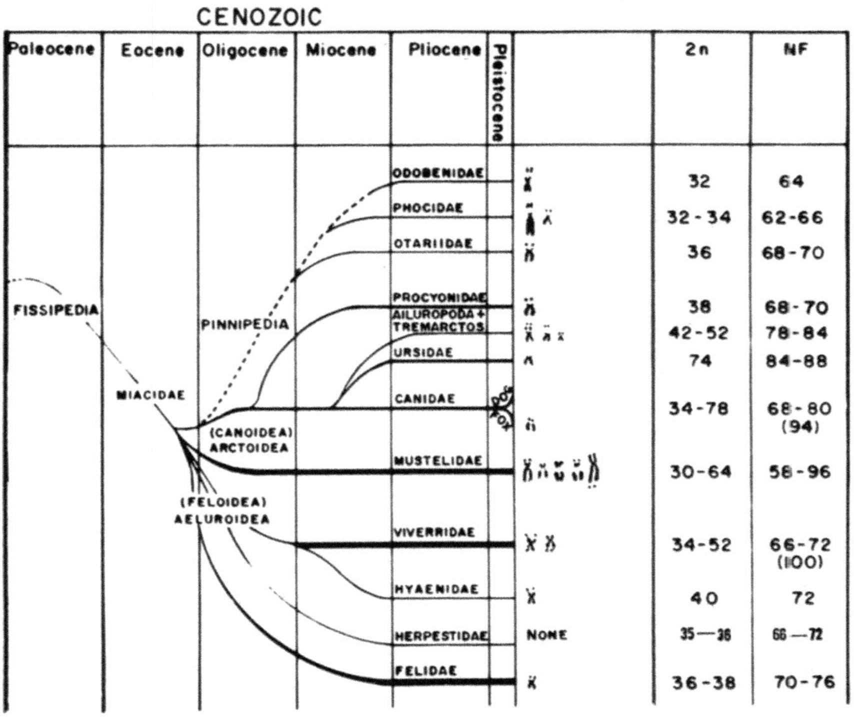

Fig. 1. This drawing depicts the carnivoran phylogeny as modified from Romer (1966) and Thenius and Hofer (1960) to accommodate the cytogenetic viewpoint. The marker chromosomes shown for each family occur in one or more species of that family.

ranges only from 66 to 88 with but 4 exceptions of 90–100 in the genet (*Genetta genetta,* 2n = 52), the striped skunk (*Mephitis mephitis,* 2n = 50), and the arctic fox (*Alopex lagopus,* 2n = 48–50), and 58 in the mink (*Mustela vison,* 2n = 30). Remarkably constant, however, throughout the Order is the presence of a marker chromosome. Each family, as seen in Figure 1, has one or more types of marker chromosomes, with the exception of Herpestid mongooses and the division of the Canidae on the branch labeled "dog." The most common type is that possessed by all the cats, a small submetacentric, satellited on the short arms, and it is this typical form that is often referred to as the "carnivore chromosome." The bears and foxes carry a different kind, which is acrocentric, and instead of being satellited, has an achromatic region in the long arm adjacent to the centromere. The following karyotypes each illustrate one or more of the types of marker seen in the diagrammatic figure: ring-tailed mongoose (*Galidia elegans,* 2n = 44), masked palm civet (*Paguma larvata,* 2n = 44), African or two-spotted

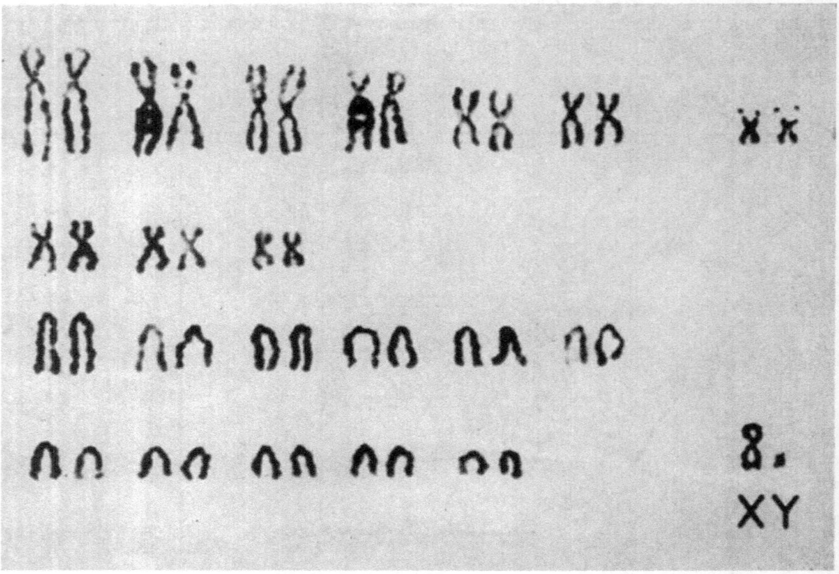

Fig. 2. Karyotype of a male ring-tailed mongoose, *Galidia elegans.*

palm civet (*Nandinia binotata,* 2n = 38), yellow-throated marten (*Martes flavigula,* 2n = 40), Chinese ferret badger (*Melogale moschata,* 2n = 38), grizzly bear (*Ursus horribilis,* 2n = 74), and the fennec fox (*Fennecus zerda,* 2n = 64) (Figs. 2–8).

It is not known whether the marker bears any significance to the relationships of the species. The persistence of its presence is, however, a mark of its stability. Since it is carried by most of the carnivoran families, as well as the Pinnipedia (Corfman and Richart, 1964; Fay *et al.,* 1966; Hungerford and Snyder, 1964), which branched off about 35 million years ago, it can be assumed that it is a primitive trait. The sea lions (Fay *et al.,* 1966; Hungerford and Snyder, 1964), like the cats, the hyenas, the Viverridae and the Procyonidae, possess the typical type of marker. Since the sea lions, the Procyonidae, the giant panda (Newnham and Davidson, 1966) and the spectacled bear (Newnham and Davidson, 1967), which are products of the canid stock, all have the typical, presumably primitive, satellited type of marker, it can be assumed that the Ursidae originated their own kind, or inverted the typical kind to form the type of marker that they carry. The Herpestidae, which I have separated into a family apart from the Viverridae, after Gregory and Hellman (1939) and Thenius and Hofer (1960), completely lack a marker in the 8 species studied so far.

The Mustelidae have the greatest variety of markers and they also have the greatest range of the fundamental number and greatest variety

Fig. 3. Karyotype of a male masked palm civet, *Paguma larvata.*

of karyotype pattern of the carnivoran families. The subfamilies of the Mustelidae that we know today were already fossil in Eurasia in the Lower Oligocene and the Miocene and have since attained a wide distribution. In contrast, the Felidae have the typical "carnivore chromosome" throughout the family and the most uniform diploid number, fundamental number and karyotype pattern of all the large families.

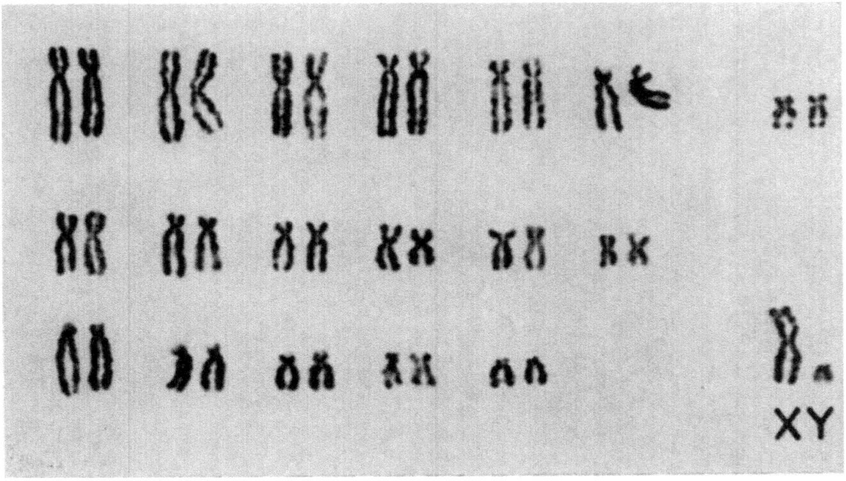

Fig. 4. Karyotype of a male African or two-spotted palm civet, *Nandinia binotata.*

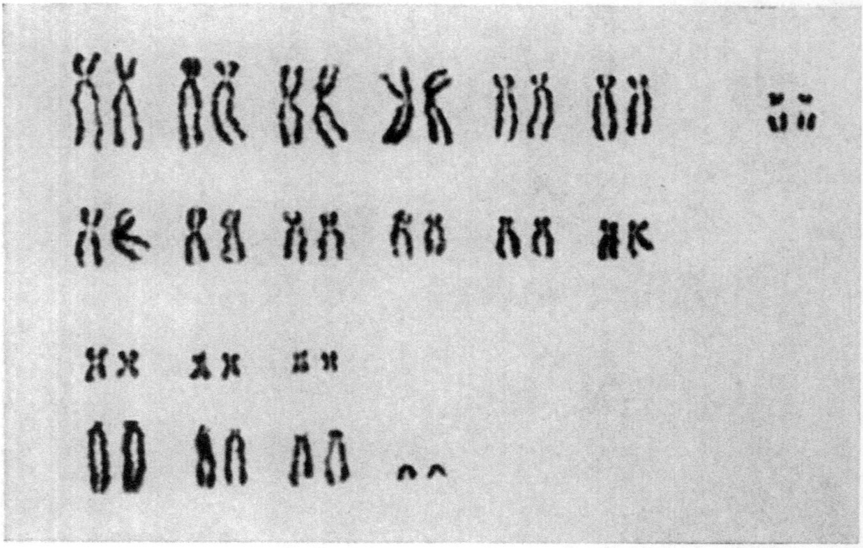

Fig. 5. Karyotype of a female yellow-throated marten, *Martes flavigula.*

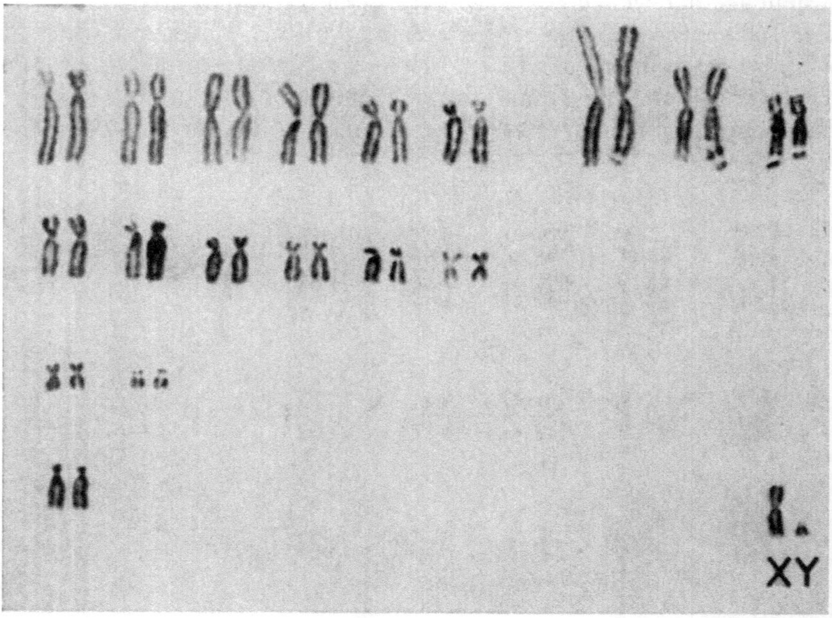

Fig. 6. Karytoype of a male chinese ferret badger, *Melogale moschata.*

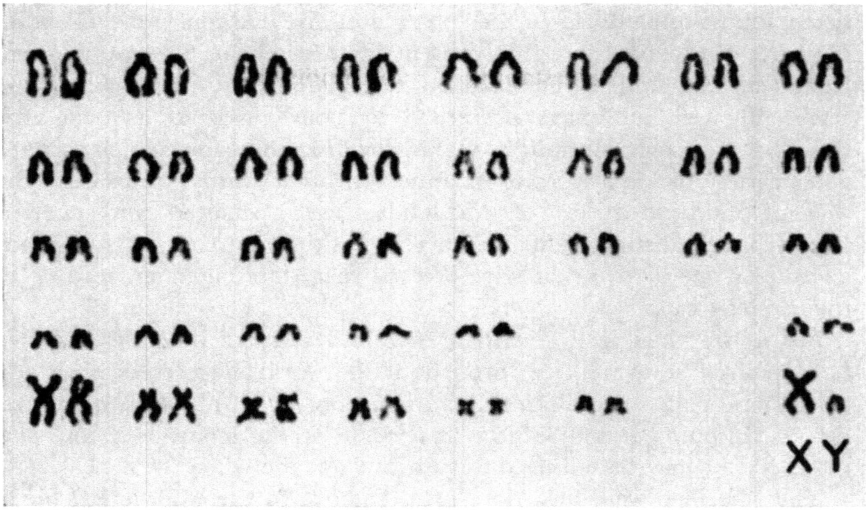

Fig. 7. Karyotype of a male grizzly bear, *Ursus horribilis*. The marker chromosomes are the small acrocentrics immediately above the sex chromosomes.

The modern cats that we know today were fossil only as recently as the Upper Pliocene and Pleistocene. We do not, of course, know anything concerning the cytogenetics of the other branches of cats in history, such as the saber-tooth cats, which became extinct along the way. From the comparison of these two families, the Mustelidae and Felidae, we can hypothesize that the branch of cats from which our modern cats have come apparently arrived at a fairly stable gene arrangement and karyotype pattern by the Upper Pliocene and this appears to have per-

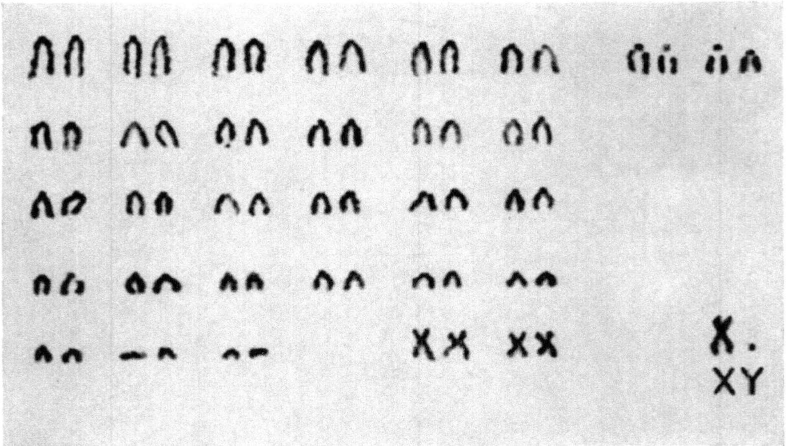

Fig. 8. Karyotype of a male fennec fox, *Fennecus zerda*.

sisted almost unmodified in the intervening five million years; all other branches of cats died out. The present-day Mustelidae, in contrast, stem from very old and well-established sub-families that have each done reasonably well and have proceeded to evolve each in its own way, thereby, after some 30 million years providing the cytogeneticist with a good variety of karyotypes to examine. In other words, we can say the ancient multiple forms of Mustelidae each persisted and evolved karyotypically distinct. And the ancient multiple forms of cats died out except for one or two that have evolved in a stable manner, leading to the modern cats.

Let us see then if the carnivoran families and subfamilies, as they have evolved and migrated throughout the world, have established any cytogenetic paths of evolution and migration. Fossil histories, of course, may be incomplete and subject to revision at the moment of any new discovery, or may be telling the truth but not the whole truth.

The Miacidae were fossil in North America in the Middle Paleocene to the Upper Eocene. They were then fossil in Eurasia from the Upper Eocene to perhaps the Lower Oligocene. This says they migrated from North America to Eurasia (see Fig. 9).

From the Miacidae in North America arose the following modern families of the Canoid branch: the Canidae, fossil in the Upper Eocene, and the sub-family Procyoninae of the Procyonidae, fossil in the Lower Miocene. The Canidae appeared simultaneously in Eurasia either by migration or parallel evolution. The Procyoninae migrated to South America appearing there by the Upper Miocene.

From the Canoid branch of the Miacidae in Eurasia, besides the simultaneous appearance of the Canidae, arose the following modern

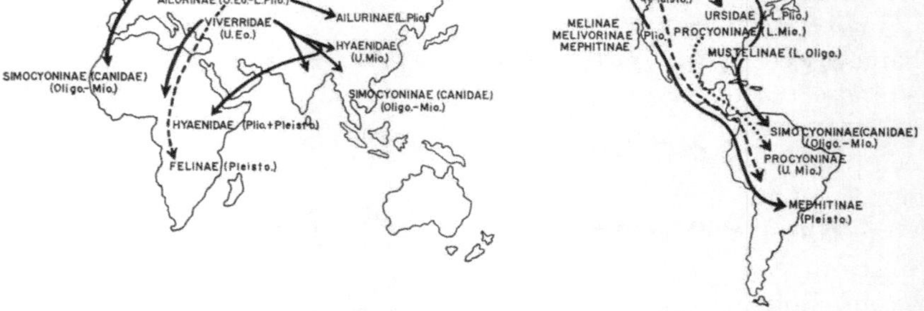

Fig. 9. This drawing depicts earliest fossil remains of Carnivoran families and subfamilies and assumed routes of migration deduced from fossil histories.

families and subfamilies: first the Ailurinae of the Procyonidae, appearing in the Upper Miocene and Lower Pliocene. The question arises whether these migrated from North America or whether they evolved independently. The Ailurinae contain two fossil genera in Europe as well as the two existing ones in Asia today represented only by the giant panda (*Ailuropoda melanoleuca,* 2n = 42) and the lesser panda (*Ailurus fulgens,* 2n = 36). From the study of their karyotypes I would propose that the giant panda is a very early offshoot of the bears and that the evolution of its karyotype is similar to that pathway of karyotype evolution being followed by the spectacled bear (*Tremarctos ornatus,* 2n = 52) which is the only bear to have reached South America. From the study of the lesser panda's karyotype I would propose that it represents an example of parallel evolution in cytogenetics in having evolved from the Miacidae in Europe in a fashion similar to the evolution of the Procyonidae from the Miacidae in North America. These karyotypes will be discussed in detail later. The Ursidae or bears were first fossil in Europe in the Middle Miocene and evidently migrated to North America appearing there by the Lower Pliocene.

Three subfamilies of the Mustelidae were fossil first in Eurasia about mid-Miocene to Pliocene and appeared only later in North America, and a fourth sub-family, the Mustelinae, was fossil earlier than the others in the Lower Oligocene in both Eurasia and North America. From the Feloid branch of the Miacidae, the Viverridae were first fossil in Europe very early, that is the upper Eocene. The first of the Hyaenidae, which is an offshoot of the Viverridae, is in Asia in the Upper Miocene. The hyenas evidently migrated only southward into southern Asia and Africa and never crossed over into North America. The remaining modern family, the Felidae, was first fossil in Europe as recently as the Upper Pliocene with their fossils appearing in Africa and America first in the Pleistocene.

This is a complex fossil history of a large order that is today geographically widespread and cytogenetically diverse. The point to note is that these carnivoran families arose and migrated back and forth in the northern hemisphere. They dispersed southward secondarily. In general, a southerly migration has tended to be accompanied, within a family, by a decreasing diploid number while the marker chromosome remains stable and shows no trend correlative with the migratory route. If one assumes a very primitive karyotype to consist of 80 or more acrocentric chromosomes, it can be seen that attainment of many metacentrics and a diploid number of about 34 to 44 is a stable arrangement for the carnivores. Of the more than 100 species of carnivores studied so far about 70% have a diploid number between 34 and 44. Those that migrated in a southerly direction from the northern hemisphere have tended to lower their diploid number, and this is particularly noticeable

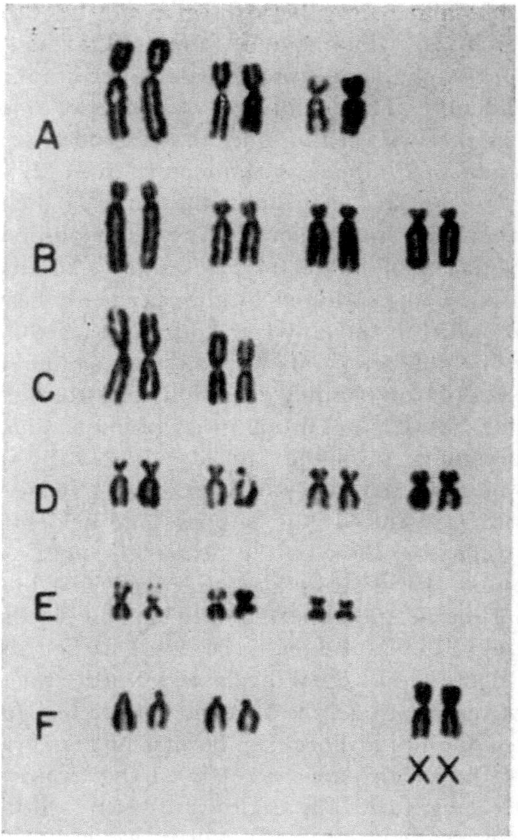

Fig. 10. Karyotype of a female Indian Golden cat, *Felis temmincki.*

in those species that migrated to South America. Although there are a number of exceptions, there are examples of this generalization in each family.

For instance, of the Felidae, all cats of the nothern hemisphere have $2n = 38$ while a number of the South American cats have $2n = 36$. These include the ocelot (*Felis pardalis*), tiger cat (*Felis tigrina*), marguay cat (*Felis wiedi*) and Geoffroy's cat (*Felis geoffroyi*). The decrease in number is often manifested in a reduced number of acrocentrics as these cats have only one pair of acrocentrics, or as in the case of the Geoffroy's cat, no acrocentrics at all. Instead, they have an extra pair of metacentrics in the C group, probably representing fusion of the F group acrocentrics. Figure 10 shows the karyotype of an Indian golden cat (*Felis temmincki,* $2n = 38$); this karyotype is indistinguishable from that of the domestic cat and is patterned after the standard cat karyotype agreed upon at the San Juan conference. Figure 11 shows the karyotype of a male Geoffroy's cat from South America with 2n = 36, no acrocentrics, and an extra pair in the C group. The ocelot (Fig. 12) has one pair of acrocentrics and three pairs in the C group but one less pair in the D group, probably indicating fusion of one pair of acrocentrics with one pair of D group chromosomes to form a new pair of C group chromosomes. Thus, they have similar karyotypes, but chromosomes in comparable positions are not necessarily homologous. The trend, however, is to reduce the acrocentric elements and thus usually the diploid number also, but not always. The jaguarondi (*Felis yagouaroundi,* Fig. 13) of South America has no acrocentrics but retains a diploid number of 38. It has an extra pair in the A group, or possibly the D group, and one extra pair in the E group. Here the acrocentric elements appear to have escaped fusion but to have undergone inversion. The trend toward the reduction of acrocentrics is also seen in the

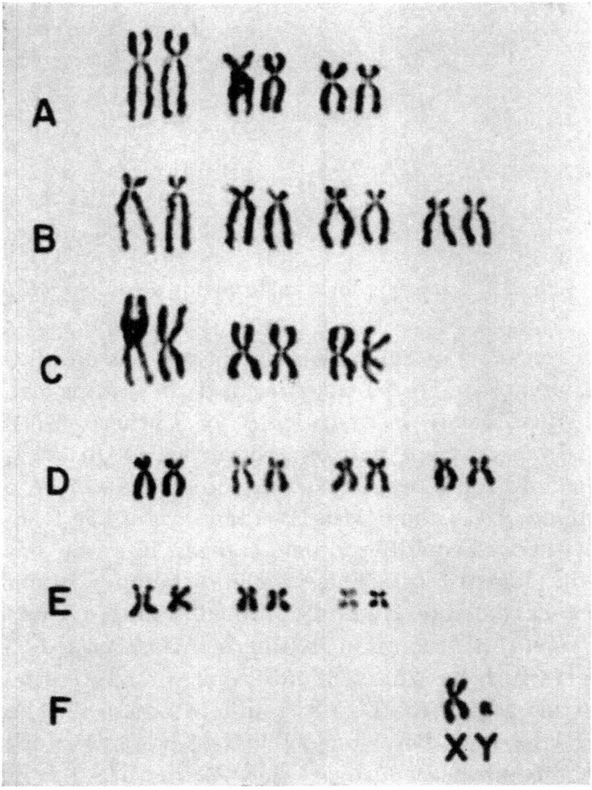

Fig. 11. Karyotype of a male Geoffroy's cat, *Felis geoffroyi.*

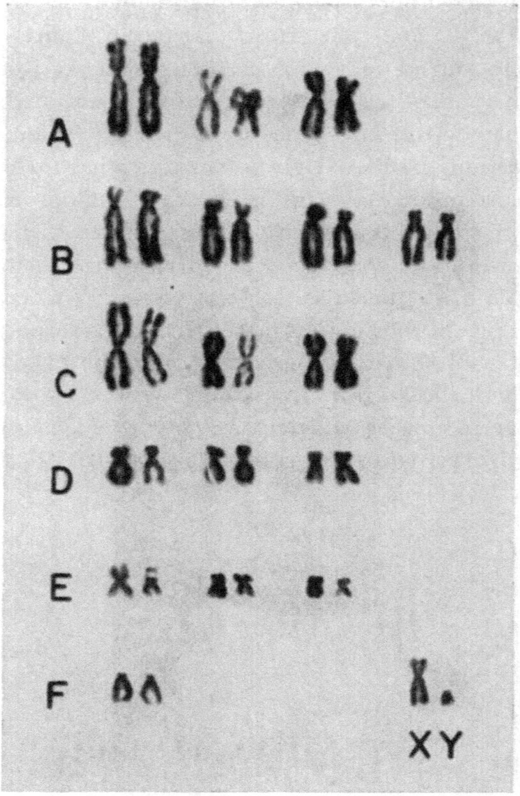

Fig. 12. Karyotype of a male ocelot, *Felis pardalis.*

puma (*Felis concolor*) of South America (as well as North America), the cheetah (*Acinonyx jubatus*) of Africa, and the fishing cat (*Felis viverrina*) and leopard cat (*Felis bengalensis*) of southern Asia, all of which have 2n = 38 but only one pair of acrocentrics. Of the 23 species of cats so far studied, 13 have two pairs of acrocentrics, 7 have one pair and three have none. All of those with less than 2 pairs live below the Tropic of Cancer. Others of equally southerly ranges, however, have karyotypes similar to the domestic cat. These include the Indian golden cat (Fig. 10) and one of the most recently studied southern cats, the clouded leopard (*Neofelis nebulosa*) of southern Asia, Borneo, Sumatra and Java (Fig. 14) which has 2n = 38 and 2 pairs of acrocentrics.

The viverrids are all dwellers in southern areas. The genet (*Genetta genetta*) and the Egyptian mongoose (*Herpestes ichneumon*) (Makino, 1951), both of southern Europe and Africa, are the northern-most viverrid residents and have the highest diploid numbers, 52 and 48 respectively. All other viverrids live more south and southeast and have diploid numbers ranging from 34 to 44. The genet is unique, however, in having 46 metacentric autosomes and only 4 acrocentric autosomes, giving it a fundamental number of 100. Again, if we assume a very primitive karyotype to be 80 or more acrocentric chromosomes, then the genet can be considered to have a karyotype that has undergone almost complete fusion of its acrocentrics, and the rest of the viverrids to have progressed even further in having undergone fusion, inversion, and further fusion in many steps, thus rendering the great cytogenetic diversity found among the Viverridae today.

Of the Canidae, the genus Canis possesses a stable 2n = 78 and uniform karyotype. All members of the genus thus far studied have identical karyotypes. Very similar in karyotype pattern but with the slightly reduced diploid number of 76 are the short-eared dog (*Atelocynus microtis*) and the maned wolf (*Chrysocyon brachyurus*) (Newnham and Davidson, 1966) of South America, again demonstrating the trend for

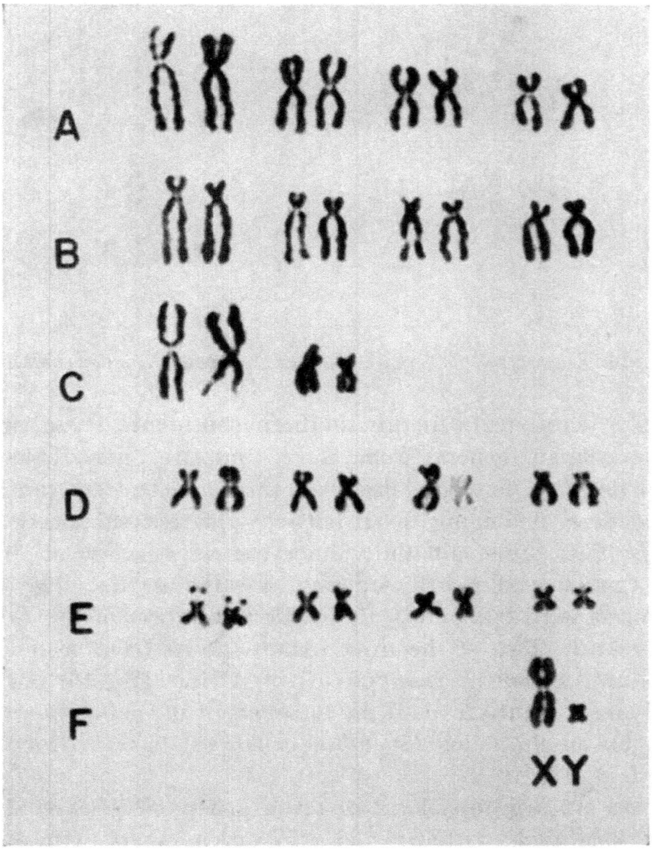

Fig. 13. Karyotype of a male jaguarondi, *Felis yagouaroundi*.

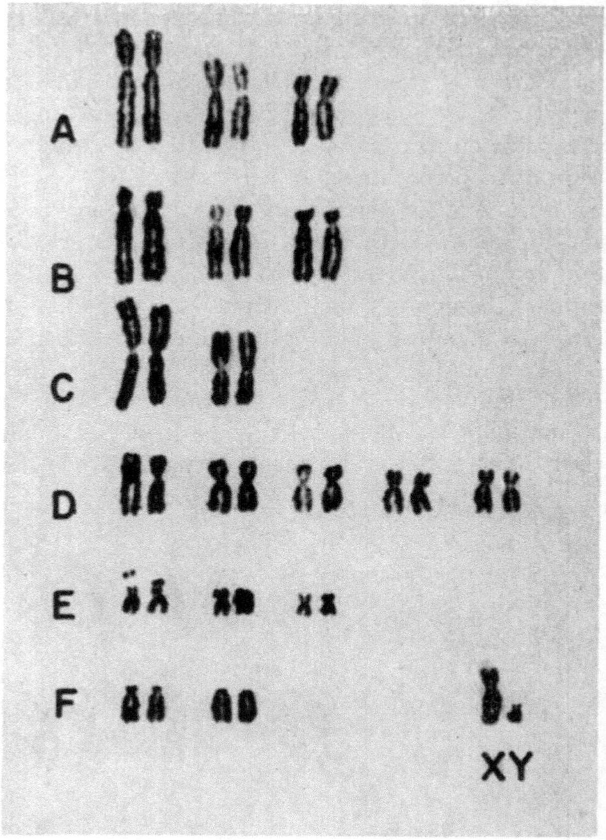

Fig. 14. Karyotype of a male clouded leopard, *Neofelis nebulosa.*

reduction of acrocentrics in this southern continent. These two species represent different genera from *Canis* but are, nevertheless, karyotypically "dogs." There are, however, three species representing three different genera, belonging to an entirely different subfamily from the other dogs. This is the subfamily Simocyoninae members of which were fossil in Europe and North America as early as the Oligocene. The ancient forms were apparently the earliest members of the canid stock that went south. Two of the three extant species have been studied so far: the hunting dog (*Lycaon pictus*) of Africa (Fig. 15) has 2n = 78 and a karyotype identical to all the members of the genus *Canis,* and the South American bush dog (*Speothos venaticus*) has a reduced diploid number of 74 (Fig. 16).

The foxes are a group of six or seven genera represented by widely divergent karyotype patterns and diploid numbers. However, their fundamental numbers, with the exception of the arctic fox, all fall

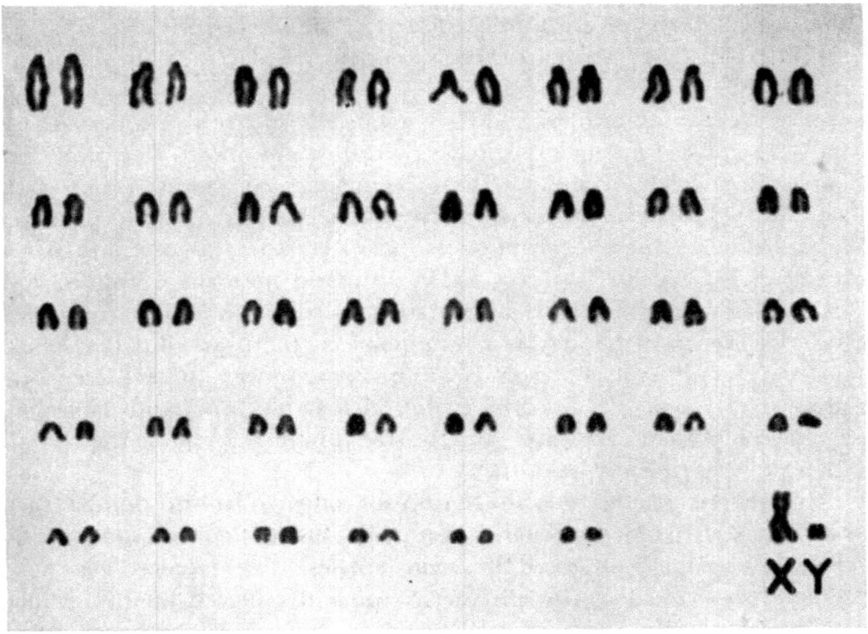

Fig. 15. Karyotype of a male hunting dog, *Lycaon pictus.*

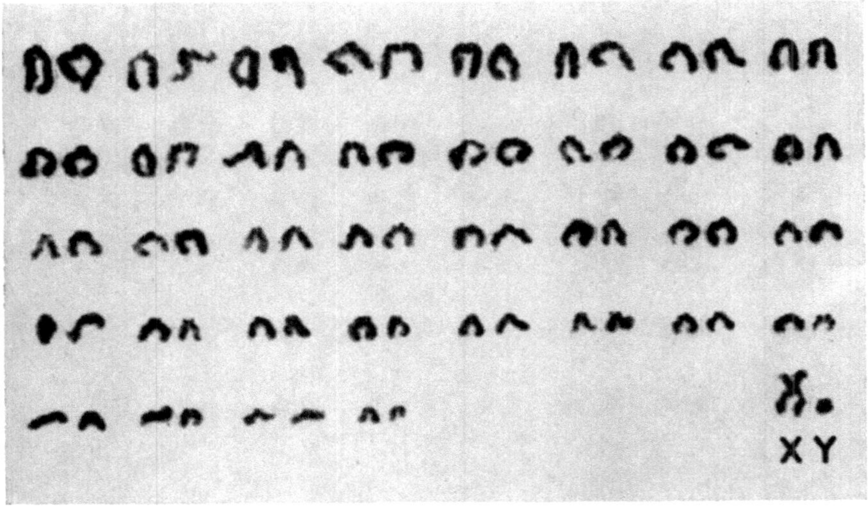

Fig. 16. Karyotype of male bush dog, *Speothos venaticus.*

within the range of 70–76, as distinct from a figure of 80 for the dogs. Some of the foxes also possess markers which do not appear in any of the dogs. The foxes are thus indicated to be an actively evolving, unstable group that, cytogenetically, should merit placement in a separate subfamily.

Most bears are northern and have essentially identical karyotypes with $2n = 74$ and a fundamental number of about 84 to 88. A typical bear karyotype is that of the sun bear (*Helarctos malayanus*) (Fig. 17) of southern Asia. There are no bears in Africa. The only bear to penetrate far into the southern hemisphere is the spectacled bear of South America (Fig. 18) which has a very reduced diploid number of only 52, but its fundamental number is 84 like that of all its northern cousins. It has, however, not the marker chromosome of the bears, but that of the raccoon family, the Procyonidae. This bear may thus represent the product of a type of karyotype evolution that the procyonids have long ago accomplished, and may also be the product of an earlier origin than the other New World ursids.

Cytogenetic studies encourage us not only to try to follow these phylogenetic parades through history but also to look at some of the difficult relationships posed by some species. The raccoon dog (*Nyctereutes procyonoides*), for instance, is presently placed in the caninae. Its diploid number was reported by Minouchi in 1929 to be 42. I can now confirm this number through the recent study of a male and female specimen. The fundamental number is 68 which is similar to that of the raccoon family and the foxes. The karyotype pattern (Fig. 19) is similar

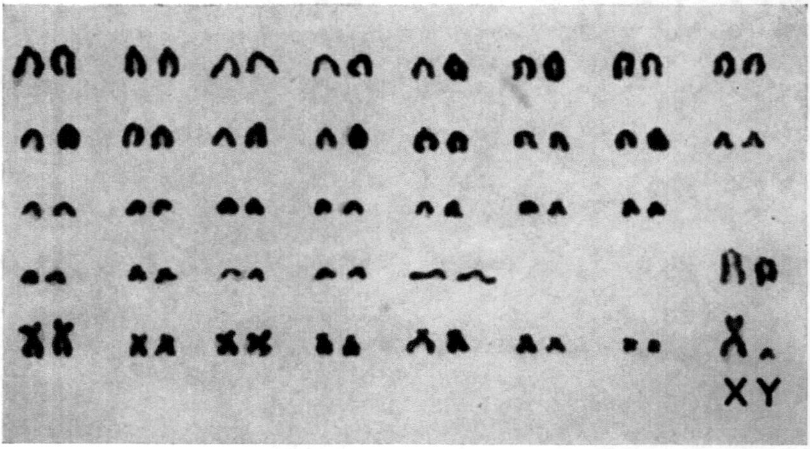

Fig. 17. Karyotype of a male sun bear, *Helarctos malayanus*. The marker chromosomes are the large acrocentrics immediately above the sex chromosomes.

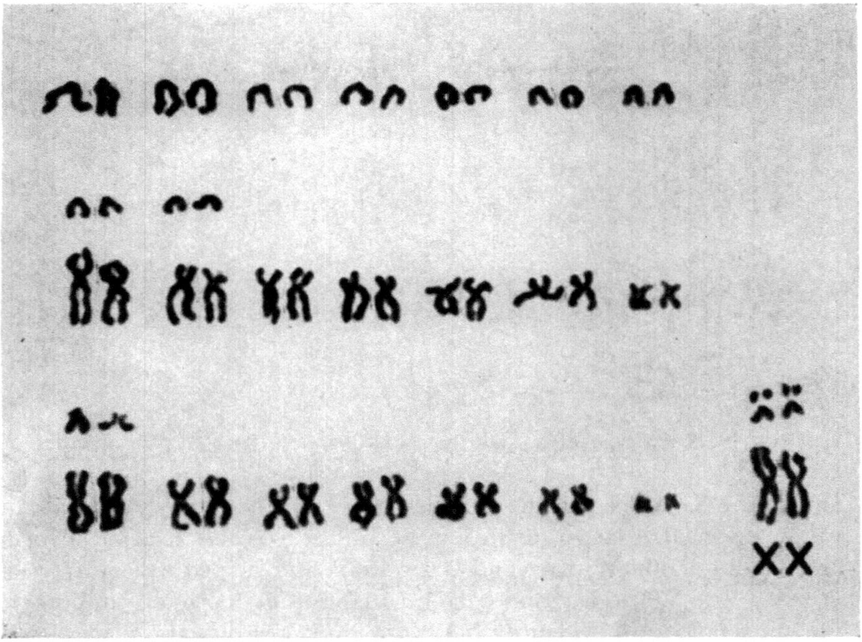

Fig. 18. Karyotype of a male spectacled bear, *Tremarctos ornatus*.

to nothing else. It has 14 acrocentric autosomes while the foxes have mainly large numbers of acrocentrics and the Procyonidae have only six to eight. If 24 of its 26 metacentric autosomes were envisaged as split into 48 acrocentrics, it would then have 62 acrocentric and 2 metacentric autosomes and, with the exception of the sex chromosomes, have a karyotype similar to the grey fox (*Urocyon cinereoargenteus*) or, if all of its acrocentric autosomes were to fuse, it would then have a diploid number and karyotype similar to that of the red fox (*Vulpes vulpes*) (Wurster and Benirschke, 1968). It has, however, a large acrocentric X chromosome and this is unlike the metacentric X chromosome of all the Canoidea, and in fact, of all the Carnivora thus far studied. The Y is a small satellited acrocentric and one of the smallest of the submetacentric autosomes and is satellited on the short arms. These features are difficult to show and are apparent in relatively few spreads. This animal has a unique karyotype. In this case the karyotype does not help us to place it taxonomically other than to indicate that it is very different from other members of its subfamily. It is easy to envisage it, however, as a fox representing karyotypically just another member of this widely divergent group.

The aardwolf (*Proteles cristatus*, 2n = 40) has been variously placed

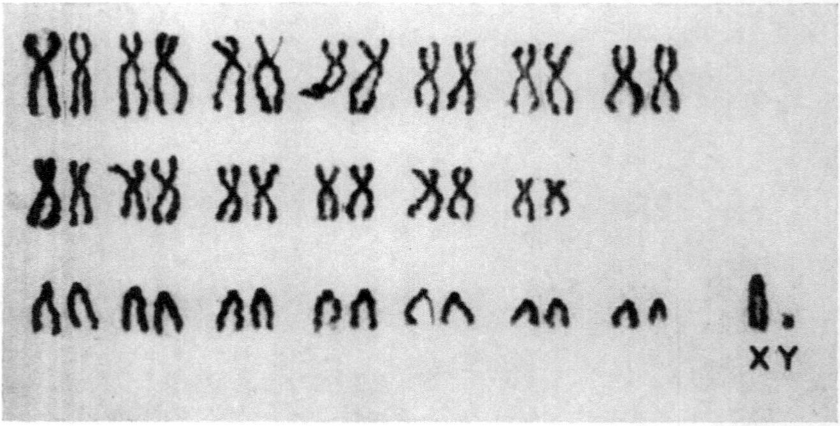

Fig. 19. Karyotype of a male raccoon dog, *Nyctereutes procyonoides.*

among the Canoidea and the Feloidea. Since its karyotype has been found to be essentially identical with that of the spotted hyaena (*Crocuta crocuta,* 2n = 40) (Wurster and Gray, 1967) and striped hyaena (*Hyaena hyaena,* 2n = 40) (Hsu and Arrighi, 1966) this is good support for the morphological evidence that it is properly placed in the family Hyaenidae. The giant panda has been variously assigned to the Ursidae and the Procyonidae. It was of interest when its karyotype (Newnham and Davidson, 1966) was found to be intermediate in character with 2n = 42, a fundamental number (78) close to that of the bears, but a marker similar to that of the procyonids. The Procyoninae were fossil in North America in the Lower Miocene. They migrated to South America but it is questionable whether any ever marched westward to Asia. The Ailurinae were fossil in Europe in the Upper Miocene and migrated southeastward into Asia. These two subfamilies of the Procyonidae are of dubious relationship and may represent cytogenetic parallel evolution from early canid or miacid stock. The lesser panda (Fig. 20) with 2n = 36 has a karyotype that is almost indistinguishable from that of the clouded leopard of Asia except that it has one pair less in the F group. This is not to suggest that it is related to cats, but to show the parallelism to the new world Procyoninae which also show karyotypic similarities to cats. It is, however, also of interest that the lesser panda has semi-retractile claws, a feline feature from which it derives its common name of cat-bear and which is possessed by no other member of the Procyonidae. From the karyotypic evidence and fossil history I would speculate that the lesser panda is more ancient than the giant panda, that the giant panda arose as an early offshoot of the ursids or from the same stock just prior to their origin, and that the two species are no more closely related than the New World Ursidae and Procyonidae and thus do not belong in the same family. Classifications, however, cannot and should not be based upon one parameter only, and I speak thus only from a cytogenetic point of view.

In summary, the Order Carnivora is large, geographically widespread, and cytogenetically diverse. It had its origin in the northern hemisphere and migration of its species occurred back and forth between the Old and New Worlds. Southward migrations were secondary and tended to be accompanied by a reduction in the diploid number, and this is especially remarkable in South American species. A marker chromosome is a presumably primitive and stable trait that exists throughout the Order and shows no modifications correlative with migratory patterns. Cytogenetic studies can be useful in determining difficult relationships and are an added parameter which may pose many new questions, in phylogenetic and taxonomic investigations.

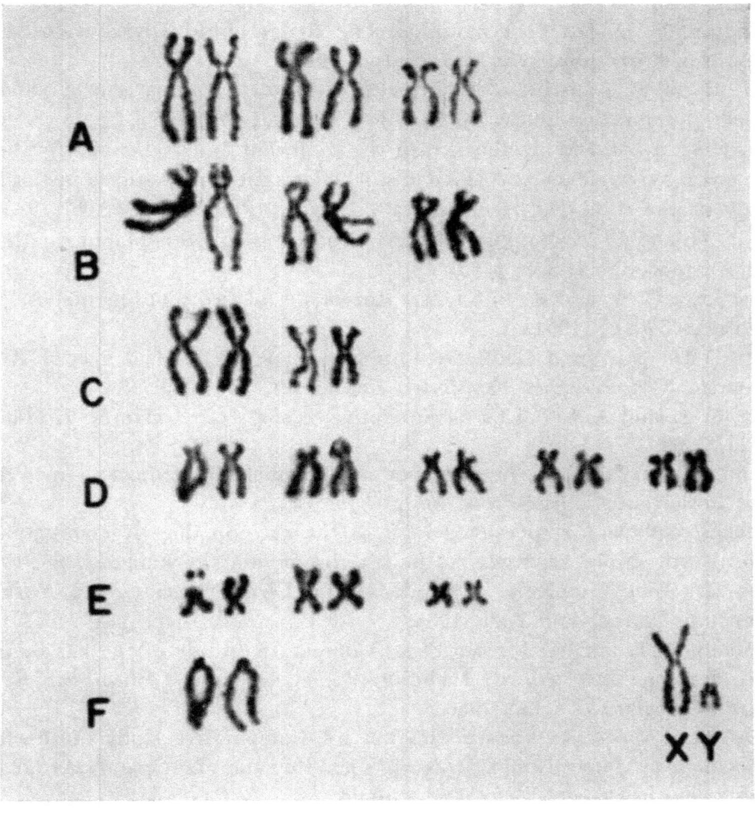

Fig. 20. Karyotype of a male lesser panda, *Ailurus fulgens*.

Acknowledgments

For skin biopsies I am grateful to Dr. Clinton Gray of the National Zoological Park in Washington, D.C., to Dr. B. W. O'Gara of the Montana Cooperative Wildlife Research Unit in Missoula, Montana, and to Mr. John Craighead, the National Science Foundation Grizzly Bear Research Project for tissues from the grizzly bear. For technical assistance I thank Mary Layton, Katalin Imredy, and Anne Babcock. This study was supported by a grant from The National Institutes of Health, GM 10210.

References

Colbert, E. H.: Evolution of the Vertebrates. New York: John Wiley & Sons, Inc., 1955.

Corfman, P. A. and R. M. Richart: Chromosomes of the ring seal. Nature *204:* 502, 1964.

Darlington, P. J., Jr.: Zoogeography: The Geographical Distribution of Animals. New York: John Wiley & Sons, Inc., 1957.

Fay, F. H., V. R. Rausch and E. T. Feltz: Cytogenetic comparison of some pinnipeds (Mammalia: Eutheria). Canad. J. Zool. *45:*773, 1966.

Gregory, W. K. and M. Hellman: On the evolution and major classification of the civets (Viverridae) and allied fossil and recent Carnivora: a phylogenetic study of the skull and dentition. Proc. Amer. Phil. Soc. *81:*309, 1939.

Hsu, T. C. and F. E. Arrighi: Karyotypes of 13 carnivores. Mammal. Chromosomes Newsletter *21:*155, 1966.

Hungerford, D. A. and R. L. Snyder: Karyotypes of two more mammals. Amer. Naturalist *98:*125, 1964.

Jones, T. C.: San Juan Conference on Karyotype of Felidae, Special Report. Mammal. Chromosomes Newsletter No. *15:*121, 1965.

Leone, C. A. and A. L. Wiens: Comparative serology of carnivores. J. Mammal. *37:*11, 1956.

Makino, S.: An Atlas of the Chromosome Numbers in Animals. Ames, Iowa: The Iowa State College Press, 1951.

Minouchi, O.: On the spermatogenesis of the raccoon dog (*Nyctereutes viverrinus*), with special reference to the sex chromosomes. Cytologia *1:*88, 1929.

Morris, D.: The Mammals. A Guide to the Living Species. New York and Evanston: Harper and Row, 1965.

Newnham, R. E. and W. M. Davidson: Comparative study of the karyotypes of several species in Carnivora including the giant panda (*Ailuropoda melanoleuca*). Cytogenetics *5:*152, 1966.

——— and ———: Author's addendum to: Comparative study of the karyotypes of several species in Carnivora, including the giant panda (*Ailuropoda melanoleuca*). Cytogenetics *6:*156, 1967.

Pauly, L. K. and H. R. Wolfe: Serological relationships among members of the order Carnivora. Zoologica *42:*159, 1957.

Romer, A. S.: Vertebrate Paleontology. Chicago and London: University of Chicago Press, 3rd ed., 1966.

Simpson, G. G.: Life of the Past. An Introduction to Paleontology. New Haven and London: Yale University Press, 1953.

Thenius, E. and H. Hofer: Stammesgeschichte der Säugetiere. Eine Uebersicht über Tatsachen und Probleme der Evolution der Säugetiere. Springer-Verlag, Berlin, 1960.

Wurster, D. H. and C. Gray: The chromosomes of the spotted hyena, *Crocuta crocuta*. Mammal. Chromosomes Newsletter *8*:197, 1967.

——— and K. Benirschke: Comparative cytogenetic studies in the order Carnivora. Chromosoma *24*:336, 1968.

CYTOGENETIC STUDIES OF SOME ARMADILLOS

K. Benirschke, R. J. Low and V. H. Ferm
Department of Pathology, Dartmouth Medical School, Hanover, New Hampshire

DEDICATED TO PROFESSOR F. A. SAEZ IN HONOR OF HIS 70TH BIRTHDAY

Introduction

Of the numerous edentates, most of which inhabit South America, only few species have been studied cytologically so far. The reason for this dearth of knowledge is partly the fact that the energies of many cytogeneticists have been absorbed by studies of the human karyotype and partly due to the unavailability of most of these species in zoos because they make difficult specimens to exhibit. For this reason relatively few edentates are kept in zoological gardens and, furthermore, these animals have rarely been used for laboratory research work. There are many interesting, biological aspects in this Order, however, that merit further study, among which are their karyotypes and genetic relationships.

Edentates have been classified in many different ways; the most notable taxonomic summaries are the works of Yepes (1928), Cabrera (1957) and Talmage and Buchanan (1954). A convenient and perhaps oversimplified grouping by Morris (1965) lists four species of anteaters, seven species of sloths, and 21 species of armadillos. The last family in particular has been of perhaps the greatest interest to taxonomists and it has also presented many problems. Indeed, the validity of specific rank for many species listed by the above authors can be questioned and no modern study, employing populations, aided by the examination of large numbers of specimens, has been undertaken to clarify what is presently a difficult group. Perhaps it is here then where cytogenetics may be of use in delineating taxonomic relationships, as has been the case with some other families.

To date, only two species of armadillos have been studied by modern cytogenetic techniques, the "Mulita" (*Dasypus hybridus* Desmarest), a seven-banded armadillo, and the nine-banded armadillo (*Dasypus novemcinctus* Linn.). The karyotype of the former, including its meiosis, was described by Saez, Drets and Brum (1964) while the nine-banded species has been the subject of numerous studies in this laboratory. It is the purpose of this paper to review these findings briefly and to describe the karyotypes of three additional species of armadillos in detail, *Chaetophractus villosus* (from Argentina), *Euphractus sexcinctus* (from Brazil) and *Cabassous centralis* (from Colombia).

Dasypus hybridus and *Dasypus novemcinctus*

These two armadillos are remarkable in that they are the only species with regular polyembryony which, coupled with their physical characteristics, suggests close taxonomic relationship. Both possess very similar chromosomes with $2n = 64$ and nearly identical karyotypes (Figs. 1 and 2). The karyotypes studied by Professor Saez were prepared from bone marrow, while we have studied embryonic, adult kidney, skin and lymphocyte cultures. The most notable karyotypic differences are:

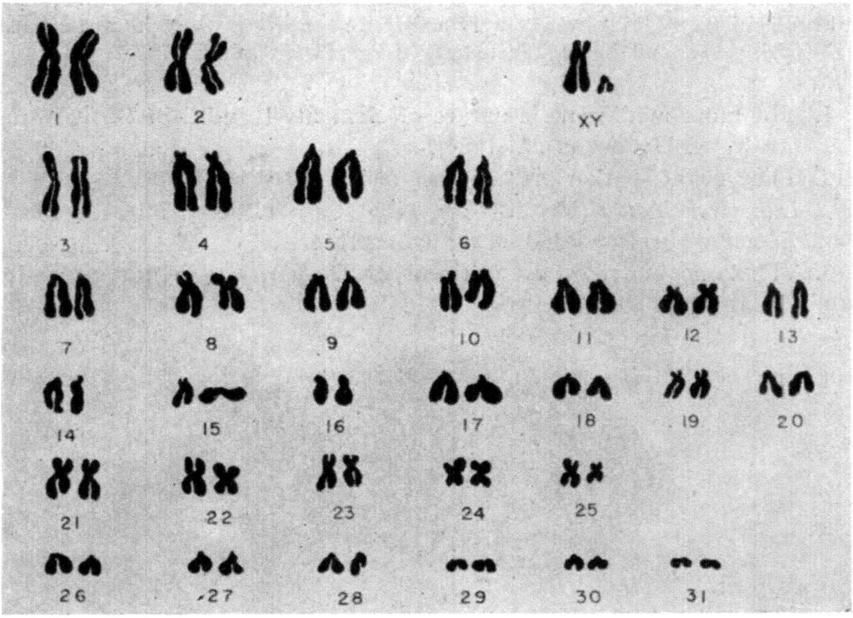

Fig. 1. Karyotype of *Dasypus hybridus* ($2n = 64$). This preparation comes from Saez *et al.* (1964) and is here reproduced with permission of Pergamon Press.

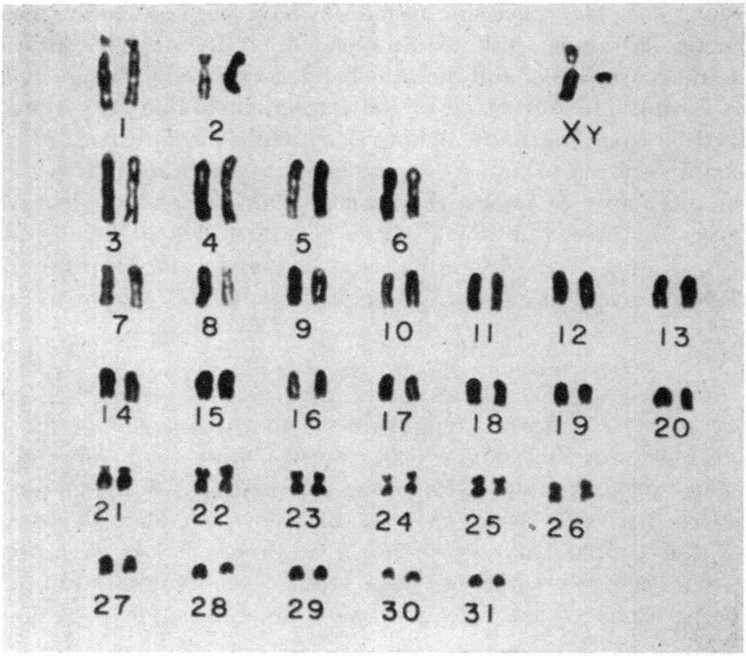

Fig. 2. Karyotype of male *Dasypus novemcinctus* from Texas (2n = 64). Kidney culture. Although faint satellites were observed at microscopy on the metacentric pair #26, these were impossible to reproduce photographically.

1. The more mediocentric nature of elements 1 and 2 in *D. hybridus* (as drawn by Saez *et al.,* 1964).
2. The possibility that one member of the small metacentric group in *D. novemcinctus* (Nos. 21–26) might be considered an acrocentric element (here counted as an acrocentric).
3. The bone marrow preparations of *D. hybridus* do not show the

Fig. 3. *Dasypus novemcinctus* (Texas). Indistinguishable also from specimen obtained from Florida, USA.

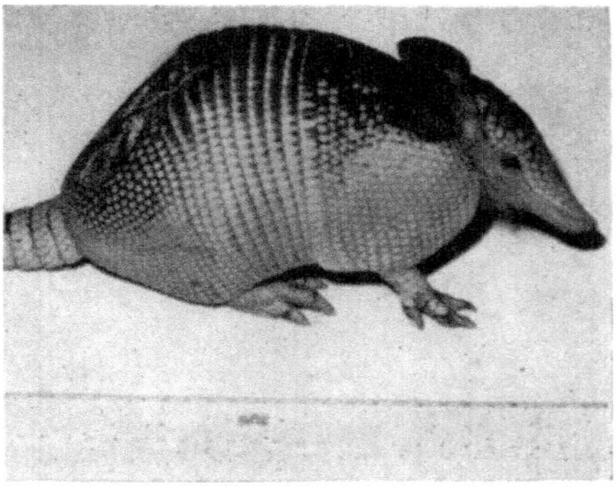

Fig. 4. *Dasypus novemcinctus* SA (Colombia). Compare pigment distribution, ear length and head: body ratio with specimen from Texas in Fig. 3.

conspicuous satellited small metacentric pair (#26) found in *D. novemcinctus*. In view of the fact, however, that this finding is also made only in our best tissue culture preparations and because bone marrow chromosomes tend to be contracted, one must reserve judgment on this discrepancy. These are details which need further study when additional seven-banded armadillos become available and when these can be studied by different techniques.

Our specimens of *D. novemcinctus* have almost all come from Texas and Florida. Between 40 and 50 such specimens have been examined cytologically in this laboratory and all have had identical karyotypes, including the typical sex chromosome dimorphism. In female cells, one X is late replicating, forms a Barr body at the interphase state and the "drumstick" of leukocytes (Grinberg *et al.,* 1966). There are remarkable phenotypic differences in this species, however, and we have been interested in cytogenetic studies of specimens from different areas. The senior author saw (at the National Museum in Rio de Janeiro) the shell of a *D. novemcinctus* shot near the Amazon river. It was twice as large as the largest specimen obtained from Texas. In contrast, four mature specimens of nine-banded armadillos obtained in 1962 from Leticia, Colombia, differed in many respects (Figs. 3–8, Table 1). Externally,

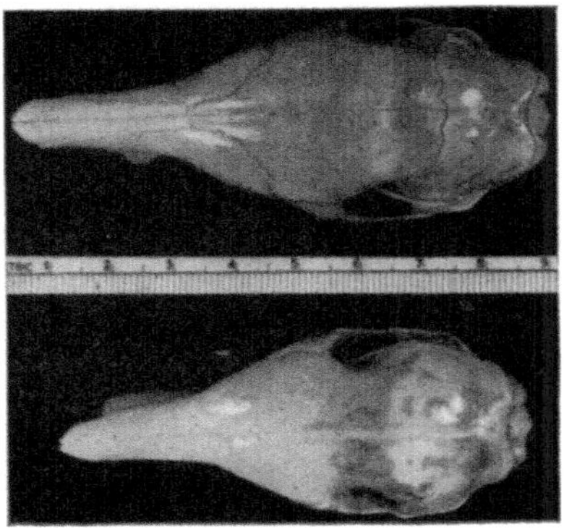

Fig. 5. Dorsal view of skulls. Above: Texas variant. Below: Colombia variant. Note frontal air sinuses and marrow cavity.

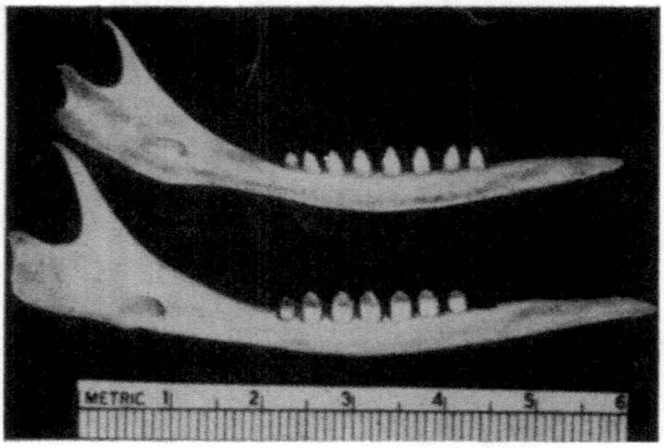

Fig. 6. Medial view of mandibles. Above: Texas variant. Below: Colombia variant. Note coronoid process, mandibular notch and tooth configuration.

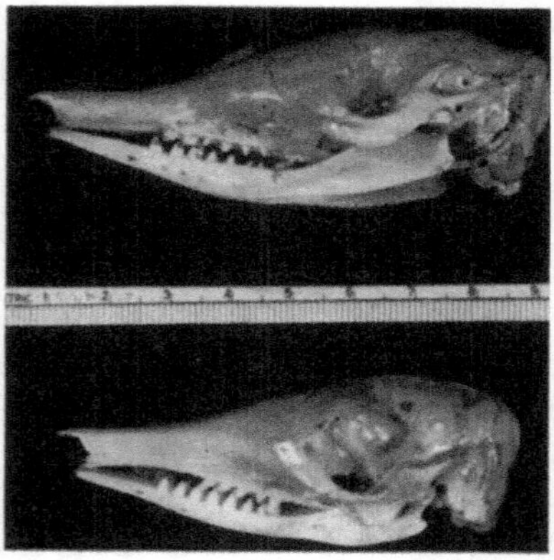

Fig. 7. Lateral view of skulls. Above: Texas variant. Below: Colombia variant. Note zygomatic arches, mandibular coronoid processes.

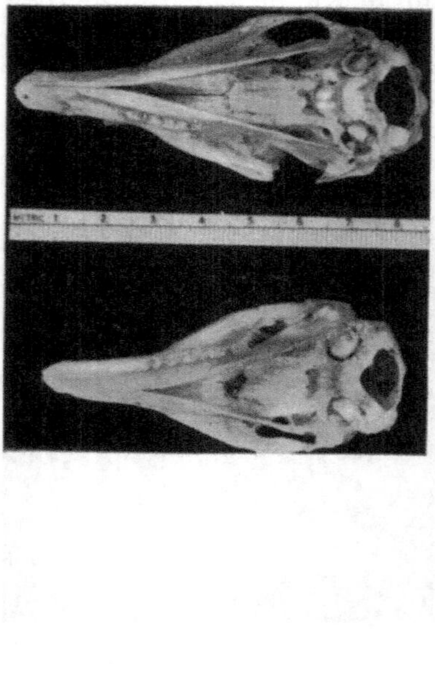

Fig. 8. Ventral view of skulls. Above: Texas variant. Below: Colombia variant.

Table 1: Comparison of Morphological Features of North and South American Nine-Banded Armadillos

Feature	North American (Texas)	Colombian
Adult Weight (Kg)	3–4	1.2–1.4
Bands	9	9
Scutes in Bands (Total)	Two groupings: 520 and 560	522–525
Pigment Distribution	Generalized distribution over entire shell, head, extremities, tail (Fig. 3)	Pigment restricted to dorsum of head, shell, tail (Fig. 4)
Head:Body Ratio	.41	.33
Ear:Head Ratio	.41	.59
Tail:Total Ratio	.47	.40
Mandibles	Prominent coronoid process and mandibular notch; medial projection of mandibular head (Fig. 6)	Posterior projection at angle of mandible; vertical projection of mandibular head (Fig. 6)
Dental Formula	$\frac{7\text{–}8}{7\text{–}8}$	$\frac{8}{8}$
Tooth Shape	Square	Cone
Frontal Bones	Air sinuses occupy 80% of frontal bone; marrow cavity reduced proportionately (Fig. 5)	Air sinuses occupy 50% of frontal bone; marrow cavity prominent (Fig. 5)
Zygomatic Arches	Short and widened at anterior margin (Fig. 7)	Uniform contour (Fig. 7)
Foramen Magnum	Round	Oval

these specimens were smaller and less pigmented and they exhibited certain differences in skull structure. Chromosome preparations from lymphocyte and kidney cultures of these four specimens also differed slightly from all Texan armadillos in that chromosome pairs 11, 18 and 27 had definite short arms, in contrast to their acrocentric counterparts of specimens from Texas (Fig. 9). While it is appreciated now that such minor structural karyotypic polymorphism (?pericentric inversions) is found in many families, they might become more important when other specimens from different South American locations are compared with their relatively isolated North American relatives.

Euphractus sexcinctus, the Six-Banded Armadillo

Two male and one female animals were studied; unfortunately only the males' skin biopsies could be processed to complete karyotypes, as the female specimen was contaminated by fungi. The remarkable difference

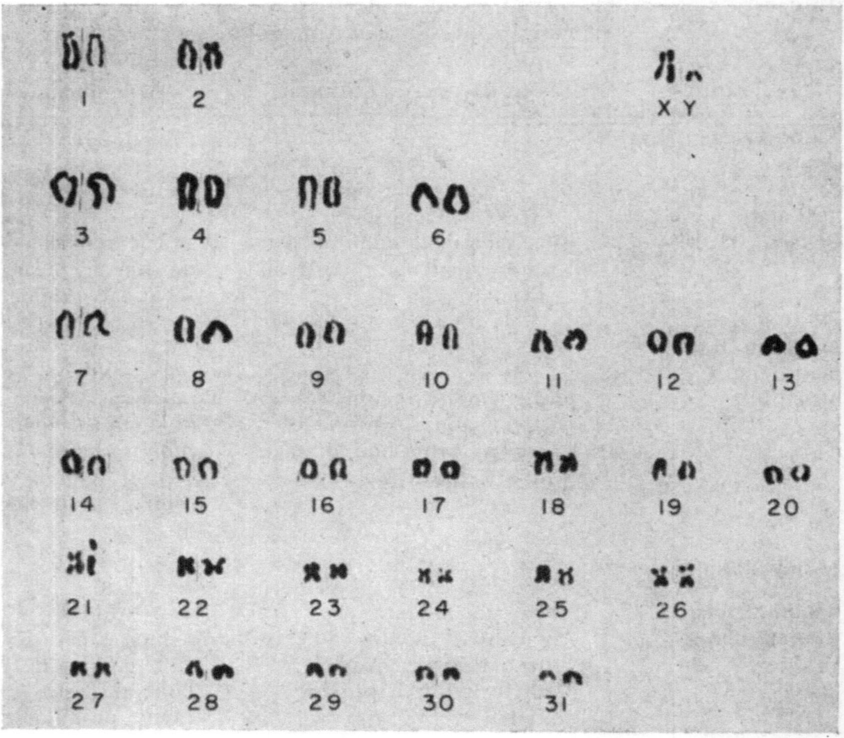

Fig. 9. Karyotype of *Dasypus novemcinctus* from Colombia of specimens shown in Figs. 3–8. 2n = 64. Chromosomes 11, 18, 27 have definite short arms as contrasted to the acrocentric elements in the Texas variant (Fig. 2).

of the bicornuate uterus of this specimen, as opposed to that of the Dasypodini with a fused, single-bodied corpus, is shown in Figure 10, a finding rarely commented upon by other authors.

The skin biopsies of the male specimens were kindly provided by Dr. A. Coimbra Filho of the Zoological Gardens of Rio de Janeiro and they were processed by the Basrur technique (1963). The modal number of chromosomes in 27 perfect metaphase plates was found to be 2n = 58. A karyotype of this species is shown in Figure 11 and it can be seen that it differs greatly from the two Dasypus species studied. Only six pairs of chromosomes are acrocentrics, the remainder being metacentrics or submetacentrics. Difficulties exist in assigning pair #22 to either submetacentrics or acrocentrics. A definite short arm can be seen in all metaphases and some authors may refer to this element as a subacrocentric chromosome. For the purpose of calculating the nombre fondamental (NF), this element is assigned two arms. The X chromosome is difficult to distinguish unequivocally and its representation in this karyo-

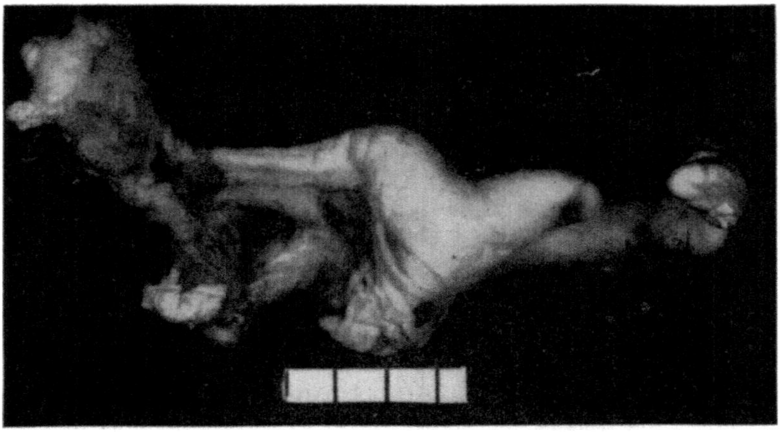

Fig. 10. Bicornuate uterus of *Euphractus sexcinctus.* Ruler is in mm.

type is tentative. It is of the "usual" variety (Ohno *et al.,* 1964). The Y chromosome is minute and, in the clearest spreads available, it appears to be a submetacentric element.

Chaetophractus villosus Desmarest, the Hairy Armadillo, or Peludo

Two female specimens were available (Fig. 12). The first was caught in the Argentinian pampas, approximately 400 km west of Buenos Aires and the best preparations became available from skin cultures of this animal. The second specimen was obtained in Uruguay through the

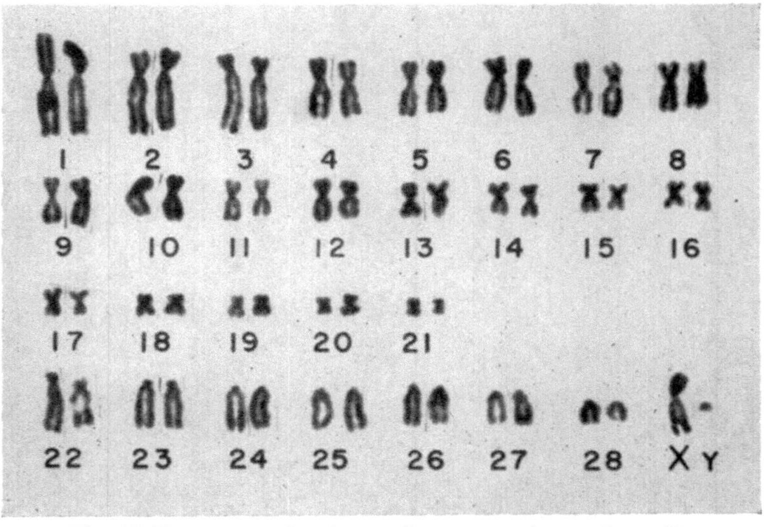

Fig. 11. Karyotype of male *Euphractus sexcinctus.* 2n = 58.

Fig. 12. *Chaetophractus villosus* (Argentina) at top, *Euphractus sexcinctus* (Brazil) at bottom.

courtesy of Dr. L. Branda, Montevideo. Numerous metaphase plates showed a uniform modal number of 2n = 60 with a karyotype as displayed in Figure 13. As can be seen, 14 pairs of chromosomes are clearly acrocentrics, the remainder being metacentrics or submetacentrics. The sex chromosomes cannot be identified as yet although a study with tritiated thymidine would be likely to delineate the X inasmuch as typical Barr bodies were found on routine histologic sections. These

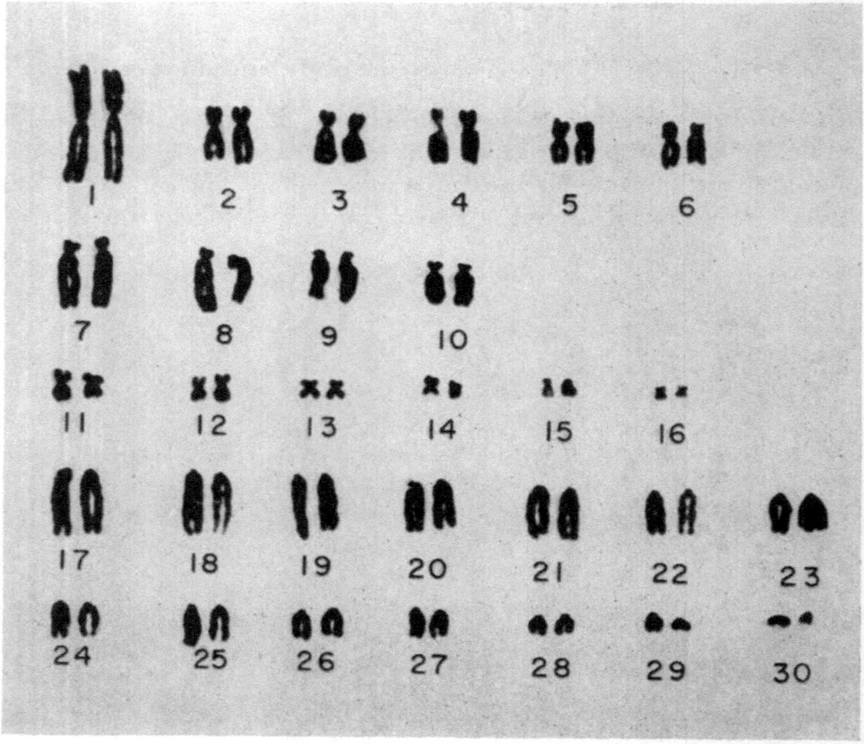

Fig. 13. Karyotype of female *Chaetophractus villosus,* 2n = 60. The X-chromosomes cannot be identified. Skin culture.

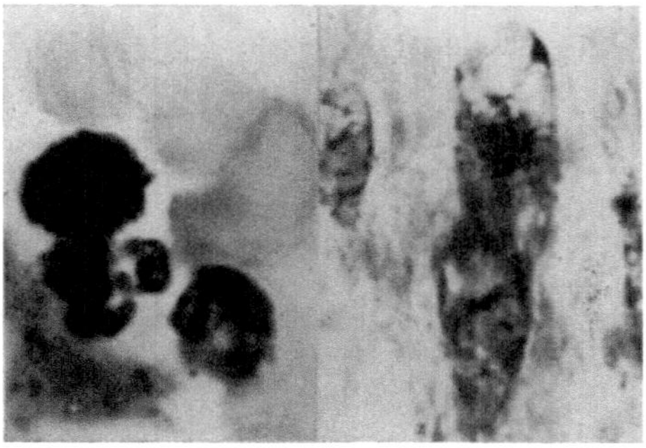

Fig. 14. Barr body (smooth muscle) and drumstick (polymorphonuclear leukocyte) in *Chaetophractus villosus* (H & E, Giemsa × 1200).

were present most prominently in smooth muscle cells, and "drumsticks" were found in a small percentage (2–3%) of polymorphonuclear leukocytes on blood smears (Fig. 14).

Cabassous centralis Miller, Soft-Tailed Armadillo

Tissue cultures of two males and one female were started from skin and kidneys after their arrival dead following shipment from Florida. One of each sex succeeded and excellent preparations were obtained which were labeled with tritiated thymidine for the identification of the X chromosome. These animals were imported from Colombia.

It is not clear from the literature whether truly more than one species of soft-tailed armadillo exists or whether the numerous separate species described in the literature and possessing very similar phenotypes should be considered subspecies. Morris (1965) lists *C. unicinctus* and *C. centralis,* while Cabrera (1957) lists many additional species, but he is not certain whether *C. centralis* is a good species. Only few specimens have been studied in detail by taxonomists and those only by their scute and band pattern, as well as their size and skull structure. Considering the origin of our specimens from Colombia, their smaller size than that assigned to *C. unicinctus,* and their bone structure (Fig. 15), our specimens appear to be best classified as *C. lugubris,* as described and depicted by Gray, 1873, and by Frechkop and Yepes (1949). Cabrera (1957) considers these as a form of *C. tatouay* and Miller and Miller (1899), who write specifically about this group, separate this smaller form as *C. centralis,* which name is here employed. The difficulties of taxonomy are especially obvious in these specimens and it is for these reasons that a photographic record is supplied. Also, it is suggested that karyologic comparisons in this group may help in simplifying the confusing and seemingly arbitrary nomenclature currently employed. Our female specimen also possessed a bicornuate uterus. All animals were mature as judged by the histologic picture of the gonads. Typical Barr bodies were present in many tissues and easiest

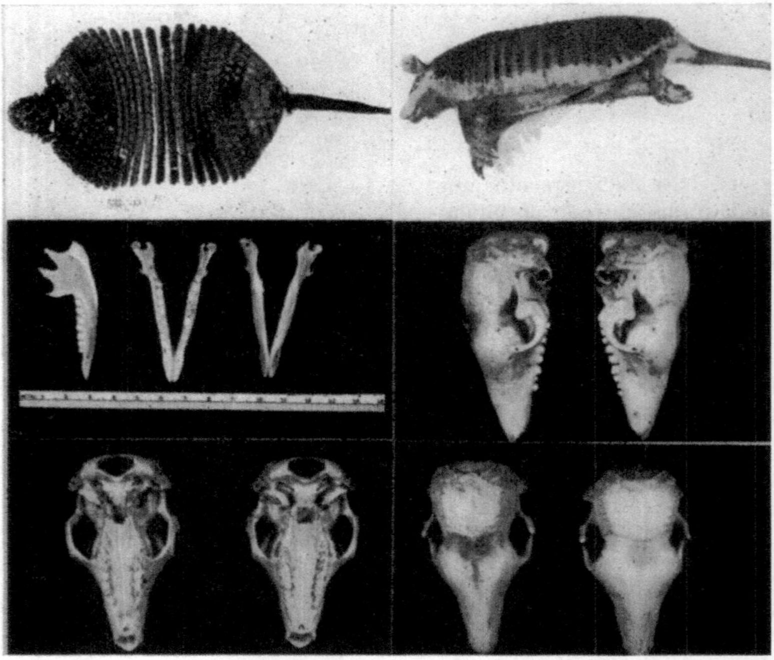

Fig. 15. Male (left) and female (right) *Cabassous centralis* from Colombia. Ruler applies only to photographs of bones.

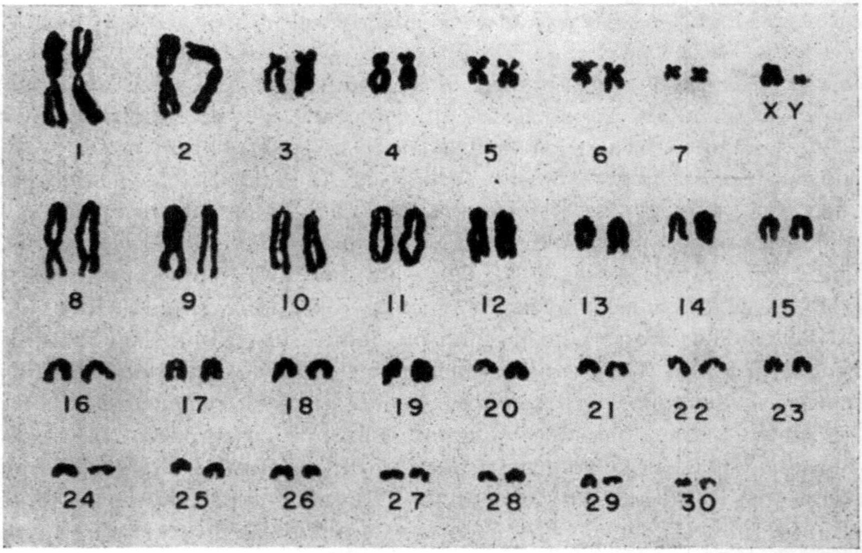

Fig. 16. Karyotype of male *Cabassous centralis*. 2n = 62. Kidney culture. Note unusual length of autosomes.

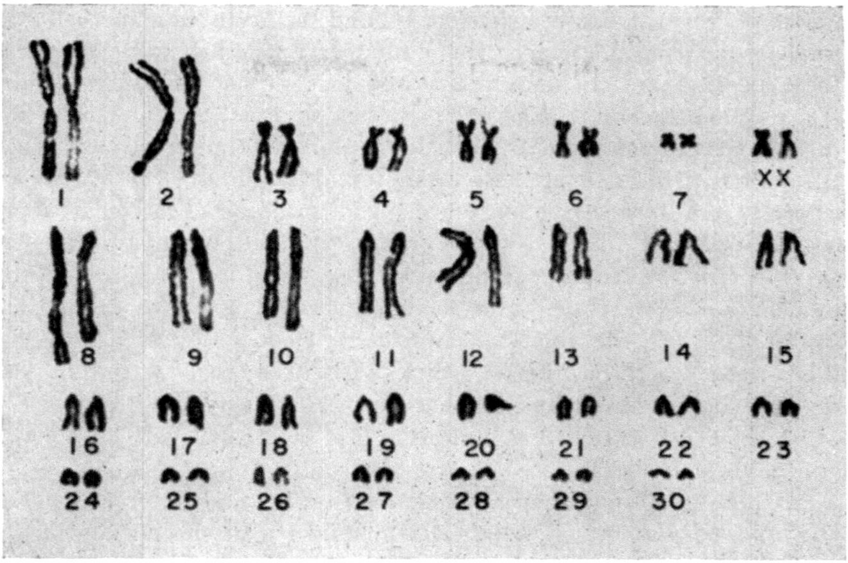

Fig. 17. Karyotype of female *Cabassous centralis.* 2n = 62.

identified in the smooth muscle cells of the uterus. The diploid number in 100 cells of both specimens was found to be 62. Of these, 14 are metacentric or submetacentric autosomes and 46 are acrocentrics. The X chromosome is a small metacentric element, the Y appears to be one of the smallest acrocentrics. No typical satellites were encountered on any chromosome and there were no unusual associations. It was striking to find these chromosomes to be of unusual length when compared with our usual preparations, and several secondary constrictions were observed, particularly among the larger metacentrics (Figs. 16 and 17).

Discussion

This study has contributed some additional knowledge on the karyology of two further tribes of Dasypodidae. The peludo and the six-banded armadillo of the tribe Euphractini were found to posses 60 and 58 chromosomes respectively and, as might be expected, their karyotypes were more similar to one another than those of other tribes. The bare-tailed armadillo of the Priodontini with 62 chromosomes has more acrocentric elements than either of the former and, for this reason, it may be considered to possess a more primitive karyotype. It will be interesting to see whether the giant armadillo has similar features when specimens become available for study.

The sex chromosomes of four of the five Dasypodidae so far studied are similar, the X being of the "original" type (Ohno *et al.*, 1964), a metacentric or submetacentric element, and the Y chromosome being the smallest element and acrocentric or submetacentric in nature. In peludo these elements are not identified as yet since we had only female specimens available. The females of all species possessed typical Barr bodies in interphase nuclei and drumsticks in polymorphonuclear leukocytes.

Satellited chromosomes have been found so far only in *D. novemcinctus*. Here they are present on one arm of the small metacentric pair of #26 and only when excellent preparations are available. Moreover, they are difficult to photograph and, often, they are bent over so that they become superimposed onto the main chromosome arms. In no other species have we been able to identify these elements but it is suspected that further search may be rewarding, particularly in *D. hybridus*. Also, we have not observed any associations of the two satellited elements which might be construed as indicating that they participate in organization of the nucleolus.

Matthey (1945) has suggested that the fundamental number (NF), i.e., the number of major chromosome arms of a female diploid set, may be useful in ascertaining taxonomic relationships of mammalian families. Such might be expected, particularly if Robertsonian fusion (centric fusion of acrocentrics to produce metacentrics) were a prominent mechanism of karyotype evolution in a particular family. Indeed, amongst Bovidae the NF is remarkably constant over a wide range of species (Wurster and Benirschke, 1967, 1968) while in Equidae this relationship is less convincing (Benirschke, 1967). The NF of the species described here is: Cabassous 78, Chaetophractus 92, Euphractus 104, and Dasypus 82–88. These figures reveal no apparent trend and further species must be examined before a decision can be made as to which of various mechanisms might be operative in the karyotype evolution of this Order.

A closer examination of the karyotype structure of the various Dasypodini suggests the possibility that pericentric inversions may be a common event in this group. All specimens have been shown to have 64 chromosomes but the Colombian nine-banded specimens had several more submetacentrics which raises their NF from 82 (of the Texan race) to 88. Furthermore, the mulita of Uruguay, studied by Saez *et al.* (1964), possesses also more submetacentrics or subacrocentrics which makes their NF approximately 86. If taxonomy of armadillos in general is confused. that of the Dasypodini is seemingly arbitrary in particular and largely guided by the variable number of movable bands and differences in size. Hamlett (1939) has attempted to clarify the situation by reviewing the existing nomenclature up to 1939. He comes to the conclusion that *D. septemcinctus* and *D. hybridus* are distinct species. He refers to Linnaeus who previously had raised the question as to whether the number of movable bands was sufficient to separate as distinct species *D. novemcinctus* from *D. septemcinctus*. As pointed out in Hamlett's paper, rather wide variations exist not only in the number of movable bands but also in the size, configuration of head and ears, and in body coloration of these various species. Nevertheless, Hamlett feels the need to separate them into five different species (*D. novemcinctus, septemcinctus, hybridus, kappleri* and *mazzai*). All authors agree that close affinities exist between these forms, not only phenotypically but also physiologically. They all possess the same unique reproductive mechanism with a globular uterus and polyembryony (with 2–14 offspring) (Ihring, 1886; Buchanan, 1957) and all have 64 chromosomes with great karyotypic similarities. Meiosis of *D. hybridus* (Saez *et al.*, 1964) is also very similar to that of *D. novemcinctus* (Devi, 1968). One can thus perhaps entertain the question again whether all the recognized Dasypodini may not be one species with numerous subspecies. Perhaps the various forms are merely races amongst which gene flow has not completely ceased but is limited because of the restrictions of habitat. The great similarities between these characters cited, particularly that of polyembryony and chromosome structure, suggest that this notion might be further tested by experimental hybridization.

To date only three other members of the Order Edentata have been studied for their karyotype, the giant anteater (*Myrmecophaga tridactyla*), the lesser anteater (*Tamandua tetradactyla*) (Hsu, 1965), and the two-toed sloth (*Choloepus hoffmanni*) (Corin-Frederic, 1968). Their diploid numbers were 60 and 54 for the anteaters respectively. In the latter species all chromosomes are biarmed, while the giant anteater possesses six pairs of acrocentrics. There is then no obvious close similarity between the karyotypes of the two Myrcemophagidae studied and the five examined Dasypodidae. The two-toed sloth has in the female 49 (X0) and in the male 49 (XY) with the Y translocated to an autosome.

Summary

Very few members of the Order Edentata have been studied for their chromosome complement by modern cytological techniques. We review here this work (two species of anteaters, one sloth and two armadillos) and report the findings on three additional species of the Dasypodidae. *Euphractus sexcinctus* has 58 chromosomes, *Chaetophractus villosus* has 60 chromosomes and *Cabassous centralis* was found to have 62 chromosomes. These findings are described in detail and all edentate chromosomes are compared with each other. In addition, possible chromosomal polymorphism is described in Dasypodini and their speciation is briefly considered.

Addendum

Since this conference the results of chromosome studies on two giant armadillos (*Priodontes giganteus* Geoffroy) have become known. R. L. Snyder and D. A. Hungerford studied a male specimen in the Philadelphia Zoo, we studied a male specimen in the Antwerp Zoo with the same results: 2n = 50, 24 acrocentrics, 26 metacentrics, NF 76, X metacentric, Y small metacentric.

Acknowledgment

We acknowledge gratefully the support by a grant from the National Institutes of Health, GM 10210, and a travel grant from the United Health Research Foundation.

References

Basrur, P. K., V. R. Basrur and J. P. W. Gilman: A simple method for short term cultures from small biopsies. Exper. Cell Res. *30*:229, 1963.

Beath, M. M., K. Benirschke and L. E. Brownhill: The chromosomes of the nine-banded armadillo *Dasypus novemcinctus*. Chromosoma (Berl.) *13*:27, 1962.

Benirschke, K.: Sterility and fertility of interspecific mammalian hybrids. In: Comparative Aspects of Reproductive Failure. K. Benirschke, ed. Springer-Verlag New York, Inc., 1967, p. 218.

Buchanan, G. D.: Variation in litter size of nine-banded armadillos. J. Mammal. *38*:529, 1957.

Cabrera, A.: Catalogo de los Mamiferos de America del Sur. Vol. I. Rev. Mus. Arg. Cien. Nat. Serie Ciencias Biologicas Bernardino Rivadavia *4:* No. 1, 1957.

Corin-Frederic, J.: L'assortiment chromosomique du Paresseux: *Choloepus hoffmanni* Peters. Etablissement et étude détaillée du caryotype. Mise en évidence d'une formule gonosomique particulière et ses implications dans le mécanisme général de "dosage-compensation" chez les mammifères. pp. 144. Thèse Sc. Zool., Liège (Belgium), 1968.

Devi, M.: Meiotic chromosomes of the nine-banded armadillo, *Dasypus novemcinctus*. In preparation, 1968.

Frechkop, S. and J. Yepes: Etude systématique et zoogéographique des Dasypodidés conservés a l'institut. Bull. Inst. Roy. Sci. Nat. Belg. *25*:1, 1949.

Gray, J. E.: Handlist of the Edentate, Thick-skinned and Ruminant Mammals in the British Museum. British Museum, London, 1873.

Grinberg, M. A., M. M. Sullivan and K. Benirschke: Investigation with tritiated thymidine of the relationship between the sex chromosomes, sex chromatin, and the drumstick in the cells of the female nine-banded armadillo, *Dasypus novemcinctus*. Cytogenetics *5*:64, 1966.

Hamlett, G. W. D.: Identity of *Dasypus septemcinctus* Linnaeus with notes on some related species. J. Mammal. *20*:328, 1939.

Hsu, T. C.: Chromosomes of two species of anteaters. Mammal. Chromosomes Newsletter *15*:107, 1965.

Ihring, H. v.: Über "Generationswechsel" bei Säugethieren. Arch. Physiol. 443, 1886.

Matthey, R.: L'évolution de la formule chromosomiale chez les vertébrés. Experientia *1*:50, 1945.

Miller, G. S. and G. S. Miller, Jr.: Notes on the naked-tailed armadillos. Proc. Biol. Soc. Wash. *13*:1, 1899.

Morris, D.: The Mammals. A Guide to the Living Species. New York: Harper & Row, 1965.

Ohno, S., W. Beçak and M. L. Beçak: X-autosome ratio and the behavior pattern of individual X-chromosomes in placental mammals. Chromosoma (Berl.) *15*:14, 1964.

Saez, F. A., M. E. Drets and N. Brum: The chromosomes of the mulita (*Dasypus hybidus* Desmarest): A mammalian Edentata of South America. In: Mammalian Cytogenetics and Related Problems in Radiobiology. New York: Pergamon Press, 1964, p. 163.

Talmage, R. V. and G. D. Buchanan: The Armadillo (*Dasypus novemcinctus*). A Review of its Natural History, Ecology, Anatomy and Reproductive Physiology. Rice Institute Pamphlet. Vol. *41*, No. 2, 1954.

Wurster, D. H. and K. Benirschke: Chromosome studies in some deer, the springbok, and the pronghorn, with notes on placentation in deer. Cytologia *32*:255, 1967.

——— and ———: Chromosome studies in the superfamily Bovoidea. Chromosoma (Berl.), *25*:152, 1968.

Yepes, J.: Los Edentata argentinos sistemática y distribución. Rev. Univ. Buenos Aires *24*:461, 1928.

ARTIODACTYL MAMMALS: THEIR CHROMOSOME CYTOLOGY IN RELATION TO PATTERNS OF EVOLUTION

K. M. Taylor, D. A. Hungerford and R. L. Snyder

Department of Biology, San Diego State College, San Diego, California; The Institute for Cancer Research, Fox Chase, Philadelphia, Penna.; Penrose Research Laboratory, Zoological Society of Philadelphia and Department of Pathology, University of Pennsylvania School of Medicine

The orders Artiodactyla, Carnivora and Primates comprise the three most successful and the most widespread groups of large mammals. There are 9 living families of artiodactyls, in which are represented more than 190 species, including such forms as pigs, camels, deer, giraffe, antelopes, cattle, sheep, and goats. This order has long occupied much of the world's land area, except for the oceanic islands and Australia. Even in the latter areas, some artiodactyls have been introduced by man. They are clearly the dominant plant-eating mammals of the present time.

Artiodactyls have had a long association with man, serving as his prey for food, clothing, and tools for thousands of years; at present they include almost all of man's domesticated animals.

There have been several major adaptive radiations within the order. Two of these are well represented in living families, the Cervidae with more than 40 species and the Bovidae with approximately 130 species, and several cytogenetic problems are available to be studied in this group.

If anything characterizes artiodactyl chromosomes, it is the evidence of widespread Robertsonian chromosome changes with reduction in chromosome numbers resulting from the union of non-homologous acrocentric chromosomes to form biarmed chromosomes. This type of change has occurred, seemingly independently, in three families, the Suidae, Cervidae, and Bovidae. In addition, there are interesting problems presented by the uniformity of karyotypes in one family, the Camelidae, and by some examples of sex chromosome evolution. Occasionally, there are some unexpected karyotypes, and there are a few major areas in which no chromosome studies have been published.

Suiformes

The modern artiodactyls which are most directly related to the earliest (Eocene) fossil artiodactyls belong to the suborder Suiformes, the pigs, peccaries and hippopotamuses. Apparently no studies have been accomplished on hippos, but two species of hogs (family Suidae) and one species of peccary (family Tayassuidae) have been karyotyped.

The wild boar and its domesticated form (*Sus scrofa*) has a diploid number (2n) of 38 and a fundamental number (NF) of 64 (see, for example, Gimenez-Martin and Lopez-Saez, 1966); the wart hog (*Phacochoerus aethiopicus*) has 2n = 34 and NF = 64 (Wallace and Fairall, 1967a). The difference between these two species is an example of probable Robertsonian change. Relevant in this connection is the work of McFee, Banner and Rary (1966) who have reported a case of Robertsonian fusion in both homozygous and heterozygous condition in a herd of wild pigs in Tennessee. These pigs were descended from European wild pigs loosed in this region about 1912. The wild pigs have 2n = 36 and 2n = 37, but the NF = 64 in both cases. The 2n = 37 pigs have one more metacentric, and the 2n = 36 pigs have two more metacentrics than the domestic pigs, some of which are feral in this area and undoubtedly interbreed with the European wild pigs. The collared peccary (*Tayassu tajacu*) is quite different from the two hogs, having 2n = 30 and NF = 48 (Pirtle, 1967). This chromosome complement cannot be related to that of the hogs by any simple fusion or fission theory of chromosome change. The difference has had a long time to become established, the suid-tayassuid separation occurring during Eocene.

Tylopoda

Another suborder, Tylopoda, can be traced in the fossil record back to early artiodactyl history, being recognizable as camelid-like animals in mid-Eocene. Most of the evolution of this group occurred in North America, but representatives died out on this continent several thousand years ago. They were possibly contemporary with early man in western North America, but today survivors occur only as two species of old world camels in Africa and Asia and two to four species of new world camelids in South America. Although these survivors at the extremes of the range of the Camelidae have been separated for upwards of one million years and have diverged morphologically, physiologically and behaviorally in many respects, their karyotypes are indistinguishable at the level of observation currently employed (Taylor *et al.*, 1968b). The raw material of evolution within the surviving species of this suborder must therefore have been gene mutations or minor chromosomal rearrangements of types which cannot be detected by the use of current techniques. This uniformity of karyotypes among all the members of a major taxonomic group is unique among the artiodactyls. It is distinctly different from the karyotypic variation to be observed among different tribes within the family Bovidae, in which the evolutionary divergence has occurred in some cases over approximately the same time span as in the camelids.

Ruminantia

By far the greatest assemblage of artiodactyl species is in the suborder Ruminantia. Primordial members of this group, ancestors of the family Tragulidae, arose in the Eocene probably in close relationship with the early camelid ancestors. They gave rise to modern tragulids, the chevrotains or mouse deer, and during Oligocene to the ancestors of Cervidae, the deer family. Unfortunately, no karyotypic descriptions of any of the chevrotains are available.

Three more families were later offshoots of this ancestral line:

1. Giraffidae were a Miocene offshoot from the Cervidae, but also show relationships to the Bovidae.
2. Antilocapridae, represented today by a single species, the pronghorn (*Antilocapra americana*) were recognizably separate from, but closely related to the Bovidae (antelopes, cattle, goats and sheep) at the time of the appearance of both families during the Miocene.
3. The Cervidae can be separated into four subfamilies; the Cervinae, which are mostly Eurasian in their present distribution, and the Odocoileinae, which are mostly North and South American in distribution. These groups include most of the modern species. The Moschinae include only one living species, and this has not been karyotyped. There are 6 species of Muntiacinae, and only one (Reeve's muntjak, *Muntiacus reevesi*) has been completely described (Wurster and Benirschke, 1968). It has 46 chromosomes, all acrocentrics except the Y, and is quite different from all other Cervidae whose karyotypes are known.

Seven species of Cervinae have been karyotyped, and they are characterized by a NF of 70 in each case. Most have $2n = 68$, but two species have diverged markedly. The sambar (*Cervus unicolor*) has $2n = 58$ and $NF = 70$, while the barasingha (*Rucervus duvauceli*) has $2n = 56$ and $NF = 70$ (Chandra *et al.*, 1967). Both are examples of extensive fusion of acrocentrics.

On the other hand, the Odocoileini, with one exception, have a NF of 74, and with two exceptions have $2n = 70$. The roe deer (*Capreolus capreolus*) has only two metacentrics among its 70 chromosomes so its

NF is 72 (Amrud and Nes, 1966). The North American moose (*Alces alces*) has been described as having 2n = 70 and a NF of 74 (Wurster and Benirschke, 1967a), as in the reindeer (*Rangifer tarandus*) (Amrud and Nes, 1966), and the mule deer (*Odocoileus hemionus*) and the whitetail deer (*Odocoileus virginianus*) (Wurster and Benirschke, 1967b). However, the same species, *Alces alces,* the elk of Europe, is described as having 2n = 68 and NF = 74, with one more pair of metacentric chromosomes than in the North American form (Amrud and Nes, 1966). The red brocket deer (*Mazama americana*) shows the same amount of Robertsonian change, with 2n = 68 and NF = 74 (Taylor *et al.,* 1968a).

The giraffe, one of the two surviving species in the family Giraffidae, has 2n = 30 with only one pair of acrocentrics and thus a NF of 58 (Wallace and Fairall, 1965; Taylor *et al.,* 1967). Although this family is considered to be a Miocene offshoot of the cervid line, its chromosome number has been reduced by Robertsonian change far more than any cervid, and its NF of 58 is very close to the commonest bovid NF of 60. Morphologically the Giraffidae show relationships both to the Cervidae and to the Bovidae. The Giraffidae separated from the cervid line in Miocene, while the early history of the bovoid line is uncertain. When the Bovidae first appeared in the fossil record in the Miocene, they were already distinct from both the Cervidae and the Antilocapridae, though clearly related to the latter family. In spite of its seemingly closer relationship to the cervids, the giraffe has a karyotype more closely resembling that of the bovids. It would be of great interest to study the chromosomes of the okapi, the only other living giraffid.

The Antilocapridae are represented by a single living species, the pronghorn (*Antilocapra americana*). Both morphologically and karyotypically the relationship to the bovids is evident; 2n is either 56 or 58 and the NF is either 58 or 60 (Wurster and Benirschke, 1968).

The Bovidae comprise such a large and varied group, and so many have been karyotyped that they cannot be treated individually in this report. However, for these very reasons, they are the most interesting group in the entire order. Much of the following discussion of bovid chromosomes is summarized from a review of the Bovidae by Wurster and Benirschke (1968) and from our own work in Philadelphia. Over fifty species and subspecies have been karyotyped more or less completely. Of these, five have a NF of 58, thirty-two have a NF of 60, and nine have a NF of 62. The number of metacentric pairs, each presumably representing the union of two acrocentric pairs, ranges from zero, as in the goat, ibex, impala and others, to all of the chromosomes, i.e., 15 pairs of metacentrics in the blackbuck, *Antilope cervicapra* (Wurster and Benirschke, 1968; Taylor *et al.,* 1968a). The distribution of members of metacentric pairs is of interest because most species have at least one such pair as is shown in Table 1.

Table 1: Number of Metacentric Pairs in Bovids

Metacentric Pairs	Species or Subspecies
0	8
1	14
2	5
3	5
4	3
5	2
6	3
7	1
9	1
10	1
11	2
13	1
14	2
15	1

One might postulate on the basis of this evidence that evolution has proceeded by fusion rather than by fission in the Bovidae. There is at present no satisfactory way of estimating the average length of time required to fix such a chromosomal fusion in a species. Therefore, the distribution in Table 1 could be interpreted as the result of the more or less random occurrence and survival of such fusions, or it could mean that some intrinsic structural factor, or possibly selection, has favored the presence of low numbers of such fusions in the surviving species of this family. Insight into the problem might come from relating these data to the evolutionary history and geographical distribution of this family. When arranged according to subfamilies, the data on NF and numbers of metacentric pairs (fusions) are as shown in Table 2.

There does not appear to be any striking difference between subfamilies with respect either to NF or to the number of metacentric pairs, except that the Bovinae are somewhat more variable than the others, and the Caprinae show no variability of NF. These two groups contain many of the domesticated species (various cattle, goats, and sheep).

When one separates the species into African and Eurasiatic groups, a

Table 2

	NF				No. of Metacentric Pairs						
Subfamily	58	60	62	Other	0	1	2	3	4	5	>5
Bovinae	4	3	7	1	1	6	2	1	1	1	3
Cephalophinae			1			1					
Hippotraginae		10	1		1	4	1		1	1	3
Antilopinae	1	5		2	2	1	1		1		3
Caprinae		14			4	2	1	4			3
TOTAL	5	32	9	3	8	14	5	5	3	2	12

Table 3

	NF				No. of Metacentric Pairs						
	58	60	62	Other	0	1	2	3	4	5	>5
African	4	16	2	2	3	6	3	1	3	1	7
Eurasiatic & North American	1	16	7	1	5	8	2	4		1	5
TOTAL	5	32	9	3	8	14	5	5	3	2	12

separation which corresponds to some extent, but not entirely, to the previous division into subfamilies and is also related to their evolutionary history, the pattern appears no more enlightening. This distribution is shown in Table 3.

Again, there is no clearcut difference between the patterns in these two major geographic categories. One is left with the opinion that Robertsonian decrease in chromosome number, with relatively little variation in the number of chromosome arms, has been an important factor in the evolution of the Bovidae. But no clue is apparent as to why this type of chromosome evolution has occurred so much more widely in the bovids than in other families of artiodactyls.

General Discussion

Stebbins (1966) argues that "aneuploid reduction" has been important in the evolution of many plant species in which outcrossing prevails and which occupy pioneer habitats. These conditions produce a high degree of heterozygosity on the one hand and require repeated colonization of new areas on the other hand. Stebbins feels that under such conditions the most successful colonizers will be those with reduced genetic variability and that in such cases "aneuploid reduction" reduces genetic variability by diminishing the number of linkage groups.

White (1968) advances what he calls a stasipatric model of speciation based on his studies of populations of wingless grasshoppers: ". . . stasipatric model, since it involves the spread of a chromosomal rearrangement which forms the basis of a cytogenetic isolating mechanism throughout a substantial part of the range of an already existing species." Each Robertsonian fusion is probably a unique event. If these events have been assumed to play a significant role in the evolution of a group, one must account for their survival and spread in populations. The evidence indicates that in some cases such chromosomal rearrangements do serve as strong isolating mechanisms by interfering with meiosis in interpopulation hybrids to a greater or lesser extent. White suggests that the spread of chromosomal fusions in such cases may be due to a segregational advantage, that is, meiotic drive, favoring the fusion chromosome in the heterozygote.

Although Stebbins based his ideas largely on studies in higher plants and White's concepts came from studies on nearly sedentary animals, White does suggest that application of such a model might also be made in some small mammals such as insectivores and rodents, whose mobility may be greater. The pertinence of such ideas in reference to artiodactyls is open to question. However, there are certain resemblancs in the cervids and bovids. Both these groups have undergone major radiations and have spread (colonized) into most of the major land areas of the world. On the other hand, the few published studies on hybrids which were heterozygous for one or more chromosome fusions (domestic-wild pig, McFee *et al.,* 1966; sheep-goat, Ilberry *et al.,* 1967; cattle-bison, Basrur and Moon, 1967) indicate no significant reproductive problems on the chromosome level. The suggestion of meiotic drive might conceivably be tested in the Caprinae where the goat (*Capra hircus,* 2n = 60, no metacentrics) has been successfully crossed with aoudad (*Ammotragus lervia,* 2n = 58, one pair of metacentrics). The hybrid of this cross can be crossed to the ibex (*Capra ibex,* 2n = 60, no metacentrics). Another possibility might occur in the argali sheep, *Ovis ammon,* in which three different subspecies have, respectively, zero, one, and three pairs of metacentric chromosomes (Wurster and Benirschke, 1968). It is interesting to note that progressive accumulation of metacentric chromosomes presumably by fusion of acrocentrics has been reported for bovid cells grown in culture by Basrur *et al.* (1964) for a bovine lymphosarcoma, by Lithner and Pontén (1966) for fetal calf lung cells, and by Nelson-Rees *et al.* (1967) for several cell lines derived from cattle, sheep, and goats, though not for a cell line from the dog.

Another interesting area of artiodactyl chromosome cytology is that of the evolution of their sex chromosomes. Most of them have a large acrocentric X chromosome which constitutes approximately 5% of the haploid chromosomal complement, the "original-type" mammalian X of Ohno, Beçak and Beçak (1964). This has been shown by Wurster and Benirschke (1968) through chromosome area measurements for several species of bovids and by Taylor *et al.* (1968a) through measurements of chromosome length for two species of deer and six species of bovids. In these eight species the X was not only about 5% of the haploid complement, but it was acrocentric. Three species with metacentric X chromo somes suggest origin of the X by fusion since these X's are larger. The nilgai (*Boselaphus tragocamelus*) X is 7–8% of the haploid complement, as is also the giraffe X, while the blackbuck (*Antilope cervicapra*) X is 12–13% of the haploid total. Wurster and Benirschke (1968) find the same thing to be true for the blackbuck X (14.96%) and for the sitatunga (*Tragelaphus spekei*) X (13.08%).

Other evidence of the occurrence of compound X chromosomes is provided by Wahrman and Horowitz (1967) in their studies on two species of gazelles (*Gazella dorcas* and *Gazella gazella*). The first of these has 2n = 34 in the female but 2n = 35 in the male, while the second species has 30 chromosomes in the female and 31 in the male. This has been termed an XY_1Y_2 sex chromosome constitution; probably the simplest interpretation is that one of the Y's is the homologue of an acrocentric (former autosome) which has been fused onto the original X.

Another interesting example of a compound sex chromosome appears to be that of the Y chromosome of the kudu (*Tragelaphus sp.*) reported by Wallace and Fairall (1966). In this case there has probably been a combination of an autosome with the original Y, as indicated by the occurrence of a multivalent composed of acrocentric X + acrocentric autosome + metacentric element in male meiosis. Most of the artiodactyl Y chromosomes, however, are very small, usually metacentric. The other exceptional Y's are in the sitatunga with a very large metacentric Y and the nilgai with a large acrocentric Y.

One more unusual observation relating to artiodactyl chromosomes is that of satellites on the long arms of acrocentric chromosomes by Amrud and Nes (1966) for European elk (*Alces alces*), the reindeer (*Rangifer tarandus*) and the roe deer (*Capreolus capreolus*), and by Chandra *et al.* (1967) for the sambar (*Cervus unicolor*) and the barasingha (*Rucervus duvauceli*), all species in the family Cervidae.

Finally, there are a few unusual karyotypes which cannot be derived from the karyotypes of related species by any simple fusion, translocation, or inversion model. Among the cervids, Reeve's muntjak (*Muntiacus reevesi*) has 2n = 46 and NF = 46 (Wurster and Benirschke, 1967b). Among the bovids, the four-horned antelope (*Tetracerus quadricornis*) has 2n = 38 and NF = 38 (Wurster and Benirschke, 1968); and Kirk's dik-dik (*Madoqua kirki*) has 2n = 46 and NF = 48 (Chandra *et al.*, 1967).

Consideration of all the chromosome reports of the entire order does not provide a solid basis for any hypothesis of an ancestral artiodactyl karyotype. However, it seems reasonable to postulate that the ancestral bovid karyotype was composed of 60 acrocentric chromosomes while that of the cervids was probably made up of 70 or 74 acrocentrics. Beyond these two evolutionary lines the evidence is too incomplete for even tentative suggestions.

Recently, much is being accomplished in the field of comparative biochemistry, and in general these works reinforce or extend and clarify classical taxonomic studies and cytogenetic studies, for example the recent work of Nadler *et al.* (1967) in comparing the serum proteins and transferrins of North American *Alces alces* and *Rangifer tarandus* with those of the European members of these two cervid species and also with those of the North American Dall sheep (*Ovis dalli*). Doolittle and his associates disagree with classical taxonomy on the basis of their biochemical studies (Doolittle, Schubert and Schwartz, 1967; Mross and Doolittle, 1967). By studying amino acid sequences of fibrinopeptides, they have characterized several species of artiodactyls, and they suggest that the Caprinae (sheep and goats) are as different from the Bovinae as are the Cervidae and Giraffidae. In this respect, cytotaxonomy conforms more closely to classical taxonomy in that the chromosomes of the Bovinae are much more similar to those of the Caprinae than to those of the deer and giraffe. However, as was pointed out earlier, the Bovinae do appear to be somewhat more variable than other bovid subfamilies, both with respect to chromosome number and to NF. Such biochemical studies should be extended to other subfamilies of bovids and to non-domesticated species of Caprinae.

Perspectives

The following gaps in cytotaxonomic information about the artiodactyls should especially repay investigation. (1) There are no published reports on the chromosomes of the Tragulidae, and these animals are descended from early tragulids which were close to the origins of both the camelid and cervid-bovid lines of evolution. (2) Apparently, no hippopotamuses have been studied. Although there are only two surviving species, this leaves an entire infraorder totally uninvestigated. (3) Because of the unusual karyotype of Reeve's muntjak, more members of the subfamily Muntiacinae should be studied. (4) The okapi, as the only surviving near relative of the giraffe, would be of interest. (5) Only one species of duiker, in the large subfamily Cephalophinae, has been studied. (6) Only two species in the large tribe Neotragini have been studied, and one of these is the karyotypically unusual Kirk's dik-dik. While there remain numerous unstudied questions in the cytotaxonomy of artiodactyls and major cytogenetic problems yet to be resolved, encouraging progress has already been achieved.

Acknowledgments

This work was supported in part by the U.S. Public Health Service: Fellowship 1-F3-CA-33594 from the National Cancer Institute; research grants CA-05903 and CA-06927 from the National Cancer Institute; and research grant HO-00543 from the National Institute of Child Health and Human Development.

The authors acknowledge the technical assistance of Gundula U. LaBadie and Marion R. Taylor.

References

Amrud, J. and N. Nes: The chromosomes of the roe (*Capreolus capreolus*). Hereditas *56*:217, 1966.

Basrur, P. K., J. P. W. Gilman and B. J. McSherry: Cytological observations on a bovine lymphosarcoma. Nature *201*:368, 1964.

——— and Y. S. Moon: Chromosomes of cattle, bison, and their hybrid, the cattalo. Am. J. Vet. Res. *28*:1319, 1967.

Chandra, H. S., D. A. Hungerford, J. Wagner and R. L. Snyder: Chromosomes of five artiodactyl mammals. Chromosoma (Berl.) *21*:211, 1967.

Doolittle, R. F., D. Schubert and S. A. Schwartz: Amino acid sequence studies on artiodactyl fibrinopeptides. I. Dromedary camel, mule deer, and cape buffalo. Arch. Biochem. Biophys. *118*:456, 1967.

Giménez-Martin, G. and J. F. Lopez-Saez: Cromosomas de mamiferos domesticos. Genet. Iberica *18*:15, 1966.

Ilberry, P. L. T., G. Alexander and D. Williams: The chromosomes of sheep × goat hybrid. Aust. J. biol. Sci. *20*:1245, 1967.

Lithner, F. and J. Pontén: Bovine fibroblasts in long-term tissue culture: Chromosome studies. Int. J. Cancer *1*:579, 1966.

McFee, A. F., M. W. Banner and J. M. Rary: Variation in chromosome number among European wild pigs. Cytogenetics *5*:75, 1966.

Mross, G. A. and R. F. Doolittle: Amino acid sequence studies on artiodactyl fibrinopeptides. II. Vicuna, elk, muntjak, pronghorn antelope, and water buffalo. Arch. Biochem. Biophys. *122*:674, 1967.

Nadler, C. F., C. E. Hughes, K. E. Harris and N. W. Nadler: Electrophoresis of the serum proteins and transferrins of *Alces alces* (elk), *Rangifer tarandus* (reindeer), and *Ovis dalli* (Dall sheep) from North America. Comp. Biochem. Physiol. *23*:149, 1967.

Nelson-Rees, W. A., A. J. Kniazeff and N. B. Darby, Jr.: Debut and accumulation of centric fusion products: An index to age of certain cell lines Cytogenetics *6*:436, 1967.

Ohno, S., W. Beçak and M. L. Beçak: X-autosome ratio and the behavior pattern of individual X-chromosomes in placental mammals. Chromosoma (Berl.) *15*:14, 1964.

Pirtle, E. C.: Chromosomes of the female peccary (*Tayassu tajacu*). Mammal. Chrom. Newsl. *8*(1):16, 1967.

Stebbins, G. L.: Chromosomal variation and evolution. Science *152*:1463, 1966.

Taylor, K. M., D. A. Hungerford and R. L. Snyder: The chromosomes of four artiodactyls and one perissodactyl. Mammal. Chrom. Newsl. *8*(3):233, 1967.

Taylor, K. M., D. A. Hungerford and R. L. Snyder. *Unpublished*, 1968a

Taylor, K. M., D. A. Hungerford, R. L. Snyder and F. A. Ulmer, Jr.: Uniformity of karyotypes in the Camelidae. Cytogenetics *7*:8, 1968b.

Wahrman, J. and A. Horowitz: Personal communication, 1967.

Wallace, C. and N. Fairall: Chromosome analysis in the Krueger National Park with special reference to the chromosomes of the giraffe. Koedoe *8*:97, 1965.

Wallace, C. and N. Fairall: The chromosomes of the warthog. S. Afr. J. Med. Sci. *32*:51, 1967a.

Wallace, C. and N. Fairall: The chromosomes of the kudu. S. Afr. J. Med. Sci. *32*:61, 1967b.

White, M. J. D.: Models of speciation. Science *159*:1065, 1968.

Wurster, D. H. and K. Benirschke: The chromosomes of twenty-three species of the Cervoidea and Bovoidea. Mammal. Chrom. Newsl. *8*(3):226, 1967a.

Wurster, D. H. and K. Benirschke: Chromosome studies in some deer, the springbok, and the pronghorn, with notes on placentation in deer. Cytologia *32*:273, 1967b.

Wurster, D. H. and K. Benirschke: Chromosome studies in the superfamily Bovoidea. Chromosoma (Berl.) *25*:152, 1968.

PRIMATES

J. Egozcue
Department of Genetics, Oregon Regional Primate Research Center, Beaverton, Oregon

The study of primate chromosomes is important for an understanding of the phylogenetic relationships and evolutionary trends within the order. The scope of genetic studies in primates has gone a long way since Painter (1922) first described the chromosomes of a "ring tail monkey" in an attempt to determine whether mammals have an X-Y sex-determining mechanism. Today the rhesus macaque (*Macaca mulatta*) is no longer called by such offensive names as rhesus macacus monkey. The cytogeneticist knows quite well that he has to rely on the taxonomist for proper species identification, without which his results are meaningless; and the taxonomist, in turn, recognizing that cytogenetics, immunogenetics, and biochemical genetics provide information essential to his work, has forgiven.

Since Bender and Chu (1963) published their review of primate chromosomes, many new species have been studied, and what was at the time a confused picture is now becoming a sharp, clear account of the evolution of primate chromosomes. Although the ancestors of living primates are no longer available for karyological studies, related species do provide many examples of the different stages through which the most highly evolved karyotypes have passed. The similarity between the chromosomal complements in other species suggests a common origin, whereas their differences permit the reconstruction of some of the steps in their evolution. However, it is necessary to point out that similarity between chromosomes of related species does not preclude homology. Meiotic studies in interspecific hybrids are necessary to prove that homology exists between look-alike chromosomal complements. Since such studies are scarce, most of the predicated relationships among primate groups are based on the assumption that homology really exists but whether or not it does is still problematic.

Progress in methodology has accelerated chromosomal studies in mammals at such a fantastic rate that a review soon becomes outdated. Fortunately, the beleaguered reader can find help in his attempts to keep pace with the flood of new information. Hsu and Benirschke (1967, 1968) and Chiarelli *et al.* (1968) have compiled atlases of mammalian chromosomes and primate chromosomes respectively and a list of primate chromosome numbers is compiled yearly by Borgaonkar (1966b; 1967).

Note on Methodology

Most authors follow the traditional classification of chromosomes into metacentrics, submetacentrics, and acrocentrics or use even more complex systems. Although we have used this type of classification in the past, we wholeheartedly agree with Klinger (1963) that "such differentiations would be difficult to make without extensive arm ratio determinations" and believe with him that "only the terms acrocentric and metacentric" should be used. Nevertheless, since the term submetacentric is more flexible than metacentric, we have chosen to retain it and to use the acrocentric and submetacentric classification except in those cases where a more complex classification is useful to emphasize similarities. Many species of Platyrrhini, for instance, share the common characteristic of two pairs of small metacentric chromosomes in their complements. Calling these chromosomes metacentrics serves to heighten the similarity of the species that have them and the dissimilarity of the species than do not. Other than such cases, all nonacrocentrics are classified in this report as submetacentrics, and chromosomal formulae are compared only when the similar formulae are accompanied by morphological similarity of the karyotypes and of the individual chromosomal pairs.

In the figures depicting evolution in the different primate groups, the disposition of the generic or specific names that accompany the chromosomal formulae does not mean that one species evolved from the other, but rather that the karyotype now found in one species probably evolved from a karyotype similar to that now found in the other.

Suborder Prosimii

This suborder is composed of three superfamilies: the Lemuroidea, the Lorisoidea, and the Tarsioidea. The Lemuroidea include the Lemuridae, the Indridae, and the Daubentonidae.

The data on Lemuridae, which have been extensively investigated, are summarized in Table 1. Studies of Chu and Swomley (1961), Chu and Bender (1961), and Chiarelli (1961a) showed that interspecific, and maybe intraspecific, variability is common in the genus Lemur. Intraspecific variability was later confirmed by Egozcue (1967). Diploid

Table 1: Chromosomes of Lemuroidea

Species	Specimens Male	Female	2n	S	A	X	Y	Common Name
Lemuridae								
Lemur mongoz	2	3	60	4	54	A	A	Mongoose lemur
L. mongoz	1	7	58	4	52	A	A	Mongoose lemur
L. fulvus fulvus	2	1	58	4	52	A	A	Brown lemur
L. f. rufus	—	2	60	4	54	A	A	Red-fronted lemur
L. f. albifrons	—	1	60	4	54	A	A	White-fronted lemur
Hybrid l. fulvus × L. f. albifrons	1	—	58	4	52	A	A	
L. fulvus fulvus	1	1	48	16	30	A	A	Brown lemur
L. variegatus	1	—	46	18	26	S	A	Ruffed lemur
L. macaco	4	4	44	20	22	A	A	Black lemur
L. sp.	1	—	52	12	38	S	A	
L. catta	2	—	56	10	44	A	A	Ringtail lemur
L. catta	—	2	56	14	40	A	—	Ringtail lemur
Hapalemur griseus griseus	1	1	54	10	42	A	A	Gray gentle lemur
H. g. olivaceus	—	1	58	6	50	A?	—	
Cheirogaleus major	—	1	66	2	64	—	—	Greater dwarf lemur
Microcebus murinus	—	1	66	2	64	—	—	Miller's mouse lemur
Indridae								
Propithecus verreauxi verreauxi	1	—	48	—	—	—	—	Verraux's sifaka
P. v. coquereli	1	2	48	—	—	—	—	

numbers of 58 and 60 have been found in *Lemur mongoz*, and of 48, 58, and 60 in *L. fulvus*. Chromosome numbers vary from 44 to 60 in lemurs, from 54 to 58 in hapalemurs. On the basis of chromosomal results, we proposed a modification of Chu and Swomley's classification of (1961) Lemurs and Hapalemurs and divided them into three groups (Egozcue, 1967).

The first group includes these species with a diploid number of 2n = 58 or 2n = 60 and a predominance of acrocentric chromosomes (*L. mongoz, L. fulvus fulvus, L. fulvus rufus*, and *L. fulvus albifrons*). The second group includes those species with a diploid number around 46 and a high nmber of submetacentric chromosomes (*L. fulvus fulvus, L. variegatus*, and *L. macaco*). The third group includes those species with a diploid number around 56 and an intermediate number of submetacentrics (*L. catta, Hapalemur griseus griseus*, and *H. griseus olivaceus*). Nevertheless, a comparison of the chromosomal formulae (Fig. 1) shows that all species in the two genera are closely related and that the classification is artificial and destroys the unity of the group. Furthermore, subspecies of *L. fulvus* are found in two different groups of the classification. Thus, the attempt to break down the Lemurs and Hapa-

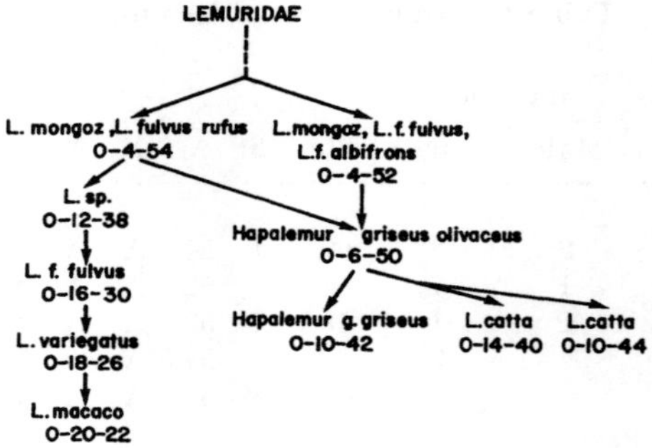

Fig. 1. Chromosomal evolution in the Lemuridae.

lemurs into groups is probably unnecessary. In the Lemurinae, the X and Y chromosomes are usually acrocentric. *L. variegatus* and *L. sp.* (Chu and Swomley, 1961) have submetacentric X chromosomes. The Lemurinae are the only primates to show microchromosomes in their complements, a characteristic they share with the class Aves and the order Squamata (Beçak *et al.,* 1964; Ohno, 1967). No information is available on the Lepilemurs.

In the subfamily Cheirogaleinae, only two species, one from each genus, have been studied. Both *Cheirogaleus major* and *Microcebus murinus* (Chu and Swomley, cited by Chu and Bender, 1962) have 66 chromosomes. The karyotypes are identical and, except for one pair of submetacentrics, include only acrocentric chromosomes. No information is available on the sex chromosomes.

Among the *Indridae,* two subspecies of *Propithecus verreauxi* have been investigated. *P. verreauxi verreauxi* and *P. verreauxi coquereli* both have 48 chromosomes (Chu, cited by Chu and Bender, 1962). No description of the karyotype has been published. There is no information on the genera Avahi and Indri. No studies have ever been conducted in the Daubentonidae.

The Lorisoidea include the Lorisidae and the Galagidae. Table 2 summarizes the available data. *Perodicticus potto* (Chu and Bender, 1961; Bender and Chu, 1963) has 62 chromosomes; the complement is composed of 24 submetacentrics and 36 acrocentrics. These results have been confirmed in our laboratory in one male potto (this report). The chromosomal complement of the potto is similar to that of *Galago crassicaudatus* (Fig. 2), a fact that provides a link between the two families of the Lorisoidea.

In the genus Nycticebus, counts of 2n = 50 have been reported for *N. coucang* (Bender and Mettler, 1958; Klinger, 1963; Egozcue and Vilarasau de Egozcue, 1967a), *N. coucang bengalensis* (Chu and Bender,

Table 2: Chromosomes of Lorisoidea

Species	Specimens Male	Female	2n	S	A	X	Y	Common Name
Lorisidae								
Perodicticus potto	2	4	62	24	36	S	A	Potto
Nycticebus coucang	3	3	50	48	—	S	S	Slow loris
Arctocebus calabarensis	—	1	52	50	—	S	—	Angwantibo
Loris tardigradus	1	—	62	34	26	S	S	Slender loris
L. tardigradus	1	1	62	38	22	S	A	Slender loris
Galagidae								
Galago senegalensis	1	—	38	24	12	S	S	Lesser bushbaby
G. senegalensis	1	—	38	22	14	S	A	Lesser bushbaby
G. senegalensis	2	2	38	30	6	S	A	Lesser bushbaby
G. crassicaudatus	2	2	62	6	54	S	A	Thick-tailed bushbaby
G. crassicaudatus	1	—	62	30	30	S	S	Thick-tailed bushbaby

1962), and *N. pygmaeus* (Chu and Bender, 1962). In these species, as well as in *Arctocebus calabarensis* (Egozcue and Vilarasau de Egozcue, 1966a), which has a diploid number of 2n = 52 and a karyotype similar to that of Nycticebus, all chromosomes are submetacentric.

Different karyotypes for *Loris tardigradus* have been published by Egozcue *et al.* (1966) and Manna and Talukdar (1968); however, the differences are minor and the results may reflect a polymorphism in the species. All Lorisidae have submetacentric X chromosomes; the Y is acrocentric in *P. potto* and in the *L. tardigradus* studied by Manna and Talukdar (1968); submetacentric in *Nycticebus* and in the *L. tardigradus* studied by us (Egozcue *et al.,* 1966). Since the only *Arctocebus calabarensis* studied was a female, no information is available on the

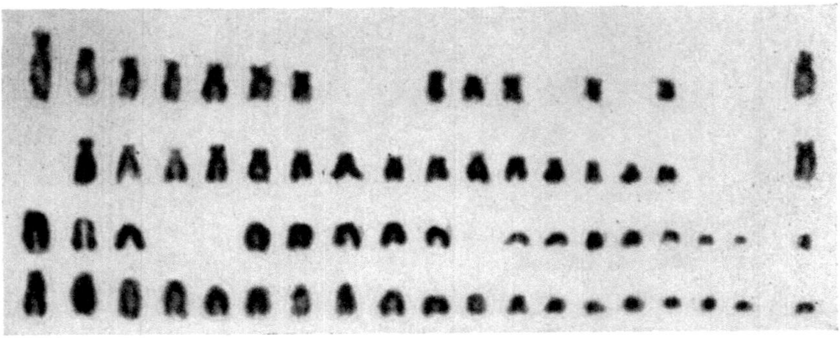

Fig. 2. Comparative haploid complements of *P. potto* (top and bottom rows) and *G. crassicaudatus* (central rows) showing the many similarities between the two karyotypes. The sex chromosomes are at the end of the series.

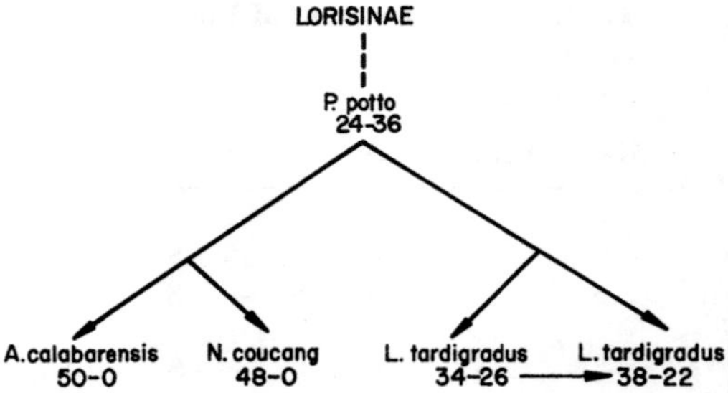

Fig. 3. Chromosomal evolution in the Lorisinae.

morphology of the sex chromosomes in this species. Figure 3 shows the evolutionary trends in the Lorisinae.

Two species have been investigated in the Galagidae with conflicting results (Table 2). *Galago senegalensis* has 38 chromosomes; the complement is composed of 24 submetacentrics and 12 acrocentrics (Matthey, 1955). Using testicular material, we described 22 submetacentrics and 14 acrocentrics in one male *G. senegalensis* (Egozcue, 1968). Chu

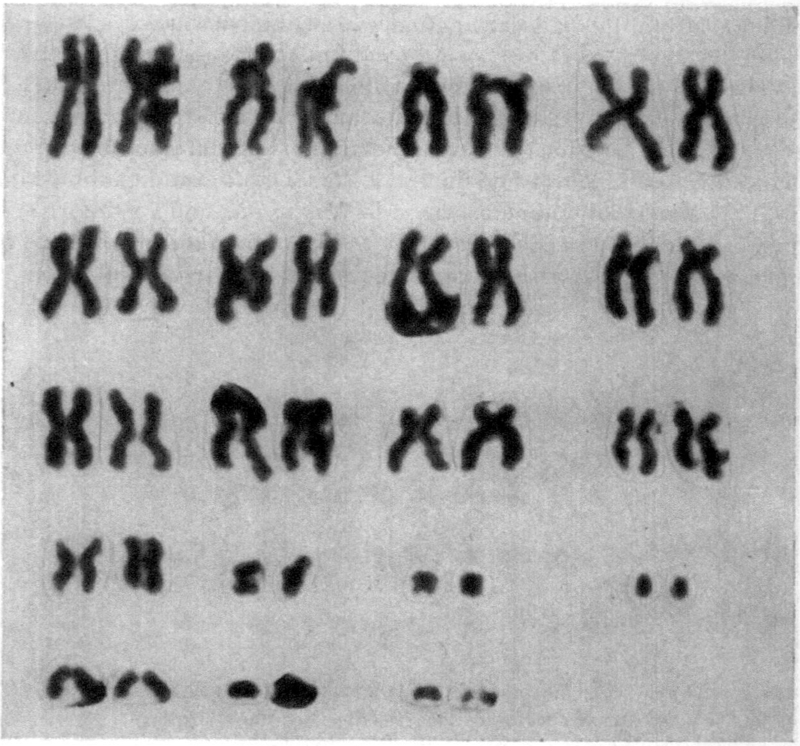

Fig. 4. Karyotype of female *Galago senegalensis*.

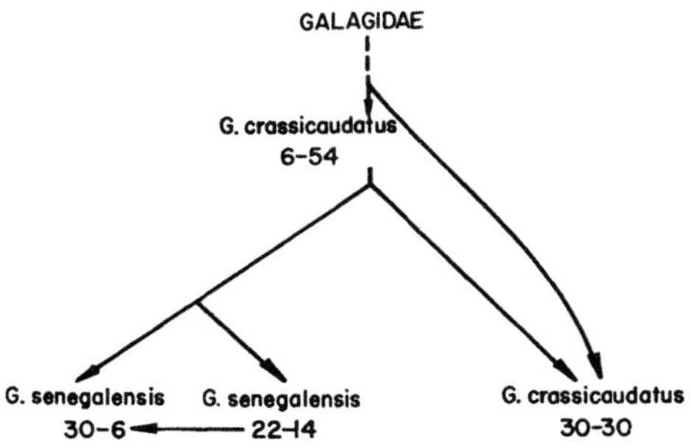

Fig. 5. Chromosomal evolution in the Galagidae.

and Bender (1961) found 30 submetacentrics and 6 acrocentrics in skin cultures of two male and two female *G. senegalensis;* we have recently confirmed their results (this report) in blood cultures of a female (Fig. 4). The similarity between our earlier results and Matthey's, and the differences between these and the results of Chu and Bender and our own suggest a polymorphism in the species.

Galago crassicaudatus has 62 chromosomes. According to Chu and Bender (1961), the complement is composed of 6 submetacentrics and 54 acrocentrics. In studies of one male *G. crassicaudatus* (this report), we found the diploid number to be 2n = 62, with 30 submetacentrics and 30 acrocentrics in the complement (Fig. 2). The situation here is similar to that in *G. senegalensis.* The chromosomal complement of *G. crassicaudatus* is similar to that of *Perodicticus potto,* a member of the Lorisinae. The X chromosome of the Galagidae is submetacentric; the Y is acrocentric, but the specimen of *G. crassicaudatus* studied by us has a submetacentric Y. Figure 5 shows the evolutionary trends in the Galagidae.

In the Tarsioidea, two species of Tarsius have been studied (Table 3). Klinger (1963) reported a diploid number of 2n = 80 in a female *T. bancanus,* with 14 submetacentrics and 66 acrocentrics in the complement. A count of 2n = 80 has also been found in testicular material of one male *T. syrichta* (Chiarelli and Egozcue, 1968), erroneously reported as *T. bancanus.* No information is available on the morphology of the sex chromosomes.

Table 3: Chromosomes of the Tarsioidea

Species	Specimens		2n	S	A	X	Y	Common Name
	Male	Female						
Tarsius bancanus	—	1	80	14	66	—	—	Horsfield's tarsier
Tarsius syrichta	1	—	80	—	—	—	—	Philippine tarsier

Suborder Platyrrhini

The family Cebidae includes 10 genera, six of which have been fairly well studied (Tables 4 and 5). *Aotus* (= *Aotes*) *trivirgatus* (Fig. 6) has either 54 (Bender and Mettler, cited by Bender and Chu, 1963; Chiarelli, 1961a; Egozcue *et al.*, 1968c) or 50 chromosomes (Chiarelli and Barberis, 1966). In the bone marrow preparations we studied, the karyotype is

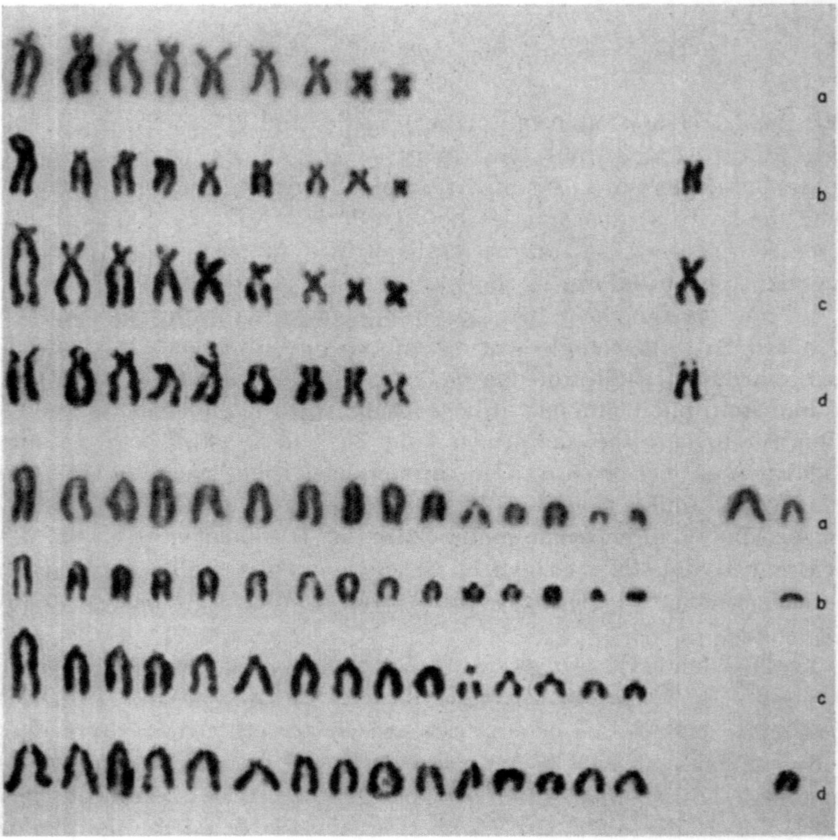

Fig. 6. Comparative haploid karyotypes of *Cebus capucinus* (a, first row), *Cebus albifrons* (b, second), *Alouatta caraya* (c, third) and *Aotus trivirgatus* (d, fourth). The sex chromosomes are not shown. The chromosomes are divided into submetacentrics (top) and acrocentrics (bottom). The long acrocentric of *C. capucinus* (extreme right) is missing in the other three species; the small acrocentric is present in *C. albifrons* and *A. trivirgatus* but not in *A. caraya*, indicating that the submetacentric (extreme right) of *C. albifrons* and *Aotus trivirgatus* probably originated by pericentric inversion of the long acrocentric, while the submetacentric of *A. caraya* originated by Robertsonian fusion of the two acrocentrics.

Table 4: Chromosomes of Cebidae

Species	Specimens Male	Female	2n	S	A	X	Y	Common Name
Lagothrix ubericola	1	–	62	–	–	–	–	Brown woolly monkey
Ateles arachnoides	–	1	34	–	–	–	–	Brown woolly spider monkey
A. paniscus	–	2	34	–	–	–	–	Black-faced spider monkey
A. paniscus chamek	1	1	34	30	2	S	A	Black-faced spider monkey
A. belzebuth	5	6	34	30	2	S	A	Golden spider monkey
A. geoffroyi cucullatus	2	–	34	30	2	S	A, S	Hooded spider monkey

Table 5: Chromosomes of Cebidae

Species	Specimens Male	Female	2n	M	S	A	X	Y	Common Name
Alouatta seniculus	1	–	44	4	12	26	A	S	Red howler monkey
A. villosa	–	1	53	–	–	–	–	–	Guatemalan howler monkey
A. caraya	2	1	52	4	16	30	S	A	Black howler monkey
Aotus trivirgatus	2	–	54	4	16	32	S	S	Owl monkey
Cebus albifrons	4	–	54	4	16	32	S	A	White-fronted capuchin
C. capucinus	1	–	54	4	14	34	S	A	White-faced capuchin
C. capucinus	1	–	52	4	16	30	A	A	White-faced capuchin
C. capucinus	3	1	54	4	20	28	A	–	White-faced capuchin
C. apella	6	1	54	4	20	28	A	A	Tufted capuchin
Callicebus moloch	2	–	46	4	16	24	S	A, M	Red titi monkey
Callicebus torquatus	–	1	20	–	10	10	–	–	
Cacajao rubicundus	2	–	46	4	16	24	S	–	Red uakari
Pithecia pithecia	–	1	46	–	–	–	–	–	Pale-headed sakiwinki
Saimiri sciureus	2	3	44	4	26	12	S	A	Squirrel monkey
Saimiri madeirae	1	6	44	4	28	10	S	A	Squirrel monkey
Saimiri boliviensis	8	1	44	4	28	10	S	A	Squirrel monkey

most similar to that of *Cebus albifrons,* with 4 metacentrics, 16 submetacentrics, and 32 acrocentrics. The X and Y chromosomes are submetacentric (Fig. 7). There are only minor differences between the karyotype of *Aotus* and those of *Cebus, Alouatta caraya, Cacajao* and *Callicebus moloch.*

Two species of the genus Callicebus have been investigated. *Callicebus moloch* (= *cupreus*) has 46 chromosomes (Bender and Mettler, 1958) with 4 metacentrics, 16 submetacentrics, and 24 acrocentrics in the complement. We have confirmed these results in blood and kidney cultures of one male *C. moloch* (this report, Fig. 8). The X is submetacentric, the Y a minute element that, in good preparations, can be identified as a submetacentric (Fig. 8, insert). The karyotype shares many characteristics with those of *Cebus, Aotus,* and *Alouatta caraya,* and is most similar to that of *Cacajao rubicundus.* Only one female *Callicebus torquatus* has been investigated (this report, Fig. 9); its karyotype is most unusual and unlike anything found in other primates. The diploid number is 2n = 20, the lowest for any primate, and the complement is composed of 10 submetacentrics and 10 acrocentrics. No information is available on the morphology of the sex chromosomes.

In the genus Pithecia, only one female *P. pithecia* has been studied (Chu and Bender, 1961). The diploid number is 46, but a description of the karyotype was not published. No information is available in the genus Chiropotes.

The only species studied in the genus Cacajao is *C. rubicundus.* Chu and Bender (1961) reported a diploid number of 46, with no information on the karyotype. A recent study of one male in kidney cultures (this report) showed the animal to have only 45 chromosomes, obviously abnormal. There are 4 metacentrics, 16 submetacentrics, and 24 acrocentrics in the complement, plus one unmatched submetacentric, pre-

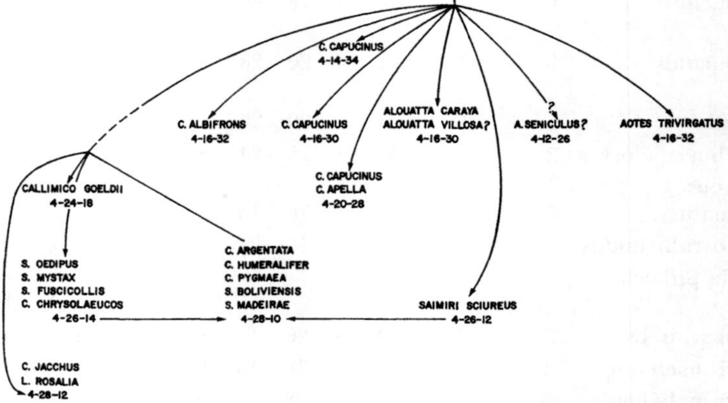

Fig. 7. Chromosomal evolution in the Cebidae and Callithricidae.

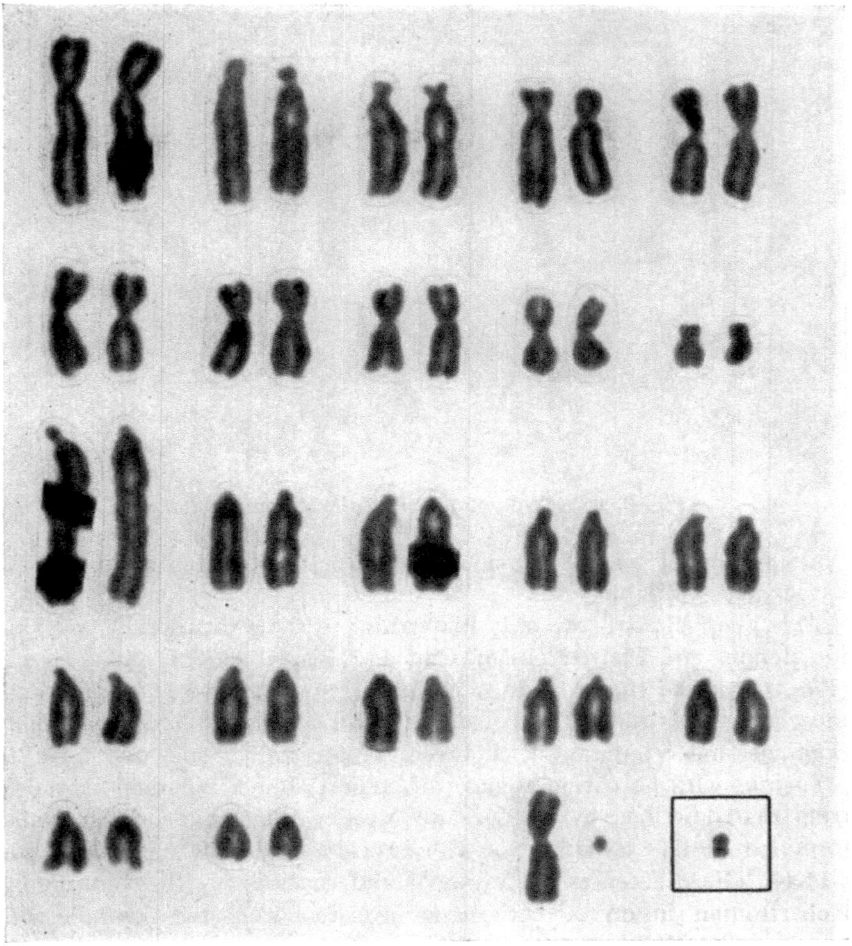

Fig. 8. Karyotype of male *Callicebus moloch.*

sumably the X chromosome. The karyotype is similar to that of *Callicebus moloch.*

Three species of Alouattinae have been studied. *Alouatta seniculus* has 44 chromosomes (Bender and Chu, 963). The complement is probably composed of 4 metacentrics, 12 submetacentrics, and 26 acrocentrics. The X chromosome is acrocentric, the Y submetacentric. *Alouatta caraya* has 52 chromosomes (Egozcue and Vilarasau de Egozcue, 1966b, 1967b). The complement (Fig. 6) is composed of 4 metacentrics, 16 submetacentrics, and 30 acrocentrics. The X is submetacentric, the Y acrocentric. The karyotype is similar to those of Cebus and Aotus and shares many common characteristics with those of *Callicebus moloch* and *Cacajao rubicundus.* Hsu (1965) studied one female *A. villosa;* the

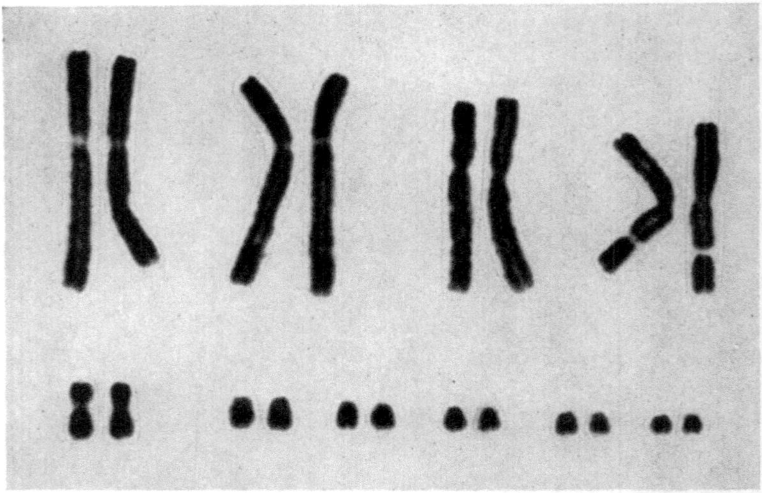

Fig. 9. Karyotype of female *Callicebus torquatus.*

diploid number was 53, obviously abnormal. No information on the karyotype is available.

Cebus capucinus is the only Platyrrhini to show intraspecific variability. Bender and Mettler (1958), Chu and Bender (1961), and Chiarelli (1961a) reported counts of 2n = 54 for *C. capucinus* and *C. apella.* Since then, we have studied one male *C. capucinus* with 52 chromosomes (Egozcue and Vilarasau de Egozcue, 1966c, 1967c) and one male *C. capucinus* with 54 chromosomes (this report) but a different karyotype from that described by Bender and Mettler (1958). The chromosomal formulae so far described for *C. capucinus* are 4-20-28, 4-16-30, and 4-14-34. The differences are minimal and consistent with evolution by Robertsonian fusion or pericentric inversion from the formula with more acrocentrics to the types with more submetacentrics (Fig. 6). Similar relationships can be found between *C. capucinus* and *C. albifrons, Alouatta caraya,* and *Aotus trivirgatus* (Fig. 6). The X chromosome of *C. capucinus* is either acrocentric or submetacentric, the Y is acrocentric. *Cebus fatuellus* has 54 chromosomes (Chiarelli, 1961a); no description of the karyotype is available. *Cebus albifrons* has 54 chromosomes (Egozcue and Vilarasau de Egozcue, 1967c, d) with 4 metacentrics, 16 submetacentrics, and 32 acrocentrics in the complement. The X is submetacentric, the Y acrocentric.

Three species of the genus Saimiri have been studied so far. *Saimiri sciureus* (Bender and Mettler, 1958) has 44 chromosomes, with 4 metacentrics, 26 submetacentrics, and 12 acrocentrics in the complement; these results have been confirmed by Chiarelli (1961a) and by us (this report). *Saimiri madeirae* and *Saimiri boliviensis* (Egozcue *et al.,* 1967) also have 44 chromosomes, but the complement is composed of 4 meta-

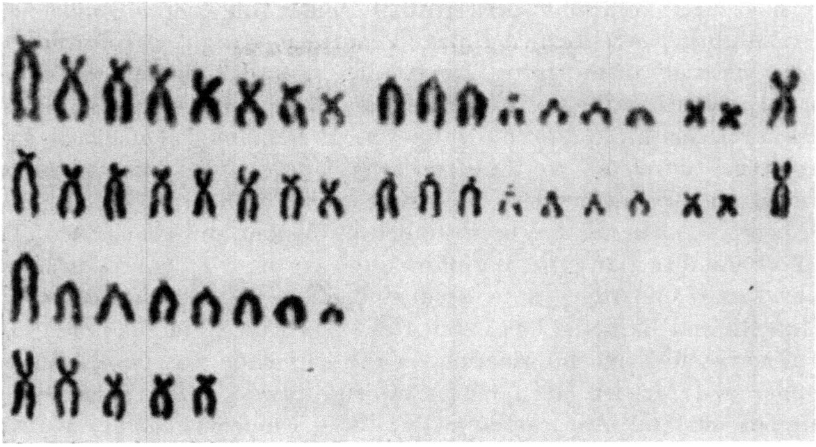

Fig. 10. Similarities (top) and dissimilarities (bottom) between the haploid karyotypes of *Alouatta caraya* (top row) and *Saguinus fuscicollis* (bottom row).

centrics, 28 submetacentrics, and 10 acrocentrics. Evolution from one type to another occurred either by pericentric inversion or centromere shift. In all three species the X is submetacentric, the Y acrocentric.

Saimiri and Callimico constitute two clear links between the Cebidae and the Callithricidae. We have seen before that *Cebus, Alouatta* and *Aotus,* and also *Cacajao* and *Callicebus moloch* are closely related from a chromosomal point of view. At the same time, they share many common characteristics with Saimiri and the Callithricidae (Fig. 10). Callimico is a controversial animal that has been assigned its own genus and family because it is difficult to decide whether to classify it with the Cebidae or the Callithricidae. It has the pes and manus of Saguinus and resembles the tamarins (Saguinus), but its dental formula is that of the Cebidae (Hill, 1957). From a chromosomal point of view, it is closer to the Callithricidae (Egozcue *et al.,* 1968d). On the other hand, Callimico as well as the Callithricidae have many common chromosomal characteristics with Cebus, Alouatta, Aotus, Cacajao and Callicebus. Finally, Saimiri belongs in the Cebidae, but the karyotypes of *Saimiri madeirae* and *S. boliviensis* are similar to those of *Callithrix argentata, C. humeralifer* and *C. pygmaea* (Fig. 11). The chromosomal similarities

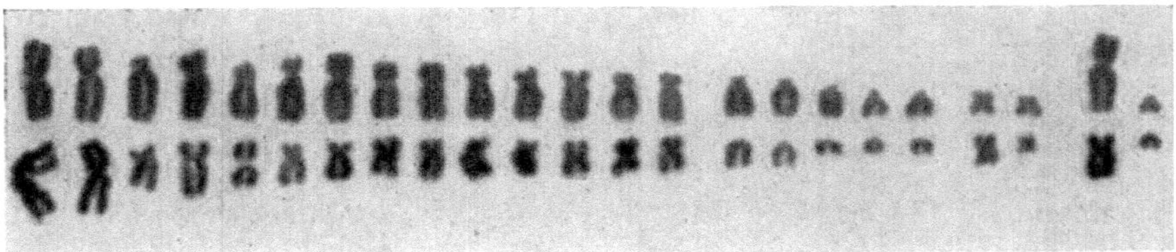

Fig. 11. Karyotypes (haploid) of *Callithrix humeralifer* (top) and *Saimiri madeirae* (bottom).

are such that a continuity between the Cebidae, the Callimiconidae, and the Callithricidae is clear. As in the Lemuridae, a rigid classification destroys the unity of the group. Nevertheless, not all Cebidae are chromosomally related to the Callithricidae, or at least there is no evidence for such a relationship. *Callicebus torquatus, Ateles,* and *Lagothrix* all have karyotypes unrelated to those reviewed above. The information on Pithecia is too fragmentary to permit comparison.

Lagothrix ubericola has been studied by Bender and Chu (1963). The diploid number is 2n = 62; no information on the karyotype is available. Four species of Ateles have been studied so far. *Ateles arachnoides* (Chiarelli and Barberis, 1966) and *Ateles paniscus* (Chiarelli, 1961) have 34 chromosomes. No information on the karyotype was published, but Bender and Mettler (1958) and Chu and Bender (1961) reported 34 chromosomes and identical karyotypes for *A. paniscus chamek, A. belzebuth,* and *A. geoffroyi cucullatus;* the karyotype is composed of 30 submetacentrics and 2 acrocentrics. The X is metacentric and the Y acrocentric. Studies of *A. geoffroyi cucullatus* (Egozcue and Hagemenas, 1967; Hsu and Benirschke, 1968) report identical results except for the presence of a submetacentric Y chromosome.

As we said before, *Callimico goeldii* is a special case and is classified by itself in the Callimiconidae (Table 6). Bender and Mettler (1958)

Table 6: Chromosomes of Callimiconidae and Callithricidae

Species	Specimens		2n	M	S	A	X	Y	Common Name
	Male	Female							
Callimico goeldii	2	2	48	4	24	18	S	A	Goeldi's marmoset
Callithrix chrysoleuca	1	2	46	4	26	14	S	S	Silky marmoset
C. jacchus	6	—	46	4	28	12	S	A	Common marmoset
C. pygmaea	9	—	44	4	28	10	S	A	Pigmy marmoset
C. argentata	2	—	44	4	28	10	S	M	Silver marmoset
C. humeralifer	1	—	44	4	28	10	S	A	White-shouldered marmoset
Saguinus oedipus	3	2	46	4	26	14	S	M	Cottontop pinché
S. fuscicollis illigeri	3	3	46	4	26	14	S	M	Red-mantled tamarin
S. nigricollis	2	—	46	4	26	14	S	M	Black-collared tamarin
S. mystax	—	—	46	4	26	14	S	M	Moustached tamarin
S. leucopus	—	—	46	—	—	—	—	—	White-handed tamarin
S. tamarin	—	—	46	—	—	—	—	—	Black tamarin
Leontideus rosalia	1	1	46	4	28	12	S	M	Golden lion-tamarin

reported counts of 48 chromosomes. These counts have been confirmed by us (Egozcue *et al.,* 1968d). The complement is composed of 4 metacentrics, 24 submetacentrics, and 18 acrocentrics. The X is submetacentric, the Y acrocentric (Fig. 7). The chromosomes of *Callimico goeldii,* like the species itself, are at a middle point between those of the Callithricidae and those of most Cebidae and are related to both groups.

The family Callithricidae includes three genera: Callithrix, Saguinus and Leontideus (Table 6, Fig. 7). For the classification of the Callithricidae we follow Hershkovitz (1958, 1966a, 1966b). Three types of chromosomal complements have been found in the group. *Callithrix chrysoleuca* (= *crysoleucos*) (Bender and Mettler, 1960), *Saguinus oedipus* (Chiarelli, 1961; Benirschke and Brownhill, 1962; Egozcue *et al.,* 1968d), *Saguinus fuscicollis illigeri* (Bender and Mettler, 1960; Egozcue *et al.,* 1968c), *Saguinus nigricollis* (Benirschke and Brownhill, 1962; Wohnus and Benirschke, 1966), and *Saguinus mystax* (Benirschke and Brownhill, 1963; Wohnus and Benirschke, 1966) all have 46 chromosomes, with 4 metacentrics, 26 submetacentrics, and 14 acrocentrics in the complement. *Callithrix pygmaea* (Benirschke and Brownhill, 1962; Wohnus and Benirschke, 1966; Egozcue *et al.,* 1968a, 1968d), *C. humeralifer* (Egozcue *et al.,* 1968d) and *C. argentata* (Egozcue *et al.,* 1968d) all have 44 chromosomes, of which 4 are metacentric, 28 submetacentric, and 10 acrocentric; their karyotype is most similar to that of *Saimiri madeirae* and *S. boliviensis* (Fig. 11). *Callithrix jacchus* (Chiarelli, 1961; Wohnus and Benirschke, 1966; Benirschke *et al.,* 1962), and *Leontideus rosalia*

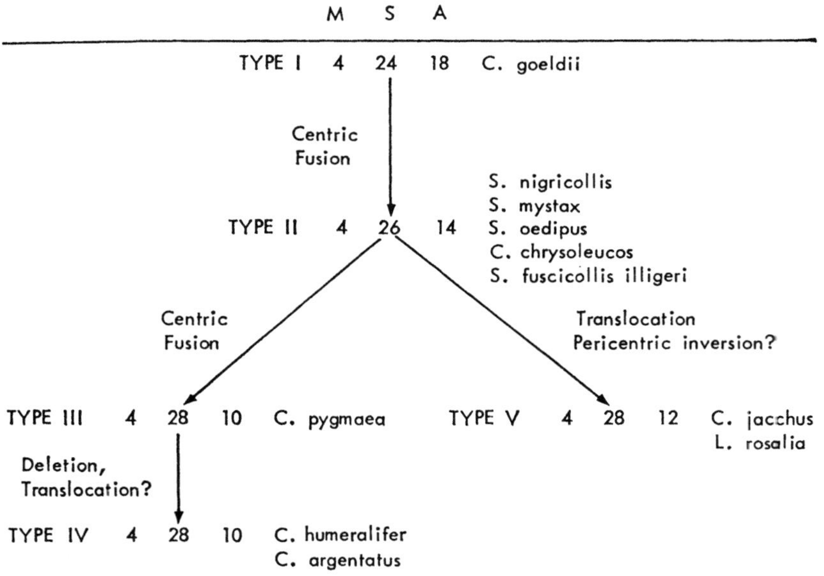

Fig. 12. Chromosomal evolution in the Callithricidae.

(Benirschke *et al.,* 1962) have a diploid number of 2n = 46, with 4 metacentrics, 28 submetacentrics, and 12 acrocentrics in the complement. The three groups are closely related (Wohnus and Benirschke, 1966; Egozcue *et al.,* 1968d). Figure 12 shows the evolutionary steps leading from one group to the other. *Saguinus tamarin* and *S. leucopus* (Benirschke and Brownhill, 1963) also have 46 chromosomes, but no description of the karyotype is available.

Suborder Catarrhini

The karyology of some genera in this group is not completely clear (see Table 9) and makes a comparative analysis rather difficult. Classification characteristics that place the genera Macaca, Papio, Theropithecus, and Cercocebus in a single subfamily, the Cercopithecinae, with the genera Cercopithecus and Erythrocebus, and the subfamily Hylobatinae in the superfamily Hominoidea with the great apes and man make the situation even more confusing. Thus, following the recommendations of Chiarelli (1966a) we will consider the genera Macaca, Papio, Theropithecus, and Cercocebus in a single subfamily, the Papinae. The subfamily Cercopithecinae will then include only the genera Cercopithecus and Erythrocebus; and the subfamilies Papinae and Cercopithecinae will constitute the family Cercopithecidae. Further, with Buettner-Janusch (1966) we will consider Theropithecus not as a different genus but as

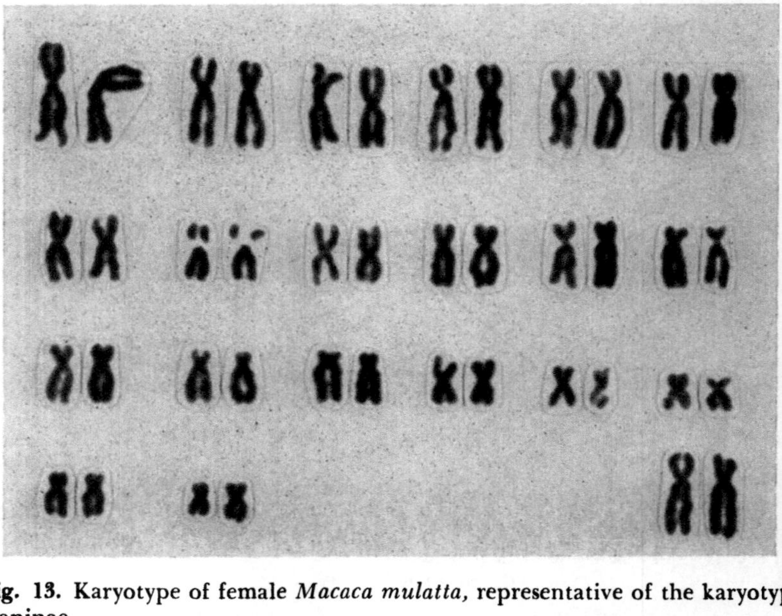

Fig. 13. Karyotype of female *Macaca mulatta,* representative of the karyotypes of Papinae.

a species of Papio, *Papio* (= *Theropithecus*) *gelada*. In accord with Chiarelli (1962a) we will consider Cynopithecus not as a genus but as a species of Macaca: *Macaca* (= *Cynopithecus*) *niger;* and with Chiarelli (1963a) we will consider Erythrocebus as a Cercopithecus species, *Cercopithecus* (= *Erythrocebus*) *patas*. Finally, following Klinger *et al.* (1963), Hamerton *et al.* (1963), and Chiarelli (1966a), we will place the Hylobatinae with the Cercopithecoidea rather than with the Hominoidea.

All Papinae have 42 chromosomes and similar karyotypes (Fig. 13); all chromosomes in the complement are submetacentric. The X chromosome is submetacentric; the Y has usually been described as an acrocentric but is really a submetacentric (Table 7, Fig. 14). The species studied in this group are listed below. The karyotypes of the Papinae are similar to those of the Colobinae and the Hylobatinae (Fig. 14, Table 8). Colobus have 44 chromosomes, all of which are submetacentric; the X is submetacentric, the Y acrocentric. Presbytis have 44

Table 7: Chromosomes of Papinae

Species	Specimens		2n	S	A	X	Y	Common Name
	Male	Female						
Macaca mulatta	46	133	42	40	—	S	S	Rhesus macaque
M. sylvana	2	—	42	40	—	S	S	Barbary ape
M. speciosa	2	4	42	40	—	S	S	Stumptail macaque
M. fuscata	3	1	42	40	—	S	S	Japanese macaque
M. assamensis	1	1	42	40	—	S	S	Assamese macaque
M. silenus	2	2	42	40	—	S	S	Liontailed macaque
M. nemestrina	7	22	42	40	—	S	S	Pigtail macaque
M. radiata	2	3	42	40	—	S	S	Bonnet macaque
M. sinica	—	1	42	40	—	S	S	Toque macaque
M. irus	3	1	42	40	—	S	S	Irus macaque or Cynomolgus macaque
M. cyclopis	1	—	42	40	—	S	S	Formosan macaque
M. maura	5	2	42	40	—	S	S	Moor macaque
M. (= Cynopithecus) niger	7	15	42	40	—	S	S	Celebes ape
Papio sphinx	2	1	42	40	—	S	S	Mandrill
P. hamadryas	4	5	42	40	—	S	S	Hamadryas baboon
P. cynocephalus	13	4	42	40	—	S	S	Yellow baboon
P. (= Theropithecus) gelada	1	1	42	40	—	S	S	Gelada baboon
Cercocebus albigena	2	—	42	40	—	S	S	Grey-cheeked mangabey
C. aterrimus	2	2	42	40	—	S	S	Black mangabey
C. galeritus	3	—	42	40	—	S	S	Crested mangabey
C. torquatus	4	—	42	40	—	S	S	
C. t. torquatus	1	—	42	40	—	S	S	Sooty mangabey
C. t. lunulatus	2	—	42	40	—	S	S	White-crowned mangabey

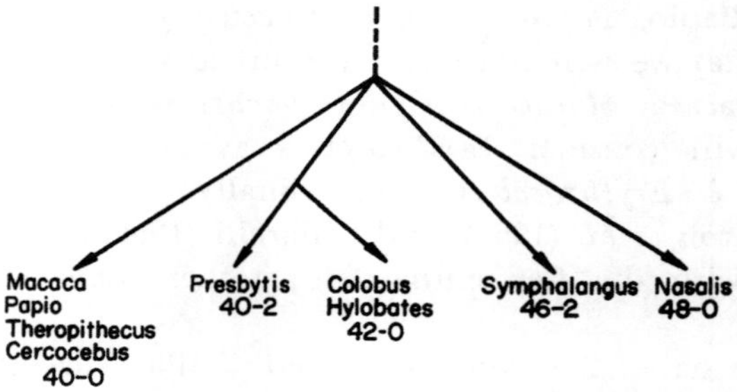

Fig. 14. Chromosomal evolution in the Papinae, Colobinae and Hylobatinae.

chromosomes, with 40 submetacentrics and 2 acrocentrics in the complement; the X is submetacentric, the Y acrocentric. Nasalis have 48 chromosomes, all of which are submetacentric; only one female has been studied. Hylobates have 44 chromosomes, with 42 submetacentrics; the X and Y are submetacentric. The karyotypes of Colobus, Presbytis, and Hylobates are quite similar and very close to those of the Papinae. Symphalangus have 50 chromosomes, of which 46 are submetacentric and 2 acrocentric; the X and Y are submetacentric. The relationships between these groups are shown in Figure 14.

The species of Papinae, Colobinae, and Hylobatinae studied are as follows:

Subfamily Papinae, genus Macaca: *Macaca mulatta* (Painter, 1924; Shiwago, 1939; Darlington and Haque, 1955; Chu and Giles, 1957;

Table 8: Chromosomes of Colobinae and Hylobatinae

Species	Specimens		2n	S	A	X	Y	Common Name
	Male	Female						
Colobinae								
Presbytis entellus	3	4	44	40	2	S	A	Indian common langur
P. obscurus	1	1	44	40	2	S	A	Dusky leaf-monkey
P. senex	1	1	44	40	2	S	A	Purple-faced langur
Colobus polykomos	—	1	44	42	—	S	—	Colobus monkey
C. badius (= Procolobus kirkii)	—	—	44	42	—	S	A	Red colobus
Nasalis larvatus	—	1	48	46	—	S	—	Proboscis monkey
Hylobatinae								
Hylobates lar	6	4	44	42	—	S	S	White-handed gibbon
H. agilis	2	2	44	42	—	S	S	Dark-handed gibbon
H. moloch	2	1	44	42	—	S	S	Sunda island gibbon
H. hoolock	—	2	44	42	—	S	S	Hoolock gibbon
Symphalangus syndactylus	1	1	50	46	2	S	S	Siamang

Rothfels and Siminovitch, 1958; Chiarelli, 1962a; Baylet and Grattepanche, 1964a; Borgaonkar, 1966a; Farber, 1966a; Egozcue *et al.*, 1968d; Manna and Talukdar, 1968); *M. sylvana* (Chiarelli, 1962a); *M. speciosa* (Chiarelli, 1962a; Egozcue *et al.*, 1968d); *M. fuscata* (Chiarelli, 1962a; Egozcue *et al.*, 1968d); *M. assamensis* (Chiarelli, 1962a); *M. silenus* (Chiarelli, 1962a; Egozcue *et al.*, 1968d); *M. nemestrina* (Darlington and Haque, 1955; Chiarelli, 1962a; Egozcue *et al.*, 1968d); *M. radiata* (Chiarelli, 1962a; Egozcue *et al.*, 1968d); *M. sinica* (Chiarelli, 1962); *M. irus* (Chu and Bender, 1961; Chiarelli, 1962a; Egozcue *et al.*, 1968d); *M. cyclopis* (Makino, 1952); *M. maura* (Chiarelli, 1962a; Egozcue *et al.*, 1968d); *M. niger* (Chiarelli, 1962a; Ushijima *et al.*, 1964; Egozcue *et al.*, 1968d). All these species have 42 chromosomes; all the chromosomes in the complement are submetacentric.

Subfamily Papinae, genus Papio: *Papio sphinx* (Chiarelli, 1962a; Bender and Mettler, 1958); *P. hamadryas* (includes *P. leucophaeus*) (Chiarelli, 1962a; Bender and Chu, 1963; Nemtseva, 1964; Egozcue *et al.*, 1968d); *P. cynocephalus* (includes *P. papio, P. anubis, P. comatus, P. ursinus,* and *P. doguera*) (Darlington and Haque, 1955; Chu and Giles, 1957; Chiarelli, 1962a; Bender and Chu, 1963; Baylet and Grattepanche, 1964a; Irsigler, 1967; Egozcue *et al.*, 1968d; *Tobias,* 1956). The characteristics of the complement are the same as for Macaca. The diploid number is 42, and all chromosomes are submetacentric.

Subfamily Papinae, genus Theropithecus: as indicated, this genus should be considered a species of Papio; thus, *Papio* (= *Theropithecus*) *gelada* (Chiarelli, 1962a).

Subfamily Papinae, genus Cercocebus: *Cercocebus albigena* (Chiarelli, 1962a; Tappen, 1960); *C. aterrimus* (Chiarelli, 1962a; Klinger, 1963); *C. galeritus* (Chiarelli, 1962a; Tappen, 1960); C. torquatus torquatus (Bender and Mettler, 1958); *C. t. lunulatus* (Chu and Giles, 1957). The chromosomal characteristics of this group are the same as for Macaca and Papio. The diploid number is 42; all chromosomes in the complement are submetacentric.

Family Colobidae, subfamily Colobinae: in this group, no information is available on the genera Pygathrix, Rhinopithecus, and Simias. Genus Presbytis: *Presbytis entellus* (Makino, 1952; Ushijima *et al.*, 1964; Egozcue *et al.*, 1968d; Sharma and Kakati, 1966); *P. obscurus* (Chiarelli, 1962b); *P. senex* (Goldstein and Birdwell, 1967). All Presbytis have 44 chromosomes, with 40 submetacentrics and 2 acrocentrics in the complement. The X is submetacentric, the Y acrocentric.

Subfamily Colobinae, genus Colobus: *Colobus polykomos* (Bender and Chu, 1963; Chiarelli, 1962b); *C. badius* (= *Procolobus kirkii*) (Kuhn, 1967). Colobus have 44 chromosomes, all of which are submetacentric. The X is submetacentric, the Y acrocentric.

Subfamily Colobinae, genus Nasalis: only one female *Nasalis larvatus* has been studied (Chiarelli, 1966c). The diploid number is 48; all chromosomes in the complement are submetacentric.

Family Hylobatidae, subfamily Hylobatinae, genus Hylobates: *Hylobates lar* (Chiarelli, 1962b; Bender and Chu, 1963; Hamerton *et al.*, 1963; Klinger *et al.*, 1963); *H. agilis* (Chiarelli, 1962b); *H. moloch* (Chiarelli, 1962b; Hamerton *et al.*, 1963); *H. hoolock* (Chiarelli, 1961b; Chu and Bender, 1961). Hylobates have 44 chromosomes, all of which are submetacentric. Their karyotype is closest to that of Colobus and quite similar to those of Presbytis, Macaca, Papio and Cercocebus.

Subfamily Hylobatinae, genus Symphalangus: *Symphalangus syndactylus* (Bender and Chu, 1963; Klinger, 1963) has 50 chromosomes, with 46 submetacentrics and 2 acrocentrics. The X and Y are submetacentric.

The subfamily Cercopithecinae, genus Cercopithecus, which, with the Papinae, is part of the family Cercopithecidae, is the most complex group of primates from a chromosomal point of view. Chromosomal variability is extremely marked in this genus (Table 9, Fig. 15). The difficulty is increased by conflicting reports on some of the species as well as by problems of classification. For instance, *Cercopithecus*

Table 9: Chromosomes of Cercopithecinae

Species	Specimens Male	Specimens Female	2n	S	A	X	Y	Common Name
Cercopithecus patas	6	6	54	38	14	S	A, S	Patas monkey
C. talapoin	1	—	54	38	14	S	A	Talapoin
C. aethiops	15	3	60	36	22	S	A, S	Green vervet monkey
C. a. sabaeus	4	3	60	36	22	S	S	African green monkey
C. a. tantalus	2	1	60	—	—	S	S	White monkey
C. diana	1	1	58	—	—	—	—	Diana monkey
C. d. roloway	1	1	60	—	—	—	—	Diana monkey
C. nigroviridis	2	1	60	36	22	S	A	
C. neglectus	2	—	(60)	—	—	—	—	DeBrassa's guenon
C. cephus	—	1	66	48	16	S	—	Moustached guenon
C. mona mona	2	5	66	48	16	S	S	Mona guenon
C. m. campbelli	1	4	66	48	16	S	A	Campbell's guenon
C. m. denti	1	—	66	—	—	—	—	
C. m. preussi	—	1	66	—	—	—	—	
C. ascanius	—	—	66	—	—	—	—	White-nosed guenon
C. petaurista	—	—	66	—	—	—	—	White-nosed guenon
C. nictitans buttikoferi	—	3	66	48	16	S	—	White-nosed guenon
C. nictitans	2	2	70	—	—	—	—	White-nosed guenon
C. l'hoesti	—	1	72	48	22	S	—	l'Hoest's guenon
C. mitis	3	3	72	48	22	S	S	Diadem guenon
C. m. mitis	2	—	72	48	22	S	A	
C. m. albogularis	1	—	72	48	22	S	A	

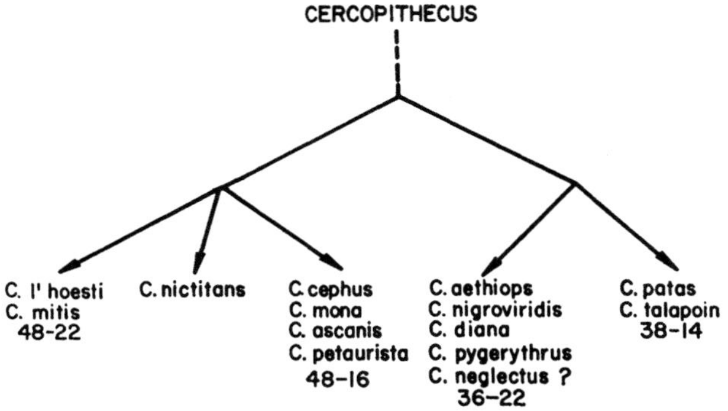

Fig. 15. Chromosomal evolution in Cercopithecus.

ascanius, C. petaurista, and *C. buttikoferi,* which were considered synonyms of *C. nictitans* or subspecies (*C. nictitans ascanius,* etc.), have 66 chromosomes, whereas *C. nictitans* proper has 70 chromosomes. A revision of the group has been attempted by Kuhn (1967) and by Chiarelli (1966a). The chromosome numbers vary between 54 and 72; nevertheless, the existence of two major groups within the Cercopithecus is clear. One includes the species with low diploid numbers, 2n = 54 and 2n = 60; the only exception is *C. diana* (Chiarelli, 1962b), which has 58 chromosomes (but *C. diana roloway* has 60; Chu and Giles, 1957). The second group includes those species with high diploid numbers, 2n = 66 and 2n = 72 (Fig. 16); the only exception is *C. nictitans,* with 70 chromosomes (Chiarelli, 1963a; Kuhn, 1967). Since description of the karyotypes is not available for many species, comparison between them is difficult.

Cercopithecus patas (Chu and Giles, 1957; Chiarelli, 1962b; Baylet and Grattepanche, 1964a; Egozcue *et al.,* 1968d); *C. talapoin* (Chiarelli, 1962b). Both species have 54 chromosomes, with 38 submetacentrics and 14 acrocentrics in the complement. The X is submetacentric; the Y has been described as either a submetacentric or an acrocentric.

Cercopithecus aethiops (= *pygerythrus,* Farber, 1966b) (Chiarelli, 1962b; Bender and Chu, 1963; Baylet and Grattepanche, 1964a; Egozcue *et al.,* 1968d); *C. aethiops tantalus* (Chu and Giles, 1957); *C. aethiops sabaeus* (= *sabaceus*) (Chu and Giles, 1957; Baylet and Grattepanche, 1964b; Ruzicska and Hsu, 1967); *C. diana roloway* (Chu and Giles, 1957); *C. nigroviridis* (Chiarelli, 1962b); and probably *C. neglectus* (Tappen, 1960; Chiarelli, 1962b) all have 60 chromosomes, with 36 submetacentrics and 22 acrocentrics. The X is submetacentric, the Y also submetacentric. *C. diana* (Chiarelli, 1962b) has 58 chromosomes; no description of the karyotype is available.

Cercopithecus cephus (Chiarelli, 1963a); *C. mona mona* (Bender and Mettler, 1958; Bender and Chu, 1963; Chiarelli, 1963a); *C. mona camp-*

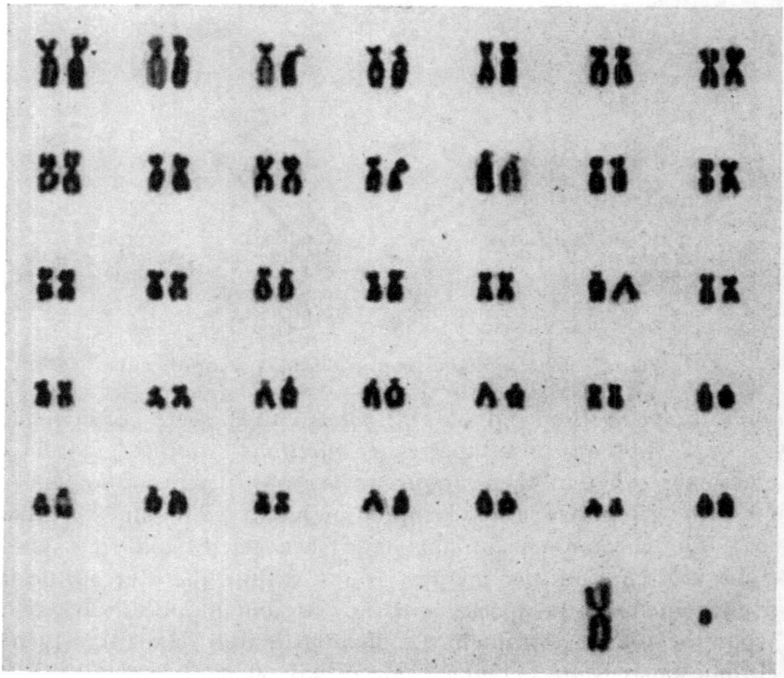

Fig. 16. Karyotype of male *Cercopithecus mitis.*

belli (Chiarelli, 1962b; Chu and Giles, 1957); *C. mona denti* (Tappen, 1960; cited by Bender and Chu, 1963); *C. mona preussi* (Chiarelli, 1962b); *C. ascanius* (Kuhn, 1967; Chiarelli, 1962b); *C. petaurista* (Kuhn, 1967); and *C. nictitans buttikoferi* (Chu and Giles, 1957) all have 66 chromosomes, with 48 submetacentrics and 16 acrocentrics in the complement. The X and probably the Y are submetacentric. Some of the results in this group reinforce the suggestion that a revision of the genus Cercopithecus is in order. *Cercopithecus ascanius, petaurista,* and *buttikoferi* are synonyms for *C. nictitans* or have been used to designate subspecies *(C. nictitans buttikoferi,* etc.). All three groups have 66 chromosomes and are probably a different species from *C. nictitans,* which has 70 chromosomes (Kuhn, 1967) although no description of the karyotypes is available for comparison. Chiarelli (1963a) also reported tentative counts of $2n = 70$ for *C. nictitans.*

Cercopithecus l'hoesti (Chu and Bender, 1961) and *C. mitis* (Bender and Chu, 1963; Egozcue and Vilarasau de Egozcue, 1967e), *C. mitis mitis* (Chiarelli, 1963a) and *C. m. albogularis* (Chiarelli, 1963a) all have 72 chromosomes, with 48 submetacentrics and 22 acrocentrics in the complement. The X is submetacentric, the Y acrocentric.

Table 10: Chromosomes of Hominoidea

Species	Specimens Male	Female	2n	S	A	X	Y	Common Name
Pongidae								
Pongo pygmaeus	5	3	48	26	20	S	S	Orangutan
Gorilla gorilla gorilla	2	4	48	30	16	S	S	Lowland gorilla
G. g. beringei	—	1	48	30	16	S	—	Mountain gorilla
Pan troglodytes paniscus	3	4	48	34	12	S	A	Pygmy chimpanzee
P. t. troglodytes	5	9	48	34	12	S	A	Northern chimpanzee
Hominidae								
Homo sapiens	Several hundred		46	34	10	S	A	Man

The superfamily Hominoidea (Table 10, Fig. 17) includes the Pongidae and the Hominidae. The Pongidae include the Hylobatinae, which as we have seen should be placed among the Cercopithecoidea, and the Ponginae, which include Pongo, Pan and Gorilla. The Hominoidea include only the Hominidae, subfamily Homininae; the only genus is *Homo*.

On the basis of immunochemical studies, Goodman (1962) suggested that Gorilla and Pan should be removed from the Pongidae and placed with Homo in the Hominidae. On chromosomal evidence, the karyotypes of Pan and Gorilla more closely resemble that of Homo than that of Pongo; on the other hand, the transition from one karyotype to the other is better understood from Pongo to Pan and Gorilla than from any of them to Homo. All species of Hominoidea (excluding the Hylobatidae) have been studied. *Pongo pygmaeus* (Chiarelli, 1961c, 1962c; Hamerton *et al.*, 1963; Klinger *et al.*, 1963) has 48 chromosomes, with 26 submetacentrics and 20 acrocentrics in the complement. The X and Y are submetacentric.

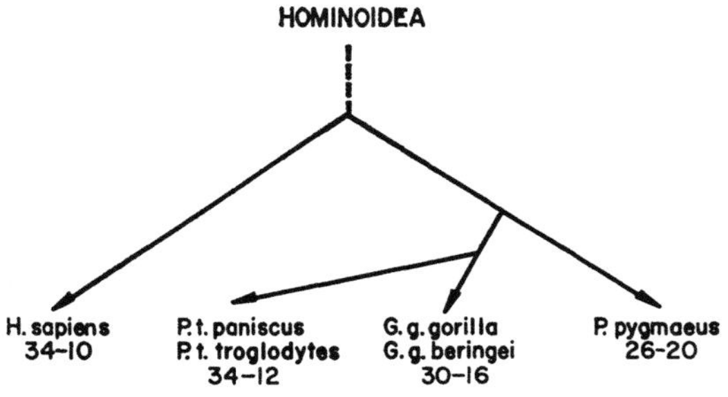

Fig. 17. Chromosomal evolution in the Hominoidea.

Two subspecies of Gorilla have been investigated. *Gorilla gorilla gorilla* (Hamerton *et al.,* 1961, 1963; Klinger, *et al.,* 1963; Egozcue and Chiarelli, 1966) and *G. g. beringei* (Hamerton *et al.,* 1963) have 48 chromosomes. The complement is composed of 30 submetacentrics and 16 acrocentrics. The X and Y are submetacentric.

In 1940, Yeager *et al.* found 48 chromosomes in testicular material of two "chimpanzees." Young *et al.* (1960) described 48 chromosomes in nine specimens of Pan troglodytes; the subspecies was not identified. Since then these counts have been confirmed in *Pan troglodytes troglodytes* (Chu and Bender, 1961; Chiarelli, 1962c; Bender and Chu, 1963; Hamerton *et al.,* 1963; Klinger *et al.,* 1963; Egozcue *et al.,* 1968d) and in *P. t. paniscus* (Chiarelli, 1962c; Hamerton *et al.,* 1963; Klinger *et al.,* 1963). The karyotypes of the two subspecies are identical but for one minor difference: one of the smaller chromosomal pairs is more metacentric in paniscus than it is in troglodytes. The complement is composed of 34 submetacentrics and 12 acrocentrics; the X is submetacentric, the Y acrocentric.

For a long time *Homo sapiens* was believed to have 48 chromosomes (Painter, 1923), but recently it has been proved that the correct diploid number is 46 (Tjio and Levan, 1956; Ford and Hamerton, 1956). The complement is composed of 34 submetacentrics and 10 acrocentrics; the X is submetacentric, the Y acrocentric.

Chromosomal Evolution

The general principles of chromosomal evolution in relation to primates have been discussed by a number of authors (Hamerton, 1958, 1963; Bender and Chu, 1963). Nevertheless, new data, some evolutionary trends and the mechanisms responsible for them indicate the need for a review. A comparative analysis of primate chromosomes shows that practically all members of a genus (and with few exceptions, of a family as well) share many common chromosomal characteristics. Evolutionary trends can be understood in all primate groups where sufficient information is available. Not only is there a clear relationship between all Lemuridae, but practically all Platyrrhini genera show common chromosomal characteristics, and even four different Papinae genera have the same karyotype. These data show the close relationship between primate groups, and prove not only that taxonomic classification is accurate but also that it may be splitting in excess. Nevertheless, we must remember here again that morphological similarity between the chromosomes of two different species does not assure homology.

It is well known that species with high chromosomal numbers have also a high number of acrocentrics, whereas species with low diploid numbers show a predominance of submetacentrics (Bender and Chu,

1963). From this, one may conclude that the tendency in the order Primates is toward a reduction in chromosomal number and has taken place mainly by the Robertsonian mechanism (Robertson, 1916), a translocation at or near the centromere, frequently and incorrectly referred to as centric fusion. This mechanism achieves a reduction in the number of chromosomes with a minimum loss of genetic material and reflects itself in the chromosomal formula as a loss of four in the number of acrocentric chromosomes and an increase of two in the number of submetacentrics. A change of this type is easily spotted (Figs. 1, 6, 7, 12): it has been found in many cases and has been frequently reported. Because of this, it has been assumed that changes of the Robertsonian type have been prevalent in the chromosomal evolution of primates.

On the other hand, let us remember with Hamerton (1958) that the importance of chromosomes "lies not in their morphological but in their genetic character," and that the real importance of the changes undergone by chromosomes during evolution can "be summed up in one word, isolation." The combination of genetic and reproductive isolation may then result in the formation of a new species and may eventually lead to phenotypic differences. Recent evidence (this report) shows that pericentric inversion, an extremely effective mechanism of genetic isolation, has been at least as important as the Robertsonian changes in the evolution of primate karyotypes and has probably been a leading factor in speciation (Fig. 18). This is particularly evident in the Platyrrhini, where the results of the independent action of Robertsonian fusion and pericentric inversion on the same ancestral karyotype is clear (Fig. 6). One of the karyotypes now found in *Cebus capucinus* is probably quite similar to the one that originated the karyotypes of *Alouatta caraya, Aotus trivirgatus,* and *Cebus albifrons.* As shown in Figure 6, the karyotype of *A. caraya* probably originated by Robertsonian fusion, whereas those of *Aotus* and *C. albifrons* did so by pericentric inversion or centromere shift. A similar mechanism is probably responsible for the differences between the karyotypes of *Lemur mongoz*

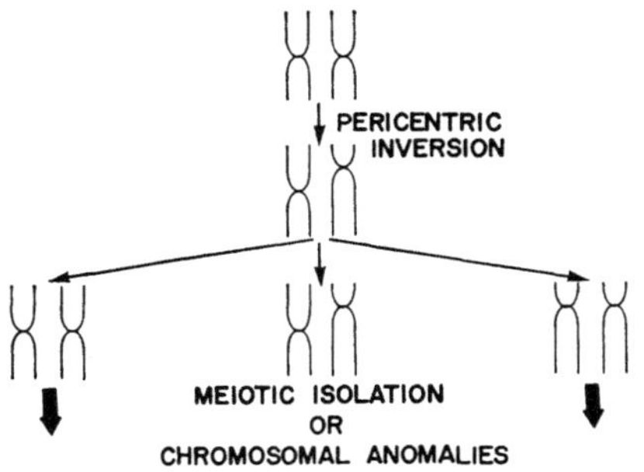

Fig. 18. Role of pericentric inversion in speciation.

and *L. fulvus rufus* and those of *L. mongoz, L. fulvus fulvus,* and *L. f. albifrons* and the differences between the karyotypes found in *L. catta* (Fig. 1); for the differences between the karyotypes described for *Loris tardigradus* (Fig. 3); perhaps for some of the differences between *Perodicticus potto* and *Galago crassicaudatus* (Figs. 2, 3, 5); for the very important differences between the karyotypes described for *Galago senegalensis* or for *G. crassicaudatus* (Fig. 5), although in this case the information is still very fragmentary; for the differences between the species of Saimiri or between *Callithrix jacchus* and *Leontideus rosalia* when compared to *C. humeralifer, C. argentata, C. pygmaea, Saimiri madeirae,* and *S. boliviensis* (Fig. 7, 12). Such a mechanism can also be responsible for many of the minor differences found between the Papinae, the Colobinae, and the Hylobatinae (Fig. 14). A situation similar to that found in Cebus, Alouatta and Aotus, in which evolution takes place in one direction by Robertsonian fusion and in the other by pericentric inversion, is probably responsible for the differences between the two Cercopithecus groups with a low chromosomal number and the differences between the two groups with a high chromosomal number (Fig. 15). Finally, the differences in the karyotypes of the great apes are also better explained by pericentric inversion (Fig. 17). In summary, similar karyotypes in related species with different chromosomal numbers can best be explained if it is kept in mind that evolution by Robertsonian fusion, compared to evolution by pericentric inversion, results in "loss" of one chromosomal pair with every change. That is, the karyotype that gains one submetacentric pair by Robertsonian fusion will have one pair of chromosomes less than the karyotype in which one submetacentric is gained by pericentric inversion.

The tendency in primates toward reduced chromosome numbers and some of the evidence available at the time led Bender and Chu (1963) to speculate that lower diploid numbers and a predominance of submetacentric chromosomes might be a characteristic of the more specialized forms, as opposed to higher diploid numbers and a predominance of the acrocentrics in the more primitive species. We know now that some very specialized forms, like the Tarsiers, have some of the highest diploid numbers and many acrocentrics, whereas species that are regarded as primitive (e.g., *Callithrix pygmaea*) have low counts and a predominance of submetacentrics. Thus, it seems that the "primitiveness" of a species has no correlation with the morphology of its chromosomes, and it is evident that species regarded as primitive have undergone more chromosomal rearrangements than other more specialized forms. In summary, there is no reason to support the contention that a high diploid number and a predominance of acrocentric chromosomes represent a primitive feature.

To explain the wide range in chromosomal numbers found in primates, which vary from 2n = 20 in *Callicebus torquatus* to 2n = 80 in *Tarsius bancanus* and *T. syrichta*, a number of theories have been suggested. One of them, polyploidy, was examined and rejected (Bender and Chu, 1963) on the grounds of its incompatibility with a sex-determining system of the XX/XY type. The argument is, of course, valid. The other objection to polyploidy was the presence in most Catarrhini of one pair of "marked" (Chiarelli, 1966b) chromosomes, consisting of a pair of acrocentric or metacentric chromosomes with a long achromatic region. If polyploidy had occurred, one would expect to find different numbers of such chromosomes; however, new evidence found in other primate groups refutes this. The importance accorded the presence of such chromosomes in the relationship between primate groups has been grossly overrated. Marker chromosomes are found in most primate species, in many mammalian species, and in large numbers of nonmammals. The fact that related species with identical or similar karyotypes (Fig. 6) show different numbers of chromosomes with achromatic regions in different places suggests that such regions (which, according to the classical notion of heterochromatin, are probably inactive) appeared after speciation occurred; that is, had polyploidy been responsible for the increase in chromosomal numbers, it would have occurred before the appearance of the achromatic regions and would have had no effect on the number of marker chromosomes. Indeed, the achromatic regions may indicate the inactivation of regions too deleterious to have in duplicate and the possibility that additive mechanisms have had a role in increasing the number of chromosomes in primates (Chiarelli, 1967). Nevertheless, it is not likely that polyploidy had a role in increasing the chromosomal number of the primitive primates.

A more appealing theory has been put forth by Todd (1967). The increase in chromosome numbers in primates would have taken place by chromosome fission (Darlington, 1958), an extensive mechanism that transforms submetacentric chromosomes into acrocentrics, acting on an original karyotype of probably 2n = 14. Processes of fission would underline the periods of explosive (Romer, 1960) or eruptive (Mayr, 1963) evolution, characterized by extensive speciation followed by adaptive radiation. The periods of sequential evolution of one form to another would be characterized by genetic stabilization by chromosomal fusion, pericentric inversion, etc. The further proposition that the sex chromosomes would not be affected is compatible with the generally accepted notion of great sex chromosome stability (Ohno, 1967). The theory interprets the karyotypes of some specialized groups with high chromosome numbers and a predominance of the acrocentrics, such as dogs (Minouchi, 1927) and bears (Low *et al.*, 1964; Benirschke, 1967), as conspicuous examples of karyotypic fissioning. As we have said, there is no reason to consider such karyotypes as primitive features, and the presence in a specialized group of a high diploid number with many acrocentrics does not need justification. On the other hand, we have seen evidence that speciation has often occurred by pericentric inversion or Robertsonian fusion independently of a previous fissioning event. Nevertheless, the theory is appealing and with necessary modifications can account for the extreme differences in chromosome numbers observed in mammals. Successive fissioning, alternating with periods in which Robertsonian changes and pericentric inversion produce new submetacentric elements, would result in the number of chromosomes needed to explain karyotypes, like those of Tarsius, after only a few fissioning events. On the other hand, fissioning would result in a progressive reduction of chromosomal size. Consistent with this is the fact that species with low chromosomal numbers tend to show extremely long chromosomes, whereas species with high diploid numbers even show microchromosomes.

On the other hand, Robertsonian changes produce some loss of genetic material. An accumulation of translocation events would result in reduction of the DNA contents in the genome of a species. The fact that, with few exceptions (Ohno, 1967), all mammalian species show similar DNA contents (Kleinfeld and Chu, 1958) seems to favor fission as an explanation, but Chiarelli has shown (1967) that the total length of the genome decreases with chromosomal number in Cercopithecus. His interpretation that this reflects the occurrence of additive mechanisms in the evolution of primate karyotypes is as valid as the suggestion that it may reflect the loss of genetic material that takes place in translocations. In any event, neither additive mechanisms nor fissioning alone can explain the numerical increase (or decrease) in chromosomal numbers in primates and the differences in DNA content or genome length. Probably both types have been operative. As of now, the trend seems to be toward a reduction in chromosomal numbers by means of Robertsonian fusions and pericentric inversions.

Summary

A review of primate chromosome complements shows that a close relationship exists among most species in a given genus and among many genera in any given family. Recent evidence indicates that the trend in primates seems to be toward a reduction in chromosomal numbers by Robertsonian changes and pericentric inversion. The importance of pericentric inversion in chromosomal evolution and speciation is evidenced in some of the primate groups. Analysis of the available data shows that fissioning, as well as additive mechanisms, may be responsible for the extreme variability in chromosomal numbers now found in primates.

Acknowledgment

Publication No. 358 of the Oregon Regional Primate Research Center, supported in part by Grant No. FR000163 of the National Institutes of Health.

References

Baylet, R. and H. Grattepanche: Sur les chromosomes des Cercopithecidae *Papio papio, Macaca mulatta, Cercopithecus aethiops, Erythrocebus patas.* C. R. Soc. Biol. *158:*1382, 1964.

——— and ———: Chromosomes de *Cercopithecus aethiops.* C. R. Acad. Sci. (Paris) *259:*3096, 1964.

Beçak, W., M. L. Beçak, H. R. S. Nazareth and S. Ohno: Close karyological kinship between the reptilian suborder Serpentes and the class Aves. Chromosoma (Berl.) *15:*606, 1964.

Bender, M. A. and E. H. Y. Chu: The chromosomes of primates. In: Evolutionary and Genetic Biology of Primates. J. Buettner-Janusch, ed. New York: Academic Press, 1963.

——— and L. E. Mettler: Chromosome studies of primates. Science *128:*186, 1958.

——— and ———: Chromosome studies of primates. II. *Callithrix, Leontocebus* and *Callimico.* Cytologia (Tokyo) *25:*400, 1960.

Benirschke, K.: Sterility and fertility of mammalian hybrids. In: Comparative Aspects of Reproductive Failure. K. Benirschke, ed. New York: Springer-Verlag New York Inc., 1967.

———, J. M. Anderson and L. E. Brownhill: Marrow chimerism in marmosets. Science *138:*513, 1962.

——— and L. E. Brownhill: Further observations on marrow chimerism in marmosets. Cytogenetics *1:*245, 1962.

——— and ———: Heterosexual cells in testes of chimeric marmoset monkeys. Cytogenetics *2:*331, 1963.

Borgaonkar, D. S.: Cytotaxonomic studies on the Tenrecoidea Simpson. Fifth Conf. Mamm. Cytol. & Som. Cell Genet., St. Simons Island, Georgia, 1966a.

———: A list of chromosome numbers in primates. J. Hered. *57:*60, 1966b.

———: Additions to the list of chromosome numbers in the insectivores and primates. J. Hered. *58:*211, 1967.

Buettner-Janusch, J.: A problem in evolutionary systematics: nomenclature and classification of baboons, genus Papio. Folia primat. *4:*288, 1966.

Chiarelli, B.: Some chromosome numbers in primates. Mammal. Chrom. Newsl. *6:*3, 1961a.

———:Cariologia e sistematica dei primati con cenni sulla origine del cariotipo umano. Comunicazione alla Soc. Italiana di Antropologia ed Etnologia. 1961b.

———: Chromosomes of the Orang-Utan (*Pongo pygmaeus*). Nature *192:*285, 1961c.

———: Comparative morphometric analysis of primate chromosomes. II. The chromosomes of the genera *Macaca, Papio, Theropithecus* and *Cercocebus.* Caryologia *15:*401, 1962a.

———: Some new data on the chromosomes of Catarrhina. Experientia *18:*405, 1962b.

———: Comparative morphometric analysis of primate chromosomes. I. The chromosomes of anthropoid apes and of man. Caryologia *15:*99, 1962c.

———: Primi resultati di ricerche di genetica e cariologia comparata in primati e loro interese evolutivo. Rivista di Antropologia *50:*87, 1963a.

———: Comparative morphometric analysis of primate chromosomes. III. The chromosomes of the genera *Hylobates, Colobus* and *Presbytis.* Caryologia *16:* 637, 1963b.

———: Caryology and taxonomy of the Catarrhine monkeys. Amer. J. Phys. Anthrop. *24:*155, 1966a.

———: Marked chromosome in Catarrhine monkeys. Folia primat. *4:*74, 1966b.

———: The chromosome complement of *Nasalis larvatus* (Wurm, 1781). Experientia *22:*797, 1966c.

———: Statistical analysis of the relation between the length of the genome and the diploid number of chromosomes in *Cercopithecus.* Experientia *23:* 672, 1967.

——— and L. Barberis: Some data on the chromosomes of Prosimiae and of new world monkeys. Mammal. Chrom. Newsl. *22:*216, 1966.

——— and J. Egozcue: The meiotic chromosomes of some primates. Mammal. Chrom. Newsl. *9:*85, 1968.

———, ——— and M. Monchietto: Caryological Atlas of Living Primates. In preparation. 1968.

Chu, E. H. Y. and M. A. Bender: Chromosome cytology and evolution in primates. Science *133:*1399, 1961.

——— and ———: Cytogenetics and evolution of primates. Ann. N.Y. Acad. Sci. *102:*253, 1962.

——— and N. H. Giles: A study of primate chromosome complements. Amer. Naturalist *91:*273, 1957.

——— and B. A. Swomley: Chromosomes of Lemurine lemurs. Science *133:* 1925, 1961.

Darlington, C. D. and A. Haque: Chromosomes of monkeys and men. Nature *175:*32, 1955.

Egozcue, J.: Chromosome variability in the Lemuridae. Amer. J. Phys. Anthrop. *26:*341, 1967.

———: The meiotic chromosomes of the lesser bushbaby (*Galago senegalensis*). Mammal. Chrom. Newsl. *9:*92, 1968.

——— and B. Chiarelli: The idiogram of the lowland gorilla (*Gorilla gorilla gorilla*). Folia primat. *5:*237, 1967.

——— and F. Hagemenas: The chromosomes of the hooded spider monkey (*Ateles geoffroy cucullatus*). Mammal. Chrom. Newsl. *8:*12, 1967.

———, B. Chiarelli and M. Sarti-Chiarelli: The somatic and meiotic chromosomes of *Cebuella pygmaea* (Spix, 1823) with special reference to the behavior of the sex chromosomes during spermatogenesis. Folia primat. *8:*50, 1968a.

———, F. Hagemenas and M. Vilarasau de Egozcue: Chromosome studies in Catarrhinae monkeys. Mammal. Chrom. Newsl. *9*:65, 1968b.

———, E. M. Perkins and F. Hagemenas: The chromosomes of *Saguinus fuscicollis illigeri* (Pucheran, 1845) and *Aotus trivirgatus* (Humboldt, 1811). Folia primat. (in press), 1968c.

———, E. M. Perkins and F. Hagemenas: Chromosomal evolution in marmosets, tamarins and pinchés. Folia primat. *9*:81, 1968d.

———, R. N. Ushijima and M. Vilarasau de Egozcue: The chromosomes of the slender loris (*Loris tardigradus*). Mammal. Chrom. Newsl. *22*:204, 1966.

——— and M. Vilarasau de Egozcue: The chromosomes of the Angwantibo (*Arctocebus calabarensis*). Mammal. Chrom. Newsl. *20*:53, 1966a.

——— and ———: The chromosome complement of the howler monkey (*Alouatta caraya,* Humboldt, 1812). Cytogenetics *5*:20, 1966b.

——— and ———: The chromosomes of *Cebus capucinus*. Mammal. Chrom. Newsl. *20*:71, 1966c.

——— and ———: The chromosome complement of the slow loris (*Nycticebus coucang*). Primates *7*:423, 1967a.

——— and ———: The somatic chromosomes. In: Biology of the Howler Monkey (*Alouatta caraya*). M. R. Malinow, ed. Bibliotheca Primatologica No. 7. Basel: S. Karger, 1967b.

——— and ———: Chromosome evolution in the Cebidae. In: Progress in Primatology. D. Starck, R. Schneider and H. J. Kuhn, eds. Stuttgart: Gustav Fischer Verlag, 1967c.

——— and ———: The chromosome complement of *Cebus albifrons* (Erxleben, 1777). Folia primat. *5*:285, 1967d.

——— and ———: The marked chromosomes of primates: origin and evolutionary significance. In: Progress in Primatology. D. Starck, R. Schneider and H. J. Kuhn, eds. Stuttgart: Gustav Fischer Verlag, 1967e.

———, ——— and F. Hagemenas: The chromosomes of two species of Saimiri: *Saimiri madeirae juruanus* and *Saimiri boliviensis nigriceps*. Mammal. Chrom. Newsl. *8*:14, 1967.

Farbar, P. A.: Chromosomes of *Macaca mulatta* leukocytes. Folia primat. *4*:409, 1966a.

———: The karyotype of *Cercopithecus pygerythrus*. Mammal. Chrom. Newsl. *22*:205, 1966b.

Ford, C. E. and J. L. Hamerton: The chromosomes of man. Nature *178*:1020, 1956.

Goldstein, P. R. and T. R. Birdwell: Cytogenetic studies in the purple-faced langur. Mammal. Chrom. Newsl. *8*:197, 1967.

Goodman, M.: Immunochemistry of the primates and primate evolution. Ann. N.Y. Acad. Sci. *102*:219, 1962.

Hamerton, J. L.: Problems in mammalian cytotaxonomy. Proc. Lin. Soc. Lond. *169*:112, 1958.

———: Primate chromosomes. Symp. Zool. Soc. Lond. *10*:211, 1963.

———, M. Fraccaro, L. de Carli, F. Nuzzo, H. P. Klinger, L. Hulliger, A. Taylor and E. M. Lang: Somatic chromosomes of the gorilla. Nature *192*:225, 1961.

———, H. P. Klinger, D. E. Mutton and E. M. Lang: The somatic chromosomes of the Hominoidea. Cytogenetics *2*:240, 1963.

Hershkovitz, P.: Type localities and nomenclature of some american primates with remarks on secondary homonyms. Proc. Biol. Soc. Wash. *71*:53, 1958.

———: Taxonomic notes on tamarins, genus Saguinus (Callithricidae, Primates) with descriptions of four new forms. Folia primat. *4*:381, 1966a.

———: On the identification of some marmosets, family Callithricidae (Primates). Mammalia *30*:327, 1966b.

Hill, W. C. O.: Primates. Vol. III, Pithecoidea. Edinburgh: University Press, 1957.

Hsu, T. C.: Foreword. Mammal. Chrom. Newsl. *15*:98, 1965.

——— and K. Benirschke: An Atlas of Mammalian Chromosomes. Vols. I and II. New York: Springer-Verlag New York Inc., 1967–1968.

Irsigler, G. B.: Studies on the chromosomes of African mammals. IV. The karyotype of female Chacma baboon (*Papio ursinus*). South Afr. J. Sci. *63*:527, 1967.

Kleinfeld, R. and E. H. Y. Chu: DNA determinations of kidney cell cultures of three species of monkeys. Cytologia *23*:452, 1958.

Kuhn, H. J.: Zur Systematik der Cercopithecidae. In: Progress in Primatology. D. Starck, R. Schneider and H. J. Kuhn, eds. Stuttgart: Gustav Fischer Verlag, 1967.

Klinger, H. P.: The somatic chromosomes of some primates (*Tupaia glis, Nycticebus coucang, Tarsius bancanus, Cercocebus aterrimus, Symphalangus syndactylus*). Cytogenetics *2*:140, 1963.

———, J. L. Hamerton, D. Mutton and E. M. Lang: The chromosomes of the Hominoidea. In: Classification and Human Evolution. S. L. Washburn, ed. Chicago: Aldine Publishing Co., 1963.

Low, R. J., K. Benirschke, J. L. Grimmer and T. G. Schneider: The chromosomes of three bears. Mammal. Chrom. Newsl. *13*:3, 1964.

Makino, S.: A contribution to the study of the chromosomes of some asiatic mammals. Cytologia *16*:288, 1952.

Manna, G. K. and M. Talukdar: An analysis of somatic chromosome complements of both sexes of two primates, the slender loris, *Loris tardigradus,* and Rhesus monkey, *Macaca mulatta.* Mammalia *32*:118, 1968.

Matthey, R.: Les chromosomes de *Galago senegalensis* Geoffroy (Prosimii-Lorisidae-Galaginae). Rev. Suisse Zool. *62* (suppl.):190, 1955.

Mayr, E.: Animal Species and Evolution. Cambridge (Mass.): Harvard University Press, 1963.

Minouchi, O.: The spermatogenesis of the dog, with special reference to meiosis. Jap. J. Zool. *1*:255, 1927.

Nemtseva, L. S.: The karyotype of *Papio hamadryas.* Citologija (Moscow) *6*:633, 1964.

Ohno, S.: Sex Chromosomes and Sex-linked Genes. New York. Springer-Verlag, New York Inc., 1967.

Painter, T. S.: The sex chromosomes of the monkey. J. Exper. Zool. *35*:286, 1922.

———: Studies in mammalian spermatogenesis. II. The spermatogenesis of man. J. Exper. Zool. *37*:291, 1923.

———: Studies in mammalian spermatogenesis. IV. The sex chromosomes of monkeys. J. Exper. Zool. *39*:433, 1924.

Robertson, W. R. B.: Chromosome studies. Taxonomic relationships shown in the chromosomes of the Tettigidae and Acrididae. V-shaped chromosomes and their significance in Acrididae, Locustidae and Gryllidae, chromosomes and variation. J. Morph. *27*:129, 1916.

Romer, A. S.: Explosive evolution. Zoologische Jahrbücher *88*:79, 1960.

Rothfels, K. H. and L. Siminovitch: The chromosome complement of the Rhesus monkey (*Macaca mulatta*) determined in kidney cells cultivated *in vitro*. Chromosoma (*Berlin*) *9*:163, 1958.

Ruzicska, P. and T. C. Hsu: The karyotype of the african green monkey (*Cercopithecus aethiops sabaceus*). Texas Reports Biol. Med. *25*:151, 1967.

Sharma, T. and S. Kakati: Chromosomes of indian common langur. Mammal. Chrom. Newsl. *20*:70, 1966.

Shiwago, P. I.: Recherches sur le caryotype du Rhesus macacus. Bull. Biol. Med. Expt. U.R.S.S. *8*:3, 1939.

Tappen, N.: Genetics and systematics in the study of primate evolution. Symp. Zool. Soc. Lond. *10*:267, 1960.

———: 1960, cited by Bender and Chu, 1963.

Tjio, J. H. and A. Levan: The chromosome number of man. Hereditas *42*:1, 1956.

Tobias, P. V.: Chromosomes, Sex Cells, and Evolution in a Mammal: Based Mainly on the Studies of the Reproductive Glands of the Gerbil, and a New List of Chromosome Numbers of Mammals. London: Lund, Humphries and Co., Ltd., 1956.

Todd, N. B.: A theory of karyotypic fissioning, genetic potentiation and eutherian evolution. Mammal. Chrom. Newsl. *8*:268, 1967.

Ushijima, R. N., F. S. Shininger and T. I. Grand: Chromosome complements of two species of primates: *Cynopithecus niger* and *Presbytis entellus*. Science *146*:78, 1964.

Wohnus, J. F. and K. Benirschke: Chromosome analysis of four species of marmosets (*Callithrix jacchus, Tamarinus mystax, Tamarinus nigricollis, Cebuella pygmaea*). Cytogenetics *5*:94, 1966.

Yeager, C. H., T. S. Painter and R. M. Yerkes: The chromosomes of the chimpanzee. Science *91*:74, 1940.

Young, W. J., T. Merz, M. A. Ferguson-Smith and A. W. Johnston: Chromosome number of the chimpanzee, *Pan troglodytes*. Science *131*:1672, 1960.

CONSIDERATIONS OF SEX CHROMOSOME ABNORMALITIES IN MAN

G. E. Sarto, J. M. Opitz and S. L. Inhorn
University of Wisconsin Medical Center, Madison, Wisconsin

Introduction

Most of the developmental aspects and the genetic control mechanisms of normal human gonadal and genital differentiation are not well understood. Advances in cytogenetic technique led to the discovery of many different sex chromosome abnormalities in man. These stimulated, anew, attempts to relate gonosomal function to processes of normal sex differentiation. We should like to present personally-studied patients and review previously published cases pertaining to a discussion of the role of the sex chromosomes in sex determination, gonadal and genital development.

Brief Review of Normal Development

The initially bipotential gonad is first recognized during the fourth week of gestation as a thickening of the coelomic epithelium along the mesonephric ridge. The primary sex cords are formed by an extension of coelomic epithelium into the mesenchymal mass of the indifferent gonad. It is now generally agreed that the primordial germ cells migrate (Witschi, 1948; McKay, *et al.*, 1953; Pinkerton, *et al.*, 1961) from an area of the endodermal yolk sac epithelium to the gonadal ridges and "settle" in the two somatic components: cortex and medulla. Testicular differentiation involves primarily the medullary portion of the gonad; during ovarian differentiation the cortex becomes the principal gonadal component. The internal genitalia are derived from either Müllerian or Wolffian duct primordia which coexist in both male and female embryos. In normal females, the Müllerian ducts extend caudally, meet in the midline and form the fallopian tubes, uterus and the upper one third of the vagina, while the Wolffian duct system regresses. In males, the

Wolffian ducts form the vas deferens, seminal vesicles and epididymis, while the Müllerian ducts regress. The external genitalia in both sexes arise from a neutral, bipotential, midline anlage; transformation into either male or female genitalia occurs normally after gonadal differentiation.

Theories and Experimental Findings Relating to Normal Sex Differentiation in Mammals

In 1914 Witschi postulated that the primordial germ cell is the primary inductor of the gonad. Experimental findings are not consistent in defining the role of the primordial germ cell in inducing gonadal differentiation; however, in man there is close temporal relationship between the arrival of the germ cells and the onset of differentiation of the gonad. This hypothesis is supported by the experiments of Hemsworth and Jackson (1963a, b) in which they destroyed the primordial germ cells early in development in rat embryos by the administration of busulphan; in such animals the gonad failed to develop and appeared histologically similar to the streak gonads described in the Ullrich-Turner syndrome (UTS).

Witschi further postulated that the primordial germ cell is a neutral bipotential cell which is capable of differentiating into either an oocyte or spermatocyte regardless of the genetic constitution. Two investigations in animals support this hypothesis: (1) The demonstration of XX germ cells in the testes of newborn bull twins of freemartin calves by Ohno *et al.* (1962), and (2) Benirschke and Brownhill's (1963) discovery of XX germ cells which were capable of undergoing spermatogenesis in the testes of male twins of female marmosets. In man, the presence of ova and of spermatogenesis in the ovo-testes of XY or XX hermaphrodites respectively may also indicate that the germ cell is bipotential. These observations also imply that the sex of the gonad is determined not so much by the gonosomal constitution of the primordial germ cell as by the genetic sex of the somatic cells of the gonadal ridge.

Differentiation of the internal and external genitalia depends upon the gonad. The presence of testicular tissue is normally associated with the development of male genitalia, while ovarian tissue or the absence of any gonadal tissue results in individuals with female genitalia. Jost (1947) castrated male rabbits before the 21st day of gestation; all of them developed apparently normal female genitalia, showing that a testis is necessary for normal male genital differentiation. Clinical observations (occurrence of normal female internal and external genitalia in some XY individuals) sufficiently demonstrate that in man, also, testes are required to produce normal male genital development.

Male sex determination in man depends on the presence of a Y chromosome. The male sex determinants on the Y chromosome are so strong that dysgenetic testes are produced even in individuals with an XXXXY constitution.

The first portion of this paper deals with the role of the Y chromosome in sex determination, and gonadal and genital differentiation in man. The following questions are of specific interest: (1) Where on the Y chromosome are the male sex determining factors located; (2) How is the male protected from his "single X-ness;" (3) How can one explain the presence of testicular tissue in individuals with an XX constitution. The last portion of the paper will consider the influence of the second X chromosome on female gonadal and somatic development.

Case Material

Case 1

D. Z. (B.D. 1-12-50) was first seen for intersex genitalia at the age of three years; he was lost to followup and was not seen again until he was 15 years old, at which time he was referred because of gynecomastia. He was now 153 cm tall and weighed 42 kg. No major somatic anomalies were found except for gynecomastia comparable to that seen in a pubescent female. He had a phallus resembling an enlarged clitoris with a urethral opening at its base. No gonads were palpable in the groin or in the redundant wrinkled labia majora. In addition, he had very hypoplastic labia minora and a dimple in the perineal body where the vaginal opening should be located. A small midline uterus was palpable rectally.

The internal genitalia were found to consist of a small uterus with a

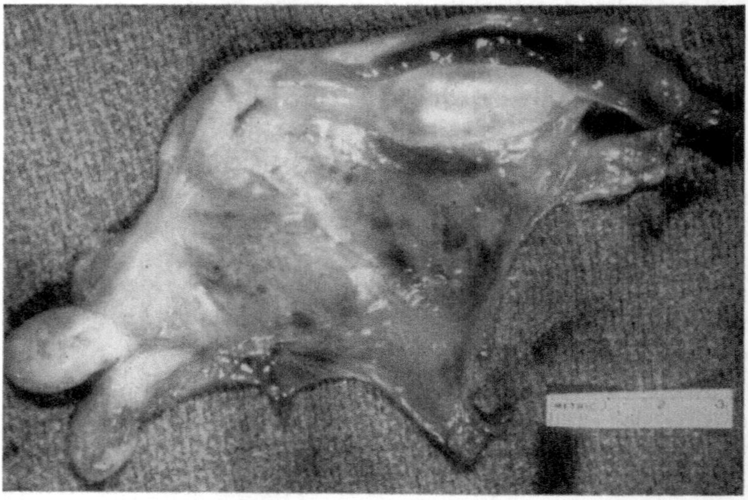

Fig. 1. Case 1. Internal genitalia of 46,XX/47,XXYr true hermaphrodite: uterus with fallopian tube and ovary on right and vas deferens and testis on left.

well-formed fallopian tube and an ovary on the right side. The tubular structure attached to the left side of the uterus ended blindly; extending from the distal end was a testis-like structure. A vagina ended blindly halfway to the perineum (Fig. 1).

Histologic Studies The right gonad was composed of ovarian tissue without testicular or unusual medullary elements. Primary follicles and oocytes were seen; however, compared to a normal ovary, the number of germ cells was greatly decreased. The left gonad consisted of testicular tissue composed of seminiferous tubules of varying degrees of differentiation. In some of the tubules spermatogonia were undergoing degeneration; in other tubules meiotic figures were observed in a few germ cells, but spermatogenesis, in general, was markedly reduced (Fig. 2).

Cytogenetic Studies The buccal smear was 40 percent sex chromatin positive. The sex chromatin body appeared to be of normal size; cells with more than one Barr body per nucleus were not observed. Chromosome preparations were made from lymphocyte cultures and fibroblasts cultured from biopsies of skin, right gonad, left gonad and from each of the tubular structures (i.e., fallopian tube and vas deferens). The cytologic results of the blood, skin and both gonads are summarized in Table 1.

Interpretation This patient is a lateral hermaphrodite, i.e., his two gonadal tissues were segregated into a testis and an ovary located respectively on the left and right sides. He has chromosome mosaicism presumably of two main cell lines (Fig. 3). One of these has an apparently normal 46,XX constitution; the other appears identical but has, in addition,

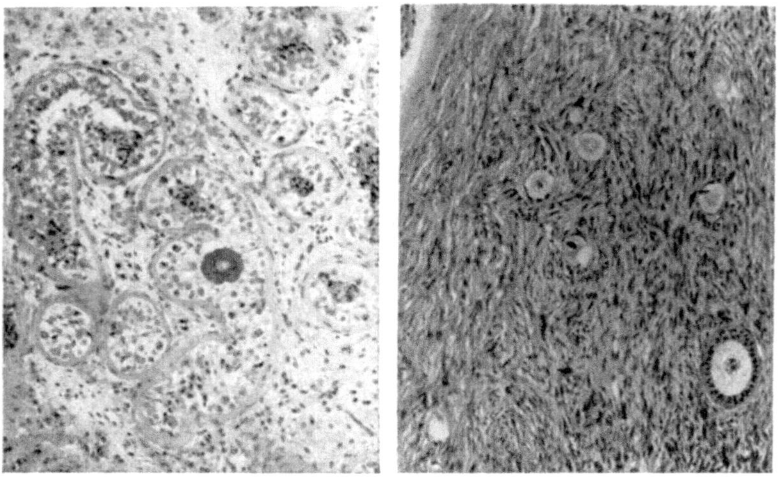

Fig. 2. Case 1. Histologic sections of ovary with follicles and oogonia; testis with seminferous tubules showing limited spermatogenesis.

Table 1: Cytologic Analyses of Tissues From a 46/XX/47, XXYr True Hermaphrodite (Case 1)

Tissue	Chromosome Number: 46,XX	47,XXYr	45	47	Other Cells: Yr+	Other Cells: Yr–	Total No. of Cells	Total No. Cells Yr+	Percent Cells Yr+
Skin	21	11	(6) 1C–, 1D*–, 2F*–, 2G–	1A*+	2	5	46	15	32.6
Ovary	40	7	(10) 5C*–, 2E*–, 2G–, 1F–	–	1	1	59	10	16.9
Testis	12	21	(8) 2C*–, 1E–, 3F–, 2G**–	1E+, 1B*+	10	4	57	35	61.4
Blood	28	22	(9) 6C****–, 3E–	1D+	–	–	60	26	43.3

* Indicates Yr present in the cell.

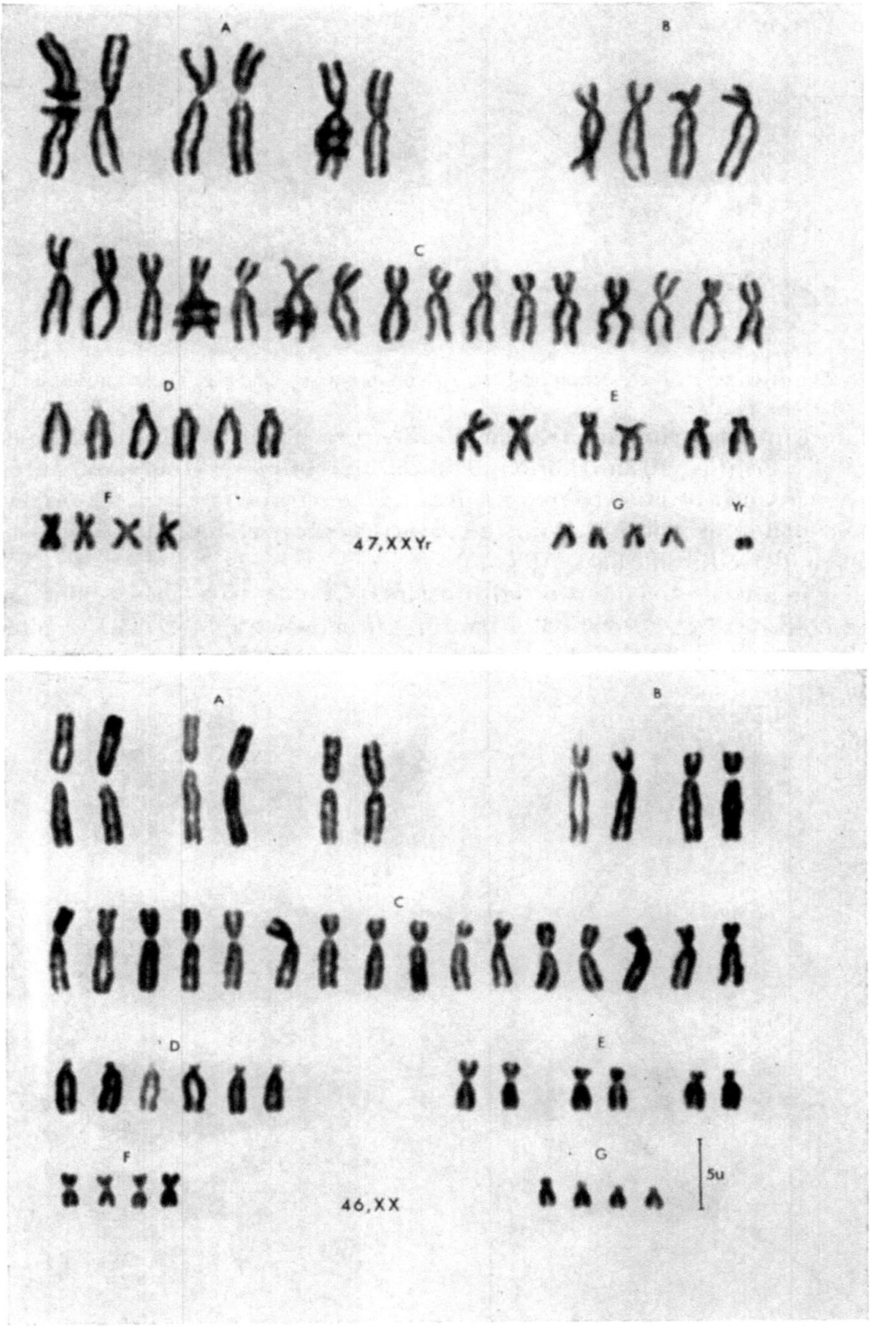

Fig. 3. Case 1. Karyotypes of two major cell lines: 46,XX and 47,XXYr.

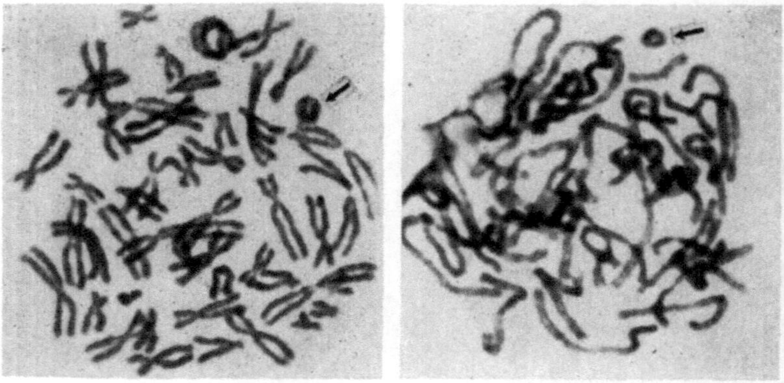

Fig. 4. Case 1. Two prometaphase plates showing the ring Y chromosome.

an extra small chromosome which is interpreted as a Y-ring chromosome (47,XXYr) (Fig. 4). In addition, both cell lines from all tissues examined were frequently observed to contain a minute centric fragment, the origin and nature of which remains unknown (possibly self-replicating centromere derived from the Y) (Fig. 5).

The presence of the two cell lines in D. Z. may have come about in several ways: (1) Formation of the ring chromosome in a 47,XXY zygote forming the 47,XXYr cell line with loss of the Yr from the original cell

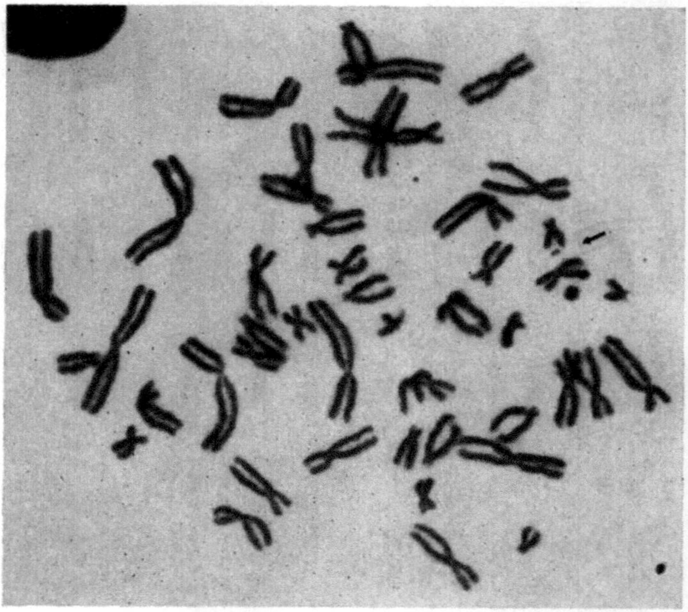

Fig. 5. Case 1. Metaphase plate showing a small centric fragment (self-replicating centromere).

line to form the 46,XX cell line; (2) Random segregation of the X and Yr in meiosis in the father leading to sperm containing both the X and Yr chromosomes; (3) A double fertilization of two maternal germ cell contributions with an X and an XYr sperm. The first hypothesis appears most plausible since it requires the least number of assumptions to explain the patient's chromosome constitution. It is not possible from the study of the patient alone to conclude which is the correct hypothesis; we were unable to perform chromosome studies in other members of the patient's family.

Cytologic studies of the gonadal fibroblast are of particular interest in that the XXYr cell line is the predominant cell line in the testis but not in the ovary (Table 1). Unless one invokes the less likely occurrence of the migration of XX and XXYr germ cells to the right and left gonadal ridge respectively, it is tentatively possible to conclude that the XX and XXYr constitutions of the somatic components of the right and left gonad were responsible for their differentiation into ovary and testis.

Case 2

J. H. (B.D. 7-25-40) was first seen at the age of 12 years because of a clitoris which had been enlarged since birth. She was reevaluated for this abnormality and for primary amenorrhea at 16 years of age. At this time she was 137 cm tall (shortness of stature) and was found to have several somatic anomalies including a highly arched palate, nevi of the face, neck and trunk, a low posterior scalp line, suggested webbing of the neck, and bilateral cubitus valgus. She had a small uterus with bilateral gonadal rudiments. There was a 1.2 cm gonad with a vas deferens on the

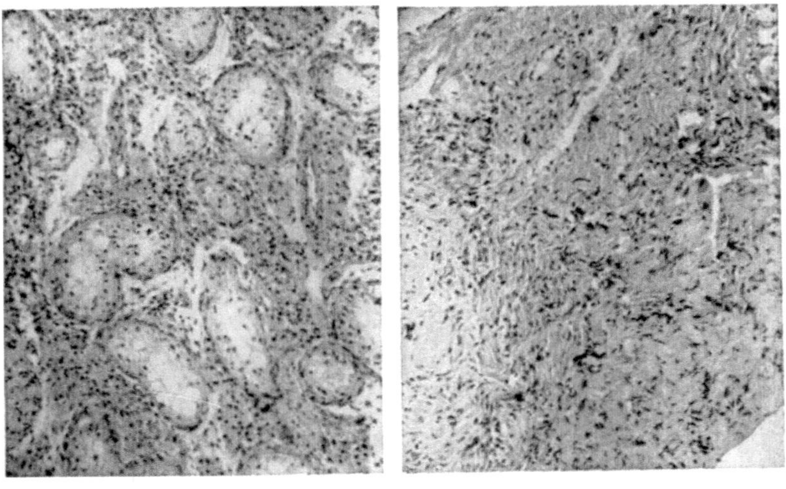

Fig. 6. Case 2. Histologic sections of fibrous "streak" gonad and dysgenetic testis with tubules containing Sertoli cells.

left, and on the right a fallopian tube with a tiny nodule of gonadal tissue. The right gonad consisted of a small streak of connective tissue entirely devoid of germ cells. The left gonad was a dysgenetic testis with Sertoli cells in some of the tubules; active spermatogenesis was not observed (Fig. 6).

Cytogenetic Examination The buccal smear was sex chromatin negative. Lymphocyte culture showed that the patient was a 45,X0/46,XY mosaic (Fig. 7).

Interpretation This patient is a genital pseudohermaphrodite with one streak gonad and a contralateral dysgenetic testis. Anaphase lag with loss of a Y in an XY zygote probably resulted in the X0 cell line. This case is described primarily to call attention to the wide phenotypic variation that can be encountered with X0/XY mosaicism. This patient presented with female genitalia with an enlarged clitoris as do many

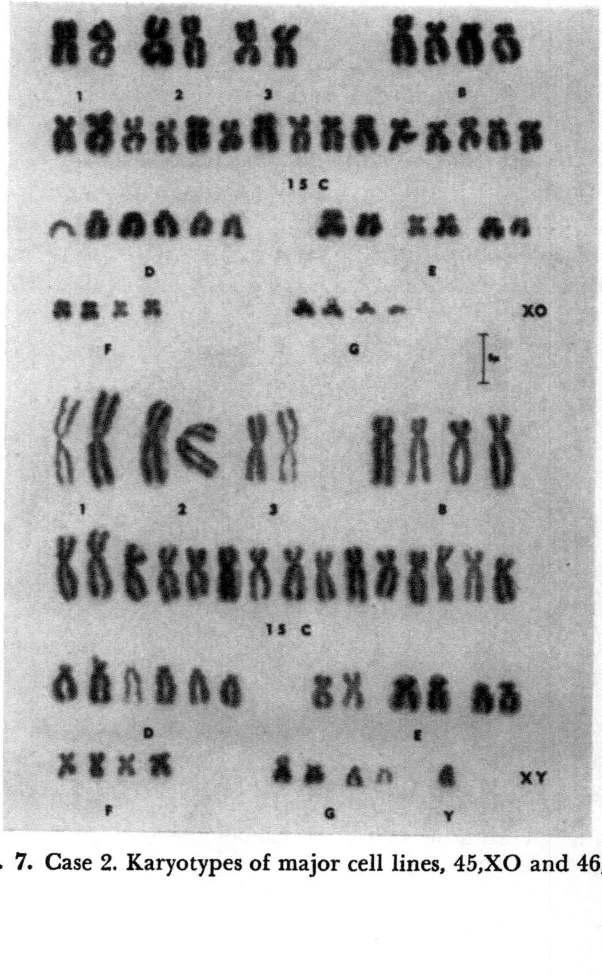

Fig. 7. Case 2. Karyotypes of major cell lines, 45,XO and 46,XY.

patients with X0/XY mosaicism. Less commonly, individuals with an X0/XY constitution may exhibit the Ullrich-Turner syndrome (Schlegel *et al.*, 1966) or may be apparently normal males with infertility (Kjessler, 1965) or males with normal fertility (Daly, 1968). The wide variation in phenotype may be related to the proportion of the two cell lines in gonadal and somatic tissues during development. If the gonadal ridge is primarily made up of somatic cells with an X0 constitution streak gonads may occur; however, should the major somatic cell line have an XY constitution a relatively normal testis with male genital differentiation may result.

Case 3

C. C. (B.D. 3-14-62) was a first child of young and healthy parents. The mother had not suffered any previous abortions and her pregnancy was uneventful and of normal duration. At birth the infant weighed 3400 gm and was 48 cm long. She was found to have multiple anomalies including asymmetry of the face, much redundant skin at the neck, a highly arched palate, a shield-like chest with widely spaced nipples, hypoplastic clavicles, and bulging metatarsal and metacarpal pads on the feet and hands. There were no obvious cardiac anomalies and the external genitalia were those of a normal female. She died at five days of age of acute respiratory distress. Examination of the internal genitalia at post mortem examination revealed a normal uterus in the midline and bilateral fallopian tubes, with the ovaries in the normal location. The right ovary was elongated, measured $2.0 \times 0.2 \times 0.3$ cm and the left ovary was of normal shape and measured $1.5 \times 0.3 \times 0.2$ cm.

Histologic Examination Serial sections of the infant's ovaries showed a symmetric reduction in size but a nearly normal organization into a cortex covered with peritoneal epithelium and a hilar region containing some residual medullary cords. Many normally appearing oocytes, presumably in the dictyotene stage of diplotene, were observed; however, when compared with a normal ovary their number was found to be greatly reduced. The amount of interstitial connective tissue was increased. Several nodules or cell nests were thought to consist of oocytes in various prophase stages of the first meiotic division (Fig. 8). Extensive invaginative processes from the surface of the ovary were observed, primarily in the juxta-hilar areas. These tubular processes contained small epithelial cells and strikingly large cells with pale cytoplasm. The former are probably derived from the surface epithelium, the nature of the latter is unknown. These large pale cells may represent either a peculiar reaction of the peritoneal surface cells to the internal cortical environment, or they may represent an attempt to invest oogonia with a coat of follicle cells.

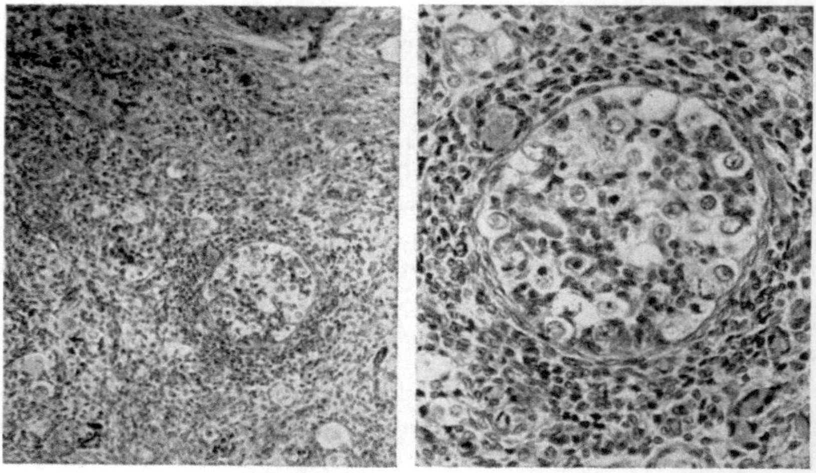

Fig. 8. Case 3. Histologic sections showing germ cells and germ cell nest (higher magnification on right) in the ovary from a 45,XO infant.

Cytogenetic Investigation The buccal smear was sex chromatin negative and an analysis of the chromosomes showed an X0 constitution.

Interpretation The infant died before extensive cytologic studies could be performed and it is therefore not possible to be certain that the degree of ovarian development which she manifested was not due to an unrecognized XX cell line. However, several recent investigations, particularly those of Singh and Carr (1966) of X0 fetuses and Carr *et al.* (1968) of infants with more or less well developed ovaries, suggest that a pure X0 constitution is compatible with the extensive degree of cortical differentiation into ovarian tissue which was observed in our patient but which may undergo degeneration before adult life.

Case 4

A. H. (B.D. 1-14-48) was first seen at 17 years of age for failure to develop secondary sex characteristics and primary amenorrhea. She had always enjoyed good health and had had no major illnesses. She was 150 cm tall and showed eunuchoidal body measurements and weighed approximately 33 kg. She had a few minor anomalies including: prominent posteriorly rotated ears, inner epicanthic folds, and an increased number of nevi (Fig. 9). The external genitalia were infantile though normal. She had a normal vagina with a small cervix and a small uterus that was palpable rectally. She had no cardiac anomalies. Bone age was 10 years for a chronologic age of 17 years. Thus far no laparotomy has been performed, but we assume she has streak gonads.

Cytogenetic Investigation Buccal smear was sex chromatin negative and a blood lymphocyte culture revealed a 46,XY chromosome constitution (Fig. 10).

Interpretation This patient represents a condition characterized by gonadal dygenesis not accompanied by the somatic abnormalities associated with UTS. This group of individuals can be separated from those with the feminizing testis syndrome because they have normal female internal and external genitalia except for "streak-like" gonads resulting in primary amenorrhea, lack of breasts, and infantile genitalia in the adult. They are of average height, often with eunuchoidal body proportions. The chromosome constitution may be either XX or XY.

Swyer (1955) first reported such a condition in two patients with male chromosomal sex and he called this a form of "male pseudo-hermaphroditism." Hoffenberg and Jackson (1957) reported an additional case which had a female nuclear sex pattern and suggested that the cases of Swyer were variants of the syndrome of gonadal dysgenesis. The term "pure gonadal dysgenesis" also has been used to describe this syndrome, but since pure gonadal dysgenesis may occur in both XX and XY individuals, it may be best to use the term Swyer's syndrome to separate the XY individuals from the XX individual with this abnormality.

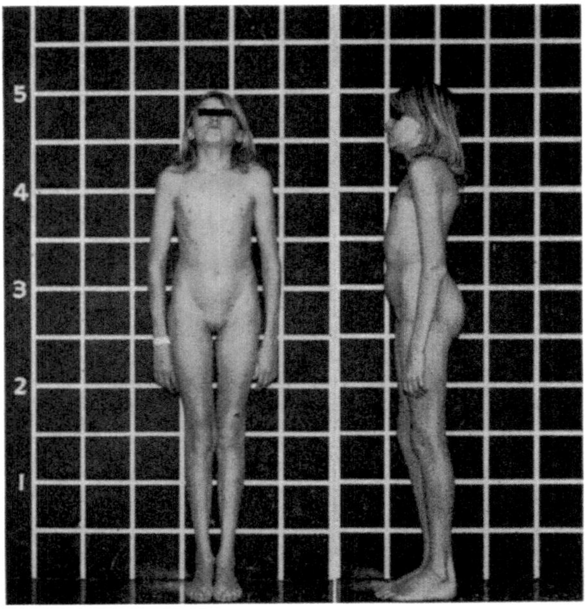

Fig. 9. Case 4. 46,XY phenotypic female with female internal and external genitalia.

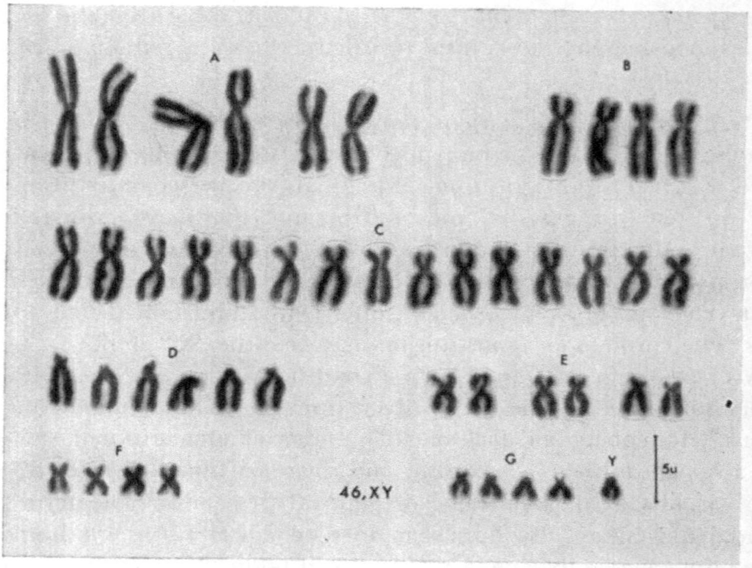

Fig. 10. Case 4. Karyotype, 46,XY, showing morphologic characteristics of the Y chromosome.

Gonadal dysgenesis without the features of UTS has been reported to occur in siblings (Cohen and Shaw, 1965) and a recent report (Sternberg *et al.,* 1968) of its presence in three sibships of a single family suggests that this condition may be gene determined. A localized defect in the gonadal ridge may prevent its developing into a gonad of either type. In familial XY gonadal dysgenesis failure of testicular development may occur at such an early stage, that only a fibrous streak results. Differentiation into a normal female in such an XY individual may be interpreted as the natural counterpart to the experiments of Jost in rabbits (1947).

There is an increased risk of neoplasm in the gonads of XY persons (Teter and Boczkowski, 1967). Frasier and Bashore (1964) reported the presence of gonadoblastomas in each of identical twin girls with an XY constitution, in addition to Fine *et al.* (1962), Brøgger and Strand (1965) and Cohen and Shaw (1965) who also reported gonadal neoplasms in their patients.

Discussion

Position of Male Sex Determining Loci on the Y Chromosome The human Y chromosome exerts a strong effect on testicular, and hence on genital differentiation. There are striking variations in the size of the Y chromosome, particularly impressive are those instances where a considerable reduction in the length of the long arm of the chromosome has not led to an impairment of fertility. Several clinically abnormal patients with structural defects of the Y chromosome have been reported (Table 2) and, depending on the at times rather difficult interpretation of similar cytologic findings, the loss has been reported to involve either the short arm or the long arm of the Y. Extreme reductions of the Y chromosome are usually accompanied by X0 mosaicism which may handicap an interpretative correlation between the clinical and cytologic findings. Finally, it must be admitted that more or less definite "masculinization" of the patient's genitalia, at times without demonstration of testicular tissue, may be the only reason for the identification of a small chromosome as Y chromosome material.

Jacobs and Ross (1966) reported four patients with Y chromosome abnormalities and concluded that the genes for the male sex determination were located on the short arm of the Y. The first two patients were phenotypic females with apparently normal female external genitalia: Case 1 had streaks and a uterus; a laparotomy was not performed and there exists no information about the internal genitalia of Case 2. Both patients had a small metacentric chromosome. The other two cases were both phenotypic males: Case 3 was an 83-year-old male who had no children; Case 4 was a 41-year-old male with Klinefelter's syndrome. Both of these men had a similar abnormality of their Y chromosome which was described as metacentric in Case 4. Jacobs and Ross interpreted the abnormal chromosome in each of their first two cases as an isochromosome of the long arm of the Y, and in their last two patients as a pericentric inversion of a Y chromosome. In the first two cases, the complete lack of masculinization was thought to be due to the deletion of the short arm of the Y chromosome with loss of the male sex determining region which was preserved in the pericentric inversion of the second two cases. Since the masculinizing effect is given as evidence for the presence of Y genetic material in other reported cases of abnormalities of the Y chromosome, the complete lack of masculinization of the first two cases of Jacobs and Ross may indicate that the abnormal chromosome in these two cases is not derived from the Y, and indeed, may be a fragment of a second X chromosome.

Lenz and Pfeiffer (1966), in reviewing the patient reported by Solomon *et al.* (1964) and personal case material with a presumed isochromosome of the long arm of the Y, arrived at the conclusion that the male sex determining region must be located on the long arm of the Y. In the patients of Lenz and Pfeiffer and Solomon *et al.* the isochromosome had been inherited from a fertile father. The evidence from our patient, D. Z., and from the cases of Lenz and Pfeiffer and Solomon *et al.*, suggests that the male sex determining "region" is located on the long arm of the Y chromosome; however, none of the cases reported thus far allows for a definite conclusion. This pertains not only to the location but also to

Table 2: Reported Cases of Structural Y Chromosome Abnormality, Review of the Literature

Authors	Description of Patient and Anomalies	Cytogenetic Abnormalities	Authors' Interpretation of the Abnormal Chromosome
Vaharu *et al.* (1961)	4½ yr. old female; short, low set ears, anti-mongoloid slanting of eyes, micrognathia. No cardiac anomalies. Genitalia: intersex; enlarged clitoris, fused labia, without palpable gonads. No laparotomy.	Sex chromatin negative. 45,X0 plus a small piece of chromosome, half the length of 22, with no visible centromere. Found in all cells examined.	Portion of a Y chromosome.
Conen *et al.* (1961)	13 year old female: short, no breast development, primary amenorrhea, deep voice. Genitalia: large labia majora with rugae, enlarged clitoris, vagina present. Laparotomy: small rudimentary uterus and fallopian tubes. Left gonad—well-formed testicular tissue with epididymis. Right gonad—fibrous streak.	Sex chromatin negative. 45,X0/46,X"y"; * shape of abnormal chromosome not discerned in all cells, but in some, had 4 equal short arms, half the length of 21 or 22.	Deletion of major portion of LA † of the Y; translocation of part of LA of Y to another chromosome not excluded.
Muldal and Ockey (1962)	Father (propositus) and two sons with muscular dystrophy, all three with hypospadias. Brother of the propositus, normal.	46,X"y".	Deletion of distal ⅓ to ½ of the Y.
Fraccaro *et al.* (1962)	61 year old female: short stature, short 4th metacarpal, short neck, coarctation of aorta, primary amenorrhea. Genitalia: enlarged clitoris. No laparotomy.	Sex chromatin? 45,X0/46,X"y"/47,XXX"y". Abnormal chromosome: metacentric, two arms similar to short arms of Y chromosome.	Deficient Y chromosome or isochromosome for SA ‡ of the Y.
Ferrier *et al.* (1962)	13½ year old female: short stature; obesity; cubitus valgus; bilateral short humeri; low posterior hair line; no webbing of neck; broad chest; phalangeal edema; concave fingernails; hirsutism; no breast	Sex chromatin negative. 45,X0/46,X"y"; abnormal chromosome: acrocentric, half the size of 22.	Partially deleted Y chromosome.

	tissue. Genitalia: infantile female; no clitoral enlargement. Laparotomy: uterus; bilateral fallopian tubes; streak gonads void of germ cells.		
Klevit *et al.* (1963)	Intersex: no features of Ullrich-Turner syndrome. Genitalia: hypospadias, empty scrotal sacs, inguinal hernias. Laparotomy: uterus, fallopian tubes. Left gonad—testis. Right gonad—dysgenetic.	45,X0/46,X"y""y". Two abnormal chromosomes, one metacentric with equal arm lengths, the other, small acrocentric.	Metacentric chromosome isochromosome for LA of Y. Acrocentric chromosome-deletion of LA of Y.
Solomon *et al.* (1964) Case of Oikawa and Blizzard (1961)	11 year old male: mental retardation; webbed neck, cubitus valgus; hypertelorism; epicanthal folds; clinodactyly 5th finger. Genitalia: infantile scrotum; gonads not palpable. Laparotomy: testes bilaterally, no ovarian tissue or Müllerian structures, no germ cells. Abnormal Y found in apparently normal father of patient.	Sex chromatin positive, large Barr bodies. 46,XXqi/47,-XXqi"y". Abnormal small chromosome resembled F group—3 cells with 48 chromosomes had additional acrocentric fragment.	Abnormal Y. Too short for LA isochromosome Y, too long for SA isochromosome Y.
Kjessler (1965)	Infertile male: hypospadias and oligospermia. Paternal uncle of propositus apparently normal but childless.	Extremely small Y in both.	Deleted Y chromosome.
Lo and Kobernick (1965)	Female: short; highly arched palate, no webbed neck. Genitalia: enlarged clitoris to 8 cm.	Sex chromatin negative. 45,X0/46,X"y" exact morphology difficult to ascertain—metacentric appearing in some cells.	Partial deletion of LA of Y chromosome.

* "y": structurally abnormal Y, usually smaller than normal.
† LA: Long arm of Y.
‡ SA: Short arm of Y.

Table 2 (Continued)

Authors	Description of Patient and Anomalies	Cytogenetic Abnormalities	Authors' Interpretation of the Abnormal Chromosome
Jones and Zourlas (1965)	Intersex female: short, micrognathia, hypertelorism, short neck, shield chest (Ullrich-Turner like). Genitalia: enlarged clitoris, no labia minora, small labial folds without palpable gonads. Laparotomy: infantile uterus and tubes, bilateral testicular tissue.	Sex chromatin negative. 45,X0/46,X"y"; very small Y chromosome.	Deleted Y chromosome.
Dewhurst *et al.* (1965)	14 mo. old female: no somatic abnormalities described. Genitalia: moderate sized phallus with hypospadias, bifid scrotum with testicular tissue on left side, vagina was present. Laparotomy: unicornuate uterus, tube and ovary on right, vas deferens on left.	Sex chromatin positive. 45,X0/46,XY/47,XX"y". Abnormal chromosome: small metacentric.	Fragment of a Y chromosome.
de Grouchy *et al.* (1966a)	18 year old female: similar to Ullrich-Turner syndrome; short, low set ears, cubitus valgus, coarctation of aorta, short 4th metacarpal, no breasts. Genitalia: enlarged clitoris. Laparotomy: uterus, rudimentary fallopian tubes and gonads.	45,X0/46,XY/46,X"y".	Partial deletion of the LA of Y chromosome.
de Grouchy *et al.* (1966b)	12 year old male: short stature, no major somatic anomalies. Genitalia: hypospadias, scrotum with testis on the left. Laparotomy: uterus, fallopian tube and fibrous streak on right.	Sex chromatin negative. 45,X0/46,X"y". Abnormal chromosome: metacentric.	Partial deletion of LA of Y or SA isochromosome of Y.
Jacobs and Ross (1966)	Case 1: 34 year old female: 155 cm, short neck, ?webbed, primary amenorrhea. Genitalia: apparently normal female. Laparotomy: uterus, streak gonads.	Sex chromatin negative. 46,X"y", small metacentric chromosome, labelling pattern same on both arms.	Isochromosome of LA of Y.

	Case 2: 62 year old female: 163 cm, bilateral choroidoretinitis, primary amenorrhea. Genitalia: apparently normal. Laparotomy: not done.	Same as above.	Isochromosome of LA of Y.
	Case 3: 83 year old male: cleft lip, no children. Genitalia: normal male, testes in scrotum normal for age. Laparotomy: not done.	Sex chromatin negative. 46,X"y", abnormal chromosome: resembled the F group, Y like appearance of chromatin material.	Y chromosome with a pericentric inversion.
	Case 4: 41 year old male: Klinefelter's syndrome. Genitalia: normal male, atrophic testes. Laparotomy: not done.	Sex chromatin positive, 47,-XX"y". Abnormal chromosome: similar to F group, Y like appearance of the chromatin material.	Y chromosome with a pericentric inversion.
Lenz and Pfeiffer (1966)	Adult male: 182 cm tall, apparently normal, 3 children, one son, 196 cm, with same chromosome abnormality. Genitalia: not described.	Apparently 46,X"y".	Isochromosome for LA of Y.
Federman *et al.* (1967)	Adult female: mental retardation, epicanthal folds, low hair line, widely spaced nipples, cubitus valgus, genito-urinary anomalies including a double collecting system bilaterally. Genitalia: fused labioscrotal folds, no gonads palpable. Laparotomy: uterus, bilateral rudimentary fallopian tubes, dysgenetic ovarian tissue on right, dysgenetic testis on left.	Sex chromatin negative. 45,X0/45, XY/D translocation.	Translocation of LA Y to D chromosome.

whether there is a single gene locus or several loci involved in male sex determination.

The Extra-Genital Function of the Y Chromosome In addition to its effect on gonadal and genital differentiation, the Y chromosome appears to exert a pervasive influence on many phases of somatic development and function in normal males. In spite of the relatively small size of the Y, its variable length and the fact that it seems to consist primarily of heterochromatin, the Y chromosome exerts the same effect as does the second X in the female in preventing the developmental defects associated with "single X-ness" (i.e., UTS in X0 individuals). In addition, normal males show somatic traits which differentiate them from normal females and which may be attributed to the Y chromosome. The maturational delay in osseous development in males probably explains the greater average adult height in men compared to that of women. Men are generally more aggressive than their female counterparts. Dermatoglyphic differences have also been documented (Penrose, 1967). XY males generally have a finger tip total ridge count lower than that of X0 females; additional Y chromosomes further decrease the total ridge count (Table 3). XYY individuals are generally taller, less intelligent and more aggressive than XY men (Jacobs *et al.*, 1965).

The question may be asked whether these extra-genital effects of the Y chromosome are developmental results of the process of male sex determination, or whether some are due to an action of the Y chromosome independent of its primary sex determining function. This problem is difficult if not impossible to resolve at the present time.

It is a generally accepted hypothesis, as reviewed by Ohno (1967), that the X and Y chromosomes developed from a homologous pair of autosomes, and that the formation of the present Y was accomplished during evolution by many structural events (translocations, inversions, deletions and so forth) which led to a great reduction in size and genetic content of the Y as compared to the present X chromosome. This does not pre-

Table 3: Sex Chromosome Complement and Mean Total Ridge Counts

Sex Chromosome Complement	No. in Sample	Mean T.R.C.	Expected Mean (E = 187 − 3X − 12Y)*
X	40	178.6	157
XY	825	145.0	145
XYY	14	133.6	133
XX	825	127.2	127
XXY	25	114.8	115
XXYY	7	106.1	103

* X,Y = number of X and Y chromosomes respectively.

clude the hypothetical possibility that the X and Y chromosomes of man contain genetic material which acts similarly during the development of XX and XY individuals because of previous homology, but which has been changed during evolution by repeated mutations so as to achieve slightly different somatic effects in the two sexes.

Testicular Tissue in XX Individuals True hermaphroditism implies the presence, in one individual, of both testicular and ovarian tissue. The most common sex chromosome constitution encountered in hermaphrodites is the 46,XX type. The presence of testicular tissue in such individuals is difficult to explain; hidden XY mosaicism can never be excluded with complete assurance. However, the finding by Rosenberg *et al.* (1963) of three XX hermaphrodites in one sibship suggests that at least some cases may indeed have a "pure" XX constitution. Though such cases might be explained on the basis of a Y/X or a Y/autosomal translocation, it is more likely that they are due to gene mutation. Sporadic XY hermaphroditism is far less common than the XX type; however, one instance of familial occurrence of the XY type has also been observed in two brothers by Milner *et al.* (1958).

The occurrence of testes and a Klinefelter-like syndrome in apparently pure XX individuals (Ferguson-Smith, 1965) is even more difficult to explain. The same mutation responsible for the (familial) occurrence of XX hermaphroditism might possibly also cause complete absence of ovarian tissue in some individuals and lead to the "XX Klinefelter's syndrome." Ferguson-Smith (1966) considers the homology of the X and Y chromosome to be so close as to permit pairing of homologous portions of these two gonosomes with occasional crossing over between the X and Y leading to transfer of the male sex determining region of the Y to the X chromosome. Published illustrations of meiosis in males still do not permit a conclusive interpretation of the sex chromosome pairing process; however, in diakinesis there appears to be end-to-end association of the short arms of the X and Y. It is conceded, however, that a terminalized association during diakinesis does not preclude more extensive pairing of the X and Y earlier during prophase. Evidence adduced to date by Ferguson-Smith to support this hypothesis consists primarily of several examples of atypical Xga inheritance in XX Klinefelter individuals (Ferguson-Smith, 1966). Rarity of such atypical cases seems to support the view, which we favor, that the human XY chromosome sex-determining mechanism is very stable and precludes pairing of homologous segments of appreciable length, so that exchanges of X and Y chromosomal material would not be expected to occur with any frequency.

Function of the Second X Chromosome in Females There have been several recent extensive reviews of the subject of X chromosome abnormalities and their clinical manifestations (Engel *et al.*, 1965; Ferguson-Smith, 1965; Inhorn and Opitz, 1968); therefore, this paper will discuss only briefly the function of the second X chromosome in female gonadal and somatic development.

It appears that the second X chromosome is essential for normal gonadal function in females, but its exact role is not known. Ovarian tissue may differentiate early in development in X0 individuals (Case 3), but it seems to degenerate before adult life. Even the loss of a portion of a second X chromosome results in gonadal dysgenesis of varying degrees.

The short arm of the second X chromosome seems to protect females from shortness of stature and the somatic abnormalities observed in the UTS. In contradistinction to individuals with deletion of the short arm of one X chromosome, patients with loss of the long arm are generally not short of stature and do not show the somatic manifestations found in UTS (Ferguson-Smith, 1965). In this way the short arm of the X seems to act similarly to the Y chromosome in males or XY females (Case 4) in protecting them from developing such stigmata. This function of the Y chromosome and the short arm of the X does not imply that there need be genes on the gonosomes affecting these characteristics directly; instead, such effects may be secondary to the developmental processes leading to normal maleness or femaleness.

Summary

Human sex determination and sex differentiation are discussed in reference to several clinical observations. We think that these observations support the hypotheses that:

1. The Y chromosome in man is the primary genetic determiner of the male sex; the exact position on the Y chromosome of the sex-determining gene(s) is still unknown; however, we would favor the hypothesis that they are located on the proximal portion of the long arm of the Y chromosome.

2. The genetic constitution of the somatic elements of the gonadal ridge is the main determinant of gonadal differentiation; the primordial germ cell may serve as an inductor of gonadal differentiation, but its genetic constitution probably does not control "sex" of the resulting gonad.

3. Loci controlling or influencing genital differentiation are located not just on the Y chromosome but are probably distributed over several of the other chromosomes as evidenced by the many autosomal mutations which produce familial abnormalities of gonadal and/or genital differentiation. These mutations include those leading to true hermaphroditism,

male and female pseudohermaphroditism and the several forms of "pure gonadal dysgenesis."

4. The X0 state does permit an extensive degree of "ovarian" differentiation prenatally; these ovaries which do not appear completely normal in gross or finer structure probably degenerate early in childhood.

Acknowledgment

These studies were supported by NIH grants #HD00995, GM08217, GM15422; the Racine Health Fund, Inc.; and the National Foundation—March of Dimes, and Children's Bureau.

References

Benirschke, K. and L. E. Brownhill: Heterosexual cells in testes of chimeric marmoset monkeys. Cytogenetics *2*:331, 1963.

Brøgger, A. and A. Strand: Contribution to the study of the so-called pure gonadal dysgenesis. Acta Endocr. *48*:490, 1965.

Carr, D. H., R. A. Haggar and A. G. Hart: Germ cells in the ovaries of X0 female infants. Amer. J. Clin. Path. *49*:521, 1968.

Cohen, M. M. and M. W. Shaw: Two XY siblings with gonadal dysgenesis and female phenotype. New Eng. J. Med. *272*:1083, 1965.

Conen, P. E., J. D. Bailey, W. H. Allemang, D. W. Thompson and E. Ezrin: A probable partial deletion of Y chromosome in an intersex patient. Lancet *2*:294, 1961.

Daly, R.: Personal communication, 1968.

Dewhurst, C. J., A. J. N. Warrack, C. E. Blank, A. M. Bishop and W. B. Heslop: Chromosome mosaicism in an hermaphrodite. J. Med. Genet. *2*:221, 1965.

Engel, E. and A. P. Forbes: Cytogenetic and clinical findings in 48 patients with congenitally defective or absent ovaries. Medicine *44*:135, 1965.

Federman, D. D., F. M. Davidoff and E. Ouellette: Presumptive Y/D translocation in mixed gonadal dysgenesis. J. Med. Genet. *4*:36, 1967.

Ferguson-Smith, M. A.: Review article: Karyotype-phenotype correlations in gonadal dysgenesis and their bearing on the pathogenesis of malformations. J. Med. Genet. *2*:93, 1965.

———: X-Y chromosomal interchange in the aetiology of true hermaphroditism and of XX Klinefelter's syndrome. Lancet *2*:475, 1966.

Ferrier, P., S. M. Gartler, S. H. Waxman and T. H. Shepard: Abnormal sexual development associated with sex chromosome mosaicism. Pediatrics *29*:703, 1962.

Fine, G., R. C. Mellinger and J. C. Canton: Gonadoblastoma occurring in a patient with familial gonadal dysgenesis. Amer. J. Clin. Path. *38*:615, 1962.

Fraccaro, M., M. G. Bott, F. M. Salzano, R. W. R. Russell and W. I. Cranston: Triple chromosomal mosaic in a woman with clinical evidence of masculinization. Lancet *1*:1379, 1962.

Frasier, S. D., R. A. Bashore and H. D. Mosier: Gonadoblastoma associated with pure gonadal dysgenesis in monozygous twins. J. Pediat. *64:*740, 1964.

Grouchy, J. de, I. Emerit and P. Corone: Mosaïque X0/XY/Xy et syndrome de Turner chez une fille de 18 ans. Ann. Genet. *9:*86, 1966a.

———, J. Lasserre, R. Maraud, P. de Soza and A. Broustet: Mosaique X0/Xy dans un cas de dysgénésie gonadique mixte. Ann. Genet. *9:*127, 1966b.

Hemsworth, B. N. and H. Jackson: Effect of busulphan on the developing gonad of the male rat. J. Reprod. Fertil. *5:*187, 1963a.

——— and ———: Effect of busulphan on the developing ovary in the rat. J. Reprod. Fertil. *6:*229, 1963b.

Hoffenberg, R. and W. P. U. Jackson: Gonadal dysgenesis in normal-looking females. Brit. Med. J. *1:*1281, 1957.

Inhorn, S. L. and J. M. Opitz: Abnormalities of sex development. In: Endocrine Pathology. J. M. B. Bloodworth, ed. Baltimore: The Williams & Wilkins Co., 1968.

Jacobs, P. A., M. Brunton, M. Melville, R. P. Brittain and W. F. McClemont: Aggressive behaviour, mental sub-normality and the XYY male. Nature *208:*1351, 1965.

——— and A. Ross: Structural abnormalities of the Y chromosome in man. Nature *210:*352, 1966.

Jones, H. W. and P. A. Zourlas: Clinical, histologic and cytogenetic findings in male pseudohermaphroditism. Obstet. Gynec. *25:*597, 1965.

Jost, A.: Récherches sur la differentiation sexuelle de l'embryon de lapin. III. Role des gonades foetales dans la différenciation sexuelle somatique. Arch. Anat. Micro. Morph. Exp. *36:*271, 1947.

Kjessler, B.: Karyotypes of 130 childless men. Lancet *2:*493, 1965.

Klevit, H. D., W. J. Mellman and W. R. Eberlein: Triple mosaicism with an isochromosome derived from a partially deleted Y in a male pseudohermaphrodite. Pediatrics *32:*56, 1963.

Lenz, W. and R. A. Pfeiffer: Die Genetik der Geschlechtsdifferenzierung beim Menschen. Münch. Med. Wschr. *108:*1, 1966.

Lo, M. T. and S. D. Kobernick: X"y"/X0 mosaicism in a phenotypic intersex—Report of a case. Amer. J. Clin. Path. *43:*251, 1965.

McKay, D. G., A. T. Hertig, E. C. Adams and S. Danziger: Histochemical observations on the germ cells of human embryos. Anat. Rec. *117:*201, 1953.

Milner, W. A., W. B. Garlick, A. J. Fink and A. A. Stein: True hermaphrodite siblings. J. Urol. *79:*1003, 1958.

Muldal, S. and C. H. Ockey: Muscular dystrophy and deletion of Y chromosome. Lancet *2:*601, 1961.

——— and ———: Deletion of Y chromosome in a family with muscular dystrophy and hypospadias. Brit. Med. J. *1:*291, 1962.

Ohno, S.: Sex Chromosomes and Sex-linked Genes. In: Monographs on Endocrinology, Vol. 1. New York: Springer-Verlag New York Inc., 1967.

———, J. M. Trujillo, C. Stenius, L. C. Christian and R. L. Teplitz: Possible germ cell chimeras among newborn dizygotic twin calves (*Bos taurus*). Cytogenetics *1:*258, 1962.

Oikawa, K. and R. M. Blizzard: Chromosomal studies of patients with congenital anomalies simulating those of gonadal aplasia: Including a case of true gonadal and sex reversal. New Eng. J. Med. *264*:1009, 1961.

Penrose, L. S.: Finger-print pattern and the sex chromosomes. Lancet *1*:298, 1967.

Pinkerton, J. H. M., D. G. McKay, E. C. Adams and A. C. Hertig: Development of the human ovary—A study using histochemical technics. Obstet. Gynec. *18*:152, 1961.

Rosenberg, H. S., G. W. Clayton and T. C. Hsu: Familial true hermaphrodism. J. Clin. Endocr. *23*:203, 1963.

Schlegel, R. J., J. Leao, R. L. Neu, E. Farias and L. I. Gardner: Testicular X0/XY mosaicism in a phenotypic female with XY sex chromosomes in other tissues. J. Clin. Endocr. *26*:847, 1966.

Singh, R. P. and D. H. Carr: The anatomy and histology of X0 human embryos and fetuses. Anat. Rec. *155*:369, 1966.

Solomon, I. L., C. W. Hamm and O. C. Green: Chromosome studies on testicular tissue cultures and blood leukocytes of a male previously reported to have no Y chromosome. New Eng. J. Med. *271*:586, 1964.

Sternberg, W. H., D. L. Barclay and H. W. Kloepfer: Familial XY gonadal dysgenesis. New Eng. J. Med. *278*:695, 1968.

Swyer, G. I. M.: Male pseudohermaphroditism: A hitherto undescribed form. Brit. Med. J. *2*:709, 1955.

Teter, J. and K. Boczkowski: Occurrence of tumors in dysgenetic gonads. Cancer *20*:1301, 1967.

Vaharu, T., R. G. Patton, M. L. Voorhess and L. I. Gardner: Gonadal dysplasia and enlarged phallus in a girl with 45 chromosomes plus "fragment." Lancet *1*:1351, 1961.

Witschi, E.: Studien über die Geschlechtsbestimmung bei Fröschen. Arch. mikr. Anat. *86*:1, 1914.

———: Migration of the germ cells of human embryos from the yolk sac to the primitive gonadal fold. Contributions to Embryology, Carnegie Institute, Washington, D. C. *32*:67, 1948.

ANOMALIES OF SEX CHROMOSOMES IN TORTOISESHELL MALE CATS

T. C. JONES
New England Regional Primate Research Center, Harvard Medical School, Southboro, Mass.

The enigma of the rare tortoiseshell male cat has intrigued scientists for many years. The explanation seems to have a cytogenetic basis, justifying a summary of present knowledge in the context of this Conference. We shall report two cases in detail, review other reported cases and discuss possible mechanisms of etiology.

The recognition of the tortoiseshell phenotype is important in this problem: errors have frequently been confusing and have appeared in published reports. The tortoiseshell cat is identified by the concurrent presence of two colors, black and orange, in the hair coat. These two colors are determined by two allelic genes, orange (*O*) and black (*O*+) which are sex-linked; therefore the tortoiseshell results from the heterozygous state at this orange locus and normally only appears in females. The hemizygous male may be orange (*O*,–) or black (*O*+,–). The designation of this coat color as orange, was agreed upon by a group of geneticists in 1967 (see Robinson, 1967), although it is often called yellow, ginger or red by cat fanciers and geneticists alike. This coat color is usually readily recognized but may be confused with a yellowish buff color which in some animals is evident around the head and lower parts of the body. This buff color is most evident in the hair between tabby markings and does not alternate with black within these tabby stripes—a point which may be helpful in distinguishing this phenotype.

The orange hair color or its alternative, black, may be modified in expression by several independently-assorting autosomal genes. The gene for Maltese dilution (*d*) when present in the homozygous state, changes the black to a grayish color called "blue" and the orange to a dilute color called "cream." Tortoiseshells with *dd* phenotype are therefore called "blue cream" tortoiseshells (or "torties"). Another series of alleles, the tabby series, change the pattern of the black or orange. The dominant of these alleles, occurs in the Abyssinian breed, is called lined or Abyssinian tabby (T^a) and results in an overall gray or mouse color without any distinctively darker markings. The next in the series in order of dominance, is the wild type, called mackerel or striped tabby ($t+$) in which the tabby markings are distinct, vertical and parallel over the sides of the body. These stripes resemble the markings of the fish from which the designation is derived. The third and recessive allele of this series, blotched tabby (t^b) is recognized by a circular pattern of tabby stripes on the sides of the abdomen and by a broad band of color on the back which usually extends downward over each shoulder.

Another autosomal gene, agouti (A) causes the hair to be "ticked" with alternate bands of white and color in the hair located in those areas of the coat between the tabby markings, this making the tabby markings recognizable. In a black cat, homozygous for non-agouti (a,a) the tabby markings are not recognizable ("self black"). With the orange coat color the markings are visible, independent of the presence of agouti. Another allelic series at the silver locus also may modify the color phenotype. This series includes silver or chinchilla (c^{ch}) which in the homozygous state changes the hair next to the skin to silver—leaving the tips black or orange. The next allele, siamese or himalayan (s^s) results in the typical Siamese color pattern with the points (head, tail, feet) approximately full color, the hair of the rest of the body much lighter in color. Burmese (c^b) is the next allele—it results in an overall golden brownish hair coat. A fifth allele, albino (c) is rare and is expressed by absence of melanin.

The coat color may also be modified by another, more frequent autosomal genetic series at the piebald or white spotting locus (S). This allelic series is probably controlled by several genes expressed by varying degrees of white coat color which in the minimal form consists of a few white hairs on the ventral midline. Increased degrees of white spotting in different animals extend over the body from the midline, in some cases nearly covering the whole body. The presence of piebald white color in association with orange and black results in such descriptions as "tortoiseshell and white," "tricolor" or "calico."

The presence of another independent, autosomal gene, dominant white (W), usually obscures all other color phenotypes except in a few instances in which a culotte of black or orange hair (or both) may occur in the top of the head. These spots in many cats tend to diminish or disappear as the cat reaches maturity. Further discussion of hair coat phenotypes may be found in Searle (1968).

Report of Two Cases

Case 1

A male, long haired, tortoiseshell and white male cat, estimated age one and one-half years, was brought to our attention by Richard M. Roberts, D.V.M., of Hanover, Massachusetts. This cat, named "Lucky," was subsequently given to us. In this animal, patches of black and orange colored hair (Fig. 1) were distributed over the dorsal aspect of the neck; back and tail and were replaced by white over about 75% of the body surface, extending from the ventral midline approximately equal distances on either side of the body. No tabby markings were evident in the black patches. The identifiable phenotype of this male cat was summarized as: long hair (*l,l,*) non-agouti (*a,a,*) piebald (75%) (*S*), orange and black (*O*+,*O*). "Lucky" had an effeminate head and his testicles were estimated to be about one fourth as large as those expected in a cat this age. No information was available on his ancestry.

Biopsy specimens of the left testicle were examined microscopically and preparations for mitotic and meiotic chromosomes were made. No cells in meiosis were found but fibroblasts yielded chromosome spreads with karyotypes essentially the same as in the skin. Cells with positive sex chromatin (single Barr bodies) were demonstrated in sections and in cultures of fibroblasts.

The surgically excised left testis was considered hypoplastic, measuring 0.5 cm in maximum dimensions. Histologic sections (Figs. 2 and 3) revealed closely packed, small seminiferous tubules lined with Sertoli cells whose cytoplasm contained large lipid vacuoles. An occasional germinal

Fig. 1. Male tortoiseshell and white cat, "Lucky," Case 1. Alternate black and orange color on back and sides indicate tortoiseshell phenotype. White areas ("piebald") and long hair are determined by autosomal genes. The orange color appears gray in this black and white photograph.

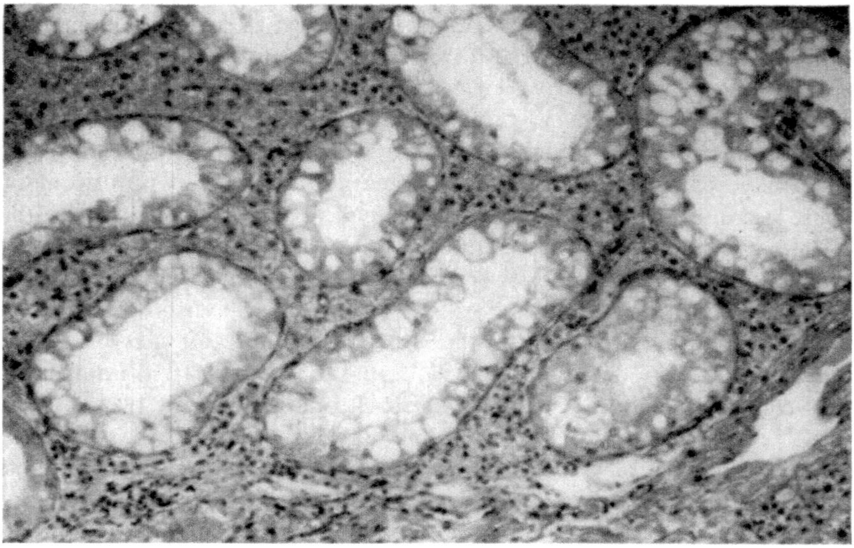

Fig. 2. Testis of Case 1, "Lucky," tortoiseshell male cat, age 2 years. Seminiferous tubules separated by closely-packed interstitial cells.

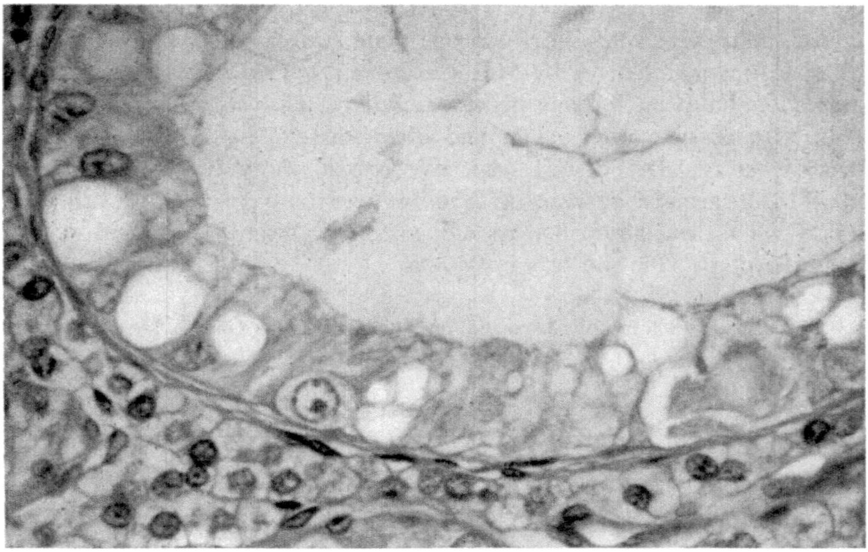

Fig. 3. Testis of Case 1, "Lucky." Higher magnification of one seminiferous tubule. The tubule is lined with Sertoli cells containing vacuoles indicating presence of lipid. A single germinal cell is present.

cell could be found but spermatogenesis was totally absent. Interstitial cells were conspicuous but probably as the result of atrophy of the seminiferous tubules—causing the interstitial cells to occupy a relatively greater amount of space in the testis. Spermatozoa were absent. Cultures of skin prepared by the method of Basrur *et al.* (1963) and of blood, leukocytes, using a modification of the technique of Moorehead (1960) were made to obtain chromosomal preparations. Cells in mitotic metaphase were photographed and karyotypes arranged according to the San Juan system for felidae (Hsu and Rearden, 1965; Smith and Jones, 1966). Chromosome numbers were determined by projecting the negative on a blackboard, then checking off and counting each chromosome.

The numbers of cells in mitotic metaphase and the karyotypes of cells from "Lucky" are depicted in Table 1. The greatest number of cells contained a diploid number of 39 chromosomes with two X's and one Y chromosome. Other populations of cells were also present, with differing numbers of sex chromosomes. The most frequent were diploid (2n 38) with XX or XY sex chromosomes and 2n 40 with XXYY sex chromosomes. Loss of chromosomes (possibly technical) and endoreduplicated cells were apparent causes of some of the low and very high chromosome numbers. It is apparent that this animal was (38XX/XY 39XXY 40XXYY) chimeric. Karyotypes are illustrated in Figures 4–8.

Case 2

An eight week old short haired, male, tortoiseshell and white cat (Fig. 9) was given to us by Mrs. Gerace of Newton. This cat was soon named "Chauncey" by one of his attendants. He was one of a litter of four kittens, two other males and one female. The siblings each had karyotypes 2n 38XY or XX respectively. "Chauncey" was much smaller than his littermates at weaning time but eventually grew into a full sized cat. His dam was about 8 years old, tortoiseshell and white, with a normal karyotype. His sire was unknown.

Table 1: Chromosomal Counts and Karyotypes of Case 1, "Lucky," Tortoiseshell Male Cat

	Chromosome Numbers								Total
2n:	35	36	37	38	39	40	41	>42	
Number of Cells:	5	2	14	21	190	14	3	14	263
Percent:	1.9	0.8	5.3	8.0	72.2	5.3	1.1	5.3	99.9

	Karyotypes					Number of Karyotypes
2n:	37	38	38	39	40	
Number:	1	4	2	19	3	27
Percent:	3.4	13.8	6.9	65.5	10.3	99.9
Sex Chromosomes:	XXY	XY	XX	XXY	XXYY	

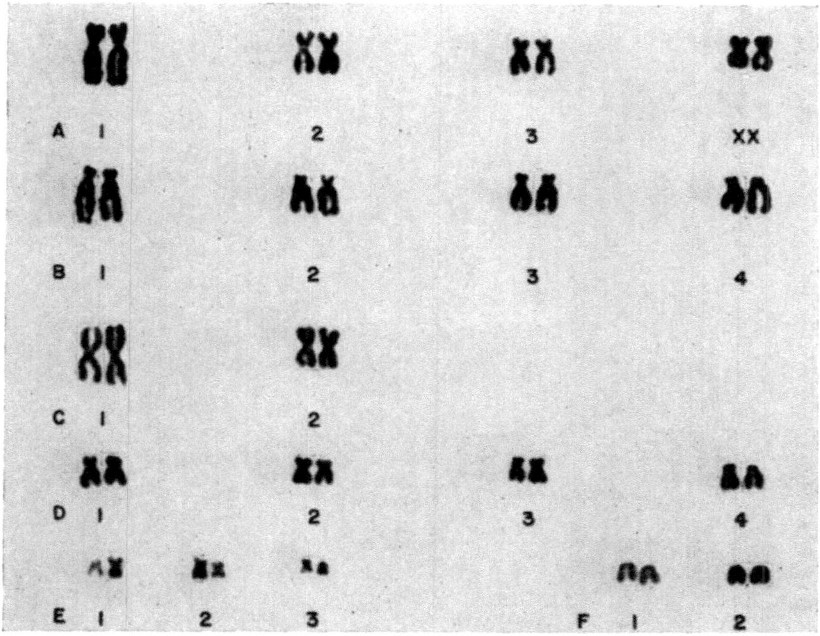

Fig. 4. Karyotype, Case 1, "Lucky," 38XX cell.

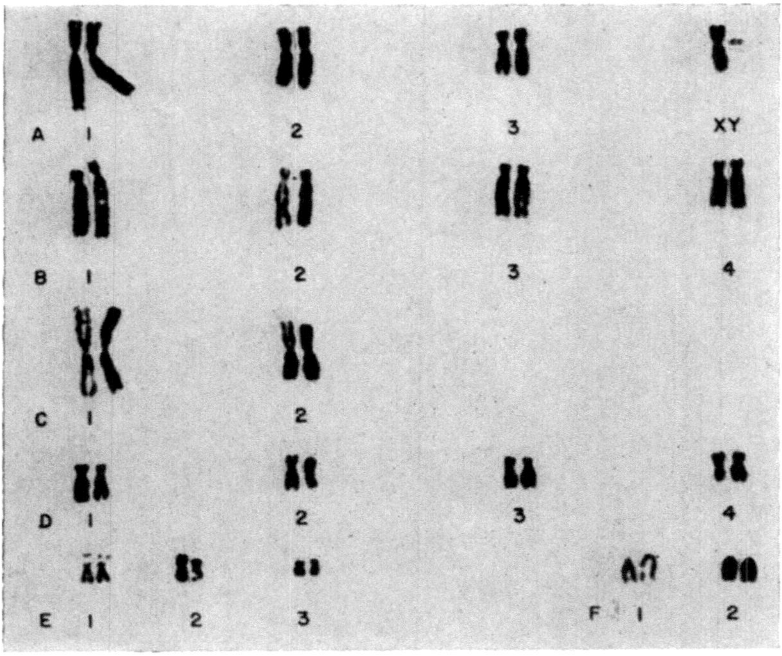

Fig. 5. Karyotype, Case 1, "Lucky," 38XY cell.

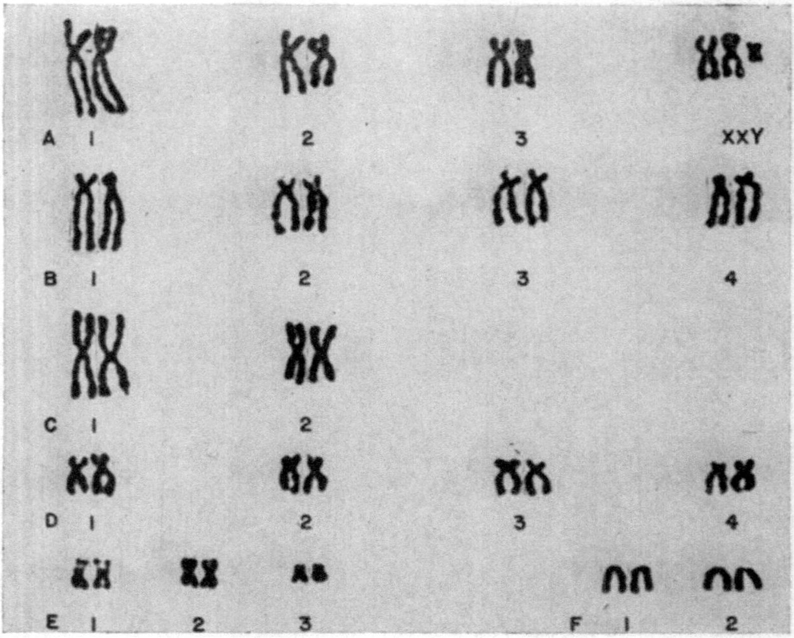

Fig. 6. Karyotype, Case 1, "Lucky," 39XXY cell.

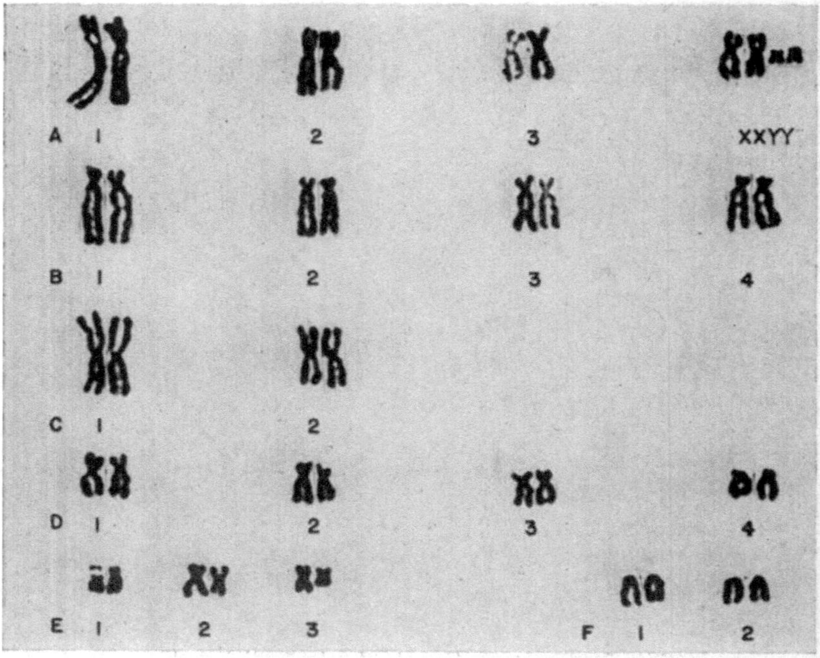

Fig. 7. Karyotype, Case 1, "Lucky," 40XXYY cell.

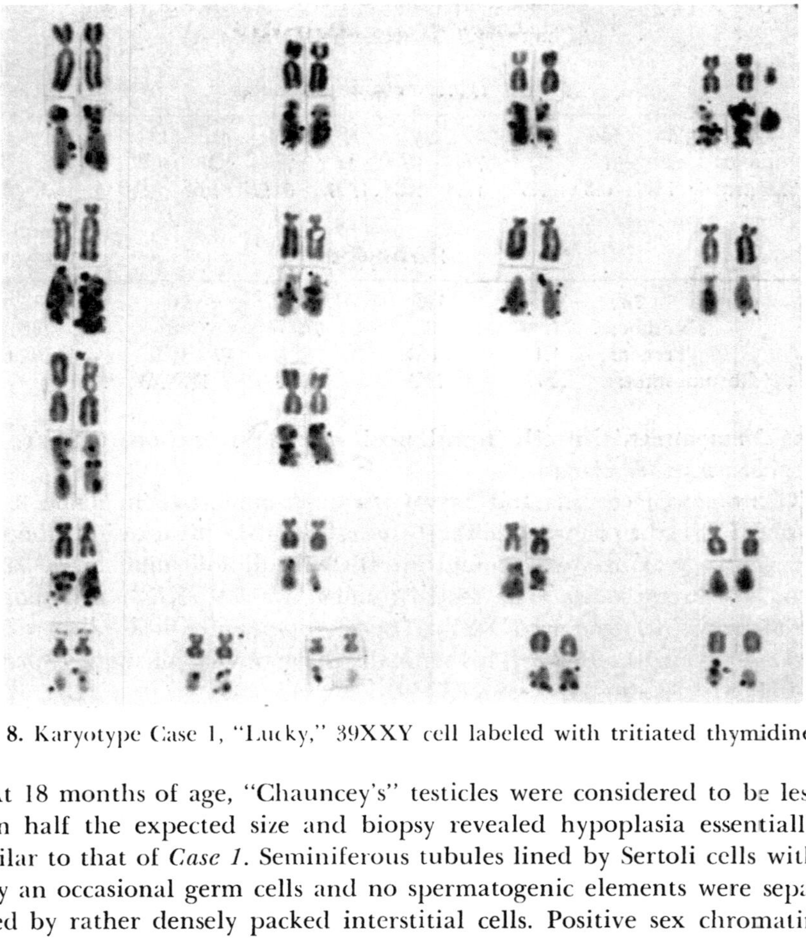

Fig. 8. Karyotype Case 1, "Lucky," 39XXY cell labeled with tritiated thymidine.

At 18 months of age, "Chauncey's" testicles were considered to be less than half the expected size and biopsy revealed hypoplasia essentially similar to that of *Case 1.* Seminiferous tubules lined by Sertoli cells with only an occasional germ cells and no spermatogenic elements were separated by rather densely packed interstitial cells. Positive sex chromatin

Fig. 9. Male tortoiseshell and white cat, "Chauncey," Case 2. The orange part of the coat appears gray. Note short hair and polydactyly.

Table 2: Chromosomal Counts and Karyotypes of Case 2, "Chauncey," Tortoiseshell Male Cat

					Chromosome Numbers						Total
2n:	<33	34	35	36	37	38	39	40	41	>42	
Number:	7	1	5	7	9	54	274	23	3	15	398
Percent:	1.8	0.3	1.3	1.8	2.3	13.6	68.8	5.8	0.8	3.8	100.5

		Karyotypes			Number of Karyotypes
2n:	*37*	*38*	*39*	40	
Number:	1	6	73	8	88
Percent:	1.1	6.8	82.9	9.0	99.8
Sex Chromosomes:	X0	XY	XXY	XXYY	

was demonstrated in cells from buccal smears, in sections of testis, and in cultures of fibroblasts.

Chromosome counts and karyotypes are summarized in Table 2. An unusual number of aneuploid cells were found in this case but the most significant numbers were found in cells with diploid numbers of 38, 39 and 40 chromosomes. The modal number was 39 (XXY) with populations of 38 (XY) and 40 (XXYY). The karyotypes of typical cells are illustrated in Figures 10–12. This animal's chromosomal phenotype may be summarized as an 38XY 39XXY 40XXYY chimera.

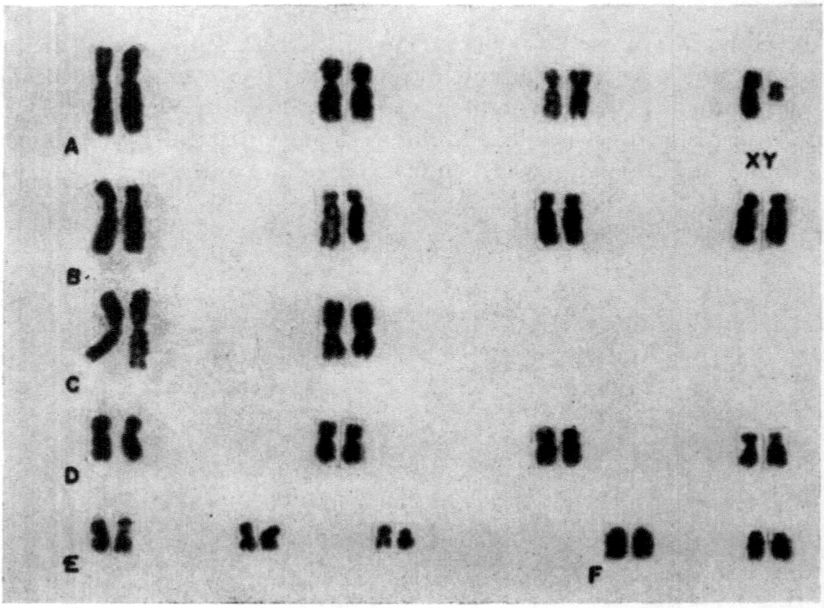

Fig. 10. Karyotype, Case 2, "Chauncey," an XY cell.

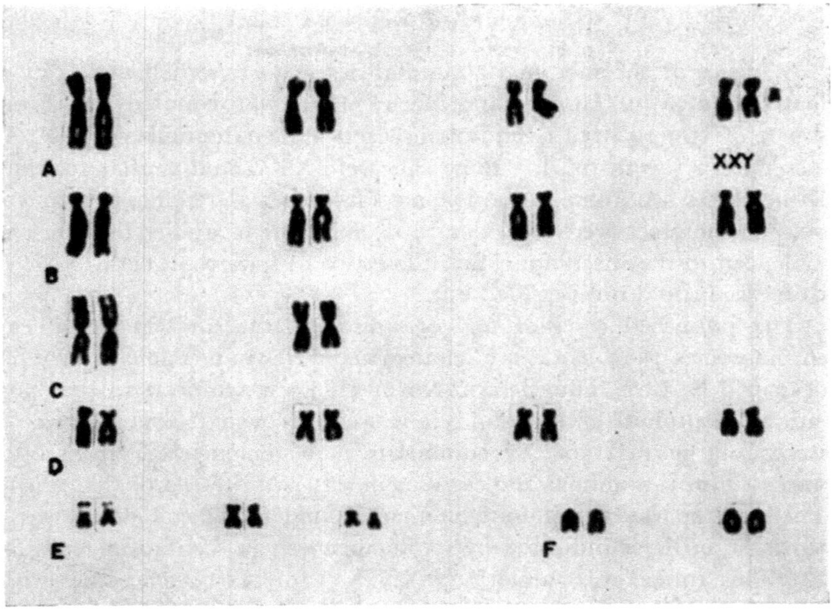

Fig. 11. Karyotype, Case 2, "Chauncey," an XXY cell.

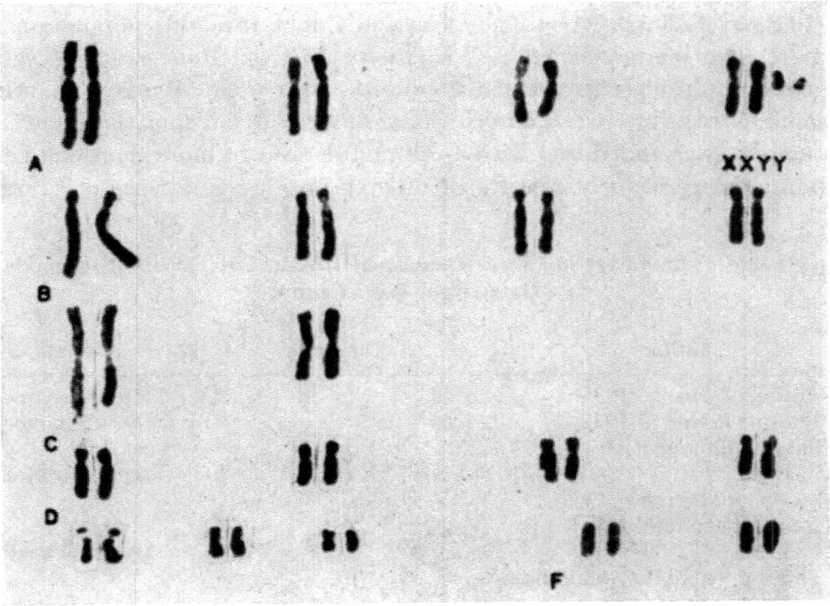

Fig. 12. Karyotype, Case 2, "Chauncey," an XXYY cell.

Summary of Reported Cases

Evidence of chromosomal abnormalities in tortoiseshell male cats was first presented by Thuline and Norby (1961) who counted the chromosomes of two such cats, finding in each a modal number of 39. The assumption was made that these cats were XXY, and similar to human Klinefelter's syndrome, although no karyotypes were prepared and the sex chromosomes were not identified. Subsequent studies on other animals lead to the speculation that these two cats were probably XXY but chimerism could not be ruled out.

The published cases of male tortoiseshell cats in which studies of chromosomes were made are summarized briefly in Table 3. The case described by Chu, Thuline and Norby (1964) was unique in that populations of diploid and triploid cells were recovered from cultures of skin. Cells positive for sex chromatin were demonstrated from buccal smears from this animal and sections of testes indicate probable sterility. The case reported by Malouf, Benirschke and Hoefnagel (1967) was potentially fertile as indicated by spermatogenesis in sections of testis (Fig. 13). This animal was chimeric (38XX/XY) for sex chromosomes with a normal diploid number. Biggers and McFeeley (1966) described a tortie male with two populations of cells, 38XY and 39XXY.

The terms "chromosome chimerism" and "mosaicism" are often used interchangeably. We have adopted the view proposed by Chu *et al.* (1964) as follows: "A *mosaic* is an individual with cell populations of more than one genotype (e.g., karyotype) derived from a single zygotic genotype through mutational or mitotic events (e.g., somatic mutation, somatic crossing over, mitotic loss, mitotic non-disjunction, etc.). A *chimera* is an individual with a cell populations of more than one genotype arising through mixture of different zygotic genotypes (e.g., trans-

Table 3: Summary Data on Tortoiseshell Male Cats, With Cytogenetic Data, from the Literature

Authors	Karyotype	Remarks
Thuline & Norby (1961)	2n 39	No karyotype
Thuline & Norby (1961)	2n 39	No karyotype
Chu, Thuline & Norby (1964)	38XX/57XXY	Diploid/triploid
Biggers & McFeeley (1966)	38XY/39XXY	
Malouf, Benirschke & Hoefnagel (1967)	38XX/XY	Possibly fertile
Thuline (cited by Malouf) (1967)	38XX/XY	
Jones (1968) Case 1	38XX/XY/39XXY/40XXYY	Sterile
Jones (1968) Case 2	38XY/39XXY/40XXYY	Sterile

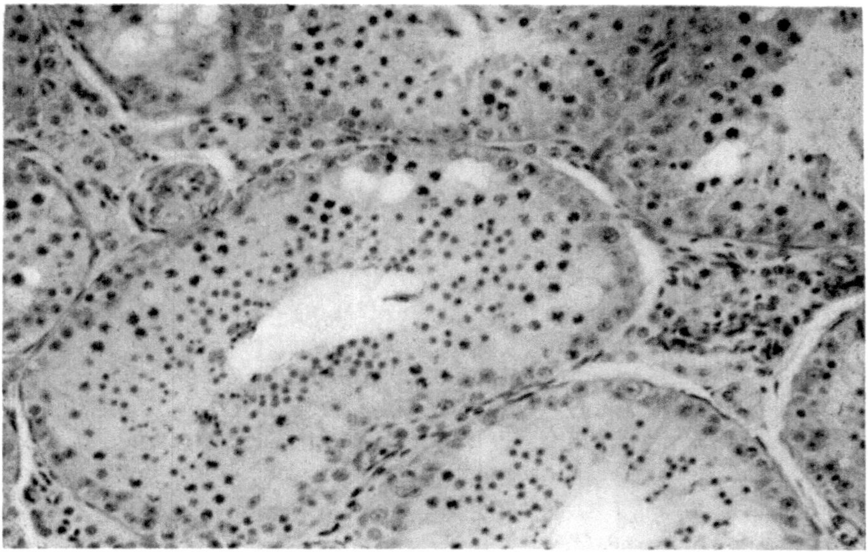

Fig. 13. Section of testis of 38XX/XY tortoiseshell cat reported by Malouf, Benirschke and Hoefnagel, 1967. Note spermatogenesis is present, in contrast to Figs. 2 and 3. Specimen courtesy of Dr. K. Benirschke.

plantation, chorionic vascular anastomoses, double fertilized meiotic products into one developing embryo, etc.)."

We have used the term "chimera" in the above sense in reference to these animals with anomalies of sex chromosomes because this seems most likely. However, the evidence is not conclusive on this point.

Theories of Pathogenesis

Several theories have been advanced to explain the mechanisms involved in these errors in sex chromosomes. A brief review of some of them seems indicated although experimental evidence is generally lacking to support any of them.

Meiotic Non-disjunction Under normal circumstances, during second meiotic division, the chromosome complement is divided equally in number between two gametes. This disjunction results in the usual haploid number of chromosomes in each gamete. Under certain circumstances it is assumed that not all chromosomes are distributed equally to the gametes. The sex chromosomes may be particularly vulnerable to this error because of replication which is non-synchronous with the autosomes. Thus, gametes formed following non-disjunction of the sex pair would have both sex chromosomes or none. In the female gonad, for example, each non-disjunctional event would result in an ovum with two X chromosomes and another with none. Fertilization of each of these gametes with Y-bearing sperm would result in an XXY and an Y0 zygote. The Y0 zygote is believed not to be viable. On the other hand, fertilization of an XX gamete by an X-bearing sperm would result in an XXX zygote. Further, a gamete without an X would produce an X0 zygote if fertilized by an X-bearing sperm. This phenomenon is outlined in Figure 14. This provides an explanation for XXY phenotypes but not for chimerism.

Polyandry; Suppression of Polar Body; Fusion Figures 15 and 16 illustrate the possible result of fertilization of a single ovum with two sperm, one carrying an X, another a Y chromosome, to produce a triploid zygote. Concomitant fertilization of a non-extruded polar body would produce a diploid zygote which could fuse with the triploid one to form a diploid-triploid chimera. Polyandry resulting in triploidy has been demonstrated in several species, including the cat (Austin, 1960).

Failure of Second Meiotic Division Another postulate which could explain the observed facts is illustrated in Figure 17. In this instance,

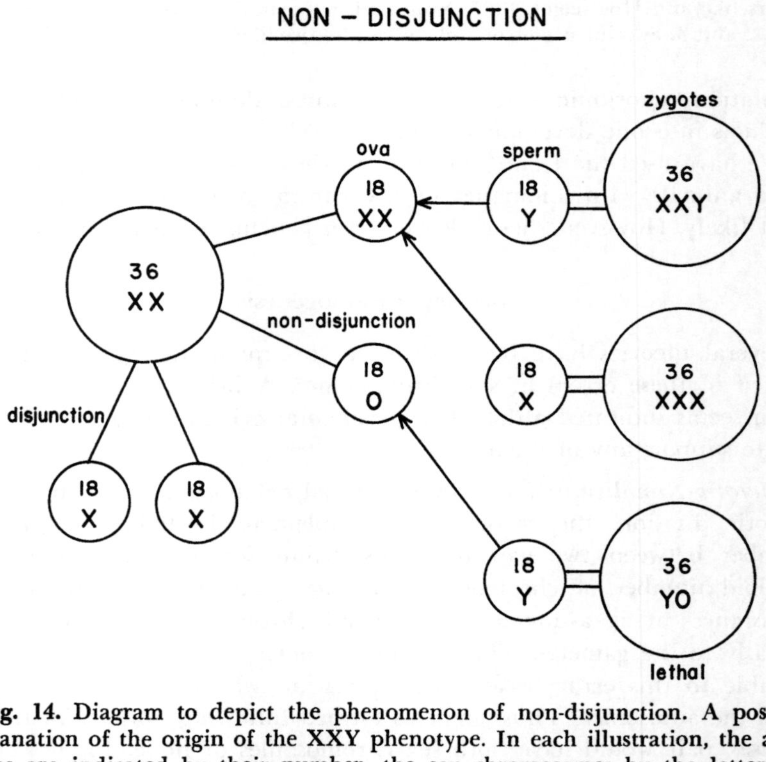

Fig. 14. Diagram to depict the phenomenon of non-disjunction. A possible explanation of the origin of the XXY phenotype. In each illustration, the autosomes are indicated by their number, the sex chromosomes by the letters (X and Y).

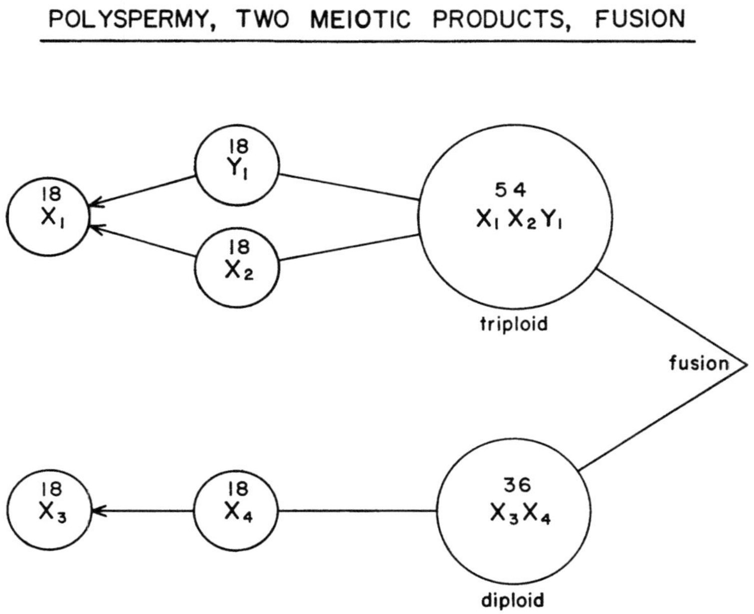

Fig. 15. The theory of polyandry, suppression and fertilization of polar body and fusion of embryos to explain diploid/triploid embryo.

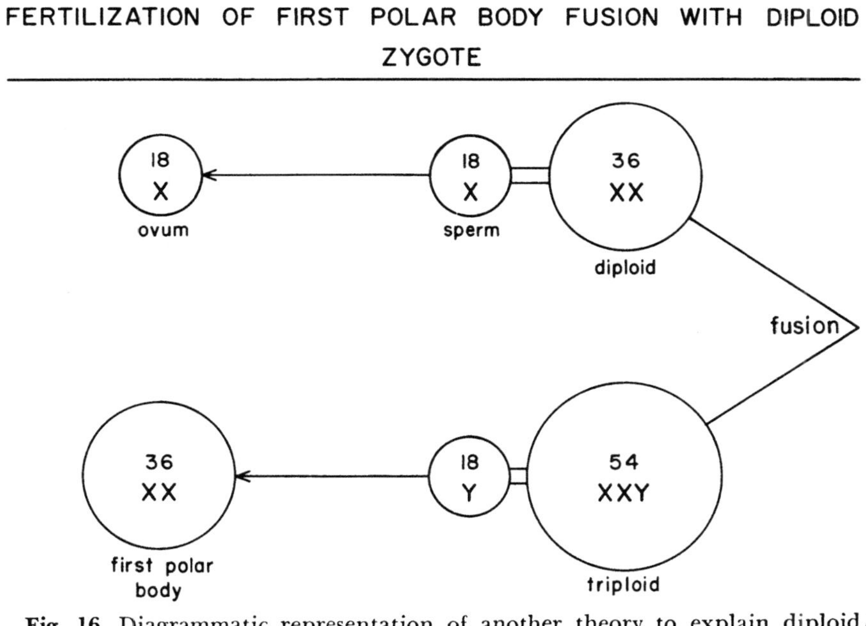

Fig. 16. Diagrammatic representation of another theory to explain diploid/triploid chimerism. Fertilization of the first (diploid) polar body followed by fusion with a diploid zygote.

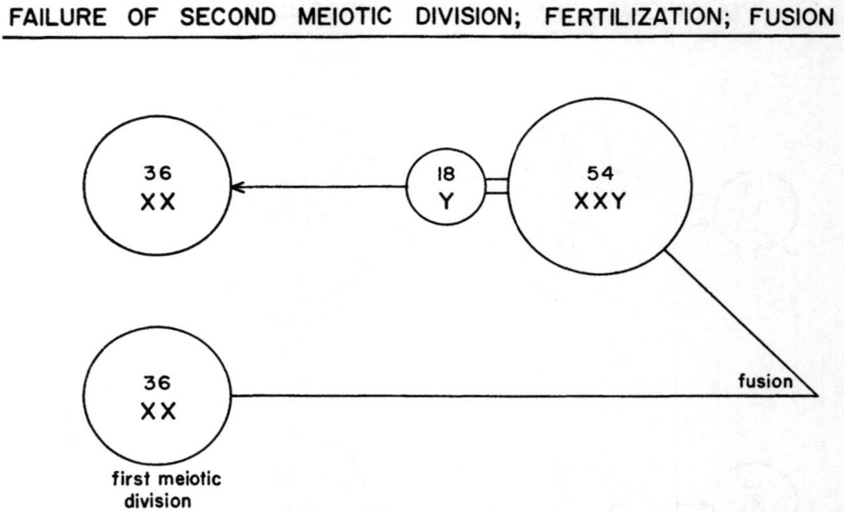

Fig. 17. Diagram to illustrate another theory concerning origin of diploid/triploid embryos. Failure of second meiotic division, resulting in fertilization of one diploid ovum and incorporation of a second by fusion into a single embryo.

second meiotic (reduction) division does not occur, leaving diploid primary oocytes susceptible to fertilization. One of the oocytes is fertilized by a Y-bearing sperm and it participates with an unfertilized diploid oocyte in forming the embryo. This rather complicated sequence of events could lead to a diploid-triploid chimeric embryo.

Abnormal Syngamy; Initial Triploidy Unusual, but possible, events at fertilization, as outlined by Austin (1960), could lead to a triploid embryo at the outset. Diploid cell lines could arise by loss of chromosomes at mitosis, perhaps through defective spindles or other mitotic factors. Chromosomes could be extruded from nuclei following these events. This theory is diagrammatically depicted in Figure 18.

Fusion of Zygotes The fusion of zygotes has been demonstrated, particularly by Mintz (1964), to be possible under experimental conditions. Early mouse blastocysts have been removed from the oviduct, the zona pellucida removed with pronase, then the embryos placed together under suitable conditions. Fusion of these blastocysts occurs and when reintroduced into a suitable uterus results in living progeny which are a chimeric medley of phenotypes from each of four parents. This phenomenon could occur spontaneously, particularly in species which extrude many ova at each ovulation. The resistance of the zona pellucida to fusion would have to be overcome in some way, such as the occurrence of two oocytes within one zona pellucida. These possibilities are outlined in Figure 19.

ABNORMAL SYNGAMY;
TRIPLOID CELLS PRODUCE DIPLOID CELLS BY LOSS AT MITOSIS

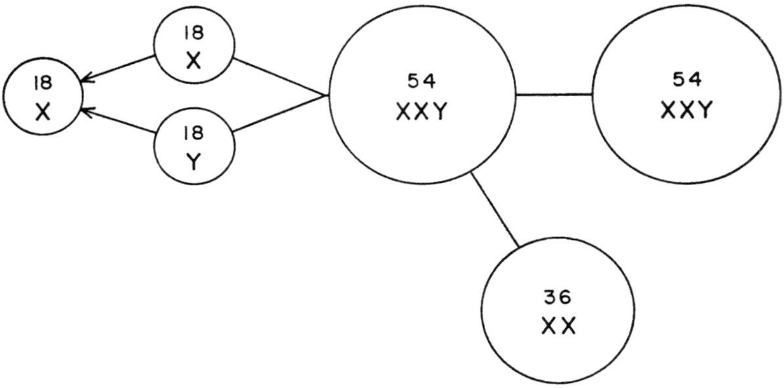

Fig. 18. Abnormal syngamy to produce triploid cells which give rise to diploid cells by loss at subsequent mitoses. In this case the phenomenon would be, by definition, called mosaicism rather than chimerism.

Exchange of Cells Between Twins Vascular anastomosis in the chorion has been shown to provide a means of exchange of hematopoietic and sex cells between twins. Chimerism of these cells have been demonstrated in cattle (Ohno *et al.,* 1962) and marmosets (Benirschke and Brownhill, 1962). Cells of other tissues, on the other hand, have

FUSION OF TWO ZYGOTES

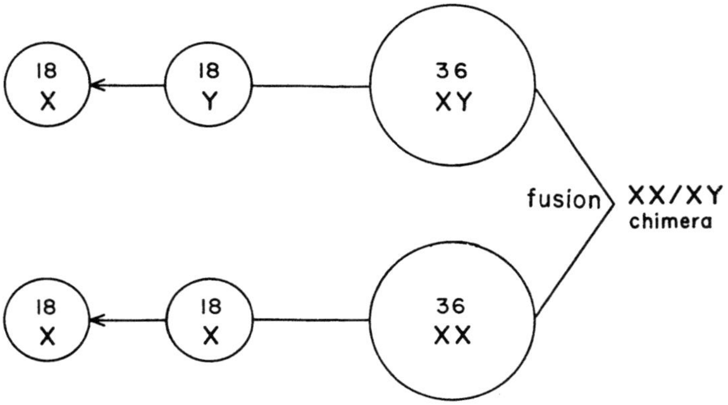

Fig. 19. One theoretical means by which a chimera might be produced by fusion of two zygotes and incorporation of their cells into one embryo.

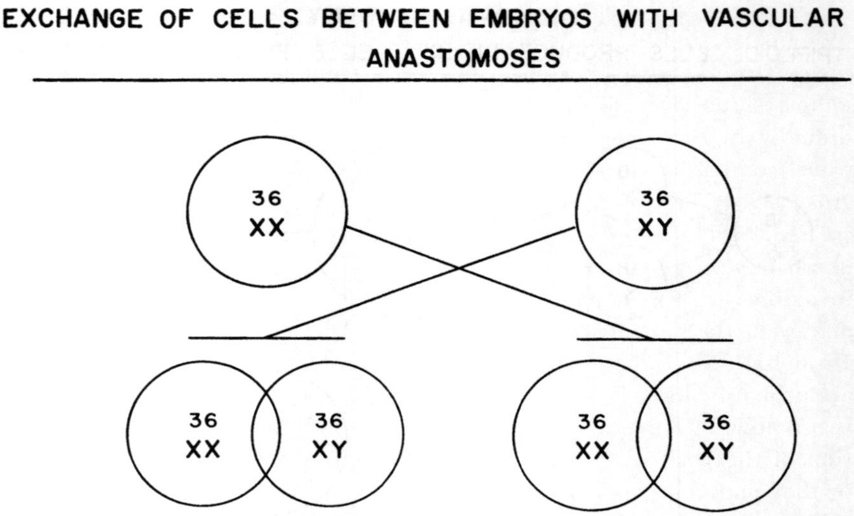

Fig. 20. Another theory concerning the origin of chimerism. Two embryos exchange cells by way of vascular anastomoses in the chorion. Such exchanges of primitive hematopoietic and sex cells have been demonstrated. Exchange of other cells has not been found.

not been shown to migrate by this means. This mechanism of interuterine exchange (Fig. 20) does not seem likely in tortoiseshell male cats.

Summary

The tortoiseshell ("tortie") coat color in *Felis catus* is the expression of the heterozygous phenotype for two sex-linked genes orange (*O*) and black (*O*+). Most tortoiseshell cats are therefore females. The recognition of the "tortie" phenotypes is sometimes made difficult by the interaction of modifying genes whose effects are described briefly. The exceptional male tortie results from the presence of two or more populations of cells with X chromosomes bearing each of the alleles at the orange locus.

Six cases of tortie males whose chromosomes had been studied were reviewed from the literature and two new cases reported. Two cases, the first reported, were shown to have 39 chromosomes and presumed to have XXY sex chromosome constitution. One case was diploid/triploid, 38XX/57XXY, with aspermatogenic testes. Two cases were diploid/diploid chimeras, 38XX/38XY and spermatogenesis was evident in their testes. One animal was chimeric, 38XX/39XXXY and believed to be sterile. The seventh case, reported here, has aspermatogenic testes and was chimeric, 38XX/38XY/39XXY/40XXYY, for sex chromosomes. The eighth cat, also sterile, was chimeric, 38XX/39XXY/40XXYY.

Possible mechanisms which could result in these chromosomal aberrations are described briefly. They are as follows: (1) meiotic non-disjunction; (2) polyandry with suppression of polar body and fusion of zygotes; (3) separate fertilization of one ovum and its heterozygous polar body followed by fusion of zygotes into one embryo; (4) abnormal syngamy-resulting in a triploid embryo which may lose haploid sets of chromosomes in subsequent mitoses; (5) fusion of two zygotes to obtain the chimeric state and (6) exchange of undifferentiated cells between heterozygous twins by means of chorionic vascular anastomoses.

Acknowledgment

This work was supported in part by research grant GM 16209 from the National Institute of General Medical Sciences, U.S. Public Health Service, Bethesda, Maryland.

References

Austin, C. R.: Anomalies of fertilization leading to triploidy. J. Cell. Comp. Physiol. *56*:Suppl. 1, 1, 1960.

Bain, A. D. and J. S. Scott: Mixed gonadal dysgenesis with XX/XY mosaicism. The evidence for the occurrence of fertilization by two spermatozoa. Lancet *1*:1035, 1965.

Bamber, R. C. and E. C. Herdman: The inheritance of black, yellow and tortoiseshell coat color in cats. J. Genet. *18*:87, 1927.

——— and ———: The problem of the tortoiseshell male cat. Verh. V. Intern. Kongr. Vererb. *1*:387, 1928.

——— and ———: The incidence of sterility amongst tortoiseshell male cats. J. Genet. *24*:355, 1931.

——— and ———: A report on the progeny of a tortoiseshell male cat, together with a discussion of his gametic constitution. J. Genet. *26*:115, 1932.

Basrur, P. K., V. R. Basrur and J. P. W. Gilman: A simple method for short term cultures from small biopsies. Exp. Cell. Res. *30*:229, 1963.

Benirschke, K. and L. E. Brownhill: Further observations on marrow chimerism in marmosets. Cytogenetics *1*:245, 1962.

Biggers, J. D. and R. A. McFeely: Intersexuality in domestic animals. Chapt. II in: Advances in Reproductive Physiology. A. McLaren, ed. New York: Academic Press, 1966.

Bissonnette, T. H.: Tortoiseshell tomcats and freemartins. J. Hered. *19*:87, 1928.

———: A case of potential freemartins in cats. Anat. Rec. *40*:339, 1928.

Chu, E. H. Y., H. C. Thuline and D. E. Norby: Triploid-diploid chimerism in a male tortoiseshell cat. Cytogenetics *3*:1, 1964.

Cutler, D. W. and L. Doncaster: On the sterility of the tortoiseshell tomcat. J. Genet. *5*:66, 1915.

Edwards, J. H., C. Yuncken, D. I. Rushton, S. Richards and U. Mittwoch: Three cases of triploidy in man. Cytogenetics *6*:81, 1967.

Ellis, J. R., R. Marshall, I. C. S. Normand and L. S. Penrose: A girl with triploid cells. Nature (Lond.) *198:*411, 1963.

Hays, F. A.: The tortoiseshell cat. J. Hered. *14:*369, 1923.

Hsu, T. C. and H. H. Rearden: Further karyological studies on felidae. Chromosoma (Berl.) *16:*365, 1965.

Ibsen, H. L.: Tricolor inheritance. III. Tortoiseshell cats. Genetics *1:*377, 1916.

Ishihara, T.: Cytological studies on tortoiseshell male cats. Cytologia *21:*391, 1956.

Jude, A. C. and A. G. Searle: A fertile tortoiseshell tomcat. Nature *179:*1087, 1957.

Komai, T.: On the inheritance of black, yellow and tortoiseshell color in cats, and the problem of the tortoiseshell male. Proc. Jap. Acad. *22:*265, 1946.

———: A new hypothesis on the origin of tortoiseshell male cat. Mem. Coll. Sci. Kyoto Univ. B. *19:*17, 1947.

———: Inheritance of common colour types and the origin of tortoiseshell males. Ann. Rep. Nat. Inst. Genet. Jap. *2:*9, 1952.

———: On the origin of the tortoiseshell male cat—a correction. Proc. Jap. Acad. *28:*150, 1952.

———: Problem of the origin of the tortoiseshell male cat and its sterility. Ann. Rep. Nat. Inst. Genet. Jap. *3:*9, 1953.

———: On the origin of the tortoiseshell male cat. Ann. Rep. Nat. Inst. Genet. Jap. *7:*12, 1957.

———: Supplementary notes on the genetics of tortoiseshell male cat. J. Fac. Sci. Hokkaido Univ. Zool. *13:*253, 1957.

——— and I. Ishihara: On the origin of male tortoiseshell cat. Am. J. Hered. *47:*287, 1956.

Lyon, M. E.: Sex chromatin and gene action in the mammalian X-chromosome. Am. J. Human Genet. *14:*135, 1962.

Malouf, N., K. Benirschke and D. Hoefnagel: XX/XY chimerism in a tricolored male cat. Cytogenetics *6:*228, 1967.

Makino, S., M. S. Sasaki and T. Fukushima: Triploid chromosome constitution in human chorionic lesions. Lancet *ii:*1273, 1964.

Matano, Y.: A study of the chromosomes for the cat. Jap. J. Genetics *38:*147, 1963.

Mintz, B.: Formation of genetically mosaic mouse embryos and early development of "lethal (t^{12}/t^{12}) normal" mosaic. J. Exper. Zool. *157:*273, 1964.

Moorhead, P. S., W. J. Mellman, D. M. Battips and D. A. Hungerford: Chromosome preparations of leukocytes cultured from human peripheral blood. Exp. Cell Res. *20:*613, 1960.

Ohno, S., W. A. Kittrell, L. C. Christian, C. Stenius and G. A. Witt: An adult triploid chicken (*Gallus domesticus*) with a left ovo-testis. Cytogenetics *2:*42, 1963.

———, J. M. Trujillo, C. Stenius, L. C. Christian and R. L. Teplitz: Possible germ cell chimeras among newborn dizygotic twin calves (*Bos taurus*). Cytogenetics *1:*258, 1962.

Robinson, R.: Standardized genetic nomenclature for the domestic cat. Carnivore Genetics Newsletter #3, 52, 1967.

Russell, L. B. and E. H. Chu: An XXY male in the mouse. Proc. Natl. Acad. Sci. U.S. *47*:571, 1961.

———: Chromosome aberrations in experimental mammals. Prog. Med. Genetics *2*:230, 1962.

——— and F. N. Woodiel: A spontaneous mouse chimera formed from separate fertilization of two meiotic products of oogenesis. Cytogenetics *5*:106, 1966.

Searle, A. G.: Comparative Genetics of Coat Colour in Mammals. New York: Academic Press, 1968.

Smith, H. A. and T. C. Jones: Veterinary Pathology. 3rd ed. Philadelphia: Lea and Febiger, 1966.

Steinberger, E., K. D. Smith and W. H. Perloff: Spermatogenesis in Klinefelters syndrome. J. Clin. Endocrinol. Metab. *25*:1325, 1965.

Tarkowski, A. K.: True hermaphroditism in chimaeric mice. J. Embryol. Exp. Morph. *12*:735, 1964.

Thuline, H. C. and D. E. Norby: Spontaneous occurrence of chromosome abnormality in cats. Science *134*:554, 1961.

Uzzell, T. M.: Natural triploidy in salamanders related to *Ambystoma jeffersonianum*. Science *139*:113, 1963.

Whiting, P. W.: The tortoiseshell cat. Am. Nat. *49*:518, 1915.

Witschi, E. and R. Laguens: Chromosomal aberrations in embryos from overripe eggs. Devel. Biol. *7*:605, 1963.

Zuelzer, W. W., K. M. Beattie and L. E. Reisman: Generalized unbalanced mosaicism attributable to dispermy and probable fertilization of a polar body. Am. J. Human Genet. *16*:38, 1964.

ANEUPLOIDY, POLYPLOIDY AND STRUCTURAL REARRANGEMENT OF CHROMOSOMES IN MAMMALS OTHER THAN MAN

RICHARD A. MCFEELY
School of Veterinary Medicine,
University of Pennsylvania, Philadelphia, Pennsylvania

The modern techniques of cytogenetics have provided scientists with a valuable tool for the investigation of many types of chromosome disorders. Although, to date, man has received the greatest degree of attention, there is an increasing number of reports of chromosome aberrations in subhuman species. Most of these studies have dealt principally with descriptions of the chromosomes and the associated clinical abnormalities. This paper will deal primarily with some of the mechanisms that might account for various types of chromosome aberrations. Although certain cases will be utilized to illustrate various points, it should be clear that this report does not encompass a large review of the literature or, in fact, that the example necessarily arose by the mechanism it serves to illustrate. It is obvious that certain aberrations may arise by more than one means and, as we view only the results of the change and not the actual occurrence, we may only speculate upon the actual series of events that have transpired in the production of a chromosome aberration. In the interest of simplicity only the more common or relatively uncomplicated mechanisms are discussed. The imaginative reader will undoubtedly be able to think of many other possibilities.

There is certain confusion in terminology among cytogeneticists. Terms used in this paper are basically those of Beatty (1957) who has provided us with an excellent monograph on chromosome aberrations in mammalian development. This report can only hope to update certain areas of Beatty's treatise. Much of the material presented here has already been reviewed comprehensively by Fechheimer (1968).

The basic number of chromosomes within a gamete is the haploid number. Fusion of the male and female gametes produces a diploid number of chromosomes in the somatic cells. Any deviation from these numbers is referred to as heteroploidy. When the chromosome number is three times the haploid number it is called triploidy; tetraploidy refers to four times the haploid set, etc. All of these are polyploid conditions. Euploidy refers to any exact multiple of the haploid set. Aneuploidy refers to any chromosome number that is not euploid. A somatic cell with three members of a particular chromosome is called trisomic. Monosomics have only one member of a particular chromosome in a somatic cell.

In addition to numerical aberrations, certain structural rearrangements can occur. Such structural changes are preceded by one or more breaks within the structure of the chromosome or chromosomes involved. A deletion involves an unhealed break with loss of the acentric fragment. A translocation occurs when there is an exchange of chromatin material between nonhomologous chromosomes. If the exchange is complete a reciprocal translocation has occurred. Two breaks within the same chromosome with the middle portion reversing itself, end for end, is called an inversion. If this involves a rotation around the centromere a special condition called a pericentric inversion results. An isochromosome is said to be the result of an abnormal splitting at the centromere with both arms of the isochromosome being equal in length and containing the same genetic material.

The effect that any of the previously described aberrations may produce depends upon the time at which the aberration takes place. They can arise at any time during the life cycle of an animal. If the cells carrying the aberration are not selected against, all of their progeny can be expected to show the aberration. However, it is quite possible that certain changes are deleterious to the affected cells and they will eventually die. It is apparent that the earlier an aberration occurs in the life cycle of an individual, the more widely the defect will be seen. The result of a meiotic error will be expected in all cells of the zygote. Likewise, errors at the time of fertilization will also be found universally in the conceptus. Chromosome abnormalities that occur early in the life cycle of an individual will generally have a greater effect upon the development of the individual because of the large number of cells that are involved. An aberration occurring late in development can be expected to have far less influence upon the individual because relatively few cells will be descended from the abnormal one.

The loss or addition of small amounts of chromatin material is generally deleterious to normal growth and development; and when larger amounts of genetic material are either lost or gained the resulting genome may be incompatible with life. Therefore, major abnormalities in the normal karyotype may exert their influence very early in the development of an individual and contribute to early death of an embryo. Although no cause and effect relationship has been demonstrated, a study of chromosomes obtained from 10-day old blastocysts from 7 litters of normal pigs revealed that about 10% of the blastocysts had grossly detectable chromosome aberrations (McFeely, 1967). Seven of the nine affected blastocysts were either triploid or tetraploid. In a more recent study reported, a blastocyst obtained on the 16th day of gestation in a cow contained a high proportion of tetraploid cells, although 11 other blastocysts from different animals were normal (McFeely and Rajakoski, 1968). Similar defects from living animals have never been reported and the assumption is that defects involving polyploidy are associated with death of the embryo. As these types of abnormalities generally arise as a result of a meiotic error or an anomaly at or about the time of fertilization, it is expected that they will be found most commonly in early pregnancy with the result that affected embryos will die early in gestation. In fact, pure polyploid mammalian adults have never been satisfactorily demonstrated.

One mechanism by which polyploid embryos can arise is through parthenogenesis. Beatty (1957) has discussed the various routes of parthenogenic development in great detail and there is very little new data to add to his review. It seems unlikely that this is an important mechanism in the production of heteroploid individuals.

The incidence of heteroploidy in mouse embryos is, in part, under the influence of the genetic composition of the parental strains (Fischberg and Beatty, 1950). Fechheimer (1961) demonstrated an increased incidence of polyploid spermatogonia in the affected strains.

Triploid embryos have been reported in several species as well as in man (McFeely, 1967; Shaver and Carr, 1967). Triploidy will probably be found to be one of the more important types of polyploid abnormalities.

Triploidy can arise by a number of mechanisms, several of which Austin (1960) has reviewed in some detail. If a normal haploid gamete is fertilized by a diploid gamete a triploid zygote arises. A diploid ovum will be produced if there is suppression of the second polar body formation. Heat treatment has experimentally produced this effect in the mouse (Fischberg and Beatty, 1952). Diploid spermatozoa have been reported (Salisbury and Baker, 1966) and presumably arise by means of some meiotic error. A second mechanism resulting in 3 sets of chromosomes comes into play if there is a fusion of the second polar body, or of a second ovum, with the fertilized egg. The first polar body contains a diploid number of chromosomes and, if fertilization of this structure should occur, a triploid zygote will result. A fourth and perhaps more common mechanism is triploidy resulting from polyspermy. If two sperms fertilize one ovum, an additional haploid set of chromosomes will be present in the embryo. There is now ample evidence to support the conclusion that delayed fertilization is associated with triploidy and other chromosome abnormalities in many species.

In pigs, Hancock (1959) found increasing numbers of trinuclear eggs the farther away from the onset of estrus the pigs were mated. He proposed that polyspermy was largely responsible for these abnormalities. Bomsel-Helmreich (1961) found 26% heteroploid embryos in pigs slaughtered on the 17th day of gestation when mating was delayed until 44 hours after the onset of estrus. In her series, the abnormalities seen were triploids and mosaics.

Hunter (1967) has demonstrated an increase of 14.1% on the mean percentage of polyspermic eggs in pigs as the ova increased in age from 4 to 16 hours at the time of sperm penetration.

More recently Shaver and Carr (1967) demonstrated a high incidence of triploidy (6 out of 43 examined) in 6-day rabbit blastocysts when fertilization was estimated to have occurred 4 to 5 hours after ovulation.

In rats, Butcher and Fugo (1967) caused a delay in ovulation for 24 to 48 hours with phenobarbital sodium and produced a variety of karyotype abnormalities including 2 triploids and 1 tetraploid. Although their control group also had 3 polyploid embryos the overall incidence of abnormal karyotypes was three times greater in the treated group. In addition, there was a marked increase in the number of degenerating embryos and implantation failures which could have been caused by chromosome aberrations, at least in part.

In domestic mammals, mating is normally restricted to the period of estrus or sexual receptivity of the female. In most species ovulation is most closely associated with the end of the estrus period. Therefore, nature seems to have designed a system whereby polyploidy resulting from delayed fertilization will be kept to a minimum. It is interesting, however, to speculate upon the adverse effect that slight endocrine imbalance, which results in delayed ovulation, might have upon the chromosome content of the resulting embryos and ultimately upon the reproductive efficiency of the animal. In species where artificial insemination is practiced, the timing of insemination takes on new importance for similar reasons.

A tetraploid embryo can also arise by several mechanisms. If there is suppression of the first cleavage division, twice the diploid chromosome number will be found in the cell. If this cell then undergoes a normal division a tetraploid embryo results.

Tetraploidy resulting from suppression of the first cleavage division has some experimental precedent. Heat treatment of fertilized ova has been shown to be effective in producing tetraploidy in mice (Fischberg and Beatty, 1950). Drugs have also been utilized to produce the same effect (Edwards, 1954).

Beatty (1957) found the incidence of spontaneous tetraploidy was 0.1% in a series of mouse embryos he analyzed. McFeely (1967) found 3 tetraploid pig blastocysts out of 88 examined.

Analysis of the sex chromosomes is useful in interpreting the origin of the tetraploid. With suppression of the first cleavage division it is apparent that there must be two sex chromosomes of maternal origin and two of paternal origin. An XXXY tetraploid must be explained in some other way. McFeely's (1967) tetraploids were either XXYY or XXXX and could be explained by failure of the first cleavage division. Butcher and Fugo's (1967) tetraploid rat embryo was an XXXX.

Polyspermy with three sperm fertilizing one ovum will also produce a tetraploid. Hunter (1967) has shown a picture of a trispermic pig egg. Shaver and Carr (1967) have reported a pentaploid blastocyst apparently occurring spontaneously in a rabbit mated at the time of an ovulation dose of gonadotropin. This study also demonstrated several mixoploids with some cells in the octaploid range.

Fusion of two fertilized meiotic products (i.e., ovum or polar body) prior to the first cleavage division could account for a tetraploid egg. If fusion occurs at a later stage of development a chimera results. Gartler *et al.* (1962) have concluded that an XX/XY human chimera resulted from the fertilization of 2 egg nuclei by 2 spermatozoa. Malouf *et al.* (1967) postulated that an XX/XY tricolored male cat arose from the fusion of 2 zygotes or through the participation of a polar body. Russell and Woodiel (1966) have described a chimeric mouse thought to have arisen by this manner. Padeh *et al.* (1965) concluded that two genetically different spermatozoa fertilized two egg nuclei to produce a horned hermaphrodite goat. And Dunn *et al.* (1968) have reported a true hermaphrodite cow likely the result of the dispermic fertilization of an egg and polar body.

A Holstein cow with two parasitic limbs attached at the level of the thoracic vertebrae and a characteristic freemartin conformation and genitalia was found to be an XX/XY chimera on leukocyte culture (McFeely, unpublished data). Available records indicate that this animal was the product of a single birth. One conclusion that can be drawn is that there was fusion of zygotes early in development with incomplete incorporation of one portion of one into the other. If this is so, additional support is given to Fechheimer's (1962) hypothesis that the sex chromosome chimerism is important in the etiology of this syndrome.

It is postulated that a diploid/tetraploid mosaic bovine blastocyst was the result of failure of an early cleavage division after chromosome replication (McFeely and Rajakoski, 1968).

Chimeras also result when there is a vascular anastomosis which develops between two separate conceptuses and an exchange of cells takes place. These may involve only circulating blood elements if the anastomoses take place very late in gestation or can involve many other types of cells, including presumed primordial germ cells (Ohno *et al.*, 1962) if the fusion occurs early in gestation. The freemartin syndrome is associated with sex chromosome chimerism which arises from fusion of the chorionic vasculature between two fetuses of different sex and an exchange of cells between them (Fechheimer *et al.*, 1962). Although this syndrome is most commonly seen in the bovine, it has been reported in sheep (Gerneke, 1965), goats (Ilbery and Williams, 1967) and swine (McFee *et al.*, 1966).

The complexity of some of these problems is illustrated by a triploid-diploid chimeric cat reported by Chu *et al.* (1964). The animal was a male tortoiseshell with a population of normal XX diploid cells and another population of XXY triploid cells. The authors postulated that this condition could have arisen by five different mechanisms as follows:

1. Fertilization of one meiotic product with one spermatozoon and fusion with another meiotic product fertilized by two spermatozoa.
2. Fusion of a fertilized ovum with a fertilized first polar body.
3. Immediate cleavage of the first meiotic division and fertilization of one primary oocyte which fused with the other which was unfertilized.
4. Triploidy arising as described previously with loss of one chromosome set during an early mitotic division.
5. Vascular anastomosis and cell exchange between diploid and triploid embryos.

A similar discussion was presented for a diploid triploid hermaphrodite mink (Nes, 1966).

With the knowledge that elevated temperature can cause heteroploidy in early embryos, it is tempting to postulate that a fever in the female at the time of ovulation and the early cleavage divisions might adversely affect development. It is known that sheep with elevated body temperatures due to high environmental temperature have an increased embryonic death rate (Ulberg and Burfening, 1967).

In addition to the polyploid conditions involving whole sets of chromosomes, anomalies involving only one or a few chromosomes can arise. Aneuploidy occurs when there is a nondisjunction during cell division. Nondisjunction is the term used to describe the phenomenon whereby a pair of homologous chromosomes do not separate into each daughter cell but rather stay together during cell division. The result is that one daughter cell has one extra chromosome and the other is lacking one. If this occurs during meiosis and the gamete later participates in fertilization the embryo will be aneuploid in all of the resulting cells. If, however, this occurs during some later mitotic division only a portion of the cells will be involved. In this case three separate cell lines would be expected, a normal diploid line, a line with one less chromosome and a line with one extra chromosome.

Aneuploidy involving the sex chromosomes has been demonstrated in several species. The tricolor male cat generally has an XXY aneuploid condition (Biggers and McFeely, 1966). Both X0 (Welshons and Russell, 1959) and XXY (Russell and Chu, 1961) mice have been reported with one less or one more chromosome than the diploid complement respectively. Recently, an XXY hyperdiploid dog has been studied (Clough, 1968). Butcher and Fugo (1967) found seven rat embryos in their delayed ovulation study that might have been either X0 or XXY. Four of these were mosaics, however. It is possible that some of them had autosomal involvement as they could not identify the sex chromosomes with certainty.

Aneuploidy of autosomes has also been reported in a number of species. Shive *et al.* (1965) reported autosomal aneuploidy in two dogs with congenital heart disease. Hare *et al.* (1967) reported aneuploidy in a dog with achondroplasia. This case was presumed to have arisen as a result of fusion of two chromosomes. Aneuploidy in a strain of European wild pigs has been reported by McFee *et al.* (1966). This polymorphism has apparently occurred spontaneously and without detriment to the affected animals. Gustavsson (1966) described aneuploidy in cattle in Sweden, and Herschler and Fechheimer (1966) have published an interesting case of aneuploidy resulting from centric fusion of chromosomes in a set of bovine triplets. Autosomal trisomy has also been reported in the mouse (Cattanach, 1964).

The phenomenon of nondisjunction in meiosis is often associated with increased maternal age in humans. In most mammals the ovum has already started meiosis before birth and remains in partial division until the time of ovulation. It has been suggested that the long delay in completing the meiotic divisions increases the chance for nondisjunction. In women, this period can be greater than 40 years. In subhuman mammals the time from birth to death would rarely involve a period even close to 40 years and thus, maternal age may be much less of a factor in subhuman mammals than in man. One study utilizing "old" mice failed to demonstrate an aging factor (Goodlin, 1965).

Another explanation, however, involves the hormonal imbalances that accompany the climacteric period. With altered hypophyseal secretions at the onset of the human menopause it is quite possible that delayed ovulation can commonly occur. Butcher and Fugo (1967) demonstrated an increase in aneuploidy with delayed ovulation in rats. This may be a better explanation for the increased incidence of nondisjunction in older women. Although there is very little evidence to support the supposition, it is thought that subhuman animals may have a lower incidence of nondisjunction based on the observation that nondisjunction has only occasionally been demonstrated. As animals generally do not go through a climacteric period, but are capable of reproduction throughout their lifetime, this mechanism may not function in subhuman species.

As mentioned earlier, structural rearrangement of chromosomal material usually involves at least one break in the integrity of the chromosome. Breaks have a tendency to repair by fusion of two broken ends. If the fusion takes place between the two segments that were originally together, little or no lasting damage is apparent. However, if the broken ends from two separate chromosomes fuse together new genetic combinations result. It is obvious that the more breaks that occur within a cell the greater will be the chance for fusion to occur between segments of nonhomologous chromosomes. Many kinds of aberrant chromosomes are formed, including inversions, translocations, dicentric and ring chromosomes and acentric fragments. Certain types of aberrations interfere with the process of cell division and are lethal, while others appear to survive without prejudice. It is possible that a great many factors have the ability to fragment chromosomes.

A case of reduced fertility in a boar associated with a translocation from one autosome to another has been reported from Sweden (Henricson and Backstrom, 1964). This defect was transmitted to a male progeny which also showed reduced fertility (Gledhill *et al.,* in press).

Spontaneously occurring pericentric inversions have been found in deer mice (Ohno *et al.,* 1966). The structural rearrangements of the chromosomes in this species apparently are not detrimental to the species and, in fact, may be beneficial. However, a presumed pericentric inversion of the X chromosome in an intersex cow apparently did interfere with the normal sexual development (McFeely *et al.,* 1967). A reciprocal translocation between an autosome and an X chromosome associated with sterility in the mouse has been described (Ford and Evans, 1964).

Ionizing radiation has been repeatedly shown to cause damage to chromosomes. Bender (1959) demonstrated a wide range of chromosome aberrations in bone marrow cells from X irradiated spider monkeys. Soukup *et al.* (1965) have seen similar structural alterations in chromosomes from rat embryos X-irradiated on the 13th day of gestation. Mice irradiated with neutrons or X-rays showed the same types of aberrations, in liver and bone marrow cells, and X-irradiation induces major aberrations in Type A mouse spermatogonia (Griffen and Bunker, 1967).

An X-autosome translocation in descendents of an irradiated mouse was identified due to suppression of sex-linked variegation (Lyon *et al.,* 1964). Many other studies, too numerous to document, support the contention that ionizing irradiation causes structural rearrangement of chromosomes.

There is an increasing list of drugs that produce chromosome rearrangements. For example, 6-amino nicotinamide has been demonstrated to cause chromosome fragmentation in mouse embryos when receiving the drug on the 11th day of gestation (Ingalls *et al.,* 1964); and the radiomimetic drugs exhibit effects similar to ionizing irradiation. An X-autosome translocation has been produced by treating male mice with triethylenemelamine and mating them to normal females (Ohno and Cattanach, 1962). Once again, there are too many reports in the literature to make a comprehensive review.

Virus-induced chromosome damage has been reviewed by Nichols (1966). It is apparent that virus infection can damage chromosome integrity and cause structural rearrangements.

An attempt has been made in this paper to review some of the major causes of chromosome abnormalities in animals. Much of the data on humans has purposely been omitted. It is obvious that all of the possible mechanisms to explain certain aberrations have not been mentioned. In fact, it would be very surprising if all of the mechanisms are even suspected at the present time. There are large gaps in our knowledge about how various aberrations occur. This is partially predicated upon the fact that we seldom have the opportunity to see the mechanism in action but primarily have retrospective studies to depend upon. However, it is interesting to speculate upon these mechanisms in the hope that a means can be found to reduce the incidence of aberrations especially where the aberrations adversely affect the development of an individual.

References

Austin, C. R.: Anomalies of fertilization leading to triploidy. J. Cell and Comp. Physiol. *56:*1, 1960.

Beatty, R. A.: Parthenogenesis and Polyploidy in Mammalian Development. Cambridge, England: Cambridge University Press, 1957.

Bender, M. A.: X-ray induced chromosome aberrations in mammalian cells *in vivo* and *in vitro*. Int. J. Rad. Biol. suppl., 103–113, 1959.

Biggers, J. D. and R. A. McFeely: Intersexuality in Domestic Mammals. In: Advances in Reproductive Physiology. A. McLaren, ed. London: Logos Press, 1966.

Bomsel-Helmreich, O.: Experimental heteroploidy in the sow. Proc. IV Int. Cong. Anim. Reprod. Hague *III:*578, 1961.

Butcher, R. L. and N. W. Fugo: Overripeness and the mammalian ova. II. Delayed ovulation and chromosome anomalies. Fertil. Steril. *18:*297, 1967.

Cattanach, B. M.: Autosomal trisomy in the mouse. Cytogenetics *4:*159, 1964.

Chu, E. H. Y., H. C. Thuline and D. E. Norby: Triploid-diploid chimerism in a male tortoiseshell cat. Cytogenetics *3:*1, 1964.

Clough, E.: Cytogenetic studies in dogs with congenital heart disease. Morris Animal Foundation: Denver, June, 1968.

Dunn, H. O., R. M. Kenney and D. H. Lein: XX/XY chimerism in a bovine true hermaphrodite: An insight into the understanding of freemartinism. Cytogenetics *7*:390, 1968.

Edwards, R. G.: Colchicine-induced heteroploidy in early mouse embryos. Nature. London. *174*:276, 1954.

Fechheimer, N. S.: Poikiloploidy among spermatogenic cells of *Mus musculus*. J. Reprod. Fert. *2*:68, 1961.

———: Consequences of chromosomal aberrations in mammals. J. Anim. Sci., *27* suppl. *1*:27, 1968.

———, M. S. Herschler and L. O. Gilmore: Sexchromosome mosaicism in unlike cattle twins. In: Genetics Today. S. J. Geerts, ed. New York: Macmillan and Co., 1963.

Fischberg, M. and R. A. Beatty: Experimentelle Herstellung von polyploiden Mausblastulae. Arch. Klaus-Stift. Vererb. Forsch. *25*:54, 1950.

——— and R. A. Beatty: Heteroploidy in mammals. II. Induction of triploidy in preimplantation mouse eggs. J. Genet. *50*:455, 1952.

Ford, C. E. and E. P. Evans: A reciprocal translocation in the mouse between the X chromosome and a short autosome. Cytogenetics *3*:295, 1964.

Gartler, S. M., S. H. Waxman and E. R. Giblett: An XX/XY human hermaphrodite resulting from double fertilization. Proc. Nat. Acad. Sci. *48*:332, 1962.

Gerneke, W. H.: Chromosome evidence of the freemartin condition in sheep—*Ovis aries*. J. S. Afr. Vet. Med. Assoc. *36*:99, 1965.

Gledhill, B. L., M. P. Gledhill and B. Henricson: An attempt to measure a chromatin distributive imbalance in spermatozoa. Hereditas (in press).

Goodlin, R. C.: Non-disjunction and maternal age in the mouse. J. Reprod. Fertil. *9*:355, 1965.

Griffen, A. B. and M. C. Bunker: The occurrence of chromosomal aberrations in pre-spermatocytic cells of irradiated male mice. III. Sterility and semisterility in the offspring of male mice irradiated in the pre-meiotic and postmeiotic stages of spermatogenesis. Canad. J. Genet. Cytol. *9*:163, 1967.

Gustavsson, I.: Chromosome abnormality in cattle. Nature *211*:865, 1966.

Hancock, J. L.: Polyspermy of pig ova. Ani. Production *1*:103, 1959.

Hare, W. C. D., J. S. Wilkinson, R. A. McFeely and W. H. Riser: Bone chondroplasia and a chromosome abnormality in the same dog. Am. J. Vet. Res. *28*:583, 1967.

Henricson, B. and L. Backstrom: Translocation heterozygosity in a boar. Hereditas *52*:166, 1964.

Herschler, M. S. and N. S. Fechheimer: Centric fusion of chromosomes in a set of bovine triplets. Cytogenetics *5*:307, 1966.

Hunter, R. H. F.: The effects of delayed insemination on fertilization and early cleavage in the pig. J. Reprod. Fert. *13*:133, 1967.

Ilbery, P. L. T. and D. Williams: Evidence of the freemartin condition in the goat. Cytogenetics *6*:276, 1967.

Ingalls, T. H., E. F. Ingenito and F. J. Curley: Acquired chromosomal anomalies induced in mice by a known teratogen. J.A.M.A. *187*:836, 1964.

Lyon, M. F., A. G. Searle, C. E. Ford and S. Ohno: A mouse translocation suppressing sex-linked variegation. Cytogenetics *3*:306, 1964.

Malouf, N., K. Benirschke and D. Hoefnagel: XX/XY chimerism in a tricolored male cat. Cytogenetics *6*:228, 1967.

McFee, A. F., M. W. Bonner and J. M. Rory: Variation in chromosome number among European wild pigs. Cytogenetics *5*:75, 1966.

———, M. Knight and M. W. Banner: An intersex pig with XX/XY leucocyte mosaicism. Canad. J. Genet. Cytol. *8*:502, 1966.

McFeely, R. A.: Chromosome abnormalities in early embryos of the pig. J. Reprod. Fert. *13*:579, 1967.

———, W. C. D. Hare and J. D. Biggers: Chromosome studies in 14 cases of intersex in domestic mammals. Cytogenetics *6*:242, 1967.

——— and E. J. Rajakoski: Chromosome studies on early embryos of the cow. Proc. VI Cong. Anim. Reprod. Paris. 1968, in press.

Nes, N.: Diploid-triploid chimerism in a true hermaphrodite mink (*Mustela vison*). Hereditas *56*:159, 1966.

Nichols, W. W.: The role of viruses in the etiology of chromosomal abnormalities. Amer. J. Human Genet. *18*:1, 1966.

Ohno, S. and B. M. Cattanach: Cytological study of an X-autosome translocation in *Mus musculus*. Cytogenetics *1*:129, 1962.

———, J. M. Trujillo, S. Stenius, L. C. Christian and R. L. Teplitz: Possible germ cell chimeras among newborn dizygotic twin calves (*Bos taurus*). Cytogenetics *1*:258, 1962.

———, C. Weiler, J. Poole, L. Christian and C. Stenius: Autosomal polymorphism due to pericentric inversions in the deer mouse (*Peromyscus maniculatus*) and some evidence of somatic segregation. Chromosoma (Berl.) *18*:177, 1966.

Padeh, B., M. Wysoki, N. Ayalon and M. Soller: An XX/XY hermaphrodite in the goat. Israel J. Med. Sci. *1*:1008, 1965.

Russell, L. B. and E. H. Y. Chu: An XXY in the mouse. Proc. Nat. Acad. Sci. *47*:571, 1961.

——— and F. N. Woodiel: A spontaneous mouse chimera formed from separate fertilization of two meiotic products of oogenesis. Cytogenetics *5*:106, 1966.

Salisbury, G. W. and F. N. Baker: Frequency of occurrence of diploid bovine spermatozoa. J. Reprod. Fert. *11*:477, 1966.

Shaver, E. L. and D. H. Carr: Chromosome abnormalities in rabbit blastocysts following delayed fertilization. J. Reprod. Fert. *14*:415, 1967.

Shive, R. J., W. C. D. Hare and D. F. Patterson: Chromosome studies in dogs with congenital cardiac defects. Cytogenetics *4*:340, 1965.

Soukup, S. W., E. Takacs and J. Warkany: Chromosome changes in rat embryos following X-irradiation. Cytogenetics *4*:130, 1965.

Ulberg, L. C. and P. J. Burfening: Embryo death resulting from adverse environment on spermatozoa or ova. J. Anim. Sci. *26*:571, 1967.

Welshons, W. J. and L. B. Russell: The Y chromosome as the bearer of male determining factors in the mouse. Proc. Nat. Acad. Sci. U.S. *45*:560, 1959.

VOUCHER SPECIMENS IN COMPARATIVE CYTOGENETIC STUDIES

RICHARD G. VAN GELDER
Department of Mammalogy, The American Museum of Natural History, New York, New York

Comparative cytogenetics is a type of taxonomic study, and it may provide the systematist with additional information concerning relationships of organisms. While many of the persons engaged in chromosome studies on a comparative basis are specifically attempting to gather cytogenetic evidence to resolve systematic problems, others are accumulating data in peripheral studies which, nonetheless, are of great interest and utility to taxonomists. The prime purpose of this paper is to present some methods by which the common material of use both to cytologists and to formal taxonomists may be made available to each group. This should help not only to improve the validity of the cytogenetic data, but also will aid the taxonomist by providing reference material for future workers in the resolution of problems that will surely arise concerning the identity of animals utilized in cytological studies.

Classical mammalian taxonomy is based largely on skins and skeletal parts, mainly skulls, that are preserved in museums. With the rise of the "new systematics" (Huxley, 1940) the breadth of characters used in systematics increased, and the new breed of taxonomists, or as they prefer to call themselves, systematists, utilize all the data that they can, at least in theory, in formulating the relationships of groups and the taxonomic identification of individuals. However, with the great reference collections of skins and skulls that exist in museums, much of mammalian taxonomy ultimately resolves itself into comparison of the characters to be seen in the skins and skulls of mammals.

The current status of the taxonomy of mammals is varied. For some groups, or for some parts of the world, identification may be quite simple and may be done by relatively untrained people with a high degree of accuracy. For other geographic areas and for other groups, the identification of individuals may be extremely difficult and beyond the ability of taxonomists who are unfamiliar with the particular group. There are some groups of mammals for which there is insufficient material for resolving the taxonomic problems at the present time, and it may be many years before a specialist becomes involved with the group or before the critical material is acquired.

However, the current taxonomic status or the ability of taxonomists to provide identifications should have no bearing on the choice of a cytogenetic problem or on the research itself provided that the cytogeneticist makes a distinct effort to provide reference material that will be available in the future when the problem is restudied or the taxa revised. To this end, it is most important that cytogeneticists provide, in their publications, detailed information about the animals that they have used, how the taxonomic determination was made, as well as the location and identifying number of voucher specimens. To this end, it would be most helpful if the editors of journals publishing data of this sort would insist on the data indicated above being presented in papers that they accept for publication.

I fully understand that the source of the material used by cytogeneticists is varied, and that often material for culture is obtained from animals that are not the property of the investigator, and that the obligations and requirements of the owner of the animal must be acceded to. However, in the majority of cases, I believe, a high degree of cooperation between the cytogeneticist and the owner of the animals may be expected.

To take the easier situation first, I would like to suggest a system of deposition of voucher specimens when the cytogeneticist is also the owner of the animal (and for ease, the animal is sacrificed at the time the cytological study is made). In suggesting this method, it must be realized, I am speaking only for the American Museum of Natural History's Department of Mammalogy, and I already know that there are some museums that will not accept such specimens. There are others that will only accept certain specimens and may make some specific requirements on even these. I also realize that there may well be specific requirements imposed on the cytogeneticist in regard to the disposition of his animals, and I will discuss some alternatives to the "ideal" situation that exists at the American Museum later.

If the animal utilized in a cytogenetic study is dead or is sacrified for the research, and if no further utilization of the specimen is planned by the investigator, it should be preserved immediately for deposition in a museum. There are numerous methods of preservation of mammal specimens, but the easiest and most adaptable from the museum curator's point of view is for the cytogeneticist to freeze the animal. For most small mammals, the freezing compartment of a refrigerator will suffice.

For various reasons, including aesthetic, it is best to wrap the animal in tinfoil or paper. Enclosed *within* the wrapper should be data concerning the specimen itself.

The more data about the animal that can be provided by the cytogeneticist, the better. Minimally, the locality where the animal was collected, the name and address of the collector, the date of collection, and the reference number specific to this individual used by the cytogeneticist should be included. Because of the effects of moisture, these data should be written with a soft pencil on good stock paper and attached to the specimen itself. Experience indicates that a label on the outside of a wrapper may readily be lost or obscured.

Frozen specimens can easily be shipped by air. We have been using plastic foam picnic containers that are cheap, lightweight, and sturdy. If the specimen is well frozen and is packed with additional ice-cubes (enclosed in plastic bags to contain melted ice), the specimen should arrive in good condition even after as much as four days in transit. For even greater security, dry ice may be packed with the specimen, and we have received specimens shipped from tropical countries still frozen after more than a week in transit.

Long before specimens are shipped, of course, the cytogeneticist should have been in contact with the curator at the museum. First he must ascertain that the museum will accept the material, and he should inform the curator in detail about the specimens and proposed manner of shipping. He should wait for a reply before sending the material. Museum people often spend several months in the field, technicians may be in the field or on vacation, or facilities for receiving may be temporarily unserviceable. Therefore, confirmation of readiness to receive the material should be awaited by the cytogeneticist. When the acceptance is received, the shipper should try to send the specimens early in the week so that they will not arrive on a weekend when they might not receive prompt attention, and should send a confirmatory telegram at the time of shipment, giving the name of the carrier, the flight number, the weight of the package, and the estimated time and place of arrival.

Arrangements for the responsibility of shipping costs should also have been made beforehand. Hopefully, the cytogeneticists will bear the costs of shipping, for scientific departments in museums have unbelievably low budgets, and if a great number of specimens are being sent in by numerous investigators, an inordinate burden will be placed on the museum. On the other hand, it is unlikely that any one cytogeneticist will be sending so much material in the course of a year that he could not afford the few dollars shipping charges that this important facet of his work will incur.

One of the items that the investigator may find that he cannot provide is locality of capture. Because much cytogenetic work is done on animals that were obtained from professional animal dealers who often do not know the actual source of their animals, care should be taken to indicate doubtful information. This should be clearly indicated on the label—that the specimen was obtained from an animal dealer who said that it came from, say, "Colombia," or even from "Leticia, Colombia." The cytogeneticist, in other words, should only present information that he has obtained firsthand, or else indicate the source of his information. Generally, a mammalian taxonomist wishes to have his data as precise as possible, and for most parts of the world, the distance and direction from the nearest town with a post office will suffice. Elevation above sea level may also be important in increasing the accuracy of an identification.

For animals that are not the property of the investigator, an attempt should be made to get the owner to save the specimen for the cytogeneticist when it dies. A high degree of cooperation in this regard is common. However, it is the responsibility of the investigator to keep tabs on the animal and to remind the owner occasionally of his agreement.

In cases at zoos where there may be a number of animals in a single enclosure, some method of identifying the individual from which the karyotype specimen was taken is needed. At the present time many zoos are interested in identifying individuals, and at the same time that the animal is restrained for collection of the cytological specimen, it may be marked by any of the standard techniques—ear tags, toe clipping, ear clipping, or tatooing. This will serve to assure that the animal that ultimately ends up in the museum is the same one from which the tissue for study was obtained.

In those cases where the owner of the animal cannot assure the investigator that the animal can be made available to him or to a museum when it dies, the cytogeneticist should obtain all of the same data that he would if he owned the animal, including sex and length of time in captivity. At the same time the investigator should take a series of photographs of the animal, especially to show the entire animal, details of the head, ears, tail, and undersides of feet. Ideally, these photographs should be taken in color. If a photograph of the mouth and teeth can be obtained, these may also be of great assistance. These photographs can eventually be helpful in obtaining an accurate identification of the animal.

To sum up the obligations of the cytogeneticist concerning the animals he is working with, he should make every attempt to obtain all of the data that a taxonomist may need, including the specimen itself.

To which museum should the material be sent? What will the museum curator do with the material and what will he provide? Specimens should be sent to a museum with a long history of care for collections and where there is a good indication that this condition will continue. Smaller collections have a way of disappearing after the death of the owner or curator, and from 1923 to 1963, 16 public and 25 private collections with a total of about 26,000 specimens "disappeared," or at least could not be located (Anderson *et al.*, 1963).

College museum collections are not immune from disposal or deterioration. Therefore, I would recommend that specimens be sent to the larger, quasi-public museums, such as the Field Museum of Natural History in Chicago, the U.S. National Museum in Washington, or the American Museum of Natural History in New York. At this writing, I can only speak for the American Museum in indicating that we will accept such material.

What will the curator do with the specimens? First of all, the cytogeneticist cannot necessarily expect an identification for the material he sends. This should not only be made clear at this time, but also should be explained by the curator during the correspondence prior to shipment of the animals. There are few mammalian taxonomists in the world, and there are even fewer institutions with reference collections sufficiently large and diverse to permit detailed studies of a majority of species of the mammals of the world. It is the prime purpose of this paper to impress upon the cytogeneticist that mammalian taxonomy is difficult and time-consuming, and that an identification is only as good as the state of classification at that time. It is to insure the opportunity to correct mistakes in the future that the deposition of voucher specimens in a museum is encouraged.

What the curator will do for the cytogeneticist is to provide him with a catalogue number of the specimen. This is a permanent number that will be attached to the specimen, hopefully in perpetuity. At any time in the future, this individual can be referred to with precision through its catalogue number. In publications about the karyotypes of this individual, the cytogeneticist should refer to the museum and catalogue number.

In many cases, of course, the curator may be able to confirm an idenification at the time of receipt of the specimens, or even more important, may be able to correct a taxonomic error on the part of the cytogeneticist. The curator also, during the correspondence that precedes shipment, may suggest a specialist to whom the specimens may be sent, and in such cases when an identification is provided by a specialist, his name and institution should be stated in publication. However, confirmed identification by a mammal taxonomist does not preclude the deposition of voucher specimens and publication of the catalogue number and institution of deposition. Taxonomists, too, are fallible.

It might be worthwhile for the cytogeneticist to deposit, at the same time as he deposits his specimens, a duplicate karyotype slide. This will be numbered by the museum with the same catalogue number as the specimen and preserved with it. For a future investigator, it may prove to be quite useful to have both the specimen and the slide at the same time. Also, as most larger museums are quite reluctant to dispose of anything, the slide may be available for study long after the cytogeneticist's personal slide collection has been disposed of by his successors.

To summarize then, all effort should be made to deposit voucher specimens of the animals from which tissues for cytogenetic studies have been taken at a museum where a permanent reference collection is maintained. The cytogeneticist must assume the responsibility for seeing to it that the animal is deposited. The curator may not assume the responsibility for identifying or confirming the identification of the specimen, but he will provide a reference number so that the identification can be made by the appropriate specialist at any time in the future. Editors of publications are urged to insist upon some means of assuring accuracy in the identification of the animals about which an author writes.

References

Anderson, S., J. K. Doutt and J. S. Findley: Collections of mammals in North America. J. Mammal. *44*:471, 1963.

Huxley, J., ed.: The New Systematics. London, England: Oxford University Press, 1940.

CONCLUDING REMARKS: WHERE DO WE GO FROM HERE?

S. OHNO
City of Hope Medical Center
Duarte, California

Kurt Benirschke has kindly assigned me the task of summarizing this five day meeting. Rising to the occasion, one is perhaps expected to praise the impressive accomplishments of the past ten years reported on at this meeting. It so happens that I believe self-criticism to be more beneficial than mutual admiration. My brief talk is entitled "Where Do We Go from Here?"

Despite the enormous amount of data presented here, it appears that we have yet to understand the role chromosomal changes played in diversification of mammals. Would karyotyping of some more rare mammalian species suddenly enlighten us with a needed clue? I think not. We have to learn little more than chromosome morphology to understand the role.

The studies on aborted fetuses and abnormally constructed individuals reported here established one important fact; that natural selection does not tolerate changes in the genome size of mammals. Triploids, tetraploids, monosomics and trisomics can not perpetuate themselves, for most of them are eliminated prenatally and those which survive are sterile. Is it a wonder that all the mammals have the same genome size (DNA content) despite immense diversity of their karyotypes? One might say that mammals are doomed because of this conservativeness. The accumulation of redundant genetic material is a prerequisite for big evolutional leaps; for new gene loci can emerge only from redundant gene loci which are not subjected to the relentless pressure of natural selection. By strictly adhering to the fixed genome size, mammals have forfeited the future.

As to chromosomal differences between related species, I have noted with acute interest that, regardless of whether he is an advocate of fusion or a supporter of fission, every speaker firmly believes in one god whose dictum is "chromosomal changes are invariably unidirectional." In the case of a mutation which is a base substitution in a DNA cistron, we take a view that mutation is a random event, and for this reason, we expect each species to accumulate about the same number of random mutations in a given number of generations. Thus, when the polypeptide X of species A differs from that of its sibling species B by a single amino acid substitution at the position Y, methionine (codon AUG) for glutamic acid (codon GAG), we deduce that the position Y of a common ancestor to both species was occupied by neither methionine nor glutamic acid but by lysine (codon AAG). In sharp contrast, when confronted with two sibling species, one with 44 chromosomes and the other with 48 chromosomes, cytogeneticists do not like to consider the possibility that a common ancestor to both may have had the diploid chromosome number of 46. Instead, a fusion advocate insists that 48 was the ancestral chromosome number; while a fission advocate is equally firm in believing the ancestral chromosome number to be 44. Both arguments are absolutely illogical unless the dictum which states that all the chromosomal changes are unidirectional is true.

The consequence of unidirectional change has to be the conservation of many original linkage relationships by diverse mammals despite apparent chromosomal rearrangements.

The cytological and genetic evidences which indicate the conservation of the original X-linkage *in toto* are now quite impressive. To the list of homologous X-linked genes listed in my paper, I might add the X-linked coat color gene *Orange* of the cat mentioned by Dr. Jones. The same X-linked gene has been found in the Syrian hamster which is a possessor of the duplicate-type X. The Syrian hamster X shares another gene *Mottled* with the mouse X.

There is an example of the conservation of a part of the autosomal linkage. The two autosomal coat color gene loci, C-locus (so-called albino locus is a structural gene locus for tyrosinase) and P-locus (so-called pink-eye locus is a structural gene locus for subunits of melanosomes), are about 16 crossing-over units apart in the mouse, and the rat as well as the deer mouse (*Peromyscus*).

Comparison of linkage relationships between diverse species of mammals is a promising avenue awaiting exploration by cytogeneticists of tomorrow. One can pursue the classical approach utilizing marker genes. Conversely, one might apply the modern technique of DNA/DNA hybridization to isolated individual metaphase chromosomes of different species.

Another promising area of research should be on the study of fossil genes the genome carries. In connection with heterochromatin, what seems to be a reasonable argument was often expressed during this meeting. The argument states that natural selection must relentlessly eliminate unnecessary genetic material from the genome. Thus, whatever is contained in the genome is needed today. I submit that this is nonsense. The winged forelimb of the Jurassic bird, *Archaeopteryx lithographia,* ended in the grasping hand which was composed of the clawed first three digits. You would think that modern birds had eliminated such an archaic trait long ago. The fact is that it takes a mere allelic mutation at a single autosomal gene locus to put back these claws on the wing of a domestic chicken.

During early development, we ourselves fashioned several pairs of gill slits to reflect our humble fish origin. At that time we were also equipped with what promised to be a respectable tail. Ontogeny recapitulates phylogeny because we have maintained most of the archaic genes. Personally, I would not be surprised if future study shows that these archaic genes, which should be suppressed at some point in the developmental sequence, reside on the so-called heterochromatic regions of chromosomes. This suppression must be a very important part of ontogeny for gills are changed to something more respectable such as thymus and the embarrassing growth of our tail is stopped in time.

It would appear that in order to understand the role chromosomal changes played in evolution, we mammalian cytogeneticists should become less of a *cyto*logist and more of a *geneticist.*

BONE MARROW PREPARATIONS FOR CHROMOSOME STUDIES

T. C. Hsu and James L. Patton

Section of Cell Biology, The University of Texas M. D. Anderson Hospital and Tumor Institute at Houston, Houston, Texas, and Department of Zoology, The University of California at Berkeley, Berkeley, California

Metaphase chromosome preparations from primary cultures are usually of better quality than those from bone marrow. However, in a number of cases bone marrow is the only practical material for cytological preparations; therefore, mastering this technique is essential in studies of mammalian karyotypes. These reasons include:

1. Not every investigator has a tissue culture laboratory at his disposal.
2. Direct bone marrow preparations reduce the cost.
3. In field trips, bone marrows can be processed with relative ease without large pieces of equipment and supply items, such as fresh growth media.
4. Bone marrow preparations should be made as an insurance in case primary cultures fail or become contaminated.
5. In case of mosaicism, bone marrow offers an excellent second material.

Direct Bone Marrow Preparations

For small animals, the specimens must be sacrificed for bone marrow preparations because of the scanty material available from each animal. Agents to arrest metaphase should be injected into the specimens prior to sacrifice. Several agents, colchicine, Colcemid and Velban (Vinblastine Sulfate), have been popular, but there is no reason why others cannot be used. Most investigators, especially beginners, fearing the lack of the drug effect, inject an overdose of the agent and prolong the treatment, thus causing an overcontraction of the chromosomes. This combination also produces a large number of C-anaphases and C-telophases with ill-defined chromosome morphology and drastically changed arm-ratios.

Probably each group of animals has its own requirement for the optimal dose and the duration of treatment. It also appears that more mitoses can be obtained in the mornings. A few tests should yield the right combination. It is also advisable to try a few animals without any agent. As a start, we suggest the use of Velban (Eli Lilly Co., Indianapolis, Indiana). Dissolve the contents of each ampule (10 mg) with 40 ml of water. Use 0.01 ml per 2 gm of body weight. Inject the solution intraperitoneally into the animal and incubate for a maximum of 2 hours.

The animal can be killed by cervical dislocation, chest pressure, or anesthesia. We find that halogenated ether (commercial brand Metofane, Pitman-Moore, Division of the Dom Chemical Co.) is very useful, especially for field trips because it is nonexplosive. Dissection of the animal should be done by making a long, straight incision in the middle of the abdomen without further damaging the specimen. Corn meal is excellent for absorbing excess fluid and keeping the animal free of blood, which aids in preparing museum specimens. The skin and skull of each specimen should be numbered, catalogued, processed, and deposited in a natural history museum. Inexperienced workers are advised to freeze the carcasses and send the frozen specimens to a museum with prior arrangement.

For animals as large as laboratory rats, one femur should provide ample marrow material. For smaller animals, it is probably wise to add the tibiae as well. Remove muscle, cut off the tip of the epiphysis of each long bone, and flush out the "marrow plug" with a small syringe (5 ml capacity) equipped with a small needle and loaded with a hypotonic solution. The size of the needle varies according to the size of the bone (#27 for very small mammals, #25 for mice, #22 for rats).

Two types of hypotonic solution have been commonly used: sodium citrate solution (1.0%) or growth medium diluted with distilled water (1:4). The hypotonicity suggested here should not be adhered to rigidly, because in some cases the cells may require a higher hypotonicity. Use 7 to 10 ml for each bone marrow sample and treat for approximately 5 minutes.

Short-Term Bone Marrow Cultures

For large and expensive specimens, it is naturally impossible to kill the animals just to get some bone marrow. Besides, it would be ridiculous to waste large amounts of colchicine or Velban. Tap small amounts of bone marrow from the sternum or iliac crest using a syringe wetted with heparin.

Place the bone marrow in approximately 10 ml of growth medium (any kind) containing 0.02 μg/ml Colcemid. Suspend the cells and leave the cell suspension in room temperature or in an incubator for one hour. At the end of the incubation period, centrifuge the cells, decant the growth medium, and resuspend the cells in a hypotonic solution for 5 minutes. A number of mitoses should be arrested by this method. This method also can be applied to animals that have died a few hours before the aspiration of the marrow.

Fixation

The investigator should decide, before fixation, the type of preparations he desires to make, because the two common procedures, the air-dried method and the squash method, require different fixatives. The former employs acetic alcohol (3 parts of ethyl or methyl alcohol and 1 part of glacial acetic acid), and the latter, 50% acetic acid (or propionic acid). Acetic alcohol should be freshly prepared each day, but 50% acetic acid can be stored.

Centrifuge the cells in conical centrifuge tubes for 3 to 5 minutes. We find the single-speed, 6-hole clinical centrifuge ideal for this purpose. The size of the cell pellet should not exceed 0.05 ml or fixation will be poor. Actually a visible speck is sufficient to yield several slides.

Decant the hypotonic solution and gently add the fixative to the tube. *Do not* suspend the cells at this time. Many cells will be broken if they are disturbed. Leave the pellet in the fixative for at least 20 minutes.

Air-Dried Preparations

The procedure for air-dried preparations is the same as that for human lymphocyte cultures (cf., Human Chromosomes Methodology, J. Yunis, ed., New York: Academic Press, 1965). Either air drying or flame drying gives satisfactory results. The advantages are:

1. It does not require much skill.
2. Bright field microscopy is sufficient since the slides must be stained.
3. It is convenient for field trips.
4. It is permanent.

Squash Preparations

Many contemporary cytologists resent making squash preparations because a large number of cells are broken. This can be avoided if the preparation is done properly. We feel that it is a useful technique, and when correctly applied, it yields superb results. The advantages of the squash procedure:

1. It is quick. One can immediately inspect a squash slide to decide whether the sample is of any use, how many slides are necessary, etc.

2. When the quality of the preparation is poor, some manipulations can be made to improve the quality to some extent.

3. When lightly stained and examined with a phase-contrast micro-

scope, a faint cellular outline gives indication as to whether the cell is broken or not. This criterion of determining intactness of a cell saves time, because the microscopist may scan the preparations with a low magnification objective to select only complete cells for analysis.

4. For autoradiography, perfect figures can be taken before the application of an emulsion.

5. Squash preparations do not have minute air bubbles over the chromosomes as seen in some air dried slides.

Prepare 1% acetic orcein by dissolving orcein powder in 50% acetic acid and heat to boil. Filter after cooling. The stain will precipitate to some extent, but no further filtration is necessary unless the stain is applied to air-dried slides. For phase contrast microscopy, overly stained chromosomes will give a golden glow and lose the resolution. For bright field microscopy, on the other hand, the stain should be as strong as possible, so that a 2% acetic orcein is recommended. When unsatisfactory staining quality occurs, one may either add stronger staining solution to intensify the stain or add 50% acetic acid to dilute the stain.

After the cell pellet is fixed for at least 20 minutes, decant and drain the fixative. If the pellet is relatively large, thorough draining of the fixative is not necessary. The amount of stain to be added varies according to the size of the cell pellet. Without experience, it is suggested that for a tiny pellet, use one drop of acetic orcein; for a larger pellet, use two or three drops. If the cell suspension is too thick, one can always add moie stain later. It is a good practice to make a tentative slide to view it microscopically to determine whether the suspension and the stain are just right. Each cell should have enough room to expand fully, but the cell density should not be too sparse so that the cytologist would have to scan large areas in order to find a mitotic figure. Nevertheless, it is a good practice to make the suspension somewhat thinner, because the quality of the slides is usually better when the cell density is low.

Use a Pasteur pipette to suspend the cells. Place a small drop on a clean slide, and place a coverslip gently over the cell suspension. We find 22 × 30 mm coverslips most ideal, but other sizes should serve equally well. If air pockets are seen under the coverslip, do not worry. Most air pockets will be driven out in the process of squashing. Even if they remain, there are plenty of areas to find good cells.

The next step is the key to squash preparations. Some persons think that squashing is an art and is not for everyone. This is wrong. Anyone can make a good squash preparation if he corrects certain mistakes which cause the cells to break. The common mistakes are:

1. Squashing too hard at the beginning.
2. Tapping.
3. Pressing only with an up-and-down motion.

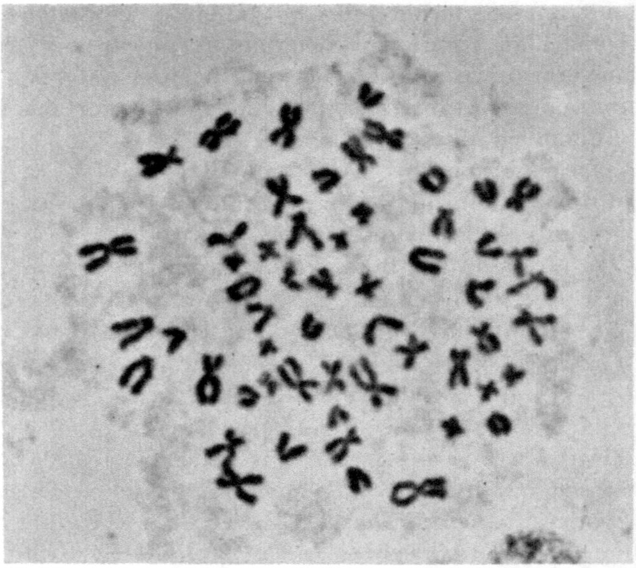

Fig. 1. A metaphase from bone marrow of a male *Microtus longicaudus,* acetic orcein squash.

The proper procedure is to place a piece of absorbent paper (e.g., Whatman #1 filter paper) over the slide. Hold the thumb at the middle of the coverslip and roll the thumb very lightly and gently along the long axis of the thumb toward one side of the coverslip, squeezing out a little stain each time. Remember when pressure is applied to the coverslip, the stain underneath the coverslip will be forced to the sides. If it is not absorbed, it will rush back to the center as the pressure is released, thereby creating a violent current which breaks the cells. If the pressure is gentle, but the liquid squeezed to the edge of the coverslip is absorbed, there is little chance for any runback. The cells will become flatter, but not broken. Putting a minute drop of cell suspension on each slide is also a good practice.

Repeat the process (from the center of the coverslip toward the sides), and each time squeeze out a little excess stain. Increase the pressure until no stain can be pressed out under very strong pressure. This is the time one inspects the slide to determine (1) whether the sample contains any mitotic figure, (2) whether the suspension is too crowded, and (3) whether the quality of the preparation is good. Use a phase-contrast microscope with an objective between 16× and 20× for scanning. If no mitosis is present, discard the entire sample. If the cells are too crowded, add one or two more drops of stain to dilute the suspension. If the quality of the metaphase spreads is good, proceed to make a few more slides.

Unlike the cells in culture, the hemopoietic cells are small. Therefore, the chromosomes have little room in which to expand. To achieve a better spread, one may borrow the botanists' technique for squashing, viz., to heat the slide. Pass the slide over an alcohol flame or gas flame with a dodging motion. The slide should be so heated that the liquid is just below its boiling point. Now quickly cover the slide with a filter paper and press with a strong pressure, rolling the thumb as before. This heating before the final press will greatly improve the quality of the metaphases in a large number of cells. Figures 1 and 2 show the same cell before and after heating. These figures are not the most dramatic demonstration of the effect of heating on the metaphase spreads, but they do show improvement. We apply the heating method to all our slides.

Seal the edge of the coverslip with Krönigs cement. Well sealed slides, when kept at refrigerated temperatures, would last for at least one year. At room temperature, they deteriorate within a few months. The slides can be made permanent by the dry ice method. Place slides on a cake of flat dry ice and leave them frozen for at least 10 minutes. It is safer to freeze the slides for 20 minutes. Use a single-edge razor blade or a surgical scalpel; take one slide at a time, and quickly scrape off the sealing cement which should be cracked. Then place the corner of the blade to one corner of the coverslip. With one quick flip, the coverslip should be detached from the slide, but the cells should remain on the slide.

Rinse the slides in two changes of commercial ethyl alcohol and dry

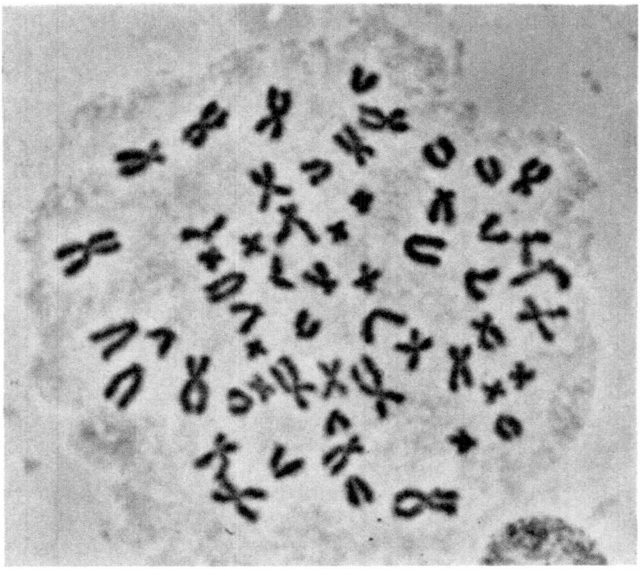

Fig. 2. The same cell after flame heating and resquash. Note improvement in the quality of spread.

them in air. The slides can be restained with Giemsa, Feulgen, carbo-fuchsin, or any nuclear stain. If the cells were treated with labeled compounds, the slides are also ready for emulsion film.

We hasten to add that pages of written recipes are often not as instructive as one demonstration. Investigators interested in learning the techniques may contact either of us (TCH, Houston, JLP, Berkeley) for personal instructions.

MEIOTIC PREPARATIONS FROM MAMMALIAN TESTES

C. E. Ford and E. P. Evans
M.R.C. Radiobiological Research Unit, Harwell, Didcot, Berkshire, England

A schedule is given below for obtaining preparations from the testes of mice suitable for stages of meiosis from late diplotene to second metaphase. It corresponds very closely with the method described by Evans *et al.* (1964) and with a few modifications, determined largely by the amount of material available, it has been applied successfully to other rodent species, to the bull, marmoset and man. The basic procedure is a direct adaptation to testicular material of the standard method of this laboratory for somatic tissues of small mammals (Ford, 1966). A similar method has been described by Benirschke and Brownhill (1963).

Schedule

1. Kill mouse and dissect out testes. Place testes in 2.2 percent (weight/vol.) trisodium citrate solution in a 2 inch diameter Petri dish.
2. Pierce tunica and expose tubules. Swirl in citrate solution to wash away fat. Time: 15 seconds.
3. Transfer to ca. 3 ml 2.2 percent citrate solution in a fresh 2 inch Petri dish. Tease tubules with curved forceps, then transfer total contents of dish to a 4 ml centrifuge tube. Pipette gently. Time: 10 minutes.
4. Spin 5 seconds to sediment larger tubule segments. Transfer supernatant to a clean 4 ml centrifuge tube. Discard sediment.
5. Spin for 5 minutes at 500 r.p.m. Discard supernatant. Resuspend pellet in minimum volume of residual supernatant.
6. Add ca. 3 ml of 1.0% trisodium citrate solution, slowly, meanwhile flicking the tube with the forefinger. Divide between two Dreyer tubes (narrow, conical tipped tubes, capacity ca. 2 ml.). Total time: 10 minutes.
7. Spin 5 minutes at 500 r.p.m. Remove supernatant. Allow tubes to

stand 1 minute, then remove fluid that has drained down from walls by a small pipette.

8. During step 7 make up fixative = 3 parts absolute ethyl alcohol:1 part glacial acetic acid plus a trace of chloroform (say, 45:15:1).

9a. Fixation, variant I. Flick tube to convert pellet to a dense suspension. Allow two drops of fixative to fall directly on to cell suspension. Flick tube vigorously. Add more fixative and continue to flick until tube is ca. ¾ full. Spin 3 minutes at 500 r.p.m. Change fixative. Spin down and change fixative twice more. Total time: 15 minutes.

9b. Fixation, variant II. Retain pellet intact. Add fixative down side of tube, carefully at first to avoid disrupting pellet, then more rapidly until the tube is full. Remove all fixative immediately. Resuspend pellet and add more fixative until the tube is ¾ full. Spin 3 minutes at 500 r.p.m. Remove fixative, resuspend pellet, add fresh fixative and leave to stand 15 minutes. Then repeat.

10. *Preparations.* The final suspension of cells in fixative should be dilute and it is best to allow it to stand for 1–3 hours before making preparations. It may then be necessary to resuspend the cells once more before proceeding.

Take up cell suspension into a small pipette. (Made from 4 mm diameter soft glass tubing; teats are made from short lengths of polythene tubing, the ends being closed by heating followed by pressure from pliers.) Allow 3 evenly-spaced drops to fall on a clean slide. Wait while the fluid spreads to its maximum extent and begins to contract. Watch in the light of a bench lamp for interference colours (Newton's rings) to appear, indicating that the remaining film of liquid is very thin. Then blow gently until dry. Repeat until there are a sufficient number of cells on the preparation, as judged by phase contrast examination. The preparation can then be stored, dry, for later examination, or stained immediately. Lactic-acetic-orcein is preferred for sharpness, toluidine blue for rapidity and intensity.

Comments

Teasing of Tubules The extent and vigour of teasing and subsequent pipetting has to be learnt by experience. Too much will give a final suspension containing many small tubule fragments; too little will result in many cells, particularly spermatogonia, remaining within the discarded tubules.

Hypotonic Pretreatment The optimum concentration and exposure time almost certainly differ from one species to another and will have to be determined empirically. Better preparations have been obtained from human testicular material, for example, by teasing in Hank's solution and then transferring to 1.5 percent trisodium citrate solution. Furthermore, it is possible that other hypotonic fluids, such as potassium chloride (Hungerford, 1965), would give improved results. Freshly prepared citrate solutions should be used.

Centrifuge Speeds and Times These are given for a centrifuge with a 7.8-inch diameter head. Since the force g operating on a revolving centrifuge tube is given by $g = 284\ RN^2 \times 10^{-7}$ where R is the radius of the head and N the revolutions per minute, the speed of revolution is the more important variable, and since variation in the diameter of the head of bench centrifuges is not very great, it is generally sufficient to quote speed (r.p.m.) alone. The object should be to make an efficient compromise between the amount of material discarded with the supernatant and the undesirability of compacting the pellet too much, with consequent risk of increasing the adherence of the cells to one another in clumps. It is a useful practice to test the normally discarded supernatants from time to time to determine just how much material is being lost.

Fixation Of all the variables implicated in the fixation process, speed of penetration is probably the most important. The maximum speed of penetration is presumably achieved when the cells are in suspension at the time the first fixative is added, as in variant I (step 9a). Conversely, if the cells are fixed as a pellet, the greater its volume, the slower the rate of penetration. This, doubtless, is the principal reason why better results are obtained with smaller pellets, the optimum perhaps being a pellet of about 2 mm depth. Still better individual cells may be obtained by employing variant I, but the total yield of cells is less, sometimes considerably less than if variant II (pellet fixation) had been employed. This must mean that many cells are disrupted completely when they are fixed in suspension, with consequent risk of differential loss of spermatocytes in the desired meiotic stages.

Preparations The empirical observation that underlies the recommended procedure is that much better spreading of the chromosomes is achieved when dilute suspensions are used than when dense suspensions are used. Since several sets of drops must be dried to provide a satisfactory final concentration of cells, a preparation may take a little longer to make than if other methods in common use were followed. In our experience, the extra time taken in preparations is more than justified by the superior results. Furthermore, the method is economical in the use of material.

The drops of fixative containing cells should spread evenly on the slide, forming a film with an even, entire (not crinkled) edge. A single gentle exhalation should suffice to dry the film. Occasionally, a crinkled edge is formed and drying is slower. When this happens the results are much less satisfactory. Factors known to influence spreading adversely include the use of siliconed glassware, some detergents used in the cleaning of slides, and some plastics including disposable syringes and even pipette teats.

Although it is recommended that the cells should remain in fixative for 1 to 3 hours before preparations are made, it is a useful practice to make one preparation immediately after the final change of fixative in order to judge the cell concentration and if necessary to adjust it.

The method of blowing dry is adapted from Tjio and Whang (1962).

Anaphases Exposure to hypotonic fluid causes metaphase arrest and therefore the progressive disappearance of spermatocytes in anaphase I or anaphase II. Anaphase figures can, therefore, be obtained only if the period of hypotonic treatment is reduced to a minimum or eliminated altogether. Immersion of tubules from a mouse testis in a 1.0 percent trisodium citrate solution for 2 minutes followed by fixation, whole, in 3 parts ethyl alcohol to 1 part glacial acetic acid, staining in Feulgen and squashing (Ford, 1962) gave reasonably satisfactory anaphases.

Spermatogonia Few spermatogonial metaphases are normally seen in preparations made from testicular material of the mouse by the method described. Their number can be increased by prior injection of Colcemid–0.01 ml of a 0.04% (weight/volume) solution per gram body weight for 1½ hours. They can also be increased by placing the tubules at step 2 into 2 ml of 0.25 percent trypsin solution in phosphate-buffered saline and pipetting gently for 5 to 6 minutes, then adding 5 ml of tissue culture medium 199 and spinning down at 500 r.p.m. for 5 minutes. The pellet is resuspended in 1.0 percent citrate solution, as in step 6 and the normal schedule followed thereafter (E. P. Evans, unpublished).

References

Benirschke, K. and L. E. Brownhill: Heterosexual cells in testes of chimeric marmoset monkeys. Cytogenetics *2*:331, 1963.

Evans, E. P., G. Breckon and C. E. Ford: An air-drying method for meiotic preparations from mammalian testes. Cytogenetics *3*:289, 1964.

Ford, C. E.: Methods in Human Cytogenetics. In: Methodology in Human Genetics. W. J. Burdette, ed. San Francisco: Holden-Day, 1962, 227–259.

Ford, C. E.: The use of chromosome markers. In: Tissue Grafting and Radiation. H. S. Micklem and J. F. Loutit, eds. New York: Academic Press, 1966, 197–206.

Hungerford, D. A.: Leukocytes cultured from small inocula of whole blood and the preparation of metaphase chromosomes by treatment with hypotonic KCl, Stain Techn. *40*:333, 1965.

Tjio, J. H. and J. Whang: Chromosome preparations of bone marrow cells without prior *in vitro* culture or *in vivo* colchicine administration. Stain Technol. *37*:17, 1962.

A COMBINED TOLUIDINE BLUE STAIN AND MOUNTING MEDIUM

G. Breckon and E. P. Evans
M.R.C. Radiobiological Research Unit, Harwell, Didcot, Berkshire, England

A simple combined stain and mounting medium has been in use in this laboratory for a number of years. It consists of an aqueous solution of toluidine blue mixed in suitable concentration with a saturated solution of a water soluble resin. The latter is dimethyl hydantoin formaldehyde resin, marketed by Rex Campbell & Co. Ltd., 7 Idol Lane, Eastcheap, London, E.C.3. It has the great advantage that viscosity hardly changes with increasing concentration. A mixture containing the stain can, therefore, be prepared that will remain fluid almost indefinitely if kept in a glass-covered jar, as used for mounting media, but which hardens slowly when exposed to air.

This resin was introduced into biology for making whole mounts of invertebrates (Steedman, 1958). It is supplied in large, almost colourless, brittle lumps which are broken up with a hammer before a solution is prepared. (Grinding is not recommended as the pieces become tacky with the heat that is generated.) Resin and water are mixed in the ratio of 70 gm to 30 ml in a suitable flask and shaken periodically, more resin being added if all has dissolved. After about a week the solution is filtered and is ready to use. A 2 percent aqueous solution of toluidine blue is then added until trial shows that a satisfactory staining intensity has been achieved.

Chromosomes in air-dried preparations of mammalian cells from all sources (somatic tissues, testicular material, blood cultures and cultured fibroblasts) and all species examined stain rapidly and intensely. Toluidine blue is a RNA stain, however, and the cytoplasm sometimes is strongly coloured. This may be rectified by hydrolysis. Immersion for 2 minutes in 5N/HCl at room temperature is sufficient. The preparation can then be dried again after rinsing in 50 percent, 95 percent and absolute alcohol (twice).

Though the first impression given by a preparation made in this way may be of clear, intensely stained chromosomes, comparison shows that there is not so much detail as in orcein stained preparations, nor such sharpness of definition as can be obtained by the Feulgen method. Furthermore, the blue colour presents difficulties for photography and the final prints are often disappointing. The advantages of the method are its speed and simplicity: preparations are ready for examination as soon as the cover glass has been applied. Its value, therefore, lies particularly where large numbers of mitotic or meiotic cells are to be scored rapidly, as in transplantation experiments using chromosome markers (e.g., Micklem *et al.*, 1966) and some experiments in radiation cytogenetics (e.g., Searle *et al.*, 1968).

Preparations left on the bench are apt to fade rapidly but commonly retain their stain for many weeks if stored away from light. Old, faded preparations can be stained again in the same way, or with a different stain such as lactic acetic orcein, and can be made permanent if desired. Adhering immersion oil is first removed by brief rinsing in a equal-parts mixture of ethyl alcohol and ethyl ether. The preparations are then put in a Coplin jar, and water at 50°–60°C allowed to run in continuously until the cover glass has separated and the resin adherent to the surface of the slide has dissolved. The preparations are then rinsed in the alcohol series, dried and restained.

References

Micklem, H. S., C. E. Ford, E. P. Evans and J. Gray: Interrelationships of myeloid and lymphoid cells: studies with chromosomally-marked cells transfused into lethally irradiated mice. Proc. R. Soc. B. *165*:78, 1966.

Searle, A. G., E. P. Evans, C. E. Ford and B. J. West: Studies on the induction of translocations in mouse spermatogonia. I. The effect of dose rate. Mutation Research (in press).

Steedman, H. F.: Dimethyl hydantoin formaldehyde: a new water-soluble resin for use as a mounting medium. Quart. J. Micr. Sci. *99*:451. 1958.

INDEX

GPSR Compliance

The European Union's (EU) General Product Safety Regulation (GPSR) is a set of rules that requires consumer products to be safe and our obligations to ensure this.

If you have any concerns about our products, you can contact us on ProductSafety@springernature.com

In case Publisher is established outside the EU, the EU authorized representative is:

Springer Nature Customer Service Center GmbH
Europaplatz 3
69115 Heidelberg, Germany

Batch number: 10620143

Printed by Printforce, the Netherlands